STATISTICAL MECHANICS OF SOLIDS

MONOGRAPHS ON THE
PHYSICS AND CHEMISTRY
OF MATERIALS #58

General Editors
R.J. Brook, Anthony Cheetham, Arthur Heuer
Sir Peter Hirsch, Tobin J. Marks, D.G. Pettifor, Manfred Rühle
John Silcox, Adrian P. Sutton, Matthew V. Tirrell, Vaclav Vitek

STATISTICAL MECHANICS OF SOLIDS

Louis A. Girifalco

2000

OXFORD
UNIVERSITY PRESS

Oxford   New York
Athens   Auckland   Bangkok   Bogotá   Buenos Aires   Calcutta
Cape Town   Chennai   Dar es Salaam   Delhi   Florence   Hong Kong   Istanbul
Karachi   Kuala Lumpur   Madrid   Melbourne   Mexico City   Mumbai
Nairobi   Paris   São Paulo   Singapore   Taipei   Tokyo   Toronto   Warsaw
and associated companies in
Berlin   Ibadan

Copyright © 2000 by Oxford University Press

Published by Oxford University Press, Inc.
198 Madison Avenue, New York, New York 10016

Oxford is a registered trademark of Oxford University Press.

All rights reserved. No part of this publication may be reproduced,
stored in a retrieval system, or transmitted, in any form or by any means,
electronic, mechanical, photocopying, recording or otherwise,
without the prior permission of Oxford University Press.

Library of Congress Cataloging-in-Publication Data
Girifalco, L.A. (Louis A.)
   Statistical mechanics of solids / Louis A. Girifalco.
     p. cm.
   Includes bibliographical references.
   ISBN 0-19-511965-7
   1. Solids—Mathematical models.   2. Statistical mechanics.
I. Title.
QC176.G54   2000
530.4'1—dc21   99-33191

9 8 7 6 5 4 3 2 1

Printed in the United States of America
on acid-free paper

Preface

Statistical mechanics is one of the greatest achievements of the human mind, combining theoretical elegance with powerful methods for studying an enormous range of specific systems and phenomena. It connects the properties of the material world to the basic constituents of matter at the deepest level. It lays bare the atomic and molecular content of thermodynamics and shows how to obtain new and important relations among the properties of matter, how to organize enormous amounts of scientific information, and how to understand it. Furthermore, it is a dynamic discipline in that it continues to grow and enhance our understanding of the physical world. Its study is essential for those who want to probe deeply into the nature of things.

Statistical mechanics can be studied from a variety of viewpoints and at different levels of theoretical sophistication. It is an important discipline for the study of matter in every form and at every level of aggregation. The number of available texts on the subject is therefore large, and they range from elementary to quite advanced treatments. However, modern texts that include topics on solids generally focus on such phenomena as phase transitions at critical points, universality, and those commonalities that permit similar treatments of solids and liquids. Aside from my earlier book (*Statistical Physics of Materials*, 1973), there has been no satisfactory text that concentrates solely on the statistical mechanics of solids from the point of view of the material properties of the solid state. This is such an important subject with such far-reaching applications that the neglect of this approach is somewhat surprising. My major objective was to fill this gap.

This book grew out of a course in statistical mechanics of solids I give to students of materials science at the University of Pennsylvania. It bears some resemblance to my previous book but includes much new and updated material. My intention is to provide a useful reference as well as a basis for course work.

This book is intended for those whose primary interests are in the *properties* of solids rather than in the statistical mechanical theory itself. One of my objectives is to aid students in developing a physical insight that relates properties to statistical models and theories. Such insight is extremely valuable for those studying the application of statistical mechanics to condensed matter. I have therefore included some discussion of the physics of the systems and phenomena presented along with the statistical mechanical developments and their physical interpretations.

Another objective is to make the subject matter accessible to as wide an audience as possible among those whose background is equivalent to the first- or second-year graduate school level. This includes students with undergraduate degrees in physics, chemistry, and materials science, and their differences must be recognized. The variation in their mathematical training, their knowledge of the solid state, and, surprisingly, their grasp of thermodynamics poses a special problem for the teacher. I address this problem by starting at what should be a common base and trying not to assume too much in the way of specialized background.

A third objective is to give a representative sample of those aspects of materials and solid state science that benefit the most from the application of statistical mechanics. The field has grown enormously since my earlier effort in 1973 and includes many new important subjects. The contents of this book reflect this.

These objectives have determined both the orientation and the content of the book. Regarding the fundamentals of statistical mechanics, I have tried to present a careful exposition of the basic concepts without excessive formalism. Thus, although statistical mechanics is developed as a tool for understanding properties, rather than for its own sake, I have given a fairly detailed discussion of the nature of ensembles and their relation to thermodynamics. In fact, chapters 1–3 should be excellent preparation for advanced study of both the fundamentals of statistical mechanics and its theoretical formalism.

The book starts with a brief review of thermodynamics despite the fact that students should have had at least one course in the subject. My experience has been that the review is needed, particularly regarding the origin and meaning of entropy and the second law. I do not agree with some of the modern approaches of first teaching thermodynamics as either being derived from statistical mechanics or as a purely axiomatic structure that connects to experiment only after considerable formal development. Both these approaches are valid and interesting, but should be preceded by a knowledge of the origin of thermodynamics from experimental data.

Chapters 4 and 5 present the statistical thermodynamics of simple crystals and include a discussion of both harmonic and anharmonic properties. The traditional theory of the harmonic crystal is still of great value and has not changed much in the past decades. But modern developments make it possible to deal with anharmonicity in a more rigorous and transparent way than before, while still maintaining much of the simplicity of the original Gruneisen method.

Chapters 6 and 7 work out the consequences of the free electron theory of metals and semiconductors, for both equilibrium and transport properties.

The emphasis on properties and physical insight has led to a sequence on cooperative phenomena in which order–disorder alloys and magnetic ordering are treated separately in chapters 8 and 9 (although the commonalities of these two kinds of phenomena are stressed). The overall theory of phase transformations in chapter 10 and of critical points in chapter 11 is taken up only after the presentation of these two topics. I believe this ordering keeps a closer connection with real materials than does starting with the more general theory.

Because of the ever-growing importance of surface phenomena, chapter 12 is devoted to the thermodynamics and statistical mechanics of surfaces.

The theory of random flight has a large number of uses and is given in chapter 13. For the study of materials, the most important applications are to polymers and diffusion. Chapter 14 presents the statistical mechanics of polymer chains, which draws heavily on the theory in chapter 13.

Chapters 15–17 develop the statistical mechanics of point defects and diffusion. These are of great importance for the properties of solids and can be given an elegant treatment, at least for simple crystals, that well illustrates the power of the statistical theory.

The range of properties that have a statistical basis is so large and so important that no selection of topics can be complete. However, it is obvious that there are some subjects that must be presented because they are so basic, and others that are part of the mainstream of modern interests in material properties. These considerations, along with my personal interests, have determined my selection of specific applications of statistical mechanics. I have included a number of appendices on topics that are important for understanding statistical mechanics in solids, but that would interrupt the flow and be out of place in the body of the book.

The exercises at the end of each chapter are designed to enhance the student's understanding of the subject matter, and to help develop physical insight. The student should do them all.

I am particularly indebted to the students who have taken my course, and who have enthusiastically hunted down errors in my notes.

Contents

1 The Basics of Thermodynamics 1

1.1 The existence of equilibrium and state functions 1
1.2 Empirical temperature scales 2
1.3 The ideal gas temperature 2
1.4 The mechanical equivalent of heat 3
1.5 Walls and the zeroth law of thermodynamics 4
1.6 Spontaneous, reversible, and irreversible processes 5
1.7 Work and the dependence of work on the path 6
1.8 The first law of thermodynamics 8
1.9 Heat capacity, energy, and enthalpy 9
1.10 The second law of thermodynamics and entropy 10
1.11 Free energies and equilibrium conditions 14
1.12 Thermodynamic potentials and Legendre transformations 16
1.13 Chemical potentials 20
1.14 Conditions of phase equilibria and stability 21
1.15 Euler's theorem and the Gibbs–Duhem equation 24
1.16 Reciprocity relations of Maxwell 26
1.17 Useful differential relations 27
1.18 Equations of state and heat capacity relations 28
1.19 Magnetic systems 31
Exercises 32

2 Principles of Statistical Mechanics 34

2.1 Definitions for statistical mechanics 34
2.2 Thermodynamic state 35
2.3 Comparison of microscopic and macroscopic state 36
2.4 The relation between microscopic and macroscopic state 37
2.5 System and environment 38
2.6 Quantum states of macroscopic systems 39
2.7 Time averages 40
2.8 Ensembles 41
2.9 The canonical ensemble 42
2.10 The canonical most probable distribution 44
2.11 Summary of definitions of probabilities 47

2.12 The canonical ensemble and thermodynamics   48
2.13 Statistical entropy and the second law of thermodynamics   54
2.14 The semiclassical approximation   56
2.15 The grand canonical ensemble   60
2.16 The pressure ensemble   63
2.17 Fluctuations   64
Exercises   67

3 Particle Statistics   69

3.1 Entropy and number of complexions   69
3.2 Particle distribution functions   71
3.3 Particle statistics and thermodynamics   74
3.4 The ideal gas   76
3.5 Particle statistics from the grand canonical ensemble   81
3.6 Representations of the density of states   84
3.7 Maxwell's velocity distribution   86
3.8 Two-dimensional ideal gas   87
3.9 Independent particles and subsystems   88
Exercises   90

4 The Harmonic Crystal   92

4.1 The harmonic model   92
4.2 The monatomic linear chain and normal mode analysis   93
4.3 Partition function and free energy of the harmonic crystal   101
4.4 General heat capacity equations   104
4.5 The Einstein model   106
4.6 Superposition of Einstein oscillators   107
4.7 The Debye model   108
4.8 Debye energy and heat capacity   112
4.9 Relation between Einstein and Debye characteristic temperatures   115
4.10 Comparison of Debye theory with experiment   116
4.11 The phonon gas   119
Exercises   121

5 Anharmonic Properties and the Equation of State   124

5.1 The crystal potential energy   124
5.2 Anharmonic properties and the Gruneisen assumption   129
5.3 Heat capacity at constant pressure   134
5.4 Debye theory and the Gruneisen assumption   134
5.5 Vibrational anharmonicity   136
5.6 Theory of the Gruneisen parameter   137
Exercises   141

6 Free Electron Theory in Metals and Semiconductors   143

6.1 Free electrons in metals   143
6.2 Statistics for the electron gas   144

- 6.3 The distribution of free electrons 145
- 6.4 Thermodynamic properties of the free electron gas 148
- 6.5 Electronic heat capacity in metals 152
- 6.6 Equation of state of the free electron gas 153
- 6.7 Thomas–Fermi theory 155
- 6.8 Review of results of band theory 159
- 6.9 Impurity levels in semiconductors 162
- 6.10 Electron distribution in intrinsic semiconductors 163
- 6.11 Electron statistics in extrinsic semiconductors 167
- 6.12 Mass action laws for extrinsic semiconductors 171
- 6.13 Relation between Fermi level and impurity concentration 173
- Exercises 175

7 Statistical-Kinetic Theory of Electron Transport 177

- 7.1 Free electrons in external fields and temperature gradients 177
- 7.2 The statistical-kinetic method 180
- 7.3 The Boltzmann transport equation 181
- 7.4 Formal flux equations 185
- 7.5 The electrical conductivity of metals 186
- 7.6 Thermal conductivity and the Wiedemann–Franz law 188
- 7.7 The isothermal Hall effect 192
- 7.8 Electrical conductivity in semiconductors 196
- Exercises 201

8 Order-Disorder Alloys 202

- 8.1 Order-disorder structures 202
- 8.2 The order-disorder transition 203
- 8.3 Description of the degree of order 204
- 8.4 The Order-disorder partition function 209
- 8.5 The Kirkwood method 213
- 8.6 The Bragg–Williams approximation 217
- 8.7 The second moment approximation 222
- 8.8 The quasi-chemical approximation 225
- 8.9 Comparison with experiment 230
- Exercises 232

9 Magnetic Order 234

- 9.1 Magnetic response 234
- 9.2 Paramagnetism of independent moments 236
- 9.3 Paramagnetism of free electrons 240
- 9.4 Ferromagnetism: mean field theory 242
- 9.5 The Ising model for ferromagnetism 245
- 9.6 Antiferromagnetism: mean field theory 248
- 9.7 Spin waves 251
- Exercises 256

10 Phase Equilibria 258

10.1 Phase equilibria in one-component systems 258
10.2 The van der Waals model 262
10.3 Sublimation 270
10.4 The liquid state 272
10.5 Communal entropy 276
10.6 Vibrations and melting 277
10.7 Melting 281
10.8 Regular solution theory of binary alloys 282
Exercises 286

11 Critical Exponents and the Renormalization Group 288

11.1 Equivalent models 288
11.2 Critical points 289
11.3 Landau theory and the Kirkwood expansion 292
11.4 Fluctuations and correlation length 295
11.5 The monatomic Ising chain 298
11.6 Renormalization of the one-dimensional Ising model 302
11.7 The Kadanoff construction 305
11.8 The renormalization group 311
11.9 Scaling and the renormalization group 313
11.10 Numbers 316
Exercises 317

12 Surfaces and Interfaces 318

12.1 Basic concepts 318
12.2 Thermodynamics of interfaces 322
12.3 Thermodynamics of adsorption on solid surfaces 325
12.4 Adhesion and cohesion 328
12.5 Critical point and critical exponent for surface tension 333
12.6 Monolayer adsorption: Langmuir isotherm 335
12.7 Monolayer adsorption: mobile layer 339
12.8 Multilayer adsorption: BET isotherm 340
12.9 Segregation of impurities at interfaces 345
Exercises 346

13 The Theory of Random Flight 349

13.1 Introduction 349
13.2 The mean square total displacement 350
13.3 Random flight on a lattice 354
13.4 Reflecting and absorbing barriers 361
13.5 The Markoff method 363
13.6 The general solution 367
13.7 Self-similarity 370
13.8 The diffusion equation from random flights 371
Exercises 373

14 Linear Polymer Chains   375

    14.1   Polymer chains and random flight   375
    14.2   Persistence length   376
    14.3   Chain length fluctuations   379
    14.4   Density in a polymer chain   381
    14.5   Partition function of a polymer chain   381
    14.6   Excluded volume   383
    14.7   The force ensemble and chain elasticity   385
    14.8   Elastomers   389
    14.9   The Flory correction   395
    14.10 Solutions and gels   395
    Exercises   401

15 Vacancies and Interstitials in Monatomic Crystals   403

    15.1   Choice of ensemble   403
    15.2   The vacancy concentration   404
    15.3   The crystal free energy   407
    15.4   Vacancies and thermodynamic functions   410
    15.5   The vacancy formation functions   413
    15.6   Vacancies, divacancies, and interstitials   418
    15.7   Some numerical results   422
    Exercises   429

16 Point Defects in Dilute Alloys   431

    16.1   General comments   431
    16.2   The statistical count for substitutional defects   433
    16.3   Defect concentration formulas for substitutional defects   435
    16.4   Internal equilibria for substitutional defects   440
    16.5   Quenched-in resistivity of dilute binary alloys   441
    16.6   Some general theory   443
    16.7   Thermodynamics of the dilute alloy   446
    Exercises   448

17 Diffusion in Simple Crystals   450

    17.1   The empirical laws of diffusion   450
    17.2   Transition probabilities and Fick's laws   452
    17.3   Atomic jumps and the diffusion coefficient   455
    17.4   The jump frequency in one dimension   457
    17.5   Many-body theory of the jump frequency   460
    17.6   The diffusion coefficient   466
    Exercises   467

Appendix 1 Combinatorial Problems in Statistical Mechanics   469

Appendix 2 The Method of Undetermined Multipliers   473

Appendix 3 Stirling's Approximation   477

Appendix 4 Sums and Integrals   479

Appendix 5 Fermi Integrals   484

Appendix 6 Kirkwood's Second Moment   488

Appendix 7 The Generalized Lattice Gas   492

Appendix 8 Dyadics and Crystal Symmetry   495

Additional Readings   505

Index   509

STATISTICAL MECHANICS OF SOLIDS

1

The Basics of Thermodynamics

1.1 The existence of equilibrium and state functions

Thermodynamics addresses those relations among the macroscopic properties of macroscopic systems that do not depend on the properties of its atomic constituents. It is a fact of experience that systems exist whose properties do not change with time, within the limits of measurement. It is also a fact that systems exist whose properties do change with time. In making such statements, the time scale must be considered. Thus, a piece of iron may have properties whose values are constant over some time period, say, one week, but over a period of months or years the iron might develop rust on its surface.

To develop thermodynamic theory, we first restrict our analysis to systems in the absence of external fields, such as gravity or magnetism, and that consist of isotropic phases. An *isotropic phase* is a homogeneous substance none of whose properties depend on spatial directions. If the system consists of just one phase, then we will call it a *simple* system. A heterogeneous system is one that consists of several simple systems. The extension of the theory to heterogeneous complex systems, and to systems in external fields, is straightforward once the thermodynamics of simple systems is understood.

A system is said to be in *internal equilibrium* if none of its properties change with time. Such properties might be the pressure exerted by the system, or its volume, chemical composition, magnetic moment, refractive index, or temperature. Experiments have shown that the values of these properties are not all independent. For example, in a one-component system the temperature, volume, and pressure are related through the equation of state. In general, the *state* of a system at equilibrium is defined as the minimum amount of information needed to determine all of the system's properties. One of the tasks of thermodynamics is to find the number and types of system properties that define state.

The measured properties of a system at equilibrium are independent of the system's history. Thus, a system could arrive at a particular composition, volume, and pressure by any of an infinite number of ways, each of which would result in the system having the same value for these properties. Not all system properties are independent of a system's history. The magnetization of a piece of iron depends on how it was magnetized; the electrical resistivity of a metal depends on the defects present, and these depend on how the metal was treated.

It is part of the definition of equilibrium that the system properties not only are constant in time but also are independent of how the system got to equilibrium. This is a definition of what may be called *absolute equilibrium*. Partial equilibrium also exists, in which the properties are time independent although the values of some of them may be history dependent.

A property of a system at equilibrium that is independent of the systems history is called a *state function*.

1.2 Empirical temperature scales

The idea of temperature requires some special comment. Other quantities can be directly measured and have their origin in mechanics and in the basic measurements of mass, length, and time. Temperature, on the other hand, has its origin in our perception of hot and cold bodies and their interactions. We note that these perceptions are accompanied by properties of the bodies that can be measured quantitatively. Thus, if a thin metal rod is exposed to hot water, we notice that its length increases. We can qualitatively sense that water gets progressively hotter as it sits in a kettle over a fire, and this sense can be given a quantitative form by noting that the length of the metal rod progressively increases as the water approaches boiling.

An empirical temperature scale can be constructed by choosing some physical property that we know is different for hot and for cold bodies, constructing an instrument that measures this property, putting the instrument in contact with a system whose temperature we wish to know, and then measuring that property. Instruments of this sort (thermometers) can be based on such properties as length (mercury or alcohol thermometers), electrical resistance (resistance thermometers), voltage (thermocouples), or volume (gas thermometers). An obvious problem is that the empirical temperature scales depend on the physical property chosen to measure it and even on the material the thermometer is made of. If this could not be overcome, a temperature scale would be arbitrary and not have any fundamental significance.

1.3 The ideal gas temperature

The gas thermometer occupies a special place in defining empirical temperature scales. The reason for this is that, if the gas is sufficiently dilute, all gas thermometers give the same temperature scale. At low pressures and high volumes, the gases approach an ideal behavior, and it is therefore possible to construct a thermometer based on the properties of an ideal gas.

Experiment shows that in the limits of high temperature and/or low pressure, the volume, pressure, and temperature of a gas are related by

$$\frac{PV}{(A+t)} = \frac{P_0 V_0}{(A+t_0)} \tag{1.3.1}$$

where (P, V) and (P_0, V_0) denote the pressure and volume at any two different temperatures t and t_0, respectively, the temperature being measured on an empirical temperature scale. This is the ideal gas law. If we adopt the centigrade scale of temperature, then the constant A is found to be 273.16. We are thus led to define a new temperature scale by the relation

$$T = t + 273.16 \tag{1.3.2}$$

T is the temperature on the ideal gas temperature scale. We expect this to have a more fundamental significance than t because the ideal gas law is valid for all gases that are sufficiently dilute and are at sufficiently high temperature.

That is, the ideal gas law holds when the intermolecular interactions are very small. In fact, it will be shown that the temperature defined by ideal gases is the same as the absolute temperature of thermodynamics.

With this definition of temperature, equation (1.3.1) becomes

$$PV = NkT \tag{1.3.3}$$

where k is Boltzmann's constant and N is the number of molecules in the gas.

A gas thermometer is just a bulb or glass tube filled with a dilute gas. The gas can be held at either constant volume or constant pressure. The pressure, or the volume, is then measured and converted to temperature via equation (1.3.3).

The ideal gas is an example of a system whose thermodynamic state is completely specified by two of the three variables (P, T, V). It is an empirical fact that if any two of these are known, all properties of the system are determined. The ideal gas equation is the simplest example of an equation of state that connects the state parameters of a system.

1.4 The mechanical equivalent of heat

A large array of experiments have shown that doing work on a system raises its temperature. Starting with Count Rumford's observation that the mechanical work in boring cannon raises the temperature of the brass, and the extensive experiments of James Prescott Joule that related the performance of many kinds of work to the temperature rise of systems, it is an experimental fact that work has important thermal effects. The results of these experiments can be summarized in the statement that work can be converted into heat and that the ratio of the work expended to the heat generated is always the same constant. This implies, of course, that a concept of heat has been developed and that a method exists for its measurement. Much of the verbal description of thermal phenomena is a holdover from the days of the caloric theory, when it was believed that heat was a substance that flows from hotter to colder bodies. While this is in some respects unfortunate, it still provides a convenient descriptive language. In modern terms, the concept of heat and its measurement can be made precise as follows.

Consider a system, which for the sake of specificity we take to consist of pure water, enclosed in a perfectly insulating envelope (this can be approximated by a Dewar flask or asbestos). Assume that we have an empirical temperature scale so that we know that any temperature change in the environment has no effect on the temperature of the system and any temperature change in the system has no effect on the temperature of the environment. The walls are said to be thermally insulating or *adiabatic*. Assume also that the system can be accessed by thermometers or other instruments in such a way that the thermally insulating property of the walls is not sensibly disturbed. Now do work on the system by turning a paddle wheel that is immersed in the water. The experimental result is that the temperature of the water goes up. We also notice that, if a certain amount of work raises the temperature of the water by one degree, twice as much work will raise the temperature by two degrees.

In general, the temperature rise is proportional to the amount of work done. (This is only approximately true if the temperature range over which measurements are made is large.) Also, the temperature rise is inversely proportional to the mass of the water. It takes twice as much mechanical work to raise the temperature by one degree if there are two kilograms of water rather than one. This law is quite exact. We are thus led to define a unit quantity of heat in terms of a unit temperature rise for a unit amount of water. In fact, the definition of a calorie is that amount of heat needed to raise the temperature of one gram of water from 14.5 to 15.5°C. It is found that the amount of work required to do this in an adiabatic system is always the same. This experiment is described by saying that the mechanical work generates heat. The ratio of the mechanical work, in mechanical units, to the heat generated, in thermal units, is called the *mechanical equivalent of heat.*

Since work is done on the system, and since we insist on the law of the conservation of energy, we associate the rise in temperature, and therefore the heat generated, with an increase in the internal energy of the system.

Different substances exhibit different temperature rises for the same amount of mechanical work, and the ratio of the two temperature rises is used to define the relative heat capacities of the substances. An enormous number of calorimetric experiments are readily rationalized and understood on the basis of this concept.

The extension of the above experiments to other kinds of systems and to other kinds of work yields consistent results. For work produced by mechanical, gravitational, electrical, or magnetic means acting on gases, liquids, or solids of any composition, the mechanical equivalent of heat is always the same.

1.5 Walls and the zeroth law of thermodynamics

We know that it is possible to effectively isolate a system from its surroundings in the sense that any changes in the environment have no effect on the values of the properties of the system. A system so isolated is said to be surrounded by a completely isolating wall. For purposes of illustration, let the system be a closed bottle of gas surrounded by an adiabatic wall and immersed in a pool of water. The gas has a fixed volume, pressure, composition, and empirical temperature, and let these be constant in time. The gas is then said to be in internal equilibrium because of the constancy of its properties and because, by definition, it is not interacting with its surroundings. If some change is effected in the pool of water, such as a rise in temperature or pressure, or a change in chemical composition, then there is no change in the physical properties of the gas in the sealed bottle. This is the meaning of a completely isolating wall.

It is possible to have a partially isolating wall. Thus, if our bottle is fitted with a piston and cylinder at its neck instead of being sealed, a change in pressure of the pool of water will effect a change in the gas because the change in pressure will cause the piston to move, thereby changing the volume of the gas in the bottle. But a change in temperature of the water, while leaving the pressure constant, will not result in a change in temperature of the gas. Such a wall is said to be *adiabatic*, and the system is said to be *thermally isolated*.

Not all walls are adiabatic. In fact, special care must be taken to thermally isolate a system. Most containers will allow the systems within them to respond to temperature changes in their surroundings sooner or later. For many walls, such as those made of metal, the response is quite rapid. A system is said to be surrounded by a *diathermic wall* if it responds to temperature changes in its environment.

Now consider systems surrounded by diathermic walls, each system being at internal equilibrium. The following statements are found to be true from experiment:

a. If two systems that are each in internal equilibrium but at different temperatures are brought into contact through a diathermic wall, they will each come to a new state of internal equilibrium. Both will then have the same temperature, whose value is between the two initial temperatures. The two systems are then said to be in thermal equilibrium with each other.
b. If a system A is in thermal equilibrium with a system B, and if B is in thermal equilibrium with a system C, then systems A and C are in thermal equilibrium with each other. That is, if two systems are each in equilibrium with a third system, they are in equilibrium with each other. This is often called the zeroth law of thermodynamics.

According to the zeroth law, for a system with diathermic walls to be in internal equilibrium, it must be in thermal equilibrium with its surroundings.

1.6 Spontaneous, reversible, and irreversible processes

For the purposes of thermodynamics it is important to distinguish among the general processes by which a system can change its state. These processes are called spontaneous, reversible, and irreversible.

A *spontaneous* process is one that takes place naturally without any intervention. Examples of spontaneous processes are the flowing of water downhill, the rusting of iron in moist air, and the dissolution of sugar in hot coffee. If a system is isolated and it is found that its properties change with time, it is undergoing a spontaneous internal process. If a system is not isolated, but immersed in an environment that is sensibly constant (such as a large pool of water at constant temperature and pressure) and its properties change with time, it, too, is undergoing a spontaneous internal process but may also be involved in an interaction with its environment. The combined system plus environment is then undergoing a spontaneous process. If the environment does not have constant properties, the details of the system–environment interaction must be analyzed to determine whether or not a spontaneous process is taking place.

Consider a system that is initially at equilibrium with given values for its properties. For specificity let the system be a one component gas with pressure, volume, and temperature given by (P_1, V_1, T_1). Now suddenly change the volume to V_2, by rapidly pushing on a piston, and wait. The system eventually comes to a new state of equilibrium with pressure, volume, and temperature given by (P_2, V_2, T_2), but the process of getting there is rather chaotic. Because the volume is changed suddenly, turbulent flow takes place in the gas as it rushes to adjust to the new volume. The frictional effect of this turbulence generates heat that raises the temperature either of the gas or of its surroundings, or both. Because of this frictional effect, if the volume is restored to its original value of V_1 by pulling up on the piston, and the system is then allowed to come to equilibrium, the original values of the pressure and temperature, (P_1, T_1), are not recovered. The process of going from state 1 to state 2 via a rapid change in volume is said to be *irreversible* because it cannot be undone by just changing the volume back to its original value. It is clear that the states the system goes through between 1 and 2 are not equilibrium states. There are an infinite number of equilibrium states of the system, but none of these occur during the irreversible process because equilibrium does not exist until we wait for the system to calm down.

The irreversible effects can be mitigated by pushing on the piston more slowly to get from V_1 to V_2. The degree of turbulence and the internal frictional effects are then less. In fact, the degree of irreversibility clearly depends on the rate at which the pressure is applied to effect the volume change. If this rate is very slow, then at any instant of time, the difference between the internal pressure and the applied pressure is very low and the system is near an equilibrium state. In the limit of zero rate of change of the pressure, the system is always at an equilibrium state. This leads us to define a *reversible* process as one in which the rate of change of all relevant parameters is so low that the system is always in an equilibrium state. Clearly, the process is truly reversible only if the process is infinitely slow. Thus, there are no real processes that are reversible, but the slower the process the more nearly it approaches reversibility. In practice, a process can be called reversible if at every stage it is in equilibrium within the accuracy of measurements made on the system. These concepts apply to any sort of system and any changes in state functions, not only to (P, V, T) changes in gases.

The difference between reversible and irreversible processes is of great physical importance. From a mathematical point of view, reversible processes connect equilibrium states. Thus, a system can exist in an infinite number of equilibrium states, each state being characterized by specific values for the state functions of the system so that if F is a state function that depends on parameters x_i,

$$F = F(x_1, x_2, \ldots) \qquad (1.6.1)$$

then two equilibrium states that are very close together and differ by an amount dF are related by

$$dF = \sum_i \frac{\partial F}{\partial x_i} dx_i \qquad (1.6.2)$$

and since F is a state function, its value for any state is independent of the system's history. In a reversible process, state functions exist throughout the process, and all differentials of state functions have the form of equation (1.6.2).

1.7 Work and the dependence of work on the path

Changes in the equilibrium state of a system can be effected by changing its temperature or by doing work. If the system has diathermic boundaries, its temperature can be changed by changing that of its environment. Work can be done on systems that have either diathermic or adiabatic boundaries. If the temperature of the system is maintained at a constant value while work is being done, the process is said to be isothermal. This can be done by surrounding the system with diathermic walls and placing it in contact with another system at the same temperature that is very large compared to the system of interest.

If the system is surrounded by an adiabatic wall while work is being done on (or by) it, the process is said to be adiabatic. In an *adiabatic process*, the temperature of the system may change, but this does not affect the temperature of any other system or of the environment.

A system at equilibrium has a definite energy that depends only on the equilibrium parameters and not on its history. The energy is therefore a state function.

The energy of a system surrounded by adiabatic walls can be changed by doing work on the system. Thus, in the bottle of gas surrounded by asbestos and fitted with a piston, as described in section 1.5, work can be done on the system by applying a pressure on the piston that is greater than the gas pressure. The piston then moves, compressing the gas until its pressure rises to equal the external pressure. Since the system is impervious to all other influences, the work done on the system increases its internal energy. Let U_i be the initial energy of the system and let U_f be the energy of the system after a certain amount of work has been done. Also, let us adopt the convention that work done *by* the system is positive and work done *on* the system is negative. Then, from the law of the conservation of energy,

$$U_f - U_i = -W_{i,f} \quad \text{(adiabatic)} \tag{1.7.1}$$

where $W_{i,f}$ is the work done by the system in going from state i to state f. Equation (1.7.1) merely states that the decrease in energy of a system resulting from an adiabatic process is equal to the work done by the system.

In general, there is a temperature change accompanying an adiabatic process such that the temperature of the final state is not the same as that of the initial state.

Equation (1.7.1) is correct whether or not the process in going from the initial to the final state is reversible because only work can get through to an adiabatic system.

Now let us shift our viewpoint and consider the system doing work rather than having work done on it. That is, the system exerts a pressure on its environment, thereby changing the volume of the system from V_i to V_f. If the process is carried out reversibly so that the difference between the internal pressure and the external pressure is always very small, a certain amount of work will be done given by $\int_{V_i}^{V_f} P dV = W_{i,f}$, the relation between P and V being given by the equation of state at every stage in the process.

What if the process is carried out irreversibly? In this case, frictional processes will be set up that create turbulent internal motion and some of the energy that would have gone into work against the external pressure is dissipated in the internal friction of the system. We thus get the result that the maximum amount of work that can be performed by an adiabatic system in irreversibly going from one state to another is less than that for the reversible process. The irreversible processes do less external work.

The same is true for work done in an isothermal process, because in a reversible process the pressure exerted by the system is infinitesimally close to the external pressure. But if the process were irreversible, the external pressure would be appreciably less than that exerted by the system, and then the work done by the expanding system would be decreased. We then have a general conclusion that work done in going from one state to another is a maximum if the process is reversible.

Since the energy is a state function, equation (1.7.1) shows that the work done in an adiabatic process is independent of the detailed path taken in going from the initial to the final state. That is, the pressure could have been applied rapidly or slowly or in varying increments; the work in going adiabatically from a given equilibrium initial state to a given equilibrium final state is always the same. This is not true for all kinds of processes, and it can be shown that the amount of work done depends on the particular way the process is carried out. It is easiest to demonstrate this for a system consisting of an ideal gas.

Let us assume that the gas is surrounded by diathermic walls and is in equilibrium with a large water bath at a constant temperature. Let the system undergo a change in state from an initial pressure, volume, and temperature of (P_i, V_i, T_i) to a final pressure, volume, and temperature (P_f, V_f, T_f). Note that the temperature is constant because the walls are diathermic.

Let us examine two reversible processes by which we can go from the initial to the final state. For process 1, we first change the volume of the gas from its initial to its final value while keeping the temperature constant. The amount of work done in this step is

$$\int_{V_i}^{V_f} PdV = NkT_i \ln\left(\frac{V_f}{V_i}\right) \qquad \text{(at } T_i\text{)} \qquad (1.7.2)$$

Next we change the temperature to its final value T_f by changing the temperature of the reservoir while keeping the pressure and volume constant. No work is done in this second step and the total work done is given by (1.7.2).

In comparison, in process 2 let the first step consist of changing the temperature to its final value while keeping the pressure and volume constant. No work is done in this step. For the second step of process 2, change the volume to its final value by increasing the pressure. The amount of work done is now given by

$$\int_{V_i}^{V_f} PdV = NkT_f \ln\left(\frac{V_f}{V_i}\right) \qquad \text{(at } T_f\text{)} \qquad (1.7.3)$$

The amount of work done by the two processes is clearly different, although both processes started at the same initial state and ended up at the same final state. The work done on or by a system is therefore *not* a state function of the system. This conclusion is not restricted to ideal gas systems. The reasoning is easily extended to any kind of work and to any equation of state.

The above considerations lead to a definition of heat that is based purely on mechanical concepts as follows: in any process, the work done by a system in going from one state to another is equal to the internal energy change only if the process is adiabatic. For all other processes, the work done is *not* equal to the change in internal energy. We define the heat absorbed by the system as the difference between the work done and the internal energy change.

1.8 The first law of thermodynamics

We are now ready to formulate the first law of thermodynamics. In fact, the last sentence in the preceding section is a statement of the first law that we have obtained as a mere definition of the heat change of a process. Written in symbols, the definition is

$$Q_{i,f} = U_{i,f} + W_{i,f} \qquad (1.8.1)$$

Again note that work done *by* the system is taken as positive, so (1.8.1) defines heat as the change in internal energy minus the work done on the system.

There are two facts that give the first law a status that goes beyond that of mere definition. The first is that the internal energy is taken to be a state function, and every consequence of its being so, that can be tested by experiment, has been verified. The second fact is the existence of the mechanical equivalent of heat, which is one of the most accurately verified experimental results in all of science. The first law is therefore nothing but a statement of the law of conservation of energy extended to processes that involve thermal changes. To emphasize this, we write (1.8.1) as

$$\delta U = \delta Q - \delta W \qquad (1.8.2)$$

where δQ is the heat absorbed by the system and δW is the work done by the system in any process for which the change in internal energy is δU. Equation (1.8.2) is true for spontaneous and irreversible processes as well as for reversible processes; we adopt the convention that δ is used to label changes that can be either reversible or irreversible, while lowercase d is used to label differential changes for reversible processes.

1.9 Heat capacity, energy, and enthalpy

Consider a system that undergoes a reversible process from a state A to a state B. We assume heat can be absorbed or liberated by the system during the process, that only PV work can be done by or on the system, and that the composition of the system is constant. (It will be clear from the following discussion that the results are easily generalized to systems for which other kinds of work can be done.) The change in the state of the system is then defined by the change in the values of pressure, volume, and temperature. The ideal gas temperature scale is used to measure the temperature.

First assume that the volume is kept constant during the entire process in going from A to B, such that no work is done on or by the system. Then, from the first law, the energy change is just equal to the heat liberated or absorbed by the system during the process. If we call this heat δQ_V, then the energy change is

$$\delta U = \delta Q_V \quad \text{(constant volume process)} \qquad (1.9.1)$$

In general, the absorption of heat is accompanied by a temperature change. The heat capacity is defined as the heat absorbed when the temperature of the system is increased by a small amount. That is, if the system absorbs an amount of heat δQ when the temperature is increased by an amount δT, then the heat capacity is defined by

$$C = \lim_{\delta T \to 0} \frac{\delta Q}{\delta T} \qquad (1.9.2)$$

If the volume is constant during the temperature increase in a reversible process, then (1.9.1) shows that (1.9.2) becomes the derivative of the energy with respect to temperature at constant volume. The constant volume heat capacity is therefore defined by

$$C_V = \left(\frac{\partial U}{\partial T}\right)_V \qquad (1.9.3)$$

where the heat is transferred to the system reversibly.

Now consider a constant pressure process in which the volume goes from V_1 to V_2. The work done during this process is

$$W = \int_{V_1}^{V_2} P dV = P(V_2 - V_1) \qquad (1.9.4)$$

From the first law, the energy change is the heat absorbed minus the work done, so if U_1 and U_2 are the energies in the initial and final states, respectively, then

$$U_2 - U_1 = \delta Q_P - P(V_2 - V_1) \qquad (1.9.5)$$

or

$$\delta Q_P = (U_2 + PV_2) - (U_1 + PV_1) \tag{1.9.6}$$

δQ_P being the heat absorbed in a constant pressure process.

Equation (1.9.6) leads us to define the function

$$H = U + PV \tag{1.9.7}$$

called the *enthalpy*. For a constant pressure process, the heat absorbed is the change in enthalpy.

Clearly, the heat capacity at constant pressure for a reversible increase in temperature is the derivative of the enthalpy at constant pressure:

$$C_P = \left(\frac{\partial H}{\partial T}\right)_P \tag{1.9.8}$$

1.10 The second law of thermodynamics and entropy

We know from experience that many processes cannot take place even if they satisfy the law of conservation of energy. Water left to itself, for example, is never seen to flow uphill, and if two bodies at different temperatures are brought into contact and then left alone, heat never flows from the colder to the hotter body. This experience is embodied in the second law of thermodynamics, which has been expressed in two basic forms. One form states that a quantity of heat extracted from a system cannot be converted entirely into work while leaving everything else unchanged. This is called the *Kelvin statement*. Another version of the second law states that heat cannot be transferred from a colder to a hotter body while leaving everything else unchanged. This is called the *Clausius statement* of the second law.

The two statements of the second law are equivalent, as can be seen by showing that if one is violated then so is the other. For example, if the Kelvin statement is false, then a quantity of heat from a system at temperature T_1 can be completely converted to work. This work could then be completely used to heat a system at a higher temperature T_2, thereby violating the Clausius statement. Thus, if the Kelvin statement is false, so is the Clausius statement.

Conversely, if the Clausius statement is false, then a quantity of heat Q can be extracted from a system at a temperature T_1 and completely transferred to a body at a higher temperature T_2. Now let the second system do an amount of work equivalent to this amount of heat. The result is that an amount of heat Q is converted completely into work, thereby violating the Kelvin statement. Thus, if the Clausius statement is false, so is the Kelvin statement. The Kelvin and Clausius statements of the second law are thus seen to be equivalent.

A principle reason we value energy is because the performance of work is so essential to so much human activity. From this anthropomorphic point of view, therefore, some forms of energy (such as the potential energy at the top of a waterfall, or the chemical energy in fuel) are more desirable than others. The operation of heat engines depends on the transfer of heat between two temperatures, so thermal energy at a uniform temperature is useless for performing work. This is often described by saying that the generation of heat results in the degradation of energy to a less useful form.

Heat engines are devices that operate in cycles to convert heat into work, and their efficiency is defined as the ratio of the work obtained from the engine to the heat absorbed by the engine from its environment during a cycle. A cycle is merely a sequence of events that starts with a system in a particular state and ends up with the system in the same state.

It is clear from the Kelvin statement of the second law that the efficiency of heat engines must be less than unity. But it turns out that there is a maximum efficiency that is the same for all heat engines, regardless of the material nature of the engine. This maximum efficiency can be found by considering a material system that can accept heat from its surroundings and use it to perform work by going through an idealized cycle called the *Carnot cycle*.

For specificity, we consider a system consisting of a gas in a cylinder with a piston that can be placed in contact with heat reservoirs. Thus, heat can flow into and out of the system and the system can do work. Our conclusions, however, are valid for any kind of system that can be surrounded by heat sources or sinks and for any kind of work, not just PV work. That this is true will be obvious from an inspection of the steps in the cycle. There are four steps:

1. Start with the system that is in thermal equilibrium with a heat bath at a temperature T_2. Label this initial state A. (For a gas, the state is determined by the pressure, volume, and temperature, P_2, V_2, T_2.) Let the system do work reversibly to bring it to a new state labeled B. The temperature is still T_2, but work has been done by the system, so an amount of heat Q_2 had to be absorbed from the reservoir. (For a gas, this is a reversible isothermal expansion.) The energy change for this process is $\delta E_1 = |Q_2| - |W_1|$. To emphasize that work done on the system is negative and heat absorbed by the system is positive, this equation was written in terms of absolute values.
2. Now take the system in the state B, surround it with insulating walls, and let it do an amount of work W_2 on its surroundings reversibly. This brings the system to a new state C. (For a gas, this amounts to a reversible adiabatic expansion.) No heat is transferred, so the energy change is just the work done by the system: $\delta E_2 = -|W_2|$. Since no heat is transferred, the temperature must fall to some new value T_1.
3. Put the system in state C in contact with a heat bath at temperature T_1 and remove the insulating wall. Now do an amount of work $-W_3$ on the system reversibly and isothermally. (For a gas, this is a reversible isothermal compression.) An amount of heat Q_1 is thereby transferred to the heat bath and the first law requires that $\delta E_3 = -|Q_1| + |W_3|$.
4. To complete the cycle, the system must be brought back to its original state. This can be done by surrounding it with an insulating wall and doing an amount of reversible work that brings it back to its original temperature T_2. (For a gas, this is a reversible adiabatic compression such that the final pressure and volume are equal to the original pressure and volume so that, because of the equation of state, the final temperature is T_2.) The energy change for this step is $\delta E_4 = |W_4|$.

The sum of the energies for each of the four steps described above step must add up to zero since the energy is a state function. Thus, $|Q_2| - |W_1| - |W_2| - |Q_1| + |W_3| + |W_4| = 0$. The total amount of work done by the system during this cycle is $|W| = |W_1| + |W_2| - |W_3| - |W_4|$, so

$$W = |Q_2| - |Q_1| \qquad (1.10.1)$$

That is, the amount of work done is equal to the amount of heat extracted from the reservoir at T_2 minus the amount given up to the reservoir at T_1.

The efficiency η of the cycle is the ratio of work done to heat taken from the reservoir. That is,

$$\eta = \frac{W}{|Q_2|} = \frac{|Q_2|-|Q_1|}{|Q_2|} = 1 - \frac{|Q_1|}{|Q_2|} \qquad (1.10.2)$$

The temperatures T_2 and T_1 of the two heat reservoirs were written as if they were from the ideal gas temperature scale. In fact, they do not appear in the final result of the Carnot cycle. But a temperature scale that is independent of any material can be defined in purely thermodynamic terms by recognizing that the ratio of the heats in (1.10.2) is a function of the temperatures of the two reservoirs. By considering two Carnot cycles, each with the same lower temperature, it can be shown that

$$\frac{|Q_1|}{|Q_2|} = \frac{f(T_1)}{f(T_2)} \qquad (1.10.3)$$

where $f(T)$ is some function of the temperature.

The absolute temperature scale is defined by requiring that the function $f(T)$ is just the thermodynamic temperature such that

$$\frac{|Q_1|}{|Q_2|} = \frac{T_2}{T_1} \qquad (1.10.4)$$

This defines a thermodynamic temperature scale that is independent of any particular material. The same symbol T is used here as in the definition of the ideal gas temperature scale. In fact, if the Carnot cycle is carried out using an ideal gas as the working substance, then the ratio of the heats is indeed equal to the ratio of the ideal gas temperatures. The thermodynamic absolute temperature is therefore equal to the ideal gas temperature. The reason for this equality is that in an ideal gas there are no intermolecular interactions and therefore nothing that identifies the differences among different substances. Note that using (1.10.4) in (1.10.2) gives the maximum thermodynamic efficiency as it is usually written: $\eta = 1 - (T_1/T_2)$. Let us rewrite (1.10.4) as

$$\frac{|Q_2|}{|T_2|} - \frac{|Q_1|}{|T_1|} = 0 \qquad (1.10.5)$$

or

$$\frac{Q_2}{T_2} + \frac{Q_1}{T_1} = 0 \qquad (1.10.6)$$

where the removal of the absolute value signs defines positive heat as being heat absorbed by the system.

Since the heat absorbed in going from one state to another depends on the path, it is not a state function. But (1.10.6) can be used to define a state function that includes the concept of heat. To do this, consider any cycle that brings a system through a succession of states that returns the system to its original state. This is not necessarily a Carnot cycle, but simple construction shows that it can be approximated by a succession of many small Carnot cycles, for each of which (1.10.6) is true. By taking the limit of an infinite number of Carnot cycles, it is easy to show that the integral of dQ/T over the cycle is zero, where dQ is the infinitesimal amount of heat reversibly transferred to the system at temperature T. Thus, if the cycle is separated into two parts such that the system first goes from state A to state B and then from state B to state A, we have

$$\int_A^B \frac{dQ}{T} + \int_B^A \frac{dQ}{T} = 0 \tag{1.10.7}$$

This means that no matter how we get from A to B, the result is always the same integral. That is, integrating over any path from A to B always gives the negative of the integral from B to A. Since both paths are arbitrary, the integrals must both be path independent. The integral is therefore a state function. This is the definition of the entropy:

$$S_B - S_A = \int_A^B \frac{dQ}{T} \tag{1.10.8}$$

The importance of the entropy arises from two basic (and related) attributes: the first is that the total entropy change in physical processes is always positive, and the second is that this fact can be used to determine the conditions required for a system to be in thermodynamic equilibrium. To show this, we start with the fact that the efficiency of the transformation of a given amount of heat into work is a maximum for a reversible Carnot cycle. The reason for this is that in a Carnot cycle the work done by the system in steps 1 and 2 is a maximum and the work done on the system in steps 3 and 4 is a minimum because the work is done reversibly. Thus, the work done by the system in a Carnot cycle is the maximum that can be done in bringing the system from its initial state to its final state.

Since the energy change of the system is zero for the cycle, the heat absorbed by the system is a minimum. Remember that this minimum is relative to all the irreversible paths in going through a cycle in which heat is absorbed and work is done in a system such that the system ends up in its initial state. Thus, the maximum thermodynamic efficiency is

$$\frac{T_2 - T_1}{T_2} = \left.\frac{Q_2 + Q_1}{Q_2}\right|_{rev} > \left.\frac{Q_2 + Q_1}{Q_2}\right|_{irrev} \tag{1.10.9}$$

where the first equality comes from equation (1.10.4). From (1.10.9), it follows that

$$\left|\frac{Q_1}{T_1} + \frac{Q_2}{T_1}\right| \leq 0 \tag{1.10.10}$$

where the equality holds for a reversible process and the inequality holds for an irreversible process. Since a cycle is irreversible, if any part of it is irreversible, it follows from (1.10.10) that

$$\oint \frac{dQ}{T} \leq 0 \tag{1.10.11}$$

where the integral is taken over the entire cycle. Now consider the cycle as the sum of two processes, the first of which takes the system from its starting state A to some state B, and the second of which takes the system from state B back to state A. Then, (1.10.11) is

$$\int_A^B \frac{dQ}{T} + \int_B^A \frac{dQ}{T} \leq 0 \tag{1.10.12}$$

Now assume that the return part of the cycle, from B to A, is done reversibly. Then the second integral in (1.10.12) is just the entropy change of the system in going from state B to state A and (1.10.12) becomes

$$S_B - S_A \geq \int_A^B \frac{dQ}{T} \tag{1.10.13}$$

Equation (1.10.13) is equivalent to

$$\delta S \geq \frac{dQ}{T} \tag{1.10.14}$$

That is, the entropy difference connecting two infinitesimally close states of a system is equal to the heat absorbed by the system divided by the temperature if the heat is transferred reversibly, and is greater than this if the heat is transferred irreversibly.

Note that the first law requires that $\delta Q = \delta U + \delta W$, so because of (1.10.14) the entropy change at constant energy and volume is

$$\delta S \geq \frac{\delta U + \delta W}{T} \tag{1.10.15}$$

This equation is the basis of the equilibrium conditions of thermodynamics. If, for example, the only work involved is pressure–volume work, then it follows from (1.10.15) that

$$\delta S_{U,V} \geq 0 \tag{1.10.16}$$

where the equality holds for a reversible process and the inequality holds for an irreversible process.

Equation (1.10.16) states that the entropy for a system with constant energy and constant volume is a maximum. Of course, if other forms of work are involved, then (1.10.16) becomes

$$\delta S_{U,\Phi} \geq 0 \tag{1.10.17}$$

where Φ represents all the extensive parameters for all the types of work done by the system.

1.11 Free energies and equilibrium conditions

From (1.10.14), the heat change in a reversible process is related to the entropy by $dQ = TdS$, so the first law can be written as

$$dU = TdS - dW_r \tag{1.11.1}$$

where dW_r is the work done *reversibly* in increasing the energy of the system by dU. For an isothermal process that brings the system from state A to state B, integration of equation (1.11.1) gives $-W_r(A \rightarrow B) = (U_B - U_A) - (TS_B - TS_A)$ (isothermal process). This leads us to define a function A by

$$A = U - TS \tag{1.11.2}$$

A is called the Helmholtz free energy, and ΔA is the maximum available work when a system goes from one state to another via an isothermal process.

The Gibbs free energy function is defined by

$$G = A + PV = H - TS \qquad (1.11.3)$$

It is related to the maximum work that can be done by the system in a constant temperature, constant volume process. To see this, take the differential of (1.11.3) at constant pressure to get

$$dG_P = dA_P + PdV \qquad (1.11.4)$$

If we also require that the change in state is isothermal, then since the Helmholtz free energy is the negative of the maximum (reversible) work done by the system in an isothermal process, (1.11.4) becomes

$$-dG_{T,P} = dW_r - PdV \qquad (1.11.5)$$

so the change in Gibbs free energy at constant temperature and volume is equal to the maximum available work done by the system, exclusive of pressure–volume work.

The Helmholtz and Gibbs free energies both provide very useful criteria for a system to be at equilibrium. The fundamental equilibrium condition is given by equation (1.10.17) [or when only P–V work is present by equation (1.10.16)]. This states that the entropy is a maximum for any process in which the energy and the extensive work parameters (such as volume) are constant. In practice, most thermodynamic processes occur at constant temperature and pressure, or at constant temperature and volume. It is convenient, therefore, to have equilibrium criteria for such processes. The free energies provide such criteria.

If we consider only reversible processes so that differentials of thermodynamic functions connect two equilibrium states, then in going from one state to the other, any heat involved is transferred reversibly. That is, $dQ = TdS$. The equilibrium condition (1.10.16) then becomes

$$dS_{U,V} = 0 \qquad (1.11.6)$$

For convenience, we restrict ourselves to processes involving PV work only; the generalization to include other types of work will generally be obvious. Then, since we consider only reversible processes, the first law becomes

$$dU = TdS - PdV \qquad (1.11.7)$$

from which it follows that

$$dU_{S,V} = 0 \qquad (1.11.8)$$

From the definition of the Helmholtz free energy given by (1.11.2), we get $dA = dU - TdS - SdT$, or, using (1.11.7),

$$dA = -PdV - SdT \qquad (1.11.9)$$

so

$$dA_{V,T} = 0 \qquad (1.11.10)$$

Now start with the definition of the Gibbs free energy given by equation (1.11.3) to get $dG = dU - TdS - SdT + PdV + VdP$, and again use the first law to reduce this to

$$dG = -SdT + VdP \qquad (1.11.11)$$

At constant temperature and pressure this gives

$$dG_{T,P} = 0 \qquad (1.11.12)$$

Equations (1.11.6), (1.11.8), (1.11.10), and (1.11.12) are all conditions for equilibrium. For completeness, the enthalpy can also yield an equilibrium condition, and in fact, it is easy to show that

$$dH_{S,P} = 0 \qquad (1.11.13)$$

1.12 Thermodynamic potentials and Legendre transformations

Since equation (1.11.7) arises directly from the combination of the first and second laws, the energy is said to be a natural function of entropy and volume, so the thermodynamic state of the system is defined as a relation among the energy, volume, and entropy:

$$U = U(S, V) \qquad (1.12.1)$$

Taking the differential of (1.12.1) gives

$$dU = \left(\frac{\partial U}{\partial S}\right)_V dS + \left(\frac{\partial U}{\partial V}\right)_S dV \qquad (1.12.2)$$

Comparing this with (1.11.7) gives

$$\left(\frac{\partial U}{\partial S}\right)_V = T, \quad \left(\frac{\partial U}{\partial V}\right)_S = -P \qquad (1.12.3)$$

The enthalpy, Helmholtz free energy, and Gibbs free energy were defined in the preceding section as natural consequences of considering simple cases of heat transfer and work done by a system. These quantities, along with the energy and entropy, are called *thermodynamic potentials*, and their importance lies in the fact that each of them is a natural function of a different pair of independent variables that can be connected to differing experimental conditions. The adjective "natural" means that the relationships between the thermodynamic potential and the independent variables arise directly from the connections of the potential to work or heat processes. It is, of course, possible to express a potential in terms of variables other than the natural ones, but the differentials of the potentials take their simplest form when expressed in terms of the natural independent variables. The notation adopted for the partial derivatives recognizes that the potentials can be expressed as functions of different sets of variables by explicitly specifying them. Thus, in equation (1.12.3), the partial derivative that defines the temperature clearly refers to the energy as a function of entropy and volume as the independent variables.

But entropy and volume are not convenient variables for describing experiments, so we are led to look for other descriptions. The Helmholtz free energy, for example, is most useful for describing processes at constant temperature and volume, while the Gibbs free energy is naturally connected to constant pressure, constant temperature conditions as shown by (1.11.9) and (1.11.11).

The natural variables for the enthalpy are readily obtained from the energy equation and equation (1.9.7) defining the enthalpy. Taking the differential of (1.9.7) and combining the result with (1.11.7) for the differential of the energy gives

$$dH = TdS + VdP \qquad (1.12.4)$$

so the natural variables for the enthalpy are entropy and pressure.

The expression of the potentials in terms of their natural variables immediately yields thermodynamic relations analogous to (1.12.3). From (1.12.4)

$$\left(\frac{\partial H}{\partial S}\right)_P = T, \quad \left(\frac{\partial H}{\partial P}\right)_S = V \qquad (1.12.5)$$

while (1.11.9) and (1.11.11) yield

$$\left(\frac{\partial A}{\partial V}\right)_T = -P, \quad \left(\frac{\partial A}{\partial T}\right)_V = -S \qquad (1.12.6)$$

$$\left(\frac{\partial G}{\partial T}\right)_P = -S, \quad \left(\frac{\partial G}{\partial P}\right)_T = V \qquad (1.12.7)$$

These equations lead to expressions for the free energies that are extremely useful. If the second equation in (1.12.6) is substituted for the entropy in the definition of the Helmholtz free energy given by (1.11.2), the result is

$$A = U + T\left(\frac{\partial A}{\partial T}\right)_V \qquad (1.12.8)$$

Similarly, if the first of equations (1.12.7) is used to replace the entropy in the definition of the Gibbs free energy, (1.11.3), then

$$G = H + T\left(\frac{\partial G}{\partial T}\right)_P \qquad (1.12.9)$$

Equations (1.12.8) and (1.12.9) are the Gibbs–Helmholtz equations for the Helmholtz and Gibbs free energies, respectively.

In statistical mechanical applications it is sometimes more convenient to express heat capacities in terms of derivatives of free energies rather than energy or enthalpy. This is easily done by using the Gibbs–Helmholtz equations. Differentiating (1.12.8) with respect to temperature at constant volume gives

$$C_V = -T\left(\frac{\partial^2 A}{\partial T^2}\right)_V \qquad (1.12.10)$$

Similarly, differentiating (1.12.9) with respect to temperature at constant pressure gives

$$C_P = -T\left(\frac{\partial^2 G}{\partial T^2}\right)_P \qquad (1.12.11)$$

While the definitions of the potentials are directly connected to work and heat changes, they are also simply described as mathematical transformations of each other that all describe the same thermodynamic information. To see this, start with the energy equation (1.11.7) for a system in which only PV work is done when a system undergoes an infinitesimal reversible change, and consider a system with a given, fixed volume whose energy can vary through heat transfer. The possible states of the system are then determined by the relation between energy and entropy and can be represented by a two-dimensional plot of energy versus entropy, as shown in figure 1.1. The curve represents the energy–entropy relation for the system, while the straight line is a tangent to the line at a given point. Clearly, if every tangent to the curve is specified, then the energy–entropy curve is determined. That is, giving the slope and intercept of the tangent to every point of the curve is fully equivalent to specifying the curve itself.

For the given point with energy U, the slope $(\partial U/\partial S)_V = T$ is given by $(\partial U/\partial S)_V = (U - b)/S$, where b is the intercept. Solving for the intercept gives

$$b = U - \left(\frac{\partial U}{\partial S}\right)_V S \qquad (1.12.12)$$

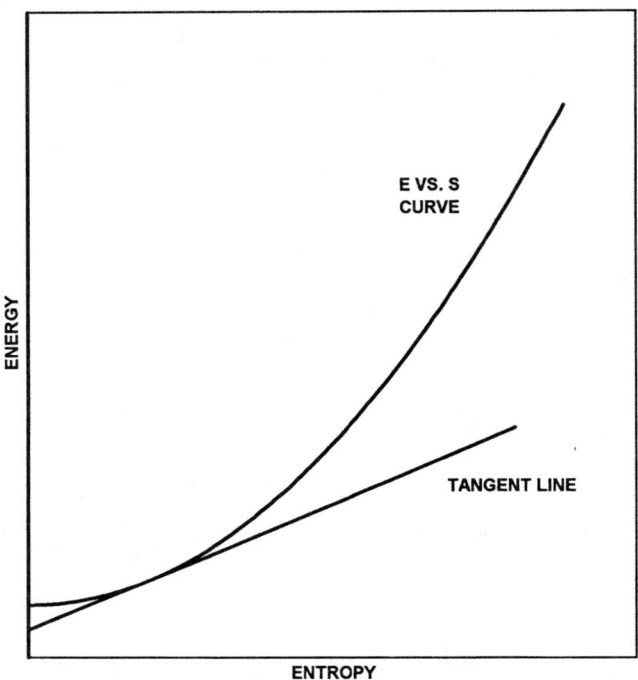

Figure 1.1. Illustration of Legendre transformation.

Equation (1.12.12) is an example of the Legendre transformation, which expresses a function in terms of its derivative. Since the derivative in (1.12.12) is the temperature, the intercept is just the Helmholtz free energy, which we see is a Legendre transformation of the energy.

The generalization to more than one variable is straightforward. In general, if $F = F(X_1, X_2, \ldots, X_t)$ is a function of t variables with derivatives $p_j = \partial F/X_j$, the Legendre transformation is defined by

$$b_F(X_1, X_2, \ldots) = F - \sum_j p_j X_j \qquad (1.12.13)$$

The transform $b_F(X_1, X_2, \ldots)$ is now the intercept of lines that are parallel to the surface defined by the function F in a t-dimensional space. It is the transform of the function F with respect to the variables $(X_1, X_2, \ldots)$. Equation (1.12.13) may be called the *complete* Legendre transformation in that all the variables were transformed. This is indicated by writing the transform as a function of all the variables. A partial Legendre transformation is one in which only some of the variables are transformed. The Helmholtz free energy, for example, is a partial transform of the energy with respect to the entropy.

The partial Legendre transform of the energy with respect to the volume is

$$b_U(V) = U - \left(\frac{\partial U}{\partial V}\right)_S V \qquad (1.12.14)$$

But the derivative in (1.12.14) is just the negative of the pressure [equation (1.12.3)], so

$$b_U(V) = U + PV \qquad (1.12.15)$$

which is the definition of the enthalpy. Finally, if a Legendre transform of the energy with respect to entropy and volume is performed, the result is

$$b_U(S, V) = U - \left(\frac{\partial U}{\partial S}\right)_V S - \left(\frac{\partial U}{\partial V}\right)_S V \qquad (1.12.16)$$

But from (1.12.3) the first derivative on the right of (1.12.16) is the temperature, and the other derivative is the negative of the pressure, so

$$b_U(S, V) = U - TS + PV \qquad (1.12.17)$$

which is just the Gibbs free energy. The four thermodynamic potentials are therefore related by Legendre transformations. Other potentials can be defined by additional Legendre transforms, but these four are the most useful. A simple scheme describing them is given in the following diagram:

$$\begin{array}{c} S + U - V \\ + \qquad - \\ H \qquad A \\ + \qquad - \\ P + G - T \end{array}$$

where each function has its independent variables on either side of it, with the appropriate plus or minus sign indicated.

1.13 Chemical potentials

Up to this point, we have assumed that our system was closed. That is, there was no transfer of matter during any change from one equilibrium state to another, so the dependence of energy on composition did not have to be made explicit. But there are many cases in which the composition changes during a thermodynamic process; the system is then an open system in that matter can pass in and out of it. The composition dependence of the energy must then be included explicitly and equation (1.12.1) must then be written as

$$U = U(S, V, N_1, N_2, \ldots, N_C) \tag{1.13.1}$$

This equation refers to a system that has c components, N_j being the number of molecules of the jth component. The differential form of (1.13.1) is

$$dU = \left(\frac{\partial U}{\partial S}\right)_{V,N_j} dS + \left(\frac{\partial U}{\partial V}\right)_{S,N_j} dV + \sum_i \left(\frac{\partial U}{\partial N_i}\right)_{S,V,N_{j\ne i}} dN_i \tag{1.13.2}$$

The first two derivatives are at constant composition, as well as constant volume and entropy, respectively, while the derivative in the ith term of the sum is at constant entropy, volume, and number of molecules of every type except for the ith. To be consistent with (1.11.7), the derivatives in the first two terms must still be given by temperature and negative pressure as in equation (1.12.3). Also, the derivatives with respect to the number of molecules are given their own symbol μ_i, so we have

$$\left(\frac{\partial U}{\partial S}\right)_{V,N_j} = T \tag{1.13.3}$$

$$\left(\frac{\partial U}{\partial V}\right)_{S,N_j} = -P \tag{1.13.4}$$

$$\left(\frac{\partial U}{\partial N_i}\right)_{S,V,j(\ne i)} = \mu_i \tag{1.13.5}$$

where μ_i is called the chemical potential of the ith component. We rewrite (1.13.2) as

$$dU = TdS - PdV + \sum_i \mu_i dN_i \tag{1.13.6}$$

The other thermodynamic potentials are easily generalized to include the composition variables and the chemical potentials. Combining (1.13.6) first with (1.12.4), then with (1.11.9), and then with (1.11.11) gives

$$dH = TdS + VdP + \sum_i \mu_i dN_i \tag{1.13.7}$$

$$dA = -SdT - PdV + \sum_i \mu_i dN_i \tag{1.13.8}$$

$$dG = -SdT + VdP + \sum_i \mu_i dN_i \tag{1.13.9}$$

Note that the chemical potential is given by partial derivatives of each of the potentials as

$$\mu_i = \left(\frac{\partial U}{\partial N_i}\right)_{S,V,N_{j \neq i}} = \left(\frac{\partial H}{\partial N_i}\right)_{S,P,N_{j \neq i}} = \left(\frac{\partial A}{\partial N_i}\right)_{T,V,N_{j \neq i}} = \left(\frac{\partial G}{\partial N_i}\right)_{T,P,N_j} \quad (1.13.10)$$

1.14 Conditions of phase equilibria and stability

Equation (1.13.6) is valid for any system; it gives the energy change when a system goes from one state to another by a reversible process, the two states differing infinitesimally in their entropy, volume, and composition. Solving for the entropy gives

$$dS = \frac{dU}{T} + \frac{PdV}{T} - \sum_i \frac{\mu_i}{T} dN_i \quad (1.14.1)$$

Since (1.14.1) is valid for any system, it is valid for a system that consists of two phases. Let one phase be labeled by a superscript A and the other by a superscript B, and consider a process in which the energy, volume, and composition of phases A and B change because of transfers of matter and energy between the two phases. Then, if we assume the total system is isolated so no heat or work is transferred to it, the total entropy vanishes and (1.14.1) becomes

$$dS = dS^A + dS^B$$
$$= \frac{dU^A}{T^A} + \frac{P^A dV^A}{T^A} - \sum_i \frac{\mu_i^A}{T^A} dN_i^A + \frac{dU^B}{T^B} + \frac{P^B dV^B}{T^B} - \sum_i \frac{\mu_i^B}{T^B} dN_i^B$$
$$= 0 \quad (1.14.2)$$

But since the system as a whole is isolated, we must also have

$$dU^A = -dU^B$$
$$dV^A = -dV^B$$
$$dN_i^A = -dN_i^B \quad (1.14.3)$$

Using (1.14.3) in (1.14.2) gives

$$\left(\frac{1}{T^A} - \frac{1}{T^B}\right) dU^A + \left(\frac{P^A}{T^A} - \frac{P^B}{T^B}\right) dV^A - \sum_i \left(\frac{\mu_i^A}{T^A} - \frac{\mu_i^B}{T^B}\right) dN_i^A = 0 \quad (1.14.4)$$

But the variations in energy, volume, and composition are independent, so each term in (1.14.4) must vanish, and therefore

$$T_A = T_B \quad (1.14.5)$$
$$P_A = P_B \quad (1.14.6)$$
$$\mu_i^A = \mu_i^B \quad (1.14.7)$$

This result can obviously be extended to any number of phases, so for internal equilibrium of a system consisting of c components distributed among p phases, the temperature must be the same in every phase (thermal equilibrium), the pressure must be the same in every phase (hydrostatic equilibrium), and the chemical potential of each component must be the same in every phase (chemical equilibrium).

From the mode of their derivation, the conditions of equality of temperatures, pressures, and chemical potentials obviously apply to a single-phase system as well as multiphase systems. For a single-phase system these conditions state that, for equilibrium, the temperature, pressure, and chemical potentials must each be constant throughout the phase.

The phase rule addresses the following question: for a multiphase system, how many thermodynamic variables are needed to specify the state of the system? Since the most useful thermodynamic independent variables for laboratory work are temperature and pressure, it is most convenient to use the Gibbs free energy to answer this question. The equilibrium condition in terms of the Gibbs free energy is that the differential given by (1.13.9) must vanish for a reversible process. The Gibbs free energy change for the Kth phase is given by an equation just like (1.13.9):

$$dG^K = -S^K dT + V^K dP + \sum_{i=1}^{c} \mu_i^K dN_i^K \qquad (1.14.8)$$

The entropy and volume are extensive quantities, and the temperature and pressure are the same for every phase, so summing (1.14.8) over all phases gives the total Gibbs free energy differential for the multiphase system. The total free energy change for the entire system must vanish because it is isolated, so

$$dG = -S dT + V dP + \sum_{K=1}^{p}\sum_{i=1}^{c} \mu_i^K dN_i^K = 0 \qquad (1.14.9)$$

This equation defines the equilibrium state of the system in that, if the temperature, pressure, and composition variables are given, the state is determined. There are c composition variables for each phase, giving a total of $cp + 2$ variables including the temperature and pressure. But these variables are not all independent. In the first place, because the system as a whole is closed, the total number of molecules of each component is fixed, so the total variation over all phases for each component must vanish. That is,

$$\sum_{K=1}^{p} dN_i^K = 0 \qquad (1.14.10)$$

There are p such equations, one for each phase so this imposes p conditions on the variables. Furthermore, the equality of chemical potentials for a given component among phases gives

$$\mu_i^1 = \mu_i^2 = \mu_i^3 \ldots = \mu_i^p \qquad (1.14.11)$$

This is a set of $(p - 1)$ equations for each component, so (1.14.11) represents an additional $c(p - 1)$ conditions that must be satisfied. If we call the difference between the number of variables and the number of conditions f, then

$$f = (cp+2) - p - c(p-1)$$
$$= c - p + 2 \qquad (1.14.12)$$

f is called the number of *degrees of freedom* of the system since it is the number of variables that may be independently varied. This is the famous phase rule of Gibbs.

Equations (1.14.5)–(1.14.7) are necessary for equilibrium, but not sufficient. They arise from the requirement that the first variation of the entropy with respect to any possible variations of the system variables is equal to, or greater than, zero. But in order for this to represent a maximum, the second differential of the entropy must be negative. That is,

$$d^2 S < 0 \qquad (1.14.13)$$

or, using (1.14.1),

$$d(dS) = \frac{\partial (dS)}{\partial U} dU + \frac{\partial (dS)}{\partial V} dV - \sum_j \frac{\partial (dS)}{\partial N_j} dN_j < 0 \qquad (1.14.14)$$

$$d^2 S = \frac{\partial}{\partial U}\left(\frac{1}{T}\right) d^2 U + \frac{\partial}{\partial U}\left(\frac{P}{T}\right) dV dU - \sum_j \frac{\partial}{\partial U}\left(\frac{\mu_j}{T}\right) dN_j dU$$
$$+ \frac{\partial}{\partial V}\left(\frac{1}{T}\right) dV dU + \frac{\partial}{\partial V}\left(\frac{P}{T}\right) d^2 V - \sum_j \frac{\partial}{\partial V}\left(\frac{\mu_j}{T}\right) dN_j dU$$
$$+ \sum_i \frac{\partial}{\partial N_i}\left(\frac{1}{T}\right) dN_i dU + \sum_i \frac{\partial}{\partial N_i}\left(\frac{P}{T}\right) dN_i dV - \sum_{i,j} \frac{\partial}{\partial N_i}\left(\frac{\mu_j}{T}\right) dN_j dN_i < 0$$
$$(1.14.15)$$

If (1.14.11) is to be true, then if the composition and volume are held constant, the first term in (1.14.15) must be less than zero. That is,

$$\left[\frac{\partial}{\partial U}\left(\frac{1}{T}\right)\right]_{V,N} < 0 \qquad (1.14.16)$$

But

$$\left[\frac{\partial}{\partial U}\left(\frac{1}{T}\right)\right]_{V,N} = -T^2 \left(\frac{\partial T}{\partial U}\right)_{V,N} = -\frac{T^2}{C_V} < 0 \qquad (1.14.17)$$

because the derivative in the middle term in (1.14.17) is the reciprocal of the heat capacity at constant volume. Thus, the condition for stable thermal equilibrium is that the heat capacity is positive:

$$C_V > 0 \qquad (1.14.18)$$

By considering the case of constant energy and composition, the condition for mechanical equilibrium is easily shown to be

$$\left[\frac{\partial}{\partial V}\left(\frac{P}{T}\right)\right]_{U,N_j} < 0 \qquad (1.14.19)$$

By keeping all compositions constant except one, as well as the energy and volume, the condition for chemical equilibrium is found to be

$$\sum_{i,j}\left[\frac{\partial}{\partial N_j}\left(\frac{\mu_i}{T}\right)\right]_{U,V,N_{i\neq j}} dN_i dN_j > 0 \qquad (1.14.20)$$

By starting with the Helmholtz free energy (1.13.8), and requiring the second differential to be positive (because the equilibrium condition requires the free energy to be a minimum), a similar derivation gives somewhat more useful forms for the conditions of mechanical and chemical equilibrium as

$$\left(\frac{\partial P}{\partial V}\right)_{T,N_j} < 0 \qquad (1.14.21)$$

$$\sum_{i,j} \frac{\partial \mu_i}{\partial N_j} dN_i dN_j > 0 \qquad (1.14.22)$$

Equations (1.14.18) and (1.14.21) are obviously true on physical grounds. If the temperature is raised, the system must gain energy, and if the pressure is increased, the volume of the system decreases (the isothermal compressibility is positive).

If the conditions arising from both the first and second differentials of the entropy (or free energy) are both satisfied, the system is in equilibrium. These are *local* conditions in that they are the result of considering changes in the system close to the maximum of the entropy function, or the minimum of the free energy. They do not guarantee that the system is in the lowest of all free energy states (or maximum of all entropy states). It is entirely possible for a system to have a number of states for which the entropy is a maximum and the free energy a minimum. There is always at least one state for which the entropy is higher, and the free energy smaller, than all the others. These are called states of *absolute*, or *stable*, *equilibrium*. The states for which all equilibrium conditions are satisfied, but which are not states of the highest entropy or lowest free energy, are called *metastable states*.

1.15 Euler's theorem and the Gibbs–Duhem equation

A major task in thermodynamics is to find the relationships among measurable physical quantities. These are needed to enable the enormous amount of possible experimental results to be described in a rational, ordered theoretical structure. Also, thermodynamics places restrictions on the possible relations among physical quantities; if experimental results violate these restrictions, they must be in error. Furthermore, a knowledge of the thermodynamic relations is necessary if any microscopic theory is to be used to understand macroscopic systems. There are a number of simple mathematical results that allow us to connect the basic laws of thermodynamics to physical properties and to obtain the desired relations among them.

The energy is a natural function of entropy, volume, and composition. These are all extensive variables, so if the number of molecules of each component in a phase increases by some fraction, the entropy volume and energy both increase by the same fraction. That is, if each composition variable is multiplied by the same number λ, then

$$\lambda U(S, V, N_1, N_2, \ldots, N_c) = U(\lambda S, \lambda V, \lambda N_1, \lambda N_2, \ldots, \lambda N_c) \qquad (1.15.1)$$

That is, the energy is a homogeneous function of its natural variables of order unity.

The Euler theorem for homogeneous functions can therefore be applied to (1.15.1) to yield very useful results. This theorem states that, given a homogeneous function of order n of variables x_j, which is defined by

$$\lambda^n f(x_1, x_2, x_3, \ldots) = f(\lambda x_1, \lambda x_2, \lambda x_3, \ldots), \tag{1.15.2}$$

then

$$nf(x_1, x_2, x_3, \ldots) = x_1 \frac{\partial f}{\partial x_1} + x_2 \frac{\partial f}{\partial x_2} + x_3 \frac{\partial f}{\partial x_3} \ldots \tag{1.15.3}$$

To prove this, differentiate (1.15.2) with respect to λ to get

$$n\lambda^{n-1} f(x_1, x_2, x_3, \ldots) = \frac{\partial f}{\partial (\lambda x_1)} \frac{\partial (\lambda x_1)}{\partial \lambda} + \frac{\partial f}{\partial (\lambda x_2)} \frac{\partial (\lambda x_2)}{\partial \lambda} \ldots$$

$$= x_1 \frac{\partial f}{\partial (\lambda x_1)} + x_2 \frac{\partial f}{\partial (\lambda x_2)} \ldots \tag{1.15.4}$$

which, upon setting $\lambda = 1$, gives (1.15.3). For $n = 1$, applying this to the energy, we get

$$U(S, V, N_1, N_2, \ldots, N_c)$$
$$= S \left(\frac{\partial U}{\partial S} \right)_{V,N} + V \left(\frac{\partial U}{\partial V} \right)_{S,N} + \sum_{j=1}^{c} N_j \left(\frac{\partial U}{\partial N_j} \right)_{V, N(\neq j)}$$

Using the relation of the derivatives to temperature, pressure, and chemical potential given by equations (1.13.3)–(1.13.5), this becomes

$$U = TS - PV + \sum_{j=1}^{c} \mu_j N_j \tag{1.15.5}$$

Thus, the Euler theorem yields an integration of the energy, equation.

The integrated forms of the enthalpy, Helmholtz free energy, and Gibbs free energy in terms of composition are easily obtained by using (1.15.5) in equations (1.9.7), (1.11.2), and (1.11.3) to get

$$H = TS + \sum_{j=1}^{c} \mu_j N_j \tag{1.15.6}$$

$$A = -PV + \sum_{j=1}^{c} \mu_j N_j \tag{1.15.7}$$

$$G = \sum_{j=1}^{c} \mu_j N_j \tag{1.15.8}$$

These equations are quite general and apply to any homogeneous phase of any number of components. From (1.15.8),

$$dG = \sum_{j=1}^{c} \mu_j dN_j + \sum_{j=1}^{c} N_j d\mu_j \qquad (1.15.9)$$

which, when combined with (1.13.9), gives

$$-SdT + VdP + \sum_{j=1}^{c} N_j d\mu_j = 0 \qquad (1.15.10)$$

This is the Gibbs–Duhem equation.

If the temperature and pressure are kept constant, then equilibrium requires the Gibbs free energy change to be constant and (1.15.10) reduces to

$$\sum_{j=1}^{c} N_j d\mu_j = 0 \qquad \text{(constant } P, T\text{)} \qquad (1.15.11)$$

Equation (1.15.11) is the Gibbs–Duhem equation at constant temperature and pressure. It is valid for any process involving an infinitesimal change in the composition of a phase but carried out at constant temperature and pressure, and is very useful for analysis of the equilibria of solutions and chemical reactions.

1.16 Reciprocity relations of Maxwell

The differentials of the thermodynamic potentials given by equations (1.13.6)–(1.13.9) all have the form

$$dF = MdX + NdY + \sum_{i=1}^{c} \mu_i dN_j \qquad (1.16.1)$$

where dF is a perfect differential and (M, N, μ_i) are all functions of (X, Y, N_i). That is, the integral of dF is independent of the path of integration, so F is a state function. The coefficients of the differentials in (1.16.1) are therefore

$$M = \left(\frac{\partial F}{\partial X}\right)_{Y,N}, \quad N = \left(\frac{\partial F}{\partial Y}\right)_{X,N}, \quad \mu_i = \left(\frac{\partial F}{\partial N_i}\right)_{X,Y,N(\neq i)} \qquad (1.16.2)$$

For simplicity of notation, let us consider a phase of constant composition, with all $dN_i = 0$, so that (1.16.1) and (1.16.2) reduce to

$$dF = MdX + NdY \qquad (1.16.3)$$

$$M = \left(\frac{\partial F}{\partial X}\right)_Y, \quad N = \left(\frac{\partial F}{\partial Y}\right)_X \qquad (1.16.4)$$

Since the order of differentiation is immaterial, differentiating the first equation in (1.16.4) with respect to Y and the second with respect to X gives the same result, so

$$\left(\frac{\partial M}{\partial Y}\right)_X = \left(\frac{\partial N}{\partial X}\right)_Y \qquad (1.16.5)$$

[Note that there is no loss of generality by considering a system of constant composition. If the composition is allowed to vary, we simply specify that the differentiation is carried out at constant composition in (1.16.5).]

If (1.16.5) is applied to the differentials of the potentials in (1.13.6)–(1.13.9), we get that

$$\left(\frac{\partial T}{\partial V}\right)_S = -\left(\frac{\partial P}{\partial S}\right)_V \qquad (1.16.6)$$

$$\left(\frac{\partial T}{\partial P}\right)_S = \left(\frac{\partial V}{\partial S}\right)_P \qquad (1.16.7)$$

$$\left(\frac{\partial S}{\partial V}\right)_T = \left(\frac{\partial P}{\partial T}\right)_V \qquad (1.16.8)$$

$$\left(\frac{\partial S}{\partial P}\right)_T = -\left(\frac{\partial V}{\partial T}\right)_P \qquad (1.16.9)$$

These are the Maxwell reciprocity relations. They have many applications in finding relations among thermodynamic quantities, such as those that are important for chemical and phase equilibria and in relating heat capacities to other physical quantities.

Since the Maxwell relations are a simple result of the fact that the order of differentiating makes no difference in the value of a second derivative, similar relations among derivatives containing composition variables also exist. They can be found by a straightforward extension of the above procedure.

1.17 Useful differential relations

The state of a thermodynamic system is completely determined by any one of the thermodynamic potentials. The Gibbs free energy, for example, specifies the state as a function of the temperature, pressure, and composition. For a single-phase system, the phase rule tells us that the number of independent variables is one more than the number of components. For a single-phase, one-component system, there are only two independent variables. That is, specifying any two variables determines all other thermodynamic quantities. Since these quantities are derivatives of the potentials, we need to know the connections among the derivatives. Finding these connections is easily accomplished with the aid of some properties of derivatives of a function of several variables, which we hereby establish.

Let z be a function of two independent variables x and y,

$$z = z(x, y) \qquad (1.17.1)$$

so that

$$dz = \left(\frac{\partial z}{\partial x}\right)_y dx + \left(\frac{\partial z}{\partial y}\right)_x dy \qquad (1.17.2)$$

Divide (1.17.2) through by dx to get

$$\frac{dz}{dx} = \left(\frac{\partial z}{\partial x}\right)_y + \left(\frac{\partial z}{\partial y}\right)_x \frac{dy}{dx} \qquad (1.17.3)$$

This equation is true for any dz, so it is true for $dz = 0$, and therefore

$$\left(\frac{\partial z}{\partial x}\right)_y + \left(\frac{\partial z}{\partial y}\right)_x \left(\frac{\partial y}{\partial x}\right)_z = 0 \tag{1.17.4}$$

Similarly, dividing (1.17.2) through by dy and applying the condition of constant z gives

$$\left(\frac{\partial z}{\partial y}\right)_x + \left(\frac{\partial z}{\partial x}\right)_y \left(\frac{\partial x}{\partial y}\right)_z = 0 \tag{1.17.5}$$

Equations (1.17.4) and (1.17.5) give

$$\left(\frac{\partial y}{\partial x}\right)_z = -\frac{\left(\frac{\partial z}{\partial x}\right)_y}{\left(\frac{\partial z}{\partial y}\right)_x} \tag{1.17.6}$$

$$\left(\frac{\partial x}{\partial y}\right)_z = -\frac{\left(\frac{\partial z}{\partial y}\right)_x}{\left(\frac{\partial z}{\partial x}\right)_y} \tag{1.17.7}$$

and from (1.17.6) and (1.17.7),

$$\left(\frac{\partial x}{\partial y}\right)_z = \frac{1}{\left(\frac{\partial y}{\partial x}\right)_z} \tag{1.17.8}$$

Now multiply (1.17.4) by the derivative of y with respect to z at constant x and use (1.17.6) to get

$$\left(\frac{\partial y}{\partial z}\right)_x \left(\frac{\partial z}{\partial y}\right)_x + \left(\frac{\partial x}{\partial y}\right)_z \left(\frac{\partial y}{\partial z}\right)_x \left(\frac{\partial z}{\partial x}\right)_y = 0 \tag{1.17.9}$$

which, because of (1.17.8), becomes

$$\left(\frac{\partial x}{\partial y}\right)_z \left(\frac{\partial y}{\partial z}\right)_x \left(\frac{\partial z}{\partial x}\right)_y = -1 \tag{1.17.10}$$

Equations (1.17.7)–(1.17.10) are very useful in manipulating derivatives of thermodynamic functions.

1.18 Equations of state and heat capacity relations

Consider a simple system consisting of a single phase of constant composition. The properties of the system are then determined by any two thermodynamic variables, and the state of the system is completely described by any of the thermodynamic potentials as a function of two variables. In this sense, the thermodynamic equations are equations of state, but this name is normally applied to the relation between pressure, temperature, and volume.

A thermodynamic equation of state is easily obtained as follows: start with equation (1.11.7), divide by dV, and then specify that the temperature is constant to get the derivative of the energy with respect to volume at constant temperature:

$$\left(\frac{\partial U}{\partial V}\right)_T = T\left(\frac{\partial S}{\partial V}\right)_T - P \qquad (1.18.1)$$

Substitution of the Maxwell relation (1.16.8) into (1.18.1) and solving for the pressure gives

$$P = T\left(\frac{\partial P}{\partial T}\right)_V - \left(\frac{\partial U}{\partial V}\right)_T \qquad (1.18.2)$$

which is a thermodynamic equation of state for the pressure as a function of volume and temperature.

Another equation of state can be derived by starting with equation (1.12.4) for the reversible differential of the enthalpy. Dividing (1.12.4) by dP and then requiring the temperature to be constant gives

$$\left(\frac{\partial H}{\partial P}\right)_T = T\left(\frac{\partial S}{\partial P}\right)_T + V \qquad (1.18.3)$$

Using the Maxwell relation (1.16.9) in (1.18.3) and solving for the volume gives

$$V = T\left(\frac{\partial V}{\partial T}\right)_P + \left(\frac{\partial H}{\partial P}\right)_T \qquad (1.18.4)$$

which is a thermodynamic equation of state for the volume as a function of temperature and pressure.

The heat capacity, thermal expansion, and bulk modulus (or its reciprocal, compressibility) are among the more important thermodynamic derivatives, not only because they can be experimentally measured and are of practical use, but also because they are readily related to atomistic theories of matter. The heat capacities at constant pressure and at constant volume have been defined in section 1.9. The definitions of the isothermal compressibility κ and the isobaric thermal expansion coefficient α are

$$\kappa = -\frac{1}{V}\left(\frac{\partial V}{\partial P}\right)_T \qquad (1.18.5)$$

$$\alpha = \frac{1}{V}\left(\frac{\partial V}{\partial T}\right)_P \qquad (1.18.6)$$

The bulk modulus is defined by

$$B = -V\left(\frac{\partial P}{\partial V}\right)_T \qquad (1.18.7)$$

and because of (1.17.8) this is just the reciprocal of the compressibility. The compressibility is the fractional decrease in volume per unit of applied pressure, and the thermal expansion coefficient is the fractional increase in volume per unit temperature increase.

30 STATISTICAL MECHANICS OF SOLIDS

Measured thermal expansions are usually reported as linear thermal expansion coefficients, which are defined as

$$\alpha_L = \frac{1}{L}\left(\frac{\partial L}{\partial T}\right)_P \tag{1.18.8}$$

where L is a linear dimension of the sample.

The relation between the volume and linear expansion coefficient is readily obtained by writing the volume as a product of the linear dimensions L_1, L_2, L_3 so that $dV = L_1L_2dL_3 + L_1L_3dL_2 + L_2L_3dL_1$ and $dV/V = dL_1/L_1 + dL_2/L_2 + dL_3/L_3$. Putting this in (1.18.6) gives

$$\alpha = \frac{1}{L_1}\left(\frac{\partial L_1}{\partial T}\right)_P + \frac{1}{L_2}\left(\frac{\partial L_2}{\partial T}\right)_P + \frac{1}{L_3}\left(\frac{\partial L_3}{\partial T}\right)_P \tag{1.18.9}$$

and if the system is isotropic, all three terms in (1.18.9) are equal, so comparing it to (1.18.8) gives

$$\alpha = 3\alpha_L \tag{1.18.10}$$

For an anisotropic system, such as an orthorhombic crystal, the linear expansion coefficient varies with direction.

A relation between the heat capacities, the thermal expansion coefficient and the compressibility is found as follows. First, write the energy as a function of volume and temperature, instead of its natural variables volume and entropy, so that its differential is

$$dU = \left(\frac{\partial U}{\partial V}\right)_T dV + \left(\frac{\partial U}{\partial T}\right)_V dT \tag{1.18.11}$$

from which $(\partial U/\partial T)_P = (\partial U/\partial V)_T (\partial V/\partial T)_P + (\partial U/\partial T)_V$. But the last term in this equation is just the constant volume heat capacity, so

$$\left(\frac{\partial U}{\partial T}\right)_P = \left(\frac{\partial U}{\partial V}\right)_V \left(\frac{\partial V}{\partial T}\right)_T + C_V \tag{1.18.12}$$

Now start with the definition of the heat capacity at constant pressure to get

$$C_P = \left(\frac{\partial H}{\partial T}\right)_P = \left[\frac{\partial(U+PV)}{\partial T}\right]_P = \left(\frac{\partial U}{\partial T}\right)_P + P\left(\frac{\partial V}{\partial T}\right)_P \tag{1.18.13}$$

From (1.18.12) and (1.18.13), the difference between the two heat capacities is

$$C_P - C_V = \left(\frac{\partial U}{\partial V}\right)_T \left(\frac{\partial V}{\partial T}\right)_P + P\left(\frac{\partial V}{\partial T}\right)_P$$

$$= \left(\frac{\partial V}{\partial T}\right)_P \left[P + \left(\frac{\partial U}{\partial V}\right)_T\right] \tag{1.18.14}$$

Because of the thermodynamic equation of state (1.18.2) the term in the square bracket is

$$P+\left(\frac{\partial U}{\partial V}\right)_T = T\left(\frac{\partial P}{\partial T}\right)_V \qquad (1.18.15)$$

so (1.18.14) becomes

$$C_P - C_V = T\left(\frac{\partial V}{\partial T}\right)_P \left(\frac{\partial P}{\partial T}\right)_V \qquad (1.18.16)$$

and using (1.17.7), this becomes

$$C_P - C_V = -T\frac{\left(\frac{\partial V}{\partial T}\right)_P^2}{\left(\frac{\partial V}{\partial P}\right)_T} \qquad (1.18.17)$$

From the definitions of the compressibility and thermal expansion coefficient given by (1.18.5) and (1.18.6), this becomes

$$C_P - C_V = \frac{\alpha^2 T V}{\kappa} \qquad (1.18.18)$$

Experimental measurements are most easily carried out under conditions of constant temperature and pressure, but the statistical mechanical theories of heat capacity are most easily developed with temperature and volume as independent variables. Equation (1.18.8) provides the needed connection between the two heat capacities.

1.19 Magnetic systems

To this point, pressure was assumed to be the only external parameter acting on the system, so it was subject only to PV work. It is obvious that the procedures used to develop the ideas of thermodynamics are the same for any kind of work. All that is needed is to treat the external force and its corresponding property of the system just as pressure and volume were treated. An important example of this in the study of solids is from magnetic materials, in which an external magnetic field H^1 acts on a system with a net magnetization M. The magnetic work accompanying a change in external field is MdH, so assuming that this is the only work (zero pressure), the thermodynamic equations for a magnetic system are obtained by replacing P with H and V with M in the equations for a PV system. Thus, the heat capacity at constant volume is replaced with the heat capacity at constant magnetization:

$$C_M = \left(\frac{\partial U}{\partial T}\right)_M = -T\left(\frac{\partial^2 A}{\partial T^2}\right)_M \qquad (1.19.1)$$

the last equality being the magnetic analog of (1.12.10). The analog of the entropy equations in (1.12.6) and (1.12.7), for a system in a magnetic field, is

$$S = -\left(\frac{\partial G}{\partial T}\right)_H = -\left(\frac{\partial A}{\partial T}\right)_M \qquad (1.19.2)$$

Instead of the compressibility, we now have the derivative of the magnetization with respect to external field, so the analog of the compressibility is the isothermal magnetic susceptibility:

$$\chi_T = \left(\frac{\partial M}{\partial H}\right)_T \tag{1.19.3}$$

For our purpose, we can take the external field as given in the same sense that we took the pressure as given. However, there is a difference between magnetic field and pressure in that the magnetic field carries energy.[2]

Exercises

1.1 Prove that

$$\left[\frac{\partial(A/T)}{\partial T}\right]_V = -\frac{U}{T^2} \tag{1.A}$$

and that

$$\left[\frac{\partial(G/T)}{\partial T}\right]_P = -\frac{H}{T^2} \tag{1.B}$$

These are alternate forms of the Gibbs–Helmholtz equations.

1.2 The Stefan–Boltzmann law for black body radiation states that the energy density of radiation in equilibrium within a cavity varies as the fourth power of the absolute temperature. Derive this law from the fact that the radiation pressure P is related to the energy density u by $P = u/3$, where $u = U/V$, U being the total energy and V being the volume of the cavity, and the energy density is a function of temperature only. (These results can be derived from the electromagnetic theory of radiation.)

1.3 The grand potential for a one-component system containing N molecules is defined by $\Psi = U - TS - \mu N$. Show that

$$N = -\left(\frac{\partial \Psi}{\partial \mu}\right)_{T,V}$$

$$P = -\left(\frac{\partial \Psi}{\partial V}\right)_{T,\mu}$$

$$S = -\left(\frac{\partial \Psi}{\partial T}\right)_{V,\mu}$$

Notes

1. The symbol for the magnetic field is the same as for the enthalpy. But we will be considering only magnetic systems at zero pressure, so there will be no need to specify constant pressure properties and the enthalpy will not

appear in the equations. I have therefore chosen to retain the traditional symbols rather than invent a new one.
2. A clear exposition of the relation of the external field to total magnetic energy is given in chapter 1 of Goodstein, David L.; 1975; *States of Matter*; chapter 1; Prentice-Hall, Englewood Cliffs, N.J.; reprinted by Dover Publications, 1985.

2

Principles of Statistical Mechanics

2.1 Definitions for statistical mechanics

The prime article of faith in scientific theory is that experimentally observed phenomena can be derived from a small number of general principles. The acceptance of this idea implies that, despite the constant changes evident in physical systems, they possess some immutable attributes from which all of their interesting properties can be calculated. A central concept in such a calculation is that of the state of an isolated system. If all of the experimentally measurable properties of a system are known at all times, then this would certainly constitute an adequate description of the state of the system. However, such a definition would be useless for theoretical purposes and is contrary to the assumption that all properties can be related to a few general laws. The most fruitful definition of state is this:

> The state of a physical system is the minimum amount of information required for a calculation of its properties.

A *theory* can be regarded as the set of concepts and mathematical apparatus needed to go from the definition of the state of a system to its observable properties.

A *system* is defined as any physical entity that can be separated from the rest of the world, at least to a good approximation. That is, the energy of interaction of the system with the rest of the world is small compared to the system's energy. No system is truly isolated since there are always some interactions with its surroundings. In fact, by definition, no measurements can be made on isolated systems, and they are therefore of limited interest. However, a system that is too strongly coupled to its environment cannot be treated as a separate entity. The systems that are of interest to us are those for which there is a small interaction with the environment so that a meaning can be assigned to its properties as an entity that is almost independent of the rest of the world. Examples of systems are an atom or molecule in a dilute gas, a gas in an insulating container, a liquid in a bottle, a crystal, and the earth with its atmosphere.

The "nature" of a system is defined by specifying its constituent particles (such as electrons, nuclei, atoms, and/or molecules), the number of different constituent particles, and the interaction potentials among these particles. Often, as a matter of convenience, we go further and specify its phases and structure (i.e., gas, liquid, solid; crystals, polymers, amorphous solid). In principle, the phases and structure can be derived from the constitution and interaction potentials, but this is a difficult task and we are often interested in systems whose phases and structure are given.

The above discussion applies to conservative systems. For nonconservative systems, the effects of the nonconservative forces must be included. But while there may be nonconservative processes at the macro level, they often reduce to conservative processes at the micro level, for example, the phenomena of friction and turbulence. Statistical mechanics is primarily concerned with *macroscopic systems*, which are defined as systems whose dimensions are large compared to atomic dimensions and whose number of constituent particles is large compared to unity.

The definition of state is different for different branches of physical theory. Given the constituent particles, their masses, and the forces acting among them, the state in classical mechanics is completely specified by the coordinates and momenta of the particles at any given instant. The reason for this is that, in classical mechanics, all properties of systems are functions of coordinates and momenta, and once the positions and momenta are given at one particular time, they can be calculated at any future (or past) time. In quantum mechanics, the state is completely defined by the wave function if the constituents, their masses, and their interaction potentials are given, since all measurable physical properties can be obtained from the wave function.

2.2 Thermodynamic state

The first and second laws of thermodynamics provide a basis for constructing relations among the macroscopic properties of physical systems. These properties include temperature, pressure, volume, energy, specific heat, compressibility, and so forth. In a one-component, one-phase system, two variables (e.g., temperature and pressure) are enough to define the thermodynamic state. All other properties are then determined by the equations of thermodynamics. For systems in equilibrium, the thermodynamic state is defined by a number of macroscopic properties determined by the phase rule given by (1.14.12).

The thermodynamic parameters of a system are of two kinds: mechanical and nonmechanical. *Mechanical variables* are those quantities that can be interpreted in mechanical terms, such as energy and internal pressure. *Nonmechanical variables* are those that have no analog in mechanics and are peculiar to thermodynamics. These include temperature and entropy.

Thermodynamic quantities can also be classified as being intensive or extensive. *Intensive parameters* are those that are independent of the size of the system, such as temperature, pressure, and concentration of components. *Extensive parameters* are those that are directly proportional to the amount of matter in the system, such as volume, energy, and heat capacity.

Also, it is useful to distinguish between the properties of the system and external parameters. The *external parameters* define the conditions under which the system exists and its interaction with the environment. These include such quantities as external fields (gravitational, electric, magnetic), the pressure of a movable piston, and the temperature of a heat bath.

Two important points must be noted. First, thermodynamics deals with equilibrium systems, so time and the concept of temporal causality do not enter into it at all. Equilibrium thermodynamic systems are macroscopically static, and the definition of state given here is incapable of describing systems whose properties are changing with time. The idea of thermodynamic equilibrium needs to be applied with care because, in thermodynamics, equilibrium is defined as being the state with the lowest free energy. But there are states whose properties do not change with time, or change so slowly that no change can be measured over laboratory time scales, and yet are not states with the lowest free energy compared to all other possible states. This can happen when the transition to a lower free energy state is hindered by extremely slow kinetic rates, as when nonequilibrium structures are quenched into solids, or when chemical reaction rates are slow, for example, the reaction between hydrogen and oxygen at room temperatures. It is often valid to apply equilibrium thermodynamics to such systems because, while the free energy may not be at an absolute minimum, it might be at a relative minimum. It is then in a metastable state.

The second point is that no actual numerical computations of the thermodynamic properties can be made from the specification of the thermodynamic state of the system. For example, if we are given the temperature and pressure of a gas, thermodynamics tells us that the volume is determined, but it cannot tell us the numerical value of this volume. To get the volume, an equation of state is needed, and this can only be obtained from some source outside of thermodynamics such as experimental data or a microscopic theory. In this sense, the specification of state in thermodynamics is incomplete.

Sometimes a macroscopic state cannot be defined. This is the case for systems that are so chaotic and changing so rapidly that their future properties cannot be determined, or even expressed, in macroscopic terms. For some nonequilibrium systems, however, it may still be possible to define a macroscopic state. For example, a system whose temperature is changing with time can be described by the equations of heat conduction if the temperature gradients are not too large. The state is then defined by including in it the thermal conductivity and the temperature as a function of position at a given time. This constitutes a definition of state for the nonequilibrium system because it allows the temperature distribution to be computed for all future times.

2.3 Comparison of microscopic and macroscopic state

It is always possible to define a microscopic state of the system since, in principle, the equations of motion (quantum or classical) of the particle constituents completely define the system. Microscopic states can be defined in classical and quantum mechanics whether or not the system is in equilibrium since the equations of motion are known.

To appreciate the task of statistical mechanics, it is useful to contrast the following characteristics of the microscopic and macroscopic definitions of state:

1. For macroscopic systems the microscopic definition of state is extremely detailed and requires knowledge about every particle in the system. Classically, it is necessary to specify all coordinates at some particular time. In quantum mechanics, all relevant properties are computed from the wave function at a particular time, the wave function being obtained from a second-order partial differential equation involving all particle coordinates. For macroscopic systems, however, there is no way to get such detailed information, and even if it were available, it would be useless.

In thermodynamics, the macroscopic state says nothing about microscopic parameters and does not even recognize their existence. Furthermore, the macroscopic state is specified by a small number of parameters (the phase rule), in sharp contrast to the enormous number of variables that enter into the microscopic state.

2. The microscopic definition of state is totally causal in the sense that the state at any time is determined by the state at some earlier time. This means that all measurable quantities are determined as a function of time. This is a precise statement even in quantum mechanics, despite the uncertainty princi-

ple, because the state is defined by the wave function whose evolution in time is precisely determined by the wave equation. (It is true that, at an atomistic level, the measurable quantities are subject to quantum mechanical probabilities, but these are determined by the wave function.) The macroscopic state, on the other hand, is only partially causal. For equilibrium systems, it is causal in the trivial sense that it does not change in time. For nonequilibrium systems, it may be causal to a good enough approximation for certain phenomenan such as diffusion and heat transfer (when gradients are not too large) in the sense that future temperature or concentration distributions can be computed from initial conditions. But for systems that are far from equilibrium, the future macroscopic state cannot always be predicted from a past macroscopic state. Note that in those cases that may be described causally, laws specific to the phenomenon must be introduced (e.g., Fick's law and Fourier's law).

3. The microscopic state is reversible with respect to time. That is, for every process evolving into the future that brings the system from a state A to a state B, there is a corresponding process evolving into the past that brings the system from state B to state A. What this means is that every temporal process is described by equations that give solutions for the properties as a function of time and valid solutions are obtained if the time is reversed in these equations. For macroscopic states, however, temporal processes are irreversible. That is, for the evolution of a macroscopic system in time, there is no corresponding process that undoes the evolution if the time is reversed.

4. If the microscopic state is known, along with some fundamental constants (such as the mass and charge of the electron, Planck's constant, and the velocity of light), then actual numerical calculations of the system's properties can be made, at least in principle (within the limitations of the uncertainty principle). Numerical values of macroscopic properties cannot be computed even in principle from the definition of macroscopic state. The equations defining macroscopic state are merely general relations among macroscopic parameters.

2.4 The relation between microscopic and macroscopic state

Our belief in the ultimate unity of scientific knowledge leads us to conclude that there must be some connection between the macroscopic (thermodynamic) and microscopic (mechanical) definitions of state. The first task of statistical mechanics is to find that connection. For the many particle systems that concern us, it is impossible to obtain an actual specification of the microscopic state and, even if such a specification were available, the equations of motion would be so complicated that they could not be solved. Also, we are not particularly interested in such detailed information. We are, however, very much interested in the possibility of interpreting thermodynamic data in microscopic terms, and in particular, it would be very desirable to obtain a method of computing macroscopic quantities from atomic properties, thereby making up for the inherent deficiency of thermodynamics.

In view of these considerations, any bridge connecting macroscopic and microscopic descriptions of material systems must have the following characteristics:

1. A great deal of microscopic information must be erased in constructing macro- from microstates so that we are not encumbered with an enormous number of microscopic parameters.
2. Enough microscopic information must be retained to enable the calculation of bulk properties from atomic properties.

3. Temporal causality and reversibility, which operate at the micro level, do not appear at the macro level. (Except insofar as introduced by special postulates such as the law of thermal conduction. The discipline of irreversible thermodynamics was developed to provide a causal description of such processes that are not too far from equilibrium. This introduces temporal causality, but not reversibility.)
4. The laws of thermodynamics must have a microscopic interpretation.

The program of statistical mechanics is to construct the bridge between micro- and macroscopic states and to use that construction to compute the bulk properties of matter.

2.5 System and environment

Consider a system containing a large number of atoms or molecules. If it were completely isolated, it would be in a definite quantum state. But no real system can be completely isolated, and no measurements can be made on an isolated system. If the interaction between the system and its environment is weak and random, then the environment is sufficiently constant, on the average, that it does not affect the measured properties of the system, at least within the accuracy of the macroscopic measurement. By a weak interaction, we mean that the changes in the system properties are very small compared to the value of those properties. By random, we mean that over the time it takes to make a measurement, there are a large number of changes in the properties of the system, that the magnitude of these changes follow the rules of probability for independent events, and that the effects of positive and negative changes are very nearly equal. Such an environment might consist of a heat bath at constant temperature that exchanges energy with the system. In addition, the system can be subject to small perturbations due to stray electric and magnetic fields, mechanical vibrations, cosmic rays, and so on. The microscopic state of the system is therefore changing in time in a random, chaotic manner, but its macroscopic state remains essentially unchanged. We will call such an environment one that is *thermodynamically constant*.

If the environment is not weak or not random, a macroscopic state and meaningful macroscopic measurements for a system can be defined only in cases where the environmental conditions are constant or changing slowly. An example is that of a metal whose different surfaces are maintained at different but constant temperatures by different heat baths. The interaction is certainly not weak because the system can exchange large amounts of energy with the environment, but a state can be defined by the use of Fourier's law.

A macroscopic system contains so many atoms that it can be mentally divided into a very large number of subsystems such that each subsystem contains an enormous number of atoms compared to unity. A particular subsystem is then in the environment provided by all the other subsystems.

The concept of a subsystem is useful for several reasons:

1. It distinguishes between internal processes and external processes. A system not at equilibrium, for example, changes with time because of interactions among its subsystems as well as interactions with its environment. A system in the environment defined by the example given above (of heat flow in a metal) changes because of its interaction with the environment, but its subsystems are also changing because of interactions among them as heat flows through the material. If all subsystems have the same values of the macroscopic parameters and these are

not changing with time, then the system is in internal equilibrium and in equilibrium with its environment. If the parameters are different for different subsystems but constant in time, the system is said to be in a *steady state*.
2. A system in equilibrium can be represented by one of its subsystems in the sense that measurements made on the subsystem will be the same as that for the entire system, if the parameter is an intensive variable, or differs only by being proportional to the volume if the parameter is an extensive variable. That is, the subsystems are in equilibrium with each other. The other subsystems then constitute the environment. Usually, subsystems are defined to have the same shape and volume so that even extensive parameters are the same for all subsystems. For the analysis of bulk properties the existence of surfaces of the system are ignored in the sense that the subsystems near the surface are assumed to be the same as subsystems in the interior. For analysis of surface properties, of course, this is not the case.
3. A system in a nonconstant or nonuniform environment can be represented by a subsystem embedded in a nonuniform environment consisting of the other subsystems. This is often necessary to construct theories of nonequilibrium systems.

These considerations are quite clear in the case of homogeneous systems, but with a bit more thought they can also be applied to nonhomogeneous systems. In a two-phase alloy at equilibrium, for example, two separate sets of subsystems can be defined, one for each phase. All the equilibrium properties (except for interfacial properties) then follow from the requirement that all subsystems be in equilibrium with each other. Alternatively, a sample of the alloy containing representative amounts of each constituent can be defined as a subsystem. This latter construct is useful if the alloy phases are very fine-grained.

2.6 Quantum states of macroscopic systems

If a subsystem is isolated by impermeable walls, it can exist in one of the quantum states defined by the solution of the Schrodinger equation. If q represents all the coordinates and t is the time, then the possible states are given by the wave function as

$$\psi_n = u_n(q)e^{-2\pi i E_n t/h} \tag{2.6.1}$$

where E_n is the energy of the nth quantum state, h is Planck's constant, and $u_n(q)$ is the space part of the wave function, which is a function of coordinates only. The system will be in one of these states.

Isolated subsystems are unrealistic, so let us consider a subsystem that can exchange energy with its surroundings. It cannot then stay in just one of the states given by (2.6.1). But, at any instant, we can expand the wave function of this subsystem in terms of those for the isolated subsystem as

$$\Psi = \sum_n c_n(t)\psi_n \tag{2.6.2}$$

This means that, at any instant, the probability of finding the subsystem in a state n is

$$p_n(t) = c_n^*(t)c_n(t) \tag{2.6.3}$$

so the subsystem is continually jumping among the states of the isolated subsystem with a probability given by (2.6.3). We will call $p_n(t)$ the *instantaneous probability function*.

Because of the uncertainty principle, the concept of an instantaneous probability function is not quite consistent with quantum mechanics. An energy change arising from the interaction of the subsystem with its surroundings can be precisely specified only if the time over which the energy change occurs is infinite. In principle, therefore, equation (2.6.2) is only approximate. However, the uncertainty principle applies to the distribution of the results of a large number of measurements of both the time and energy of interaction and refers to averages of these measurements. Also, we are normally not interested in specifying the interaction times. At any rate, this difficulty will be ignored by taking (2.6.2) to be a postulate of statistical mechanics. This is justified by the self-consistency of statistical mechanics and its agreement with experiment.

2.7 Time averages

For time differences that are small, that is, for time intervals that are comparable to the fluctuation times of the subsystems, $p_n(t)$ is a rapidly varying function of time. But since the interactions, and therefore the fluctuations, are random, the fluctuations over a period of time that is large compared with the interaction times are distributed about some mean. Therefore, let us consider a time interval $(t_2 - t_1)$ that is long compared to the interaction time but short relative to macroscopic times. That is, if the macroscopic properties of the subsystem are changing with time, then $(t_2 - t_1)$ is short with respect to the time it takes for the macroscopic properties to change appreciably. This means that the macroscopic properties must be changing slowly relative to the times of fluctuations at the micro level.

Now take a time average of equation (2.6.3) to get

$$\overline{p_n(t)} = \frac{1}{t_2 - t_1} \int c_n^*(t) c_n(t) dt \qquad (2.7.1)$$

The integral is taken over the time interval $(t_2 - t_1)$, and the time t on the left-hand side of (2.7.1) is the midpoint of the time interval $(t_2 - t_1)$; At the macroscopic level, the average defined by equation (2.7.1) is very nearly continuous for subsystems whose macroscopic properties are stationary or changing slowly with time.

During the time interval $(t_2 - t_1)$, the subsystem spends a time t_n in the nth quantum state. Therefore, the probability that it is in the nth state at the time interval centered on t is

$$\overline{p_n(t)} = \frac{t_n}{(t_2 - t_1)} \qquad (2.7.2)$$

The connection between microscopic and macroscopic mechanical variables is now easily made. The macroscopic variable is just a time average of the microscopic variable. For example, if $U(t)$ is the macroscopically measured energy of the subsystem, then

$$U(t) = \frac{1}{(t_2 - t_1)} \sum_n t_n E_n \qquad (2.7.3)$$

The summation is taken through all states that the subsystem passes through in the time $(t_2 - t_1)$.

2.8 Ensembles

Macroscopic properties can be computed from microscopic properties from equations like (2.7.3) only if the time ratios $t_n/(t_2 - t_1)$ are known. But there is no way of computing these ratios, and time averages cannot be used to construct a practical statistical mechanics. An alternative to time averages is provided by the concept of ensembles.

Imagine an enormous number of systems all of which are replicas of the physical system under consideration and all subject to the same external conditions as the physical system. Now place imaginary walls around each system and stack them together to form one huge, continuous mass. Such a collection is called an *ensemble*, and each replica is a member system of the ensemble.

The nature of the boundaries between the systems determines the type of ensemble:

1. If the boundaries are impermeable and each system is completely isolated, there are no interactions among the systems and the ensemble is called *microcanonical*.
2. If the walls are permeable to energy but not to anything else, only energy can be transferred among systems and the ensemble is called *canonical*.
3. If both energy and matter can be exchanged among systems, but the walls are impermeable to all other influences, the ensemble is called *grand canonical*.

In the above three ensemble types, it is assumed that the volume of each member of the ensemble is the same. However, it is possible to define ensembles in which the volumes of the member systems fluctuate. We then can define the following ensembles.

4. If energy can be exchanged among systems whose volumes can fluctuate but the walls are impermeable to all other influences, we have a *canonical pressure* ensemble.
5. If energy and matter can be exchanged among systems whose volumes can fluctuate, but the walls are impermeable to all other influences, the ensemble is called a *grand canonical pressure* ensemble.

The utility of the concept of ensembles lies in the fact that averages of physical quantities can be obtained by averaging over the member systems of the ensemble. These ensemble averages are then identified with experimentally observed properties. In essence, the ensemble averages are equated to the time averages defined in the preceding section. This sidesteps the need to compute time averages and provides a consistent scheme to arrive at a statistical theory of macroscopic properties. The equality of time and ensemble averages is called the *ergodic hypothesis*, and its ultimate justification lies in the agreement of the results of statistical mechanics with experiment.

Clearly, in all ensembles except microcanonical ones, fluctuations will occur in at least some values of the physical variables of the member systems. Essentially, what is done in constructing an ensemble is to replace fluctuations caused by external influences, by fluctuations caused by internal interactions among various parts of the system. In so doing, the ensemble is taken to be a supersystem isolated from the rest of the world. Different kinds of ensembles consider fluctuations of certain properties and neglect others. For example, in the pressure grand canonical ensemble, fluctuations in volume, number of atoms, and energy are all taken into account, while in a canonical ensemble only energy fluctuations are considered.

2.9 The canonical ensemble

Consider a physical system in equilibrium (both internally and with its surroundings) that can exchange only energy with its environment. The corresponding ensemble is then canonical, and each of its members can exist in one of the quantum states characteristic of the isolated system.

Let the number of systems that are in the jth quantum state be n_j. Then the set of numbers $\{n_j\}$ defines the state of the ensemble, and the possible sets $\{n_j\}$ must be considered. That is, we need to determine which $\{n_j\}$ are legitimate possible states of the ensemble. But the only restrictions on the ensemble are the values of the macroscopic parameters in its definition. For the canonical ensemble, the energies of the member systems may differ from one another, but they must always add up to the total energy of the ensemble, which is a constant. There are clearly many sets $\{n_j\}$ that are consistent with a total constant energy of the ensemble, and there is a large number of quantum states for a given set. A particular set $\{n_j\}$ will be called a *state distribution* of the ensemble, while a particular quantum state of the ensemble will be called a *complexion*. An ensemble can have many state distributions, and each distribution can have many complexions.

Number the ensemble members from 1 to X and consider a given state distribution $\{n_j\}$. One way of realizing this distribution is to have the first n_1 members in state 1, the next n_2 members in state 2, and so on. This is one complexion of the state distribution. But an equally valid complexion is obtained if one of the systems in the first group is exchanged with one in the second group. In fact, the number of different complexions is just the number of ways of arranging all possible ensemble members among all possible states such that n_1 members are in state 1, n_2 are in state 2, and so forth.

Because the members of the ensemble are macroscopic, they can be distinguished from each other, so the number of complexions for a given state distribution is just the number of ways of putting marbles in boxes such that n_1 are in the first box, n_2 in the second box, and so on. (The member systems correspond to marbles, while the quantum states correspond to the boxes.)

The number of complexions for a given state distribution is therefore (see appendix 1)

$$W\{n_i\} = \frac{X!}{\prod_j n_j!} \qquad (2.9.1)$$

and the total number of all possible complexions is

$$C = \sum W\{n_j\} \qquad (2.9.2)$$

where the sum is over all the possible distributions of the systems among the quantum states.

It is a fundamental assumption of ensemble theory that, in the absence of any constraints, all complexions are equally likely. This is the assumption of equal a priori probabilities and is assumed to hold for the microcanonical ensemble that represents the supersystem of canonical ensembles. For the canonical ensemble of member systems, however, the possible energies of the systems are constrained by the requirement that the energy of the ensemble as a whole and the number of member systems are both fixed. For the canonical ensemble, therefore, the fundamental assumption is that all complexions are equally probable subject to the number of systems and total ensemble energy being constant. Complexions subject to such constraints are called *allowed*, or *accessible, complexions*.

Because of this assumption, the ensemble energy is just the number of allowed complexions for a given distribution times the energy of that distribution divided by the total number of allowed complexions. That is, if U is the average energy (total ensemble energy divided by the number of members of the ensemble), then

$$U = \frac{\sum W_a\{n_i\}E\{n_i\}}{XC_a} \tag{2.9.3}$$

where $E\{n_i\}$ is the ensemble energy for the distribution and the summations are taken over all possible distributions. The subscript a indicates that $W_a\{n_i\}$ is the number of accessible (allowed) complexions for the distribution $\{n_i\}$, and C_a is the total number of accessible complexions.

An alternate way of writing the average energy U is in terms of the probability $f\{i\}$ that the ensemble has a distribution $\{n_i\}$ with energy $\{n_i\}$. This probability is clearly given by

$$f\{i\} = \frac{W_a\{n_i\}}{C_a} \tag{2.9.4}$$

and (2.9.3) can then be written as

$$XU = \sum_i f\{i\}E\{i\} \tag{2.9.5}$$

All the sums in equations (2.9.2)–(2.9.5) are taken over all accessible complexions.

Calculation of $f\{i\}$ from (2.9.4) is not practical, but its functional form can be obtained in a simple way. We first note that because the $f\{i\}$ determine macroscopic parameters [as in equation (2.9.5)], they cannot depend on such microscopic variables as particle coordinates or momenta. However, there are mechanical properties that arise directly from the micro-level mechanics but refer to the system as a whole. In classical mechanics these are the total energy, the components of total linear momentum, and the components of total angular momentum, which are the integrals of motion that are constant for an isolated system. These quantities play the same role in quantum mechanics since the operators for energy, linear momentum, and angular momentum all commute. These are the only properties that survive an average over microscopic times and that have a macroscopic as well as a microscopic interpretation. Therefore, it is assumed that the canonical probability distribution function for an equilibrium system is a function only of the total energy, linear momentum, and angular momentum. But the dependence on the momenta is trivial because a coordinate system can always be chosen for which the total linear and angular momenta are zero (at least for terrestrial problems). The probability that the ensemble has a particular distribution is therefore a function only of the energy of that distribution.

To get this functional form, consider two independent, identical ensembles. Then, since the two ensembles are statistically independent, the joint probability that one of the ensembles has a distribution $\{i\}$ and the other has a distribution $\{j\}$ is the product of the separate probabilities. That is,

$$f\{ij\} = f\{i\}f\{j\} \tag{2.9.6}$$

or

$$\ln f\{ij\} = \ln f\{i\} + \ln f\{j\} \tag{2.9.7}$$

But since energies are additive, the energy of the two distributions is the sum of the energy of each:

$$E\{ij\} = E\{i\} + E\{j\} \tag{2.9.8}$$

The simplest function that can satisfy both (2.9.7) and (2.9.8) is an exponential. That is,

$$\ln f\{i\} = -\alpha - \beta E\{i\}$$
$$\ln f\{j\} = -\alpha - \beta E\{j\} \tag{2.9.9}$$

where α and β are constants. This is obvious if we take the logarithm of $f\{j\}$ and expand it in a power series in $E\{j\}$. The logarithms for combined systems must then be additive, because of (2.9.7), and the energies must be additive, so only terms up to the first order in the energy can be retained in the series expansion.

What about systems that are not in equilibrium? In this case, the definition of state must include gradients (i.e., in temperature, concentration, electric fields, etc.). At any instant, the energies are still additive but the gradients are not, so the dependence on the gradients cannot be obtained in the simple manner that led to (2.9.9).

2.10 The canonical most probable distribution

For the canonical ensemble, the total energy of the ensemble is a constant and the interactions among its member systems are weak. This means that the energy for most of the member systems cannot be very far from the average energy U and there must be some distribution $\{n_i\}$ for which the number of complexions is very large relative to that for other complexions. That is, there must be a distribution that maximizes the number of complexions, subject, of course, to the conditions that the number of systems in the ensemble and the total ensemble energy are both constant. The n_i that maximize $W\{n_i\}$ subject to constant X and constant UX can be found by the method of undetermined multipliers (see appendix 2). It is more convenient to apply this method to $\ln W\{n_i\}$ than to $W\{n_i\}$. Accordingly, the equations to be solved are

$$\delta \ln W\{n_i\} = \delta\left(\ln X! - \sum_i \ln n_i!\right) = 0 \tag{2.10.1}$$

$$\delta(UX) = \delta \sum_i n_i E_i = 0 \tag{2.10.2}$$

$$\delta X = \delta \sum_i n_i = 0 \tag{2.10.3}$$

where (2.10.1) follows from (2.9.1).

Note that we have adopted the continuum notation of the calculus of variations even though the n_i can change only in discrete steps. The number of systems can be taken to be so large that this is an excellent approximation. This is often the case in the statistical mechanics of macroscopic systems, and we will use either the discrete or the continuum notation as convenient.

Equations (2.10.1)–(2.10.3) are readily simplified by using Stirling's approximation (see appendix 3),

$$\ln Y! = Y \ln Y - Y \qquad (2.10.4)$$

so that equation (2.10.1) becomes

$$\delta \left[X \ln X - X - \sum_i (n_i \ln n_i - n_i) \right] = 0 \qquad (2.10.5)$$

or, since $X = \Sigma n_i$ and $\delta X = \Sigma \delta n_i = 0$,

$$\sum_i \left(\ln \frac{X}{n_i} \right) \delta n_i = 0 \qquad (2.10.6)$$

Now multiply (2.10.2) and (2.10.3) by the constants $-\beta$ and $-\alpha$, respectively, and add the results to (2.10.6). Remembering that only the n_i are being varied and not the energy levels, the result is

$$\sum_i \left(\ln \frac{X}{n_i} - \alpha - \beta E_i \right) \delta n_i = 0 \qquad (2.10.7)$$

The Lagrangian multipliers have been given the same symbols as the constants in equation (2.9.9) in anticipation that we will arrive at a similar result. The multipliers make each term in the sum in equation (2.10.7) independent, so each term must be zero, and finally

$$\ln \left(\frac{X}{n_i} \right) - \alpha - \beta E_i = 0 \qquad (2.10.8)$$

or

$$\frac{n_i}{X} = e^{-\alpha - \beta E_i} \qquad (2.10.9)$$

This is the distribution function for the canonical ensemble.

The method of Lagrangian multipliers only guarantees that this result is an extremum. But it is easy to show that it is a maximum and not a minimum by taking the second derivative of the number of complexions $W\{n_i\}$ with respect to n and showing that it is always negative in the vicinity of the extremum (see appendix 2). Equation (2.10.9) therefore gives the probability that a member system is in quantum state i for the most probable distribution of states among the ensemble. Note that it has the same form as equation (2.9.9), which gives the probability that an ensemble has a specific distribution of systems among states, not the most probable. The reason that these two approaches are equivalent is that the number of complexions for the most probable distribution is very much larger than for any other distribution. Therefore, if all calculations are made for the most probable distribution, the results are extremely close to what would be obtained by a full analysis including all possible distributions. This is entirely reasonable because the interaction energies among member systems of the ensemble are very weak compared to the system energy: the ratio of interaction energy to system energy is of the order of the ratio of the number of atoms in the system surface to the total number of interactions. The system is therefore never pushed very far from the most probable distribution. This is fortunate because it is much easier to deal with just the most probable distribution rather than all possible distributions. Later, when we examine fluctuations, this statement will be given a quantitative form.

Therefore, the probability that a system is in a particular state i will be written as

$$f_i = \frac{n_i}{X} = e^{-\alpha - \beta E_i} \qquad (2.10.10)$$

even though it is understood that (2.10.10) is the probability that the system is in state i when the ensemble is in the most probable distribution. A rigorous notation would embellish all f_i, n_i, and E_i with some symbol or superscript to identify the fact that we are using the most probable distribution, but this would be an unnecessary proliferation of symbols.

Note that f_i is the probability of finding the system in the *state* i when the member systems of the ensemble are distributed among the quantum states in the most probable way. It equals the probability that the system has energy E_i only if there is just one system eigenfunction per energy level so that there is no degeneracy. But if there is degeneracy and we wish to describe the distribution in energy, the states of the same energy must be grouped together. Thus, if there are ω_i eigenfunctions each with the same energy E_i, then denoting the energy distribution by p_i, we get

$$p_i = \omega_i e^{-\alpha - \beta E_i} \qquad (2.10.11)$$

for the probability that the system has energy E_i. The constants can be determined formally from the requirement that the probabilities in (2.10.10) must sum to unity and from the definition of the average energy of the system.

Performing the sum over i and solving for e^α gives

$$e^\alpha = \sum_i e^{-\beta E_i} \qquad (2.10.12)$$

Therefore, (2.10.10) becomes

$$f_i = \frac{e^{-\beta E_i}}{\sum_j e^{-\beta E_j}} \qquad (2.10.13)$$

The average energy is therefore

$$U = \sum_i f_i E_i = \frac{\sum_i E_i e^{-\beta E_i}}{\sum_i e^{-\beta E_i}} \qquad (2.10.14)$$

The ensemble constants α and β are completely determined by (2.10.12) and (2.10.14) in terms of the energy levels and the average energy of the system.

The sum on the right-hand side of (2.10.12) turns out to have special importance for thermodynamics. It is called the *partition function* and is given its own symbol Z:

$$Z = \sum_i e^{-\beta E_i} \qquad (2.10.15)$$

The probability distribution function (2.10.13) is then conveniently written as

$$f_i = \frac{1}{Z} e^{-\beta E_i} \qquad (2.10.16)$$

2.11 Summary of definitions of probabilities

To clarify the meaning of the definitions of complexions and distributions, consider an ensemble of eight member systems distributed among four quantum states. Two of the possible distributions are illustrated in figures 2.1 and 2.2. In figure 2.1 we consider the distribution {1}, which is defined as having three systems in state 1, none in state 2, two systems in state 3, and three in state 4. This is shown in the first row of numbers in the figure. The rows under this top row show that there are a number of ways of realizing this distribution. The second row of numbers states that the distribution can be realized by putting systems 1, 2, and 3 in state 1; putting no systems in state 2; putting systems 4 and 5 in state 3; and putting systems 6, 7, and 8 in state 4. The next row shows that another realization of the distribution is obtained by putting systems 1, 2, and 4 in state 1, none in state 2, systems 3 and 5 in state 3, and systems 6, 7, and 8 in state 4. Each of these represents a complexion, and the

$$W\{1\} = 560, \qquad W\{2\} = 1680$$

Distribution {1} = {3,0,2,3}

	State 1	State 2	State 3	State 4
{1}	3	0	2	3
	1,2,3	0	4,5	6,7,8
	1,2,4	0	3,5	6,7,8
	1,2,5	0	3,4	6,7,8
	3,6,8	0	1,2	4,5,7
	.....	...		.....

Figure 2.1. A possible distribution of an ensemble of eight member systems distributial among from quantum states.

Distribution {2} = {2,2,1,3}

	State 1	State 2	State 3	State 4
{2}	2	2	1	3
	1,2	3,4	5	6,7,8
	1,3	2,4	5	6,7,8
	2,4	1,3	5,6,8	7

Figure 2.2. A second possible distibution of the ensenble shown in figure 2.1.

number of complexions for the distribution {1} is clearly the number of ways of putting marbles in boxes such that there are three in box 1, none in box 2, two in box 3, and three in box 4, all marbles being regarded as distinguishable from each other.

Figure 2.2 represents a distribution {2} defined by having two systems in state 1, two systems in state 2, one system in state 3, and three systems in state 4. Again, this is shown in the first row of numbers in the figure. Three of the possible complexions are shown in the subsequent rows. The a priori, unconstrained number of complexions for each distribution is $W\{1\} = 560$, $W\{2\} = 1680$.

It is important to distinguish among the following distribution functions:

$f\{j\}$ = the probability that the *ensemble* has a *distribution* $\{j\}$, defined as having given numbers of systems in each of the possible quantum states of the system subject to the constraints that the total energy and the total number of systems of the ensemble are both constant. A longer but more specific way of denoting the distribution is as $\{j\} = \{j_1, j_2, j_3, \ldots\}$, where j_1 is the number of systems in state 1, j_2 is the number of systems in state 2, and so on. The $j_1, j_2, \ldots$ are called the *occupation numbers* for the states 1, 2, and so forth.

f_j = the probability that a *member system* of the ensemble is in *quantum state j* subject to the condition that the ensemble energy and number of systems is constant, *when the ensemble has its most probable distribution.*

p_j = the probability that the *system* has energy E_j corresponding to the quantum state j of the system, when the ensemble has its most probable distribution. This may differ from f_j because of degeneracy such that a number of quantum states may have the same energy.

2.12 The canonical ensemble and thermodynamics

The E_j in the canonical distribution function are mechanical parameters of the individual member systems of the ensemble and can be computed, at least in principle, from quantum theory. However, the parameters α and β refer to the ensemble as a whole and cannot be obtained from quantum theory alone, even in principle. These parameters, in fact, provide the connection between the microscopic theory of quantum mechanics and the macroscopic theory of thermodynamics and are related to the Helmholtz free energy and the temperature. The connection is made by first showing that β has the property of a function of temperature for systems in equilibrium with each other and then by identifying some statistical mechanical equations with corresponding equations of thermodynamics.

Consider two physical systems A and B in thermal equilibrium with each other. Since each system treated separately is in equilibrium, each has a canonical distribution given by

$$f_i^A = \frac{1}{Z_A} e^{-\beta^A E_i^A} \tag{2.12.1}$$

$$f_j^B = \frac{1}{Z_B} e^{-\beta^B E_j^B} \tag{2.12.2}$$

The combined system is also in equilibrium, so it, too, must have a canonical distribution function:

$$f_{ij}^{AB} = \frac{1}{Z_{AB}} e^{-\beta^{AB} E_{ij}}, \tag{2.12.3}$$

where

$$E_{ij}^{AB} = E_i^A + E_j^B \tag{2.12.4}$$

and

$$Z_{AB} = \sum_i \sum_j e^{-\beta^{AB}(E_i^A + E_j^B)} \tag{2.12.5}$$

Equation (2.12.3) gives the probability that the combined system is in a state such that system A is in state i and system B is in state j. But a joint probability of independent events, is the product of the probabilities of the separate events, so we must have

$$f_{ij}^{AB} = f_i^A f_j^B \tag{2.12.6}$$

Therefore, multiplying (2.12.1) and (2.12.2) together and setting the result equal to (2.12.3) gives

$$\frac{1}{Z_A Z_B} e^{-(\beta^A E_i^A + \beta^B E_j^B)} = \frac{1}{Z_{AB}} e^{-\beta^{AB}(E_i^A + E_j^B)} \tag{2.12.7}$$

Rearrange (2.12.7) to give

$$(\beta^A - \beta^{AB}) E_i^A + (\beta^B - \beta^{AB}) E_j^B = \ln\left(\frac{Z_{AB}}{Z_A Z_B}\right) \tag{2.12.8}$$

The right-hand side of (2.12.8) is a constant, so the left-hand side must also be constant. But (2.12.8) is true for any pair of the infinite number of energies E_i^A and E_j^B, both of which are positive, so the left-hand side can be true for all values of the subscripts i, j, only if it is zero. This means that the coefficients of the energies must be zero. That is,

$$\beta^A = \beta^B = \beta^{AB} \tag{2.12.9}$$

and

$$Z^A Z^B = Z^{AB} \tag{2.12.10}$$

Therefore, if two systems are in equilibrium, the ensemble parameter β must be the same for both systems. But this is also the property of temperature in thermodynamics, so the first connection between statistical mechanics and thermodynamics is to assume that β is a function of temperature T.

To find the full connection between statistical mechanics and thermodynamics, consider an infinitesimal change in the state of a system. According to (2.10.14) the resulting change in energy can be produced by a change either in the probability distribution or in the energy levels. That is,

$$dU = \sum_i (E_i df_i + f_i dE_i) \tag{2.12.11}$$

From thermodynamics, the energy change is given by

$$dU = TdS - dW, \tag{2.12.12}$$

where dS is the entropy change and dW is the work done on the system during the energy change dU.

When work is done on the system, its energy levels must change. Therefore, if the statistical mechanical equation (2.12.11) is to be identified with the thermodynamic equation (2.12.12), the second term in (2.12.11) must be the work term. The first term must then be the entropy term. That is,

$$dW = -\sum_i f_i dE_i \tag{2.12.13}$$

and

$$TdS = -\sum_i E_i df_i \tag{2.12.14}$$

Now, from equation (2.10.16) for the canonical distribution, solve for the energy level to get

$$E_i = -\frac{1}{\beta}\ln Z - \frac{1}{\beta}\ln f_i \tag{2.12.15}$$

Put this into (2.12.14) and solve for dS, using the fact that the df_i sum to zero, to get

$$dS = -\frac{1}{\beta T}\sum_i \ln f_i df_i \tag{2.12.16}$$

The entropy is a state function so dS is a perfect differential. The right-hand side of (2.12.16) must then also be a perfect differential if the statistical method is valid. That this is so can be shown by writing the sum in (2.12.16) as

$$dx \equiv d\left(\sum_i f_i \ln f_i\right) = \sum_i \ln f_i df_i \tag{2.12.17}$$

and rewriting (2.12.16) as

$$dS = -\frac{1}{\beta T}dx \qquad (2.12.18)$$

The term dx is a perfect differential, so (2.12.18) can be integrated if βT is not a function of x. That this is actually the case can be shown by writing $-(\beta T)^{-1} = f(x)$ and using the additive property of the entropy, as follows.

For two systems A and B, write (2.12.18) as

$$dS_A = f(x_A)dx_A \qquad (2.12.19)$$

$$dS_B = f(x_B)dx_B \qquad (2.12.20)$$

For the combined system, AB (2.12.18) gives

$$dS_{AB} = f(x_{AB})dx_{AB} \qquad (2.12.21)$$

From (2.12.17) it is easy to show that

$$dx_{AB} = dx_A + dx_B \qquad (2.12.22)$$

and since dx is a perfect differential,

$$x_{AB} = x_A + x_B \qquad (2.12.23)$$

Also, entropy is additive and therefore

$$dS_{AB} = dS_A + dS_B \qquad (2.12.24)$$

Using (2.12.19)–(2.12.24) gives

$$f(x_{AB})dx_{AB} = f(x_A + x_B)(dx_A + dx_B) = f(x_A)dx_A + f(x_B)dx_B$$

or

$$[f(x_A) - f(x_A + x_B)]dx_A + [f(x_B) - f(x_A + x_B)]dx_B = 0.$$

Since dx_A and dx_B are independent differentials, this equation can only be true if the coefficients of the differentials are zero. That is,

$$f(x_A + x_B) = f(x_{AB}) = f(x_A) = f(x_B) \qquad (2.12.25)$$

But the two systems are completely arbitrary, so (2.12.25) can only be correct if f is not a function of x. This is the result we were after because it shows that (2.12.18) is a perfect differential, which can be integrated to give

$$S = -\frac{1}{\beta T}x = -\frac{1}{\beta T}\sum_i f_i \ln f_i \qquad (2.12.26)$$

except for an additive constant, which we take to be zero. Now solve (2.12.15) for $\ln f_i$ and put the result in (2.12.26) to get

$$S = \frac{1}{T}\sum_i f_i E_i + \frac{1}{\beta_i}\sum_i \ln Z \qquad (2.12.27)$$

or

$$S = \frac{U}{T} + \frac{\ln Z}{\beta T} \qquad (2.12.28)$$

From thermodynamics, the Helmholtz free energy is

$$A = U - TS \qquad (2.12.29)$$

so

$$S = \frac{U}{T} - \frac{A}{T} \qquad (2.12.30)$$

Comparing (2.12.28) and (2.12.30) gives

$$A = -\frac{\ln Z}{\beta} \qquad (2.12.31)$$

Again from thermodynamics,

$$S = -\left(\frac{\partial A}{\partial T}\right)_{N,V} \qquad (2.12.32)$$

so the entropy can be obtained from (2.12.31) by a differentiation. The condition of constant N is automatically satisfied, and the condition of constant volume can be satisfied by keeping all E_j constant.

Differentiating (2.12.31) with respect to temperature gives

$$S = \ln Z \frac{d}{dT}\left(\frac{1}{\beta}\right) + \frac{1}{\beta}\frac{d \ln Z}{dT} \qquad (2.12.33)$$

From the definition of the partition function, equation (2.10.15), we get

$$\frac{dZ}{dT} = -\sum_i E_i e^{-\beta E_i} \frac{d\beta}{dT}$$

so, since $f_i = e^{-\beta E_i}/Z$,

$$\frac{1}{Z}\frac{dZ}{dT} = -\sum_i E_i f_i \frac{d\beta}{dT} = -U \frac{d\beta}{dT}$$

Putting this in (2.12.33) gives

$$S = \ln Z \frac{d}{dT}\left(\frac{1}{\beta}\right) - \frac{U}{\beta}\frac{d\beta}{dT} \qquad (2.12.34)$$

Now compare this to the expression for the entropy in equation (2.12.28). Clearly, $(1/\beta)(d\beta/dT) = -(1/T)$ and $(d/dt)(1/\beta) = 1/\beta T$. When integrated, each of these equations gives

$$\beta = \frac{1}{kT} \qquad (2.12.35)$$

where k is some universal constant whose value is to be determined. β^{-1} is the statistical mechanical temperature, and as is evident from the way it appears in the formula for the canonical distribution function, it has the units of energy per degree. The constant k is just a conversion factor that connects the energy units of temperature in statistical mechanics to the absolute temperature units of thermodynamics. The value of k is obtained by working out the statistical thermodynamics of a specific system and comparing it to experiment. When we do this for the ideal gas (section 3.4), we will find that k is just Boltzmann's constant. Since k is a universal constant, it is the same for all systems.

The foregoing has not only yielded the interpretation of β, but also has actually provided the complete connection between statistical mechanics and thermodynamics. The entropy (2.12.26) now is

$$S = -k \sum_i f_i \ln f_i \qquad (2.12.36)$$

and using (2.12.35) in (2.12.31) gives

$$A = -kT \ln Z \qquad (2.12.37)$$

with the partition function given by

$$Z = \sum_i e^{-E_i/kT} \qquad (2.12.38)$$

Also, from (2.10.12) and (2.10.15), equation (2.12.37) gives the constant α as

$$\alpha = -\frac{A}{kT} \qquad (2.12.39)$$

and the canonical distribution function becomes

$$f_i = e^{(A-E_i)/kT} \qquad (2.12.40)$$

or, equivalently [see (2.10.16)],

$$f_i = \frac{1}{Z} e^{-E_i/kT} \qquad (2.12.41)$$

The statistical mechanical interpretation is now complete because all thermodynamic functions can be derived from the Helmholtz free energy. The program of statistical thermodynamics is to find the energy levels of the system of interest and to use these to compute the partition function so as to get the free energy, which then yields all the thermodynamic functions. We have thus arrived at a microscopic interpretation of thermodynamics that permits the calculation of thermodynamic properties from microscopic theory.

From the point of view of fundamental principles it should be noted that what has actually been done here has been to derive a set of statistical mechanical equations that have the *form* of thermodynamic equations, and that the connection with thermodynamics was made by identifying these equations and the terms in them with corresponding equations and terms in thermodynamics. The ultimate justification for this procedure is that its consequences are in complete agreement with experiment in every case that has been tested.

2.13 Statistical entropy and the second law of thermodynamics

The identification of entropy in terms of the probability distribution function is one of the fundamental results of statistical mechanics. For the canonical ensemble, this identification is given by equation (2.12.36). This relation is regarded as being so fundamental that an alternate approach to statistical mechanics simply assumes (2.12.36) to be the entropy for any distribution of f_i and uses it to maximize the entropy, subject to the conditions of constant average energy and constant number of ensemble members. This will then give the most probable distribution, and all of the results we obtained above can then be derived.

From thermodynamics alone, we know that the entropy of a substance increases as it is heated and goes from the solid to the liquid to the vapor phase. We also know that the system becomes more "random" in this progression in the sense that the molecules become less localized and can be found in any part of an increasing volume. In this sense, thermodynamics tells us that there is a connection between entropy and randomness.

It takes statistical mechanics in the form of equation (2.12.36) to make this idea precise and give it quantitative form. A totally nonrandom system is one in which the system is in one particular quantum state so that all f_i are zero except for one f_m, which is unity. Then (2.12.36) gives $S = 0$. A random system, on the other hand is one in which a great many of the quantum states can be occupied with equal probability. That is, there are many f_i with about equal values. These values must be small because there are so many of them and they must add up to one. If all f_i are equal, we have $f_i = 1/X$, and (2.12.36) gives

$$S = -k\sum_i \frac{1}{X}\ln\frac{1}{X} = k\ln X$$

Since X is extremely large, the entropy is very large. (In fact, the results of statistical mechanics are rigorously exact only in the limit as X becomes infinite.)

The entropy is also related to the number of complexions. To see this, replace f_i by n_i/X in (2.12.36) and use Stirling's approximation as follows:

$$S = -k\sum_i \frac{n_i}{X}\ln\frac{n_i}{X} \qquad (2.13.1)$$

which is easily rewritten as

$$XS = -k\sum_i n_i \ln n_i + kX\ln X \qquad (2.13.2)$$

and from Stirling's approximation, $n_i \ln n_i = \ln n_i! + n_i$, and $X \ln X = \ln X! + X$, so (2.13.2) becomes

$$XS = k\ln\frac{X!}{\prod_i n_i!} \qquad (2.13.3)$$

But the argument of the logarithm in (2.13.3) is just the number of complexions for the most probable distribution, so we have

$$XS = k\ln W\{n_i\} \qquad (2.13.4)$$

That is, the entropy of the ensemble is proportional to the logarithm of the total number of complexions for the distribution of systems among quantum states.

PRINCIPLES OF STATISTICAL MECHANICS

Let (2.12.36) be the definition of entropy for nonequilibrium as well as equilibrium systems. From (2.13.1), the time dependence of the entropy is given by

$$\frac{dS}{dt} = -k \sum_i \frac{df_i}{dt} \ln f_i \tag{2.13.5}$$

Now define a conditional transition probability, $\Gamma_{ij}dt$, as the probability that if a system is in state i at time t, it will be in state j at time $t + dt$. If the probability that a system is in state i is $f_i(t)$ at time t, then the change in its value during time dt is the probability that some system in state j jumps to state i in the time dt minus the probability that the system in state i jumps to some other state j in time dt, summed over all systems. That is,

$$f_i(t+dt) - f_i(t) = \sum_j (f_j \Gamma_{ji} dt - f_i \Gamma_{ij} dt) \tag{2.13.6}$$

[There is no need to omit the $i = j$ term in (2.13.7) since nothing is changed by adding this to the first term and subtracting it in the second term of the sum.] From (2.13.6) the time derivative is

$$\frac{df_i}{dt} = \sum_j (f_j \Gamma_{ji} - f_i \Gamma_{ij}) \tag{2.13.7}$$

Putting (2.13.7) into (2.13.5) gives

$$\frac{dS}{dt} = -k \sum_{i,j} (f_j \Gamma_{ji} - f_i \Gamma_{ij}) \ln f_i \tag{2.13.8}$$

The subscripts i and j can be interchanged since they are dummies. Doing this, adding the result to (2.13.8), dividing by 2, and rearranging terms gives

$$\frac{dS}{dt} = \frac{k}{2} \sum_{i,j} (\ln f_i - \ln f_j)(f_i \Gamma_{ij} - f_j \Gamma_{ji}) \tag{2.13.9}$$

Equation (2.13.9) shows that equilibrium, $(dS/dT = 0)$, is assured if

$$f_i \Gamma_{ij} = f_j \Gamma_{ji} \quad \text{(equilibrium)} \tag{2.13.10}$$

This is called the *principle of detailed balance*.

At equilibrium, the great majority of states are close together since fluctuations from the most probable state are very small. This means that most of the f_i have values that are very close together and only a few have values that are much different. For most states, therefore, we can take f_i to be nearly equal to f_j, and to a good approximation (2.13.11) then gives

$$\Gamma_{ij} = \Gamma_{ji} \tag{2.13.11}$$

This states that the conditional transition probabilities for transitions between two given states are equal. We have given a plausibility argument for the validity of (2.13.11) for a system at equilibrium. However, in kinetic theory this relation is taken to be true even for systems not at equilibrium. It is called the *principle of microscopic reversibility*. For transitions that can be described by perturbation theory, quantum theory shows that microscopic reversibility is indeed correct. We therefore adopt (2.13.11) as true for nonequilibrium systems, at least when the systems are not too far from equilibrium. Using (2.13.11) in (2.13.9), we get

$$\frac{dS}{dt} = \frac{k}{2}\sum_{i,j}\Gamma_{ij}(\ln f_i - \ln f_j)(f_i - f_j) \tag{2.13.12}$$

The reason for putting the rate of change of entropy in this form is to show that the entropy of a system not at equilibrium always increases. This follows from (2.13.12) because if $f_i > f_j$, then if $\ln f_i > \ln f_j$, and if $f_i < f_j$, then $\ln f_i < \ln f_j$, so every term in the sum is always positive. This is a statistical expression for the second law of thermodynamics for systems changing in time.

2.14 The semiclassical approximation

If the constituent particles of a system follow the laws of classical mechanics, the state of the system is then completely described by the coordinates and momenta of the particles. If the system contains N particles, the state is defined by the $3N$ coordinates $(q_1, q_2, q_3, \ldots, q_{3N})$ and the $3N$ momenta $(p_1, p_2, p_3, \ldots, p_{3N})$. These $6N$ quantities can be thought of as defining a $6N$-dimensional space. The set of $6N$ "coordinates," which we will designate as $\{q, p\}$, is then a point in that space. The $6N$-dimensional space so defined is called the *phase space* for the system; the set of $6N$ variables $\{q, p\}$ is called the *phase* of the system and is a point in the phase space that defines the state of the system.

A statistical mechanics can be constructed from this classical picture by applying the concept of ensembles to the classical systems, and this was in fact done by Gibbs before the advent of the quantum theory. Since the classical theory is a limiting case of the quantum theory, classical statistical mechanics is not developed in this book. However, there are many cases for which a classical description of the particle energy is valid to a high degree of accuracy, such as in the theory of gases and liquids when the temperature is high enough and/or the molecules are massive enough that quantum effects are small. It is therefore instructive, as well as convenient, to show how distribution functions and partition functions can be expressed in terms of coordinates and momenta in the classical limit.

Let us start with the distribution function and partition function for the canonical ensemble as given by (2.12.41) and (2.12.38). In these equations, the index i refers to the ith quantum state of the system. But in a classical system, the state is defined by a point in the phase space and the energy is a function of the coordinates and momenta of all the particles. That is,

$$E = E(p, q) \tag{2.14.1}$$

Clearly, in the classical limit, the sum over quantum states must be replaced by an integral over coordinates and momenta. We therefore write the classical limit of the partition function as a $6N$-dimensional integral over the $3N$ coordinates and the $3N$ momenta:

$$Z_{cl} = C\int e^{-E(p,q)/kT} dp\,dq \tag{2.14.2}$$

where $dp\,dq = dp_1 dp_2 \ldots dp_{3N} dq_1 dq_2 \ldots dq_{3N}$ and the subscript cl denotes the classical limit.

The proportionality constant C had to be introduced into (2.14.2) because the partition function is dimensionless whereas the integral has the dimensions of the $3N$th power of coordinate times momentum through the differentials $dp\,dq$. Each $dp_i dq_i$ pair has the dimensions of action, so the constant C has the dimensions of $(\text{action})^{-3N}$.

The classical limit of the probability distribution function (2.12.41) is now the probability that a phase point lies in the phase volume $dpdq$ and is therefore given by

$$f(p,q)dpdq = \frac{C}{Z_{cl}} e^{-E(p,q)/kT} dpdq \qquad (2.14.3)$$

The proportionality constant in (2.14.3) must be the same as in (2.14.2) because, when the probability distribution function is integrated over all possible ps and qs, the result must be unity.

All the equations connecting thermodynamics to statistical mechanics remain unchanged. The Helmholtz free energy is still given by

$$A_{cl} = -kT \ln Z_{cl} \qquad (2.14.4)$$

so that all the thermodynamic functions of a classical system can be computed from the classical partition function.

At this point, let us rename the proportionality constant C by defining a constant h such that

$$C = \frac{h^{-3N}}{N!} \qquad (2.14.5)$$

where h is a number with the dimensions of action. The choice of the symbol h reflects the fact that it turns out to be Planck's constant, and the inclusion of the $N!$ takes into account the fact that the molecules are indistinguishable. The reason the $N!$ is needed is that the partition function is defined as a sum over states whereas (2.14.2) is an integral over energies. That is, the partition function contains $N!$ states that differ from one another only because of the assignment of energy states to particular molecules. Dividing by $N!$ corrects for this. If we work out the theory of the ideal gas using both the quantum and the classical description, they agree completely if C is taken to be given by (2.14.5) with h being Planck's constant. (Note that we are presuming that our system consists of a set of identical molecules. If several different kinds of molecules are present, then $N!$ is replaced by $N_1!, N_2!, \ldots$ because two unlike molecules are distinguishable and their interchange does not leave the system unchanged.)

The presence of Planck's constant is readily understood on physical grounds by dividing the phase space into cells of volume h^{3N}. The uncertainty principle tells us that the coordinates and momenta of a particle can be measured simultaneously with a maximum accuracy limited by the uncertainty relation

$$\delta p_i \delta q_i = h \qquad (2.14.6)$$

where δp_i and δq_i are the limits of accuracy in the simultaneous measurement of any coordinate–momentum conjugate pair p_i and q_i. This means that all phase points in a phase cell of volume h^{3N} must be thought of as representing the same classical state. Therefore, for a volume element in phase space defined by

$$(\Delta p \Delta q)_k = (\Delta p_1 \Delta q_1 \Delta p_2 \Delta q_2 \ldots \Delta p_{3N} \Delta q_{3N})_k \qquad (2.14.7)$$

The total possible number of *distinct* states is

$$\frac{(\Delta p \Delta q)}{h^{3N}} \tag{2.14.8}$$

The partition function (2.12.38) can be rearranged by taking all states that are in the range $(\Delta p \Delta q)_k$ and lumping them together. Since (2.14.8) gives the number of such states, the partition function becomes

$$Z = \sum_k e^{-E_k/kT} \frac{(\Delta p \Delta q)}{h^{3N}} \tag{2.14.9}$$

This can be approximated by an integral by replacing $(\Delta p \Delta q)_k$ by differentials and dividing by $N_1! N_2! \ldots$ to account for indistinguishability to get, for a system of c components,

$$Z_c = \frac{1}{N_1! N_2! \ldots N_c! h^{3N}} \int e^{-E(p,q)/kT} dp\, dq \tag{2.14.10}$$

which is the same as (2.14.2) except that it now explicitly displays Planck's constant.

If the coordinates are expressed in a Cartesian reference system, the classical energy of the system is

$$E = \sum_i \frac{p_i^2}{2m_i} + \phi(q) \tag{2.14.11}$$

where p_i is the magnitude of the momentum of the *ith* particle and m_i is its mass, and $\phi(q)$ is the potential energy of the system as a function of all the particle coordinates. Now the partition function can be written as

$$Z_{cl} = \frac{1}{N_1! N_2! \ldots N_c! h^{3N}} \int e^{-\sum_i (p_i^2/2m_i kt)} dp \int e^{-\phi(q)/kT} dq \tag{2.14.12}$$

The momenta now appear explicitly in (2.14.12). Since a sum of exponentials is the product of each separate exponential, (2.14.12) can be written as

$$Z_{cl} = \frac{1}{N_1! N_2! \ldots N_c! h^{3N}} \prod_j \int e^{-p_j^2/2m_j kT} d\mathbf{p}_j \int e^{-\phi(q)/kT} dq \tag{2.14.13}$$

Note that, as a matter of notational convenience, we now attach a subscript to each *particle* instead of to each coordinate.

Each of the momentum integrals are identical and are given by

$$\int e^{-p_i^2/2m_i kT} d\mathbf{p}_i = (2\pi m_i kT)^{3/2} \tag{2.14.14}$$

This follows from the fact that

$$p_i^2 = p_{ix}^2 + p_{iy}^2 + p_{iz}^2 \tag{2.14.15}$$

$$d\mathbf{p}_i = dp_{ix} dp_{iy} dp_{iz} \tag{2.14.16}$$

and

$$\int_{-\infty}^{\infty} e^{-ax^2} dx = \sqrt{\frac{\pi}{a}} \tag{2.14.17}$$

[see equation (A.4.17) in appendix 4]. Putting (2.14.14) into (2.14.13) gives the classical partition function as

$$Z_{cl} = \frac{1}{h^{3N}} \prod_i (2\pi m_i kT)^{3/2} Z_q \qquad (2.14.18)$$

where Z_q is called the configurational partition function and is defined by

$$Z_q = \frac{1}{N_1! N_2! \ldots N_c!} \int e^{-\phi(q)/kT} dq \qquad (2.14.19)$$

In the classical approximation, the problem of statistical thermodynamics therefore reduces to that of evaluating the configurational partition function from the potential energy of interaction of all the particles.

If the particles are identical, such that their masses are all the same, (2.14.18) simplifies to

$$Z_{cl} = \frac{Z_q}{\Lambda^{3N}} \qquad (2.14.20)$$

where

$$\Lambda \equiv \frac{h}{(2\pi m kT)^{1/2}} \qquad (2.14.21)$$

is called the *thermal wavelength*.[1]

Note that in the special case of identical independent particles, the potential energy is zero, so $Z_q = V^N/N!$ and the partition function becomes

$$Z_{cl} = \frac{1}{N!}\left(\frac{V}{\Lambda^3}\right)^N \qquad (2.14.22)$$

This is just the partition function for an ideal gas of N particles (since we are in the classical approximation and the particles do not interact). Section 3.4 shows that the same result is obtained from particle quantum statistics and that the constant h is indeed Planck's constant.

It is sometimes useful to have an expression for the probability that a system has a set of coordinates or of momenta in a particular range. This is readily obtained by noting that the constants in front of the right-hand sides of (2.14.3) and (2.14.12) are the same, and by using equations (2.14.11), (2.14.12), and (2.14.3) to get

$$f(p, q) = \frac{e^{\sum_i p_i^2/2m_i kT} e^{-\phi(q)/kT}}{\prod_i (2\pi m_i kT)^{3/2} \int e^{-\phi(q)/kT} dq} \qquad (2.14.23)$$

The probability that a particle has a momentum or a coordinate in a particular range is obtained by integrating over all variables in (2.14.23) except the one of interest. For example, if (2.14.23) is integrated over all coordinates, the probability that the particles have the momenta $\{\mathbf{p}_1, \mathbf{p}_2, \ldots, \mathbf{p}_i, \ldots\}$ in the range $d\mathbf{p}_1 d\mathbf{p}_2 d\mathbf{p}_3 \ldots$ is

$$f(p) d\mathbf{p}_1 d\mathbf{p}_2 d\mathbf{p}_3 \ldots = \frac{e^{-\sum_i p_i^2/2m_i kT}}{\prod_i (2\pi m_i kT)^{3/2}} d\mathbf{p}_1 d\mathbf{p}_2 d\mathbf{p}_3 \ldots \qquad (2.14.24)$$

The probability that the *jth* particle has a momentum of magnitude p_j is obtained by integrating (1.14.24) over all momenta except that for the *jth* particle to get

$$f(p_j)d\mathbf{p}_j = \frac{e^{-p_j^2/2m_jkT}}{(2\pi m_j kT)^{3/2}} d\mathbf{p}_j \tag{2.14.25}$$

and the probability that the x-component of the momentum of the *jth* particle is in a range dp_{jx} is the integral of (2.14.25) over all components of momentum except p_{jx}:

$$f(p_{jx})dp_{jx} = \frac{e^{-p_{jx}^2/2m_jkT}}{(2\pi m_j kT)^{1/2}} dp_{jx} \tag{2.14.26}$$

2.15 The grand canonical ensemble

All the results of statistical mechanics, including the interpretation of thermodynamics in statistical terms, can be obtained from the canonical ensemble. But the canonical ensemble is most convenient for systems that can only exchange energy with other systems. If we want to describe a system for which fluctuations in the number of molecules, as well as energy, are allowed, a generalization of the canonical ensemble is useful. This generalization is the grand canonical ensemble.

For simplicity, assume that our system consists of just two components 1 and 2, and let us form an ensemble of systems that can exchange molecules and energy. Let

X = total number of systems in the ensemble,

$n_j(N_1, N_2)$ = number of member systems that contain N_1 molecules of type 1, N_2 molecules of type 2, and are in the *jth* quantum state, and

$E_j(N_1, N_2)$ = energy of the *jth* quantum state of a system containing N_1 molecules of type 1 and N_2 molecules of type 2.

Note that there is a different spectrum of quantum states for each system composition (N_1, N_2).

The total number of systems in the ensemble is

$$X = \sum_{j,N_1,N_2} n_j(N_1, N_2) \tag{2.15.1}$$

The ensemble averages for the number of molecules and for the energy are

$$\overline{N_1} = \frac{1}{X} \sum_{j,N_1,N_2} N_1 n_j(N_1, N_2) \tag{2.15.2}$$

$$\overline{N_2} = \frac{1}{X} \sum_{j,N_1,N_2} N_2 n_j(N_1, N_2) \tag{2.15.3}$$

$$U = \frac{1}{X} \sum_{j,N_1,N_2} E_j(N_1, N_2) n_j(N_1, N_2) \tag{2.15.4}$$

The most probable distribution of member systems among the states characterized by the number of molecules N_1, N_2 and quantum states j is found by a procedure completely analogous to that for the canonical ensemble. The combinatorial problem is similar to that for the canonical ensemble, and the number of complexions $W\{j, N_1, N_2\}$ is

$$W\{j, N_1, N_2\} = \frac{X!}{\prod_{j,N_1,N_2} n_j(N_1, N_2)!} \qquad (2.15.5)$$

Now take the variations of $\ln W(j, N_1, N_2)$, $\overline{N_1}, \overline{N_2}$, and U with respect to variations in the $n_j(N_1, N_2)$ and set them equal to zero. With the help of Stirling's approximation, the result is

$$\sum_{j,N_1,N_2} \ln \frac{X}{n_j(N_1, N_2)} \delta n_j(N_1, N_2) = 0 \qquad (2.15.6)$$

$$\sum_{j,N_1,N_2} N_1 \delta n_j(N_1, N_2) = 0 \qquad (2.15.7)$$

$$\sum_{j,N_1,N_2} N_2 \delta n_j(N_1, N_2) = 0 \qquad (2.15.8)$$

$$\sum_{j,N_1,N_2} E_j \delta n_j(N_1, N_2) = 0 \qquad (2.15.9)$$

$$\sum_{j,N_1,N_2} \delta n_j(N_1, N_2) = 0 \qquad (2.15.10)$$

Multiply equations (2.15.7)–(2.15.10) by the undetermined multipliers $-\gamma_1$, $-\gamma_2$, $-\beta$, and $-\alpha'$, respectively, add the results to (2.15.6), and set each term equal to zero, just as in finding the canonical distribution. The result is

$$f_j(N_1, N_2) \equiv \frac{n_j(N_1, N_2)}{X} = e^{\{-\alpha' - \beta E_j(N_1, N_2) - N_1\gamma_1 - N_2\gamma_2\}} \qquad (2.15.11)$$

This is the grand canonical distribution function and gives the probability that the system contains N_1 molecules of type 1 and N_2 molecules of type 2 and is in the *j*th quantum state with energy $E_j(N_1, N_2)$. It is obvious that a completely analogous derivation could have been made for a system containing any number of components and that the result would be just like (2.15.11) except that there would be Lagrangian multipliers $\gamma_1, \gamma_2, \ldots, \gamma_s$ for each of the components and the energy levels would be a function of all $N_1, N_2, \ldots, N_s$, where N_j is the number of molecules of the *j* component.

At this point, we will simplify the notation by using the index r to specify the composition, so that each r corresponds to a particular set of N_j. Equation (2.15.11) then becomes

$$f_j^r = \frac{n_j^r}{X} = e^{\left(-\alpha' - \beta E_j^r - N_1\gamma_1 - N_2\gamma_2\right)} \qquad (2.15.12)$$

It is easy to show that the β in (2.15.12) is equal to $(kT)^{-1}$ just as in the case of the canonical ensemble. To do this, we use the fact that a grand canonical ensemble is just the sum of a set of canonical ensembles. That is, if we focus only on all member systems of the grand canonical ensemble that have a specific composition r, then those systems constitute a canonical ensemble with a most probable distribution given by

$$\frac{n_j^r}{X_r} = \frac{1}{Z_r} e^{-E_j^r/kT} \qquad (2.15.13)$$

where Z_r is the partition function for the canonical ensemble containing X_r members, all having the same composition r. If n_j^r from (2.15.13) and from (2.15.12) are set equal to each other, the result is easily rearranged to give

$$\left(\beta - \frac{1}{kT}\right)E_j^r = \ln\left(\frac{XZ_r}{X_r}\right) - N_1\gamma_1 - N_2\gamma_2 - \alpha' \qquad (2.15.14)$$

But the right-hand side depends only on the composition of the canonical ensemble and is constant with respect to changes in the quantum state j. The left-hand side must then also be constant, and this can be so only if it is zero, because the energy levels can take on any of an infinite set of values. Therefore,

$$\beta = \frac{1}{kT} \qquad (2.15.15)$$

and β is related to the thermodynamic temperature just as for the canonical ensemble.

The next step to make the connection to thermodynamics is to express the entropy in terms of probabilities. Again, we use the fact that a grand canonical ensemble is a collection of canonical ensembles. Thus, if we list all systems with the same composition r, they form a canonical ensemble whose entropy, according to (2.13.4), is

$$X_r S^r = k \ln W_r\{n_i^r\} \qquad (2.15.16)$$

where X_r is the number of systems in the ensemble of composition r and $W_r\{n_i^r\}$ is the number of ways of distributing the X_r systems among the states i.

The total entropy of all systems in the grand canonical ensemble is obtained by summing (2.15.16) over all possible complexions to get

$$\sum_r X_r S^r = k \sum_r \ln W_r\{n_i^r\} = k \ln \prod_r W_r\{n_i^r\} \qquad (2.15.17)$$

But the product of the number of complexions over all canonical ensembles (one for each composition) is just the number of complexions for the entire grand canonical ensemble, and the left-hand side of (2.15.17) is just the total entropy of the grand canonical ensemble, so if S is the mean entropy of a system, then (2.15.17) is

$$XS = k \ln W\{n_i^r\} \qquad (2.15.18)$$

where $W\{n_i^r\}$ is the number of complexions for the grand canonical ensemble. Using Stirling's approximation on the factorials in the number of complexions just as in section 2.13, this gives the entropy of the system as

$$S = -k \sum_{j,r} f_j^r \ln f_j^r \qquad (2.15.19)$$

Now solve (2.15.11) for $E_j(N_1, N_2)$, substitute the result into (2.15.19), and use the fact that $\beta = 1/kT$ and the definitions of average energy and average number of molecules. Then solve for the average system energy to get

$$U = TS - kT\alpha' - kT\gamma_1\overline{N_1} - kT\gamma_2\overline{N_2} \qquad (2.15.20)$$

From the thermodynamics of open systems, the energy is given by equation (1.15.5) as

$$U = TS - PV + \mu_1\overline{N_1} + \mu_2\overline{N_2} \qquad (2.15.21)$$

μ_1 and μ_2 being the chemical potentials of the two components. Comparison of (2.15.20) with (2.15.21) shows that the statistical mechanical and thermodynamic equations agree if we make the following identifications:

$$PV = kT\alpha' \qquad (2.15.22)$$

$$kT\gamma_1 = -\mu_1 \qquad (2.15.23)$$

$$kT\gamma_2 = -\mu_2 \qquad (2.15.24)$$

The grand partition function is defined in a manner similar to the definition of the canonical partition function as

$$Q = \sum_{j,N_1,N_2} e^{-\left(E_j^r - \mu_1 N_1 - \mu_2 N_2\right)/kT} \qquad (2.15.25)$$

so by summing equation (2.15.11) and remembering that the sum of the probabilities is unity, we get

$$\alpha' = \ln Q \qquad (2.15.26)$$

and from (2.15.21),

$$PV = kT \ln Q \qquad (2.15.27)$$

This completes the identification of the grand canonical ensemble with thermodynamics.

It is useful to note that the grand canonical ensemble is an ensemble of canonical ensembles and that the grand canonical partition function is a weighted sum of canonical partition functions. That is, (2.15.25) can be written as

$$Q = \sum_{N_1,N_2} Z(N_1 N_2) e^{(\mu_1 N_1 + \mu_2 N_2)/kT} \qquad (2.15.28)$$

where $Z(N_1 N_2)$ is the canonical partition function for a system containing N_1 molecules of component 1 and N_2 molecules of component 2.

2.16 The pressure ensemble

To describe a system in which the number of molecules is fixed but the volume and energy can fluctuate, we construct the canonical pressure ensemble and derive the most probable distribution function just as in the case of the grand canonical ensemble. The result is

$$f_j(V) = \frac{1}{Z_P} e^{-[E_j(V) + PV]/kT} \qquad (2.16.1)$$

where $f_j(V)$ is the probability that the system has a volume V and is in the quantum state j with energy $E_j(V)$. Z_p is the pressure partition function and is given by

$$Z_P = \sum_{j,V} e^{-[E_j(V)+PV]/kT} \qquad (2.16.2)$$

where the sum is over all possible volumes and quantum states. The pressure partition function is related to thermodynamics through the Gibbs free energy by

$$G = -kT \ln Z_P \qquad (2.16.3)$$

Clearly, an analogous ensemble can be defined for a system subject to any kind of work as well as for pressure–volume work. For example, as discussed in section 1.19, for a magnetic system in a magnetic field H the analog of pressure is the external magnetic field and the analog of volume is the magnetization. Instead of (2.16.2) we then have

$$Z_M = \sum_{j,V} e^{-[E_j(M)+HM]/kT} \qquad (2.16.4)$$

the Gibbs free energy now being given by

$$G = -kT \ln Z_M \qquad (2.16.5)$$

2.17 Fluctuations

For macroscopic systems in equilibrium, the measured thermodynamic quantities are constant to a very high degree of accuracy. This means that if the connection between statistical mechanics and thermodynamics is valid, then the magnitude of the fluctuations of the thermodynamic quantities computed from statistical mechanics must be small. Ensemble theory does indeed lead to fluctuations that are very small for equilibrium systems, now shown here to be the case for fluctuations in energy as computed from the canonical ensemble.

The average deviation of the energy from its mean value is not a good measure for energy fluctuations because this measure is identically zero, as is evident from writing it out:

$$\overline{E-U} = \overline{E} - U = \sum_j f_j E_j - U = U - U = 0 \qquad (2.17.1)$$

We therefore choose the root mean square deviation as a measure of the spread of the energy distribution. This is always positive even though there are just as many negative as positive deviations from the mean. The mean square deviation is defined by

$$\overline{(E-U^2)} = \overline{(E^2 - 2EU + U^2)} = \overline{E^2} - U^2 \qquad (2.17.2)$$

For the canonical distribution, the average energy is

$$U = \frac{\sum_j E_j e^{-E_j/kT}}{\sum_j e^{-E_j/kT}} \qquad (2.17.3)$$

PRINCIPLES OF STATISTICAL MECHANICS

Now differentiate this with respect to temperature to get the heat capacity at constant volume. The result is

$$C_V = \left(\frac{\partial U}{\partial T}\right)_V = \frac{1}{kT^2}\left(\overline{E^2} - U^2\right) \tag{2.17.4}$$

This follows from the definitions of U and $\overline{E^2}$:

$$\overline{E^2} = \frac{\sum_j E_j^2 e^{-E_j/kT}}{\sum_j e^{-E_j/kT}} \tag{2.17.5}$$

The relative mean square deviation of the energy from the average is therefore

$$\frac{\left(\overline{E^2} - U^2\right)}{U^2} = \frac{\overline{(E-U)^2}}{U^2} = \frac{C_V kT^2}{U^2} \tag{2.17.6}$$

and the square root of this is the root mean square deviation, which is the expected magnitude of the energy fluctuations from the mean:

$$\left(\frac{\overline{(E-U)^2}}{U^2}\right)^{1/2} = \left(\frac{C_V kT^2}{U^2}\right)^{1/2} \tag{2.17.7}$$

Both the energy and the heat capacity of a physical system are proportional to the number of particles (atoms or molecules) in the system. Equation (2.17.7) therefore shows that the relative root mean square deviation of the energy is inversely proportional to the square root of the number of particles. Since this is ordinarily very large (of the order of Avagadro's number), the root mean square deviation is very small, regardless of the specific values of the energy and heat capacity. As will be shown in section 3.4, the energy and heat capacity of an ideal gas are given by

$$U = \frac{3}{2}NkT \tag{2.17.8}$$

$$C_V = \frac{3}{2}Nk \tag{2.17.9}$$

so in this case (2.16.7) becomes

$$\left(\frac{\overline{(E-U)^2}}{U^2}\right)^{1/2} = \sqrt{\frac{2}{3N}} \tag{2.17.10}$$

For a sample of gas containing one billionth of a mole, the root mean square deviation is about 10^{-7}, and this becomes smaller for larger amounts of gas.

The above calculation was for an ensemble in which the distribution of systems among energy states was the most probable since the n_j were obtained by maximizing the number of complexions for an ensemble with a fixed number of systems and a fixed energy. The fluctuations therefore refer to the probability of finding a system *in this most probable distribution* whose properties differ from the most probable value. But there is another aspect about fluctuations that needs to be considered. That is, what is the effect of distributions of the number of systems among quantum states that are *not* the most probable? So far in this book, only plausibility arguments have been used to show that the effect of all distributions except the most probable can be neglected. Let us investigate this further.

Our procedure will be to write down the number of complexions of the ensemble for both the most probable and an arbitrary distribution. Then, since the most probable distribution represents a maximum, we expand the arbitrary distribution as a Taylor series about the most probable value. From this, we will be able to show that the number of complexions for the sum of all possible distributions is very close to that for just the most probable distribution, if the number of systems is large.

To carry out this calculation, we first use Stirling's approximation to rewrite equation (2.9.1) for the total number of complexions as

$$\ln W = X \ln X - \sum_j n_j \ln n_j \qquad (2.17.11)$$

Equation (2.17.11) is true for any distribution of the n_j as well as for the most probable distribution. Up to this point, we have been using only the most probable distribution, so, for the sake of convenience, we did not label the n_j to reflect the fact that they were the most probable values. But now we want to distinguish between most probable and arbitrary distributions, so we will label the most probable distribution with a superscript and let W^* be the number of complexions for the most probable distribution in which n_j^* systems are in the state j. For the most probable distribution, we therefore write (2.17.11) as

$$\ln W^* = X \ln X - \sum_j n_j^* \ln n_j^* \qquad (2.17.12)$$

Now expand $\ln W$ about the most probable value to the second order in a Taylor series:

$$\ln W = \ln W^* + \frac{1}{2} \sum_j \left(\frac{\partial^2 \ln W}{\partial n_j^2} \right)^* (n_j^* - n_j)^2 \qquad (2.17.13)$$

the derivatives being evaluated at the most probable distribution. The linear term in the expansion vanishes because the first derivative is zero at the most probable distribution, and the mixed second derivatives vanish because $\ln W$ is a sum of terms, one for each j.

[Note that it seems as if this expansion does not take into account the fact that the total number of systems and the total energy of the ensemble remain fixed. But this lack is only apparent and not actual, as can be seen by recognizing that including the conditions of constant number of systems and constant energy means that, instead $\ln W$ in the above, we should use the function $(\ln W - \alpha X - \beta U X)$, since the variation of this function gives the most probable distribution. But it is easy to verify that expanding this in a Taylor series leads to (2.17.13) because the number of systems and the ensemble energy are both linear in the n_j. See appendix 2.]

Evaluating the second derivatives in (2.17.13) gives

$$\ln W = \ln W^* - \sum_j \frac{(n_j^* - n_j)^2}{2n_j^*} \qquad (2.17.14)$$

But this is just a Gaussian form, which we write as

$$\frac{W}{W^*} = e^{-\sum_j (n_j^* - n_j)^2 / 2n_j^*} \qquad (2.17.15)$$

(Note that since the second derivative is negative, the most probable number of complexions is a maximum, not a minimum.) For large n_j^*, each factor in this ratio is extremely close to unity when n_j is close to n_j^* but rapidly falls to zero for increasing deviations of the n_j from the most probable values. Equation (2.17.15) is the quantitative justification for using the most probable distribution and shows that fluctuations away from the most probable values are very small.

Equation (2.17.14) immediately gives the second law of thermodynamics because, if we write (2.13.5) for an arbitrary distribution and for the most probable distribution, (2.17.14) shows that the entropy of the most probable distribution is always greater than that for an arbitrary distribution. That is, the entropy of a system in statistical equilibrium is a maximum with respect to all other possible state distributions of the system with the same energy.

While the above derivation was for the canonical ensemble, it is easily extended in an obvious way to other types of ensembles. Statistical mechanics can be used to compute equilibrium thermodynamic properties because it yields fluctuations in thermodynamic properties that are very small.

Exercises

2.1 For the illustrative example of the distribution of states in an ensemble of eight members distributed among four quantum states shown in section 2.11, verify that the number of complexions for the distributions $\{1\} = \{3, 0, 2, 3\}$ and $\{2\} = \{2, 2, 1, 3\}$ are 560 and 1680, respectively.

2.2 Write the canonical partition function as a sum over *energies* instead of a sum over states. Approximate this partition function by using only the most probable term, and within this approximation show that the relation between the partition function and the Helmholtz free energy is $A = -kT \ln Z$, just as in the ordinary theory. Why does this procedure work?

2.3 Derive the equations for the distribution function of the pressure ensemble, for the pressure ensemble partition function, and for the thermodynamic functions.

2.4 Assume that a system has a Gaussian distribution in energy such that the probability that a system is in a state with an energy E in the range dE is given by $f(E)dE = (1/\sqrt{2\pi\sigma^2})e^{-(E-\bar{E})^2/2\sigma^2}dE$ where $\bar{E}$ is the average energy and σ is the standard deviation of the distribution. Assume the energy can have any positive or negative value. Show that the root mean square deviation of the energy is equal to the standard deviation and that the entropy is given by $S = (k/2)[\ln(2\pi\sigma^2) + 1]$. Note that the larger the standard deviation, the larger the entropy.

2.5 What is the root mean square deviation of the energy, as a percentage of the total energy, of an ideal gas containing 1000 molecules, one million molecules, and one millionth of a mole of molecules?

2.6 Let n_r^*, the number of systems in state r, in a most probable distribution, be 10^8. Now consider a distribution $\{\Delta n\}$ in which the number of systems in the state r differs from the most probable number by $+\Delta n$ while the number of systems in a neighboring state differs from the most probable number by $-\Delta n$. Assume that the number of systems in the neighboring state is very close to that in the r state and can be taken to also equal 10^8. Write the formula for the ratio of the number of complexions for the distribution $\{\Delta n\}$ to the number of complexions for the most probable distribution. Plot this ratio as a function of $\Delta n/n = 0$ to 0.0002.

Note

1. This name arises from the fact that $(2\pi mkT)^{1/2}$ is the momentum contribution to the partition function from one degree of freedom. In quantum mechanics, the momentum divided into Planck's constant is a wavelength, hence the designation "thermal wavelength."

3

Particle Statistics

3.1 Entropy and number of complexions

Consider a system composed of N identical particles that interact with each other very weakly. That is, they can exchange energy, but the energy of each individual particle is otherwise independent of the other particles. This situation is completely analogous to that for a canonical ensemble in the sense that a physical system of nearly independent particles can be thought of as an "ensemble" with each particle being a member "system" of the ensemble. This is the same as simply renaming the terms used in the canonical ensemble theory. Then, the ensemble canonical distribution function becomes the particle distribution function. That is,

$$f_i = \frac{N_i}{N} = \frac{e^{-\varepsilon_i/kT}}{\sum_i e^{-\varepsilon_i/kT}} \qquad (3.1.1)$$

where f_i is now the probability that a particle is in a quantum state i with energy ε_i, N_i is the number of such particles, and N is the total number of particles. Equation (3.1.1) is the Maxwell–Boltzmann distribution law for nearly independent particles.

But there is a basic flaw in the Maxwell–Boltzmann distribution function. Remember that in ensemble theory, the distribution function is derived from the number of complexions of the system, and since the systems are macroscopic, they are distinguishable. This fact is reflected in the combinatorial expression for the number of complexions, so (3.1.1) can be valid only if the nearly independent particles are distinguishable from each other. But we know from quantum theory that this is not the case. At the micro level, similar particles are indistinguishable. That is, if the positions of two electrons (or protons or identical atoms) are interchanged, the system is completely unaltered and there is no way of telling that the exchange has taken place. This has an important effect on the way the number of possible complexions of the system is counted. The theory leading to (3.1.1) must be replaced by a theory that takes into account the indistinguishability of particles.

Since the particles are indistinguishable, the probability of finding a particle in a particular place must be unchanged if the positions of particles are interchanged. This means that the square of the system wave function must be invariant with respect to an exchange in position of the particles. Therefore, after an interchange of particles, the value of the wave function must be the same as before the exchange except for a possible difference in sign. If Ψ is the system wave function for particles with coordinates $r_1, r_2, \ldots$ then we must have

$$\Psi(r_1, r_2, \ldots) = \pm \Psi(r_2, r_1, \ldots) \quad (3.1.2)$$

for any exchange of two coordinates. The positive sign defines symmetric wave functions, while the negative sign refers to antisymmetric wave functions.

The most general symmetric wave function is a linear combination of products of the particle wave functions. This shows that there is no limit to the number of particles that can be in any particular particle quantum state. Any distribution of particles among states corresponding to a valid product of particle wave functions therefore constitutes a legitimate quantum state for the system of particles. In counting the number of complexions, therefore, any number of particles can be in any quantum state.

The most general antisymmetric wave function for the system is a linear combination of determinants formed from the particle wave functions. Since a determinant is zero if any two columns or rows are identical, any distribution of particles among quantum states in which two particles are in the same state cannot exist. This leads to the requirement that at most only one particle can occupy a given particle state. This condition is a form of the Pauli exclusion principle and must be satisfied in constructing the number of possible complexions for antisymmetric particles.

Let us denote the occupation numbers of particles in states by

$$\{N_i\} = \{N_1, N_2, N_3, \ldots\} \quad (3.1.3)$$

This is a shorthand notation for saying that N_1 particles are in state 1, N_2 particles are in state 2, N_3 particles are in state 3, and so on. For particles whose wave functions are symmetric, the occupation numbers can take on any integral values. Such particles are called *bosons*, and the most important examples for us are photons and phonons. For particles with antisymmetric wave functions, the occupation numbers can have only the values zero or unity. Such particles are called *fermions*, and our most important example is the electron.

The connection between indistinguishability and ensemble theory is easily made in terms of entropy. The entropy of a canonical ensemble of X systems is given by ensemble theory as

$$XS = k \ln W \quad (3.1.4)$$

Since entropy is additive, the total ensemble entropy is the sum of the entropies of the individual systems, so (3.1.4) can be written as

$$XS = \sum_j S_j = k \sum_j \ln w_j \quad (3.1.5)$$

where S_j is the entropy of the jth system and w_j is the number of complexions for the jth system. Restricting our attention to a particular system, the subscript can be dropped and the entropy of the system of particles is

$$S = k \ln w \quad (3.1.6)$$

where w is the number of complexions for the individual system. That is, w is the number of different wave functions for a system of energy U containing N independent particles.

All that is needed to construct a statistical mechanics of particles is to maximize w with respect to variations in the number of systems occupying the particle quantum states subject to the conditions that the total energy and total number of particles are each constant. But the number of complexions depends on whether or not the particles are distinguishable and on the possible occupation numbers, and this is different for bosons and fermions.

An elementary calculation of the number of complexions starts with the fact that, for a system of N identical particles in a fixed volume, the energy levels of the system are very close together. This means that it is always possible to arrange the energy levels in a set of contiguous groups such that the jth group has energy in a range $\varepsilon_j + d\varepsilon_j$, and there are a great many energy levels in the group. That is, $d\varepsilon_j$ can always be chosen such that it is small with respect to ε_j and still contains a number of levels, ω_j, that is large with respect to unity. It is therefore possible to characterize the system by the following scheme.

Group number	$1, 2, 3, \ldots, j, \ldots$
Number of possible states in the group	$\omega_1, \omega_2, \omega_3, \ldots, \omega_j, \ldots$
Number of particles in the group	$N_1, N_2, N_3, \ldots, N_j, \ldots$
Energy of a particle in the group	$\varepsilon_1, \varepsilon_2, \varepsilon_3, \ldots, \varepsilon_j, \ldots$

Note that the index j now refers to a *group* of states all having energies that are very nearly equal. The energy ε_j of a particle in the jth group is taken to be an average or midpoint energy of all the states in the group.

The distribution of particles among particle states is defined by the set of numbers $\{N_j\}$.

3.2 Particle distribution functions

For a given distribution of particles $\{N_j\}$, the number of complexions is the number of states of the entire system, so it is just the number of ways of arranging the particles such that N_1 are in state 1, N_2 are in state 2, and so on. This number depends on whether or not the particles are distinguishable and on how many particles can be in the same energy state. Therefore, three different cases must be considered.

3.2.1 The Fermi–Dirac case

In this case the particles are indistinguishable and not more than one particle can be in a given state.

First consider one group of levels with energy ε_j. The number of ways of putting N_j indistinguishable particles into ω_j levels such that no more than one particle is in a given energy state is (see appendix 1)

$$\frac{\omega_j!}{N_j!(\omega_j - N_j)!} \tag{3.2.1}$$

The product of all terms similar to this, for all j, is the number of ways of arranging the particles among all states such that N_1 are in one of the states in group 1, N_2 are in one of the states in group 2, and so on. For the Fermi–Dirac case, the number of complexions is therefore

$$w_{FD} = \prod_j \frac{\omega_j!}{N_j!(\omega_j - N_j)!} \tag{3.2.2}$$

3.2.2 The Bose–Einstein case

In this case the particles are indistinguishable and there is no limit to the number of particles that can be in a given energy state.

For the jth group, the number of ways of putting N_j particles among the ω_j levels with no restrictions on the number of particles per level is (see appendix 1)

$$\frac{(\omega_j + N_j - 1)!}{(\omega_j - 1)! N_j!} \tag{3.2.3}$$

For all groups, the number of ways of arranging the particles such that there are N_1 in group 1, N_2 in group 2, and so forth, is just the product of terms like (3.2.3), so the Bose–Einstein number of complexions is

$$w_{BE} = \prod_j \frac{(\omega_j + N_j - 1)!}{(\omega_j - 1)! N_j!} \tag{3.2.4}$$

3.2.3 The Maxwell–Boltzmann case

In this case the particles are distinguishable and there is no limit to the number of particles that can be put in any level. The number of complexions for this case is (see appendix 1)

$$w_{MB} = N! \prod_j \frac{\omega_j^{N_j}}{N_j!} \tag{3.2.5}$$

The Maxwell–Boltzmann distribution is inconsistent with quantum theory since it presumes distinguishable particles. It is included here because of its historical significance and to illustrate the far-reaching effects of the concept of indistinguishable particles. The Maxwell–Boltzmann distribution was derived before the advent of quantum mechanics and was particularly useful for the kinetic theory of gases. The factor $N!$, which arises from the distinguishability of particles, gave some trouble that was resolved in ad hoc ways that turned out to be consistent with quantum theory.

Note that in the number of complexions for the Fermi–Dirac case every state was assumed to have a different energy. For electrons, this is not the case since a spin-up electron has the same energy as a spin-down electron. This is easily taken into account at a later stage by inserting a factor of 2 in appropriate places.

There is yet no theory that tells us whether a particle will be a fermion or a boson. All that quantum theory tells us is that both types of particles are possible. It is a fact of nature that both types exist.

The most probable distribution of particles among energy states is obtained by solving the Lagrangian multiplier problem just as in the case of the canonical ensemble. Only the Fermi–Dirac case is treated here since the mode of derivation is similar for all three cases.

As usual, we maximize the logarithm of the number of complexions and assume that there are so many particles that it is a very good approximation to use continuum notation in taking variations. From (3.2.2) the logarithm of the number of complexions for the Fermi–Dirac case (using Stirling's approximation) is

PARTICLE STATISTICS 73

$$\ln w_{FD} = \sum_j \{\omega_j \ln \omega_j - N_j \ln N_j - (\omega_j - N_j)\ln(\omega_j - N_j)\} \quad (3.2.6)$$

and the requirement of constant total energy and number of particles gives

$$N = \sum_j N_j \quad (3.2.7)$$

$$U = \sum_j N_j \varepsilon_j \quad (3.2.8)$$

The maximization equations are therefore

$$\delta \ln w_{FD} = \sum_j \frac{\partial \ln w_{FD}}{\partial N_j} \delta N_j = 0 \quad (3.2.9)$$

$$\delta N = \sum_j \delta N_j = 0 \quad (3.2.10)$$

$$\delta U = \sum_j \varepsilon_j \delta N_j = 0 \quad (3.2.11)$$

Multiply (3.2.10) and (3.2.11) by the parameters $-a$ and $-b$, respectively, and add the results to (3.2.9). The result is a sum all of whose terms are independent, and since the sum is zero, each term is zero. That is,

$$\delta \ln w_{FD} - a\delta N - b\delta U = 0 \quad (3.2.12)$$

which leads to

$$\frac{\partial \ln w_{FD}}{\partial N_j} - a - b\varepsilon_j = 0 \quad (3.2.13)$$

Using (3.2.6) to evaluate the derivative and solving for N_j gives the Fermi–Dirac distribution function in terms of the undetermined multipliers as

$$N_j = \frac{\omega_j}{e^{(a+b\varepsilon_j)} + 1} \quad \text{(Fermi–Dirac)} \quad (3.2.14)$$

A completely analogous treatment for the Bose–Einstein and the Maxwell–Boltzmann statistics gives

$$N_j = \frac{\omega_j}{e^{(a+b\varepsilon_j)} - 1} \quad \text{(Bose–Einstein)} \quad (3.2.15)$$

$$N_j = N\omega_j e^{(-a-b\varepsilon_j)} \quad \text{(Maxwell–Boltzmann)} \quad (3.2.16)$$

Note that if the exponential term in the Fermi–Dirac and Bose–Einstein cases are very large, then both distributions have the same limiting form, namely,

$$N_j = \omega_j e^{(-a-b\varepsilon_j)} \quad \text{(semiclassical)} \quad (3.2.17)$$

Since the exponential term in the denominators of the distribution functions (3.2.14) and (3.2.15) are very large in this limit, then its reciprocal in (3.2.17) is very small. Therefore, $N_j/\omega_j \ll 1$. But the number of particles is constant, so this means that the available particles are spread out over a very large range of levels. This is just the condition to be expected at high temperatures because the high thermal energy can boost particles into a large number of high-energy levels. We therefore expect (3.2.17) to be valid at high temperatures. Also, (3.2.17) is identical to the Maxwell–Boltzmann distribution except for the factor N. The appearance of N in the Maxwell–Boltzmann distribution is a direct result of treating the particles as distinguishable. If we correct for indistinguishability by dividing the Maxwell–Boltzmann number of complexions through by $N!$ and then carry out the method of Lagrangian multipliers, the result is identical to (3.2.17), which is a distribution function that has been partially corrected for quantum effects in that it takes indistinguishability into account but does not deal with the symmetry properties of the wave function. It can be used at high temperatures because the levels are so sparsely populated that there is not much chance of there being more than one particle per level no matter what the symmetry of the wave function. The distribution function (3.2.17) is called semiclassical for obvious reasons.

Just as for the canonical ensemble, it can be shown that these distribution functions represent maxima and not minima by taking second variations of the logarithm of the number of complexions.

3.3 Particle statistics and thermodynamics

The thermodynamic interpretation of particle statistics is easily obtained because we already have the statistical formula for the entropy. For each of the particle distribution functions, the starting point of our analysis was a variational equation of the form given by equation (3.2.12). Since the variational equation is also valid for variations connecting two equilibrium states, it can be written in terms of ordinary differentials as

$$d\ln w - adN - bdU = 0 \tag{3.3.1}$$

But $S = k\ln w$, so (3.3.1) gives

$$dS = bkdU + akdN \tag{3.3.2}$$

Because the volume of our system is fixed, (3.3.2) gives

$$\left(\frac{\partial S}{\partial U}\right)_{V,N} = bk \tag{3.3.3}$$

$$\left(\frac{\partial S}{\partial N}\right)_{V,U} = ak \tag{3.3.4}$$

The thermodynamic formulas [see (1.14.1)] for these derivatives are

$$\left(\frac{\partial S}{\partial U}\right)_{V,N} = \frac{1}{T} \tag{3.3.5}$$

$$\left(\frac{\partial S}{\partial N}\right)_{V,U} = \frac{\mu}{T} \tag{3.3.6}$$

where μ is the chemical potential. Comparison of (3.3.5) and (3.3.6) with (3.3.3) and (3.3.4) shows that particle statistics is consistent with thermodynamics if

$$b = \frac{1}{kT} \tag{3.3.7}$$

$$a = -\frac{\mu}{kT} \tag{3.3.8}$$

Just as in ensemble theory, the undetermined multiplier for the energy is related to the temperature. That is, the parameter b in particle statistics has the same interpretation as β has in ensemble theory. However, the multiplier a for the number of particles is related to the chemical potential, that is, the Gibbs free energy per particle, whereas in ensemble theory α was related to the Helmholtz free energy.

Equation (3.3.1) applies to any system of identical particles regardless of the particular expression for the number of complexions, so the parameters a and b have the same interpretation for Fermi–Dirac, Bose–Einstein, Maxwell–Boltzmann, and semiclassical statistics. The distribution functions (3.2.14)–(3.2.17) therefore are

$$N_j = \frac{\omega_j}{e^{(\varepsilon_j - \mu)/kT} + 1} \quad \text{(Fermi–Dirac)} \tag{3.3.9}$$

$$N_j = \frac{\omega_j}{e^{(\varepsilon_j - \mu)/kT} - 1} \quad \text{(Bose–Einstein)} \tag{3.3.10}$$

$$N_j = N\omega_j e^{(\mu - \varepsilon_j)/kT} \quad \text{(Maxwell–Boltzmann)} \tag{3.3.11}$$

$$N_j = \omega_j e^{(\mu - \varepsilon_j)/kT} \quad \text{(semiclassical)} \tag{3.3.12}$$

Each of these distribution functions is valid for different kinds of systems, but first note that the Maxwell–Boltzmann distribution is not really valid for any system. It is included in our list to stress the importance of indistinguishability and to compare it to the semiclassical distribution function, which is valid for any kind of particle if the temperature is high enough. The Fermi–Dirac distribution function applies to particles with antisymmetric wave functions and half-integral values of spin. It therefore must be used for systems of nearly independent electrons. The Bose–Einstein distribution applies to particles with symmetric wave functions and integral values of spin (including spin = 0). Important applications are to helium at low temperatures the phonons associated with crystal lattice vibrations, and to photons.

Since the temperature now appears explicitly in the distribution functions, the condition for the validity of the semiclassical approximation can be given a more explicit form. Equations (3.3.9) and (3.3.10) reduce to (3.3.12) when

$$e^{(\mu - \varepsilon_j)/kT} \gg 1 \tag{3.3.13}$$

The accuracy of the semiclassical approximation therefore depends on the chemical potential as well as on the temperature, so specific calculations of the validity of (3.3.13) must be made for different systems. It turns out that (3.3.13) is valid for all ordinary dilute gases at ordinary temperatures and the quantum corrections are important only for hydrogen and helium at low temperatures. For electrons, however, the quantum effects are important at even quite high temperatures, so it is necessary to use the Fermi–Dirac distribution function.

3.4 The ideal gas

An ideal gas is defined as a many-particle system of nearly independent identical particles that can be described by semiclassical statistics. The particles have no internal structure, so their energy levels are those of the quantum mechanical particle-in-a-box. When this definition is used, the semiclassical distribution function leads to the usual thermodynamic and phenomenological equations for the ideal gas.

Let us start by obtaining some useful statistical mechanical results for the semiclassical system of particles. If (3.3.12) is summed over all j, the result is

$$N = e^{\mu/kT} \sum_j \omega_j e^{-\varepsilon_j/kT} \qquad (3.4.1)$$

Solving this for μ gives

$$\mu = kT \ln N - kT \ln z \qquad (3.4.2)$$

where we have defined a particle partition function z by

$$z = \sum_j \omega_j e^{-\varepsilon_j/kT} \qquad (3.4.3)$$

in analogy with the partition function of canonical ensemble theory. Since the chemical potential of a one-component system is the Gibbs free energy per particle, multiplying (3.4.2) by N gives an expression for the Gibbs free energy in terms of the particle partition function:

$$G = N(kT \ln N - kT \ln z) \qquad (3.4.4)$$

Now the entropy can be written in terms of z because, from equation (1.12.7), the entropy is given by the derivative

$$S = -\left(\frac{\partial G}{\partial T}\right)_P \qquad (3.4.5)$$

Therefore, differentiating (3.4.4) with respect to temperature gives the entropy as

$$S = -Nk(\ln N - \ln z) + NkT\left(\frac{\partial \ln z}{\partial T}\right)_P \qquad (3.4.6)$$

Just as in the ensemble statistics, the partition function provides the essential connection between statistical mechanical and thermodynamic equations. All that needs to be done to work out the theory of the ideal gas is to evaluate the particle partition function and use it in the appropriate statistical thermodynamic equations.

To compute the partition function, the energy levels must be known. These are available from the quantum theory of the particle-in-a-box, which gives, for a box of volume V, the energy levels as

$$\varepsilon_j = \frac{h^2}{8mV^{2/3}}(j_1^2 + j_2^2 + j_3^2) \qquad (3.4.7)$$

where h is Planck's constant, m is the mass of the particle, and j_1, j_2, j_3 are positive integers that can take any values from zero to infinity. The set of three integers defines the state of a particle, and there is just one wave function for each set j_1, j_2, j_3. The energy levels are so close together that the sum defining the partition function can be replaced by an integral as follows. Treat the energy as if it were a continuous variable ε and let $\omega(\varepsilon)d\varepsilon$ be the number of levels with energy between ε and $(\varepsilon + d\varepsilon)$. Then the sum (3.4.3) becomes

$$z = \int_0^\infty \omega(\varepsilon)e^{-\varepsilon/kT}d\varepsilon \qquad (3.4.8)$$

(Note that replacing the sum by an integral is not strictly valid for all energy levels. What counts is the ratio of the energy difference between two closely spaced levels to the energy level itself. For low-lying states this ratio is not far from unity and the continuum approximation fails. However, as we go to higher quantum levels this ratio rapidly becomes very small, and it is trivial to show that for a macroscopic sample of gas the number of states for which the ratio is extremely small is overwhelmingly large. Only a totally insignificant percentage of the particles have energies so low that they cannot be treated continuously.)

The integral (3.4.8) can be evaluated only if the density of states $\omega(\varepsilon)$ is known. Equation (3.4.7) can be used to get $\omega(\varepsilon)$ by constructing a three-dimensional simple cubic lattice with the edge of the unit cell being of unit length. Each lattice point corresponds to three integers, which give its coordinates from one of the points taken as origin. This construction yields a j-space containing a lattice in which each point of the positive octant represents an energy level of our particle-in-a-box. Let j be the magnitude of the vector from the origin to one of the lattice points. Then,

$$j^2 = j_1^2 + j_2^2 + j_3^2 \qquad (3.4.9)$$

$$\varepsilon_j = \frac{h^2}{8mV^{2/3}}j^2 \qquad (3.4.10)$$

Because the energy levels are closely spaced, we can construct a spherical shell about the origin with a thickness dj that is small compared to the radius of the sphere but still contains a very large number of lattice points. The volume of the shell is $4\pi j^2 dj$, and the number of points in the positive octant of the shell is one eighth of this. Since each lattice point in the positive octant corresponds to a quantum state, the number of states contained in the shell is

$$\frac{1}{2}\pi j^2 dj \qquad (3.4.11)$$

To convert this to the number of states in a given energy range $d\varepsilon$, we need a relation between $d\varepsilon$ and dj. But this is readily obtained from (3.4.10) if we again treat ε_j and j as continuous variables. Then,

$$d\varepsilon = \frac{h^2}{8mV^{2/3}}2j\,dj \qquad (3.4.12)$$

If equation (3.4.12) is solved for dj and the result put into (3.4.11), we get the number of levels in the range of energy $d\varepsilon$, which is just the density of states in energy $\omega(\varepsilon)$ in equation (3.4.8):

$$\omega(\varepsilon)d\varepsilon = \frac{2\pi m V^{2/3}}{h^2} j d\varepsilon \qquad (3.4.13)$$

We want $\omega(\varepsilon)$ to be a function of energy only, so we solve (3.4.10) for j and put the result in (3.4.13) to get

$$\omega(\varepsilon)d\varepsilon = \frac{4\pi\sqrt{2}m^{3/2}V}{h^3}\sqrt{\varepsilon}d\varepsilon \qquad (3.4.14)$$

This is the density-of-states in energy and is important in all theories of a large number of independent particles.

We are now equipped to go back and compute the particle partition function by substituting (3.4.14) into (3.4.8) to get

$$z = \frac{CV}{2}\int_0^\infty \sqrt{\varepsilon}e^{-\varepsilon/kT}d\varepsilon \qquad (3.4.15)$$

For convenience, C has been defined as the collection of constants

$$C = \frac{8\pi\sqrt{2}m^{3/2}}{h^3} \qquad (3.4.16)$$

The integral is a standard form given by [see equation (A.4.19) in appendix 4]

$$\int_0^\infty \sqrt{2}e^{-\varepsilon/kT}d\varepsilon = \frac{kT}{2}\sqrt{\pi kT} \qquad (3.4.17)$$

Putting (3.4.17) into (3.4.15) and collecting constants gives the particle partition function as

$$z = V\left(\frac{2\pi mkT}{h^2}\right)^{3/2} = \frac{V}{\Lambda^3} \qquad (3.4.18)$$

where Λ is the thermal wavelength defined by (2.14.21). Note that (3.4.18) is the same as the partition function for a canonical ensemble of systems of independent particles in the semiclassical approximation derived in chapter 2.

Now all the thermodynamic functions are easily obtained in terms of the number of particles, the volume, and the temperature. The Gibbs free energy is the result of substituting (3.4.18) into (3.4.4):

$$G = NkT\left[\ln N - \ln V - \frac{3}{2}\ln\left(\frac{2\pi mkT}{h^2}\right)\right] \qquad (3.4.19)$$

Differentiating this equation according to equation (3.4.5) gives the entropy as

$$S = -Nk\left[\ln N - \ln V - T\left(\frac{\partial \ln V}{\partial T}\right)_P - \frac{3}{2} - \frac{3}{2}\ln\left(\frac{2\pi mkT}{h^2}\right)\right] \qquad (3.4.20)$$

The energy can be obtained from the thermodynamic relation $G = U - TS + PV$, but a simple expression for the energy is readily obtained directly from the distribution function (3.3.12), which in the continuum notation is

$$N(\varepsilon)d\varepsilon = \omega(\varepsilon)e^{(\mu-\varepsilon)/kT} \qquad (3.4.21)$$

The total energy is therefore

$$U = \int_0^\infty \varepsilon N(\varepsilon)d\varepsilon = e^{\mu/kT}\int_0^\infty \omega(\varepsilon)\varepsilon e^{-\varepsilon/kT}d\varepsilon \qquad (3.4.22)$$

Using the density of states as given in (3.4.14), this becomes

$$U = e^{\mu/kT}\frac{4\sqrt{2}\pi m^{3/2}}{h^3}V\int_0^\infty \omega^{3/2}e^{-\varepsilon/kT}d\varepsilon \qquad (3.4.23)$$

According to equation (A.4.20) in appendix 4, the integral has the value $3\sqrt{\pi}(kT)^{5/2}/4$, so (3.4.23) gives

$$U = e^{\mu/kT}\frac{3\sqrt{2}(m\pi)^{3/2}V}{h^3}(kT)^{5/2} \qquad (3.4.24)$$

The chemical potential is easily eliminated from this equation by dividing (3.4.19) by N to get μ and then solving for $e^{\mu/kT}$ to get

$$e^{\mu/kT} = \frac{N}{V}\left(\frac{2\pi mkT}{h^2}\right)^{-3/2} = \frac{N}{V}\Lambda^3 = \frac{P}{kT}\Lambda^3 \qquad (3.4.25)$$

where Λ is the thermal wavelength defined by equation (2.14.21) and the last equality follows by using the ideal gas law. Combining this with (3.4.24) reduces the energy to

$$U = \frac{3}{2}NkT \qquad (3.4.26)$$

To get the equation of state, start with the energy in terms of the distribution function in the discrete form:

$$U = \sum_j N_j\varepsilon_j \qquad (3.4.27)$$

The energy differential connecting two equilibrium states is therefore

$$dU = \sum_j N_j d\varepsilon_j + \sum_j \varepsilon_j dN_j \qquad (3.4.28)$$

If only pressure–volume work is done in the interaction of the system with its environment, then the first law of thermodynamics is

$$dU = TdS - PdV \qquad (3.4.29)$$

Just as was done for ensemble theory, the statistical equation (3.4.28) is compared to the thermodynamic equation (3.4.29), and the work term PdV in (3.4.29) is identified with the first term in (3.4.28) because doing work on the

80 STATISTICAL MECHANICS OF SOLIDS

system changes its energy levels. The second term in (3.4.28) must then be identified with the entropy term in (3.4.29). Therefore,

$$PdV = -\sum_j N_j d\varepsilon_j \tag{3.4.30}$$

and

$$P = -\sum_j N_j \frac{\partial \varepsilon_j}{\partial V} \tag{3.4.31}$$

The derivative in (3.4.31) is readily obtained from equation (3.4.10) as

$$\frac{\partial \varepsilon_j}{\partial V} = -\frac{2}{3}\frac{h^2 j^2}{8mV^{5/3}} = -\frac{2\varepsilon_j}{3V} \tag{3.4.32}$$

so (3.4.31) becomes

$$P = \frac{2}{3V}\sum_j N_j \varepsilon_j = \frac{2U}{3V} \tag{3.4.33}$$

Combining this with (3.4.26) gives the equation of state of the ideal gas:

$$PV = NkT \tag{3.4.34}$$

Since the enthalpy $H = U + PV$, it can now be obtained from (3.4.34) and (3.4.26) as

$$H = \frac{5}{2}PV = \frac{5}{2}NkT = \frac{5}{3}U \tag{3.4.35}$$

The heat capacities at constant volume and at constant pressure are easily obtained from their definitions as temperature derivatives of the energy and enthalpy, respectively, by using (3.4.26) and (3.4.35):

$$C_V = \frac{\partial U}{\partial T} = \frac{3}{2}Nk \tag{3.4.36}$$

$$C_P = \frac{\partial H}{\partial T} = \frac{5}{2}Nk \tag{3.4.37}$$

With the equation of state for the ideal gas in hand, we can now get rid of the derivative in (3.4.20) and get an explicit expression for the entropy as a function of temperature and pressure:

$$S = -Nk\left[\ln\left(\frac{N}{V}\right) - \frac{5}{2} - \frac{3}{2}\ln\left(\frac{2\pi mkT}{h^2}\right)\right] \tag{3.4.38}$$

This is the Sacker–Tetrode equation for the entropy of an ideal gas. [Note that equation (3.4.33) is true for any system of independent particles, not only those obeying semiclassical statistics. This is so because equations (3.4.27)–(3.4.32) are also correct for Bose–Einstein and Fermi–Dirac particles.]

Let us pause to remark on what has been accomplished here. First, the ideal gas law has been derived from atomistic considerations, and the constant k

arising in statistical mechanics has been identified as Boltzmann's constant (the gas constant per molecule). This is valid for *any* system since k is a universal constant. Second, the atomistic conditions for the validity of the ideal gas law have been determined. The ideal gas equation was originally obtained as a phenomenological equation to describe the experimental results for the relation between pressure, volume, and temperature for dilute gases. It was found that the more dilute, the gas, the better the ideal gas equation fit the data, and that as the gas gets denser (high pressures, low temperatures), there are increasingly large deviations from ideal gas behavior. This can be understood in a general way even without statistical mechanics by recognizing that the effect of intermolecular forces increases as the density of the gas increases. But the statistical mechanical derivation completely clarifies and quantifies this concept by showing that the validity of the ideal gas law requires that the interactions among the gas particles must be very weak. Not only has the phenomenological equation been given a theoretical foundation, but equations have been obtained that permit the explicit calculation of the thermodynamic functions in terms of temperature and pressure from microscopic properties. For the ideal gas, all that is needed in addition to the values of fundamental constants (Boltzmann's and Planck's constants) is the mass of the particle.

The theory of the ideal gas presented here is strictly true only for dilute monatomic gases. However, if the theory is carried out for polyatomic systems, it turns out that the ideal gas equation of state is still recovered because interactions among the molecules are still ignored but the expressions for energy, heat capacities, and entropies are different because of the contributions of the internal structure to the partition function.

Let us note that, once the particle partition function was evaluated, the ideal gas equation of state could have been derived in an almost trivial manner by using the relation between the Helmholtz free energy and the system partition function and applying the definition of the pressure as the negative derivative of the free energy with respect to volume. However, this would not have clearly displayed the relation between the pressure–volume work and the energy levels.

3.5 Particle statistics from the grand canonical ensemble

In arriving at the particle distributions in section 3.2, each particle was treated as if it were a member system in a canonical ensemble. Because of this, the energy levels had to be arbitrarily divided into groups with nearly the same energies, and then combinatorial analysis was applied to each group to get the number of complexions. This procedure works because there are many closely spaced levels for a system containing a large number of independent particles. While it has the advantage of showing how to deal with the statistics of particles directly, it does suffer from a degree of inelegance. The arbitrary grouping of levels can be avoided by use of the grand canonical ensemble, which permits the derivation of the particle distribution functions in a straightforward, easy manner.

For a one-component system of identical noninteracting particles, the grand canonical distribution function is

$$f_j = \frac{1}{Q} e^{-[E_j(N) - \mu N]/kT} \qquad (3.5.1)$$

where the energy of the system in the jth state and containing N particles is just the sum of the energies of the individual particles and is given by

$$E_j(N) = \sum_k n_k \varepsilon_k \tag{3.5.2}$$

Note that here we are using the single index j as an abbreviation for the state of the system of N particles. n_k is the number of particles with energy ε_k, so

$$\sum_k n_k = N \tag{3.5.3}$$

The summations in (3.5.2) and (3.5.3) are taken over all particle states.

The grand canonical partition function for our one-component system is

$$Q = \sum_{N,j} e^{-[E_j(N)-\mu N]/kT} \tag{3.5.4}$$

The state of the system is specified by the integers n_k, so the sum over N and j can be replaced by a sum over all possible values of the set of integers $\{n_k\} = (n_1, n_2, n_3, \ldots)$. Performing this replacement and using equation (3.5.2) for the total energy in terms of the particle energies, equations (3.5.1) and (3.5.4) become

$$f(n_1, n_2, n_3, \ldots) = \frac{1}{Q} e^{-[n_1(\varepsilon_1-\mu)+n_2(\varepsilon_2-\mu)+\ldots]/kT} \tag{3.5.5}$$

$$Q = \sum_{n_1,n_2,\ldots} e^{-[n_1(\varepsilon_1-\mu)+n_2(\varepsilon_2-\mu)+\ldots]/kT} \tag{3.5.6}$$

The right-hand side of (3.5.6) is just a product of independent sums, one for each particle, and can be written as $Q = \sum_{n_1} e^{-n_1(\varepsilon_1-\mu)/kT} \sum_{n_2} e^{-n_2(\varepsilon_2-\mu)/kT} \ldots$ or, in a briefer notation,

$$Q = \prod_k \sum_{n_k} e^{-n_k(\varepsilon_k-\mu)/kT} \tag{3.5.7}$$

Equation (3.5.5) gives the probability that n_1 particles are in state 1, n_2 particles are in state 2, and so on. We are after the particle distribution function. That is, we want the mean number of particles that are in a particular state, for example, state i with energy ε_i. To get this, we first calculate the probability that n_i particles are in state i by summing (3.5.5) over all particle states except the ith. Doing this gives

$$f(n_i) = \frac{1}{Q} \prod_{k \neq i} \sum_{n_k} e^{-n_k(\varepsilon_k-\mu)/kT} e^{-n_i(\varepsilon_i-\mu)/kT} \tag{3.5.8}$$

where the product is over all energy states except for the ith.

But the product of sums in (3.5.8) is the same as the product of sums in (3.5.7), except that the sum over n_i is missing. Canceling the sums that appear in both the numerator and denominator, (3.5.8) reduces to

$$f(n_i) = \frac{e^{-n_i(\varepsilon_i-\mu)/kT}}{\sum_{n_j} e^{-n_j(\varepsilon_j-\mu)/kT}} \tag{3.5.9}$$

This is the probability that a particle is in state i. To get the mean number of particles in state i, just multiply $f(n_i)$ by n_i and sum over all possible values of i. That is, the mean number of particles in state i, $\bar{n}_i$, is given by

$$\bar{n}_i = \sum_i n_i f(n_i) = \frac{\sum_i n_i e^{-n_i(\varepsilon_i-\mu)/kT}}{\sum_i e^{-n_i(\varepsilon_i-\mu)/kT}} \tag{3.5.10}$$

All sums in (3.5.10) are carried out over all possible values of n_i.

Up to this point, the derivation has been valid for either bosons or fermions. It is in the evaluation of the sums in (3.5.10) that the difference between Fermi–Dirac and Bose–Einstein statistics appears. For Fermi–Dirac particles, the evaluation of the sums is trivial because a state can only be empty or hold one particle. This means that n_i can only take on the values 0 and 1 and (3.5.10) reduces to

$$\bar{n}_{i\,FD} = \frac{1}{e^{(\varepsilon_i-\mu)/kT}+1} \tag{3.5.11}$$

For the Bose–Einstein case, any number of particles can be in a given state, so the n_j can take on all values from zero to infinity. The sums in (3.5.10) therefore have well-known values because they have the form

$$\sum_{j=0}^{\infty} x^j = \frac{1}{1-x} \tag{3.5.12}$$

$$\sum_{j=0}^{\infty} jx^j = \frac{x}{(1-x)^2} \tag{3.5.13}$$

[See equations (A.4.1) and (A.4.2)]. Applying these formulas to (3.5.10) by identifying x as $x = e^{-(\varepsilon_i-\mu)/kT}$ gives the mean number of particles in state i for the Bose–Einstein statistics as

$$\bar{n}_{i\,BE} = \frac{1}{e^{(\varepsilon_i-\mu)/kT}-1} \tag{3.5.14}$$

Equations (3.5.11) and (3.5.14) are equivalent to (3.3.9) and (3.3.10) because, if a group of ω_j states are chosen that have energies closely clustered around ε_j, then the number of particles in these states is $N_j = \omega_j \bar{n}_j$, so $\bar{n}_j = N_j/\omega j$.

The Fermi–Dirac and Bose–Einstein distribution functions look a lot alike and differ only in the sign of unity in their denominators, but this difference has far-reaching implications. For example, since the mean number of particles must always be positive, (3.5.14) shows that

$$e^{(\varepsilon_i-\mu)/kT} > 1 \quad \text{(Bose–Einstein)} \tag{3.5.15}$$

This means that $\varepsilon_i > \mu$ for all possible values of the energy, and since the energy can be zero, the chemical potential is always negative for bosons. The condition (3.5.15) also follows from the requirement that the sums (3.5.12) and (3.5.13) converge to a finite sum, so we must have $x < 1$.

For fermions, (3.5.11) shows that the mean number of particles can be positive for both positive and negative values of the chemical potential. The difference implied by (3.5.15) and (3.5.16) is a clue that we can expect very

3.6 Representations of the density of states

In the continuum notation, $\omega(\varepsilon)$ is the number of particle states per unit energy range. It is sometimes convenient to express the distribution functions in terms of some parameter related to the energy, such as velocity or momentum, rather than in terms of the energy itself. It is then necessary to express the density of states in terms of that parameter. This section shows how to write the density of states in terms of wave numbers, momentum, and velocity as well as energy.

Remember that the particle moves in a box of zero potential, so its energy is entirely kinetic and is given by

$$\varepsilon = \frac{\mathbf{p}_2}{2m} = \frac{1}{2}m\mathbf{v}^2 \tag{3.6.1}$$

where $\mathbf{p}$ is the particle momentum and $\mathbf{v}$ is its velocity. This is a particle description of the energy, but from the de Broglie relation we know that the particle can also be treated as a wave whose wavelength λ is connected to the particle momentum by

$$\mathbf{p} = \frac{h}{\lambda}\mathbf{e} \tag{3.6.2}$$

where $\mathbf{e}$ is a unit vector in the direction of the particle motion. We now define a vector $\mathbf{k}$ by

$$\mathbf{k} = \frac{2\pi}{\lambda}\mathbf{e} \tag{3.6.3}$$

This is the wave number vector, and it describes the wavelength and the direction of motion of the particle. In terms of $\mathbf{k}$, the momentum in (3.6.2) becomes

$$\mathbf{p} = \hbar\mathbf{k} \tag{3.6.4}$$

where $\hbar$ is Planck's constant divided by 2π. The factor 2π is included in the definition of the wave number vector in (3.6.3) because it is convenient to do so in the mathematics of wave phenomenan.

Using (3.6.4), the energy (3.6.1) in wave number language becomes

$$\varepsilon = \frac{\hbar^2 k^2}{2m} \tag{3.6.5}$$

k being the magnitude of the wave number vector. The wave number vectors can be treated as quasi-continuous just as we treat the energy. In fact, however, they form a discrete set whose values are easily obtained by combining equations (3.4.7) and (3.6.5). The result is

$$k^2 = \frac{\pi^2}{V^{2/3}}(j_1^2 + j_2^2 + j_3^2) \tag{3.6.6}$$

Let k_x, k_y, and k_z be the x, y, and z components of the wave number vector so that

$$\mathbf{k} = k_x \mathbf{i}_1 + k_y \mathbf{i}_2 + k_z \mathbf{i}_3, \qquad (3.6.7)$$

$\mathbf{i}_1$, $\mathbf{i}_2$, $\mathbf{i}_3$ being the three orthogonal unit vectors in a Cartesian coordinate system. The components of the wave number vector are the square roots of the j_1, j_2, j_3 terms in (3.6.6).

We are now able to get the density of states in terms of wave number vectors, velocities, or momenta. First note that the energy is quadratic in all of these parameters and depends only on their magnitudes, not their direction. Also, the number of states in a given range must be the same whether that range is defined in terms of energy, wave number, velocity, or momentum. Therefore,

$$\omega(\varepsilon)d\varepsilon = \omega(v)dv = \omega(p)dp = \omega(k)dk \qquad (3.6.8)$$

which gives

$$\omega(k) = \omega(\varepsilon)\frac{d\varepsilon}{dk} \qquad (3.6.9)$$

$$\omega(p) = \omega(\varepsilon)\frac{d\varepsilon}{dp} \qquad (3.6.10)$$

$$\omega(v) = \omega(\varepsilon)\frac{d\varepsilon}{dv} \qquad (3.6.11)$$

The $\omega(\varepsilon)$ is given by equation (3.4.14) and is expressed in terms of k, p, or v by using equations (3.6.1) and (3.6.5). These latter equations also yield the derivatives in (3.6.9)–(3.6.11). Doing the small amount of algebra involved, equations (3.6.9)–(3.6.11) now become

$$\omega(k) = V\frac{k^2}{2\pi^2} \qquad (3.6.12)$$

$$\omega(p) = V\frac{p^2}{2\pi^2\hbar^2} \qquad (3.6.13)$$

$$\omega(v) = V\frac{m^3 v^2}{2\hbar^3\pi^2} \qquad (3.6.14)$$

It is sometimes necessary to count the number of particles with momenta in a given range moving in a particular direction. To get the corresponding density of states, construct a spherical polar coordinate system in momentum space with radial component p, colatitude θ, and azimuthal angle ϕ, and let $\omega(p, \theta, \phi)d\Omega_p$ be the number of states for which the momentum vectors terminate in a volume element $d\Omega_p$, which is given by

$$d\Omega_p = p^2 \sin\theta \, d\theta \, d\phi \, dp \qquad (3.6.15)$$

The function $\omega(p)dp$, already given by (3.6.13), is just the integral of (3.6.15) over the angular coordinates, so

$$\omega(p)dp = V\frac{p^2}{2\pi^2\hbar^3}dp = \int_0^\pi \int_0^{2\pi} \omega(p, \theta, \phi)\sin\theta d\theta d\phi p^2 dp \qquad (3.6.16)$$

Because the distribution of states in momenta is isotropic, the number of states is the same for a given p regardless of the direction of the momentum vector. This means that the density of states function in the integral in (3.6.16) can be taken outside the integral, and then (3.6.16) becomes

$$\omega(p, \theta, \phi)\int_0^\pi \int_0^{2\pi} \sin\theta d\theta d\phi dp = \frac{V}{2\pi^2\hbar^3} \qquad (3.6.17)$$

Evaluating the integral and solving for $\omega(p, \theta, \phi)$, we get

$$\omega(p, \theta, \phi) = \frac{V}{h^3} \qquad (3.6.18)$$

The number of states in a volume element of momentum space does not depend on the choice of coordinate systems, and therefore $d\Omega_p = dp_x dp_y dp_z$ and $\omega(p, \theta, \phi)d\Omega_p = \omega(p_x, p_y, p_z) dp_x dp_y dp_z$, where $\omega(p_x, p_y, p_z)dp_x dp_y dp_z$ is the number of states with momentum components (p_x, p_y, p_z) in the range $dp_x dp_y dp_z$. Combining this with (3.6.18) gives

$$\omega(p_x, p_y, p_z) = \frac{V}{h^3} \qquad (3.6.19)$$

In a similar way, it is easy to show that the density of states for particles with velocity components (v_x, v_y, v_z) is

$$\omega(v_x, v_y, v_z) = \frac{Vm^3}{h^3} \qquad (3.6.20)$$

and the density of states in terms of the number of states with wave number vectors having components (k_x, k_y, k_z) is

$$\omega(k_x, k_y, k_z) = \frac{V}{8\pi^3} \qquad (3.6.21)$$

3.7 Maxwell's velocity distribution

The distribution function for the velocities of molecules in a gas is easily obtained from equation (2.14.25), which we rewrite for a monomolecular gas as

$$f(\mathbf{p})d\mathbf{p} = \frac{e^{-p^2/2mkT}}{(2\pi mkT)^{3/2}}d\mathbf{p} \qquad (3.7.1)$$

Since the number of molecules in a given momentum range must equal the number in the corresponding velocity range, we have

$$f(\mathbf{v})d\mathbf{v} = f(\mathbf{p})d\mathbf{p} = f(\mathbf{p})m^3 d\mathbf{v}$$

so

$$f(\mathbf{v})d\mathbf{v} = \left(\frac{m}{2\pi kT}\right)^{3/2} e^{-p^2/2mkT} d\mathbf{v}$$

$$= \left(\frac{m}{2\pi kT}\right)^{3/2} e^{-mv^2/2kT} d\mathbf{v} \qquad (3.7.2)$$

Expressing the velocity in polar coordinates, this is

$$f(\mathbf{v})d\mathbf{v} = \left(\frac{m}{2\pi kT}\right)^{3/2} e^{-mv^2/2kT} v^2 \sin\theta\, d\theta\, d\phi\, dv \qquad (3.7.3)$$

This is the probability that a molecule has a velocity whose magnitude is v in the direction given by the angles θ, φ. Integrating over the angles gives the probability that a molecules has a speed v irrespective of direction as

$$f(v)dv = 4\pi\left(\frac{m}{2\pi kT}\right)^{3/2} v^2 e^{-mv^2/2kT} dv \qquad (3.7.4)$$

This is the Maxwellian distribution of speeds in a gas.

3.8 Two-dimensional ideal gas

The theory of the ideal gas in two dimensions follows the same path as that for three dimensions given in section 3.4. In fact, the theory depends on dimensionality only through the energy levels and density of states, as can be seen by going over the derivation and identifying those steps that depend on the number of dimensions. In particular, equation (3.4.2) for the relation between the chemical potential and the partition function is still valid, and we can write

$$\mu = kT(\ln N - \ln z) \qquad (3.8.1)$$

except that now the partition function must be computed for the two-dimensional case. To do this, we start with the appropriate energy levels.

In two dimensions, the energy levels for a free particle of mass m confined in a square area σ are

$$\varepsilon_j = \frac{h^2}{8m\sigma}(j_1^2 + j_2^2) \qquad (3.8.2)$$

This is analogous to equation (3.4.7) for the three-dimensional energy levels of a particle-in-a-box. The density of states is obtained from (3.8.2) just as (3.4.14) was obtained from (3.4.7) except that the number of states with integers between j and $j + dj$ is the number of points in one quarter of a ring of thickness dj at radius j in a two-dimensional j-space. The result is that the two-dimensional density of states in energy is

$$\omega(\varepsilon) = \frac{2\pi m\sigma}{h^2} \qquad (3.8.3)$$

so equation (3.4.8) for the particle partition function becomes

$$z = \int_0^\infty \frac{2\pi m}{h^2} e^{-\varepsilon/kT} d\varepsilon \qquad (3.8.4)$$

which integrates to

$$z = \sigma\left(\frac{2\pi mkT}{h^2}\right) \quad (3.8.5)$$

The similarity to (3.4.18) is obvious. In fact, it is trivial to show that every translational degree of freedom has a partition function associated with it given by

$$z_{tr} = L\left(\frac{2\pi mkT}{h^2}\right)^{1/2} = \frac{L}{\Lambda} \quad (3.8.6)$$

L being the length of the cubic box containing the system.

The surface pressure of the two-dimensional gas is defined from thermodynamics as

$$\pi = -\left(\frac{\partial G_{2D}}{\partial \sigma}\right)_{T,P} \quad (3.8.7)$$

G_{2D} being the Gibbs free energy of the two-dimensional phase. Thus, if we use (3.8.1) in (3.8.7), remembering that the chemical potential is the Gibbs free energy per particle, we get

$$\pi\sigma = NkT \quad (3.8.8)$$

This is the equation of state for a two-dimensional perfect gas.

3.9 Independent particles and subsystems

The systems we have considered above are gases of atoms or molecules that do not interact with each other. These are examples of systems composed of independent subsystems. Because the molecules are independent subsystems, the partition function of the system can always be written as a product of partition functions of the subsystems thereby simplifying both notation and calculations.

Let us consider the general case in which the subsystems may be atoms, molecules, groups of molecules, or macroscopic entities. The only requirement we impose is that the subsystems be independent. But it is necessary to take into account the fact that some collections of subsystems are distinguishable while others are not. The particle statistics developed above applies to indistinguishable subsystems. Atoms or molecules that are identical and freely interchangeable generally must be treated as indistinguishable particles. Macroscopic subsystems and subsystems that are not identical are always distinguishable.

With this in mind, let us consider a system made of independent distinguishable subsystems. If $E(j_1)$, $E(j_2)$, $E(j_3)$, ..., $E(j_r)$ are the energy states of the r subsystems, then the total energy of the system, when it is in the state specified by the states $j_1, j_2, \ldots, j_r$ of the subsystems is

$$E(j_1, \ldots, j_r) = E(j_1) + E(j_2) + E(j_3) + \ldots \quad (3.9.1)$$

so the partition function for the total system is

$$Z = \sum_{j_1, j_2, \cdots} e^{-[E(j_1) + E(j_2) + E(j_3) + \cdots]/kT} \qquad (3.9.2)$$

$$= \sum_{j_1} e^{-E(j_1)/kT} \sum_{j_2} e^{-E(j_2)/kT} \cdots \qquad (3.9.3)$$

The partition function for the nth subsystem is defined by

$$z_n = \sum_{j_n} e^{-E(j_n)/kT} \qquad (3.9.4)$$

so the total partition function (3.9.3) becomes

$$Z = \prod_r z_r \quad \text{(distinguishable)} \qquad (3.9.5)$$

If all N subsystems are identical, this reduces to

$$Z = z^N \qquad (3.9.6)$$

What if the subsystems are not distinguishable? To see how (3.9.5) must be modified for this case, consider a system that has three identical subsystems. An energy term of the form $E_1(j) + E_2(k) + E_3(l)$ appears in the partition function for the total system. This is for a member of the system ensemble in which the subsystems labeled 1, 2, 3, are in subsystem states j, k, l, respectively. But there are also systems in the ensemble for which the subsystems 1, 2, 3, are in energy states k, j, l, respectively, and corresponding terms appear in the ensemble. In fact, all permutations of the subsystems among their states appear in the partition function, and if there are N subsystems, there are $N!$ such permutations. But since the subsystems are indistinguishable, each of these permutations represents the same state, so to get the correct number of states the partition function, (3.9.5) must be divided by $N!$. Therefore,

$$Z = \frac{1}{N!} \prod_n z_n \quad \text{(identical indistinguishable)} \qquad (3.9.7)$$

If the system consists of subsystems some of which are identical but others are not, then the $N!$ in the denominator of (3.9.7) is replaced by a product of $N_j!$s, N_j being the number of subsystems of the jth kind. For a system of identical molecules, (3.9.7) reduces to

$$Z = \frac{z^N}{N!} \qquad (3.9.8)$$

A further simplification can be achieved by noting that the separation of the partition function into products was made possible by the fact that the energy could be separated into independent parts. For a system of molecules the energy can be written as a sum of terms that include translational, vibrational, and rotational energies. That is, the total energy for the system in a state k is

$$E_k = E(\text{con}) + \sum_r [E_r(tr) + E_r(vi) + E_r(ro)]_k \qquad (3.9.9)$$

The state labeled k is a composite of the translational, vibrational, and rotational states of the molecules whose energies are given by the three terms in the sum. These constitute the kinetic energy states of the system. The energy term $E(\text{con})$ is the potential energy of interaction among the molecules and is generally not separable into contributions from each molecule. It is called the *configurational energy* because it is a function of the positions of the molecules. When (3.9.9) is put into the definition of the partition function, the result is

$$Z = Z_c \prod_r z_{tr}^r z_{vi}^r z_{ro}^r \qquad (3.9.10)$$

z_{tr}^r, z_{vi}^r, z_{ro}^r being the translational, vibrational, and rotational partition functions for the rth molecule, respectively, and the product is taken over all molecules. If all molecules are alike, this reduces to

$$Z = Z_c (z_{tr} z_{vi} z_{ro})^N \qquad (3.9.11)$$

The molecular partition functions are defined by

$$z_{tr} = \sum_t e^{-\varepsilon_t/kT} \qquad (3.9.12)$$

$$z_{vi} = \sum_t e^{-\varepsilon_v/kT} \qquad (3.9.13)$$

$$z_{ro} = \sum_t e^{-\varepsilon_r/kT} \qquad (3.9.14)$$

The sums are over all translational, vibrational and rotational states with energies ε_t, ε_v, and ε_r, respectively.

The configurational partition function is just that defined by equation (2.14.19). The translational partition function for a molecule is that for a particle-in-a-box and is given by equation (3.4.18). The vibrational partition function for a system with a set of frequencies n_j is evaluated in chapter 4 in connection with the theory of the heat capacity of solids.

Exercises

3.1 For a classical one-dimensional oscillator with energy states given by $E = p^2/2m + Kq^2/2$, where m is the mass, K is the force constant, p is the momentum, and q is the coordinate defining the instantaneous amplitude of the vibration, compute the classical partition function for the oscillator and show that the energy is $U = kT$.

3.2 For a monatomic ideal gas, use the pressure ensemble to show that the Gibbs free energy per atom is given by $G/N = -kT \ln kT/P\Lambda^3$, where the Λ is the thermal wavelength defined by $\Lambda = h^2/2\pi mkT^{1/2}$. (*Hint*: Use the fact that the number of volume states in a range dV is dV/V; i.e., the "density of volume states" is equal to $1/V$.)

3.3 Using semiclassical statistics, show that the mean kinetic energy per molecule of an ideal gas is $3kT/2$. Get the formulas for the root mean square speed and for the most probable speed of a molecule in an ideal gas. What is the value of the root mean square speed of molecules in molecular oxygen gas at 0, 25, and 1000°C?

3.4 Given a column of a monatomic gas of uniform cross-sectional area in a constant gravitational field, use semiclassical statistics to find the formula for the concentration of gas as a function of height h, relative to the concentration at $h = 0$. Take the zero of gravitational energy at $h = 0$.

3.5 Given a system of N independent particles such that each particle can assume one of the energy values $\pm\varepsilon$, ε being a positive constant, let N_+ be the number of particles with energy $+\varepsilon$ and N_- be the number of particles with energy $-\varepsilon$, so that the total energy of the system is $E = N_+\varepsilon - N_-\varepsilon$. Write the number of possible states and from this get the entropy as a function of α where $\alpha \equiv E/N\varepsilon$. From the thermodynamic relation between temperature and entropy, get the temperature as a function of α and plot this function, taking the temperature to be in units of $2\varepsilon/k$ for $1 < \alpha < 1$. What is the significance of the negative temperatures obtained for negative α?

3.6 Consider a one-dimensional line (dislocation) that has N_s sites at which an impurity can be attached. Assume the impurity has zero energy when in solution and an energy ε when attached to a site. Show that the number of impurity atoms N_l, attached to the line at temperature T, is given by $N_l/N_S = 1/(e^{\varepsilon/kT} + 1)$, and that the Helmholtz free energy of the dislocation line is $A = A° + N_s[\varepsilon - kT\ln(1 + e^{\varepsilon/kT})]$, where $A°$ is the free energy of the dislocation line when no impurities are attached to it.

Why is this formula like that for the Fermi–Dirac distribution? Why does it not contain a chemical potential? (*Note*: $\varepsilon < 0$ because we are assuming that the impurity binds to the dislocation.)

3.7 Given a fictitious system of independent particles in which each energy level ε_i can be occupied by zero, one, or two particles, use the grand canonical ensemble to get the formula for the mean number of particles in state i.

3.8 Given a gas of noninteracting particles with energy levels $\varepsilon_1, \varepsilon_2, \ldots, \varepsilon_k \ldots$ where each state can be occupied by any number of particles from 0 to P, use the grand canonical ensemble to show that the most probable occupation for the nth state is $\overline{n}_k = [e^{(\varepsilon_k-\mu)/kT} - 1]^{-1} - (1+P)[e^{(1+P)(\varepsilon_k-\mu)/kT} - 1]^{-1}$. Show that in the limit of $P = 1$ this yields Fermi–Dirac statistics and that in the limit of $P \to \infty$, it yields Bose–Einstein statistics. (Hint: The sums you need are from the geometric progression.)

3.9 What is the density of states for independent particles in a one-dimensional line of length L? For two dimensions in a square of side L?

3.10 Given a system of N noninteracting particles, each of which can exist in one of two nondegenerate states such that the energy of state 1 is ε_1 and the energy of state 2 is ε_2,

  A. write the partition function of the system,
  B. write the formulas for the average energy and the square of the average energy for the system, and
  C. derive the formula for the heat capacity of the system and show that it satisfies the equation $C_v = (N/kT^2)(\overline{\varepsilon^2} - \overline{\varepsilon}^2)$, where $\overline{\varepsilon^2}$ is the mean square energy and $\overline{\varepsilon}^2$ is the square of the mean energy.

4

The Harmonic Crystal

4.1 The harmonic model

The atoms in a monatomic crystal continually vibrate around their mean positions, and it is these vibrations that give the crystal its thermal energy. A rigorous quantum theory of the crystal should include the dynamics of all the electrons and the nuclei explicitly. However, the electrons are much lighter than the nuclei and therefore have a much faster reaction time to any disturbance. That is, if a nucleus is given a small displacement, the electrons will readjust their positions in response to the forces between the nucleus and the electrons in a very short time. This is the basis for the *adiabatic approximation*, which states that the readjustment time is so short compared to the period of vibration of the nuclei that it can be taken as zero. This means that the electrons are in a state that is determined by the instantaneous position of the nuclei.

The adiabatic approximation allows the nuclear and electronic motions to be treated separately because all the electronic information can be absorbed into a potential energy function that is a function of the nuclear coordinates alone. When the atoms are at their mean (equilibrium) positions, the potential energy is a minimum, and it increases as the atoms are displaced from equilibrium. Since the motion of the atoms is constrained to be in the vicinity of their equilibrium positions, the displacements are not extremely large and the potential energy can be expanded in a Taylor series in the displacements. If the displacements are small enough, the expansion can be truncated at the quadratic terms. This is called the *harmonic approximation* since it is a generalization of the simple harmonic motion of a mass at the end of a spring obeying Hooke's law. Note that the terms of first order in the displacements do not appear because they must vanish by virtue of the requirement that the potential energy is a minimum at equilibrium.

The above discussion applies equally well to polyatomic crystals, such as ionic solids and compound semiconductors, although the details of analyzing the vibrations are different than for monatomic crystals. For molecular crystals, the vibrations of the atoms within the molecules must be considered as well as the vibrations of the molecules as whole units. The intramolecular vibrations generally have a much higher frequency than the intermolecular vibrations because the chemical bonds holding the atoms together in the molecule are much stronger than the physical interactions that hold the molecules together in the crystal.

The harmonic model provides a good basis for the theory of crystals. It accounts for a good bit of the thermodynamics of crystals remarkably well,

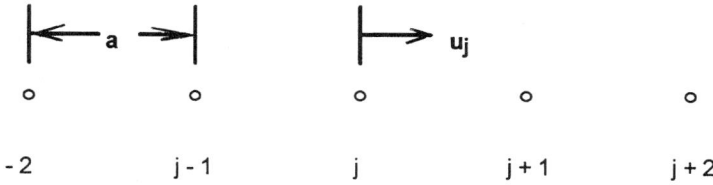

Figure 4.1. The monatomic chain.

being especially successful in the theory of the heat capacity. But for some purposes it requires extension and addition. The theory of thermal expansion, for example, requires that third- and fourth-order terms must be included in the expansion of the potential energy. Also, the theory of the heat capacity of metals at low temperatures must include the contribution of the "free" electrons as well as that of the atomic vibrations. In general, however, the harmonic model serves as one of the basic pillars of solid state theory.

4.2 The monatomic linear chain and normal mode analysis

The most important facts concerning vibrations in crystals are most easily introduced by considering a fictitious one-dimensional crystal composed of identical atoms of mass m arranged in a line. When the system is in mechanical equilibrium, the atoms are all equally spaced, adjacent atoms are an equal distance a apart, and the potential energy is a minimum. This minimum potential energy is taken to be zero. The atoms will be labeled by an index j that runs from 0 to N, and the total number of atoms in the chain is $(N + 1)$. This monatomic chain system is illustrated in figure 4.1.

The displacement of the jth atom from its equilibrium position will be denoted by u_j, and in keeping with the harmonic model, the potential energy V of the chain will be written as

$$V(u_0, u_1, \ldots, u_n) = \frac{1}{2} \sum_{i,j} C_{ij} u_i u_j \qquad (4.2.1)$$

The force constants C_{ij} are defined by

$$C_{ij} = \left(\frac{\partial^2 V}{\partial u_i \partial u_j} \right)_0 \qquad (4.2.2)$$

the derivative being evaluated at equilibrium where all displacements u_i are zero. The force constants C_{ij} that couple the ith and the jth atom depend only on the distance between them and not on the absolute position of the atoms in space. That is, if $|i - j| = |m - n|$, then $C_{ij} = C_{mn}$. Also, $C_{ij} = C_{ji}$.

The equations of motion of the chain are easily obtained by a simple application of Newton's second law in the form

$$-\frac{\partial V}{\partial u_i} = m \frac{d^2 u_i}{dt^2} \qquad (4.2.3)$$

where t is the time and the negative derivative of the potential energy is just the force on the jth atom. From (4.2.1) this force is

94 STATISTICAL MECHANICS OF SOLIDS

$$-\frac{\partial V}{\partial u_i} = -\sum_j C_{ij} u_j \qquad (4.2.4)$$

so

$$m\frac{d^2 u_i}{dt^2} + \sum_j C_{ij} u_j = 0 \qquad (4.2.5)$$

The form of (4.2.5) is very suggestive of that for the simple harmonic oscillator and would be identical to it if all force constants in the sum were zero except for C_{ii}. We will therefore try out a harmonic oscillator solution in (4.2.5) and see if it works. For the displacement of the jth atom, the trial solution is

$$u_j = v(j) e^{-i\omega t} \qquad (4.2.6)$$

where ω is an angular frequency and v depends only on j. Taking the second derivative of (4.2.6) gives

$$\frac{d^2 u_j}{dt^2} = -\omega^2 v(j) e^{-i\omega t} \qquad (4.2.7)$$

Substituting (4.2.6) and (4.2.7) into (4.2.5) gives

$$\sum_j C_{ij} v(j) - m\omega^2 v(i) = 0 \qquad (4.2.8)$$

This shows that a set of harmonic oscillator solutions to the equations of motion exist if a set of $v(i)$ is chosen that satisfies (4.2.8).

We can go further by making use of the translational symmetry of the chain. If i and j are replaced by $i + 1$ and $j + 1$ in (4.2.8), the equation is still valid and we have

$$\sum_j C_{i+1,j+1} v(j+1) - m\omega^2 v(i+1) = 0 \qquad (4.2.9)$$

But since the force constants depend only on the distance between the atomic sites, C_{ij} must equal $C_{i+1,j+1}$. This means that the solutions to (4.2.8) and (4.2.9) can differ by at most a constant factor. That is,

$$v(j+1) = B(1) v(j) \qquad (4.2.10)$$

where $B(1)$ is a constant. The argument (1) identifies it as the constant that connects two solutions that are related by a unit translation. It is clear that a similar relation holds for any lattice translation, say, by r units, so

$$v(j+r) = B(r) v(j) \qquad (4.2.11)$$

Also, for two translations r and r', we have

$$v(j+r+r') = B(r') v(j+r) = B(r') B(r) v(j) \qquad (4.2.12)$$

But treating $(r + r')$ as a single translation gives

$$v(j+r+r') = B(r+r') v(j) \qquad (4.2.13)$$

and therefore

$$B(r+r') = B(r)B(r') \tag{4.2.14}$$

Since B for the sum of two translations taken together is the product of the Bs for the translations taken separately, it is convenient to define a parameter σ that allows us to write $B(r)$ in exponential form. Accordingly, we define σ by

$$B(r) = e^{ir\sigma} \tag{4.2.15}$$

so that (4.2.14) now becomes

$$e^{i(r+r')\sigma} = e^{ir\sigma}e^{ir'\sigma} \tag{4.2.16}$$

It is clear that if we knew the values of the parameter σ, we would have an explicit expression for the harmonic oscillator solutions to our problem. These values can be obtained by considering the ends of the chain. If the chain is long enough, what happens to the ends of the chain cannot have any measurable effect on its physical properties. It can, in fact, be proven mathematically that the boundary conditions at the ends (surfaces for three-dimensional lattices) can have no physical effects if the number of atoms is large. We will simply take this result to be physically reasonable and treat the ends of the chain on the basis of convenience. Accordingly, we treat the chain as if its two ends were superimposed by requiring that

$$u_0 = u_N \tag{4.2.17}$$

That is, we imagine that the chain forms a circle that now contains N atoms. For a large enough number of atoms, the properties of this circle cannot differ appreciably from those of a linear chain. This is the periodic boundary condition, which when applied to (4.2.6) gives

$$v(N) = v(0) \tag{4.2.18}$$

But from (4.2.11) and (4.2.15),

$$v(N) = B(N)v(0) = e^{iN\sigma}v(0) \tag{4.2.19}$$

It therefore follows that

$$e^{iN\sigma} = 1 \tag{4.2.20}$$

or, equivalently,

$$N\sigma = 2\pi n \tag{4.2.21}$$

where n is any integer, $(0, \pm 1, \pm 2, \pm 3, \ldots)$.

The periodic boundary conditions have allowed us to determine a set of σs relating $v(r)$ to $v(0)$ for any r by

$$v(r) = e^{ir\sigma}v(0) \tag{4.2.22}$$

where σ is limited to the set of values

$$\sigma = \frac{2\pi n}{N} \tag{4.2.23}$$

Now we can go back to (4.2.6) and write the proposed trial solutions as

$$u_m = v(0)e^{i(m\sigma - \omega t)} \quad (4.2.24)$$

where $v(0)$ is a constant. Let us define a new parameter q by $\sigma = aq$ and write (4.2.24) as

$$u_m = v(0)e^{i(amq - \omega t)} \quad (4.2.25)$$

This is just the form of a plane wave of frequency ω with am being the distance parameter and q being 2π times the wave number. Because of (4.2.23) the qs can only take on the values

$$q = \frac{2\pi n}{Na} \quad (4.2.26)$$

Because of the periodic boundary condition, the integers 0 and N in (4.2.26) yield identical waves. Therefore, there are only N independent solutions of the form of (4.2.25) and only N independent values of q, so there are as many solutions as there are atoms in the chain. Also the integer n produces the same wave as the integer $n + N$, so the solutions are periodic and the range of q from $-\pi/a$ to $+\pi/a$ covers all N solutions.

Because q has the interpretation of a wave number, it must be closely related to the frequency ω. This relationship can be found by substituting (4.2.22) into (4.2.8) to get

$$\sum_r C_{rj} e^{ir\sigma} - m\omega^2 e^{ij\sigma} = 0 \quad (4.2.27)$$

or, solving for $m\omega^2$,

$$m\omega^2 = \sum_r C_{rj} e^{i(r-j)\sigma} \quad (4.2.28)$$

It has already been pointed out that the force constant C_{rj} depends only on the magnitude of the difference of the atomic position indices $|r - j|$. Therefore, the right-hand side of (4.2.28) can be written as a sum over $|j - r|$, and when the summation is performed, neither j nor r will appear in the result. Equation (4.2.28) is therefore an equation for ω^2 as a function of σ or, as it is usually stated, a function of q. The ωs form a discrete set because the qs can only take on the values determined by (4.2.26), and there are precisely N of them, one for each degree of freedom of the chain. From now on, we will label the frequency to emphasize this fact and write it as $\omega(q)$.

We now have a set of special solutions to the equations of motion whose number is just equal to the number of degrees of freedom. The general solution is just a linear combination of the special solutions, so the displacement of the rth atom becomes

$$u_r = \sum_q A_q e^{i[arq - \omega(q)t]} \quad (4.2.29)$$

where the $v(0)$ has been absorbed into the constant A_q.

Equation (4.2.29) shows that the atomic displacements are linear combinations of plane waves. It is useful to write (4.2.29) as

THE HARMONIC CRYSTAL 97

$$u_r = \sum_q \eta_q e^{iarq} \tag{4.2.30}$$

where η_q is defined by

$$\eta_q = A_q e^{-i\omega(q)t} \tag{4.2.31}$$

Equation (4.2.30) can be regarded as connecting two different sets of coordinates, the u_r and the η_q. The advantage of this is that the equations of motion for the chain turn out to be separable into N equations, one for each degree of freedom when expressed in terms of the η_q. (This is not the case when the atomic displacements themselves are taken as the coordinates.) To show this, substitute (4.2.30) into the equation of motion (4.2.5) to get

$$m \sum_q \frac{d^2 \eta_q}{dt^2} e^{iarq} + \sum_{j,q} C_{rj} \eta_q e^{iajq} = 0 \tag{4.2.32}$$

or

$$m \sum_q \frac{d^2 \eta_q}{dt^2} + \sum_{j,q} C_{rj} \eta_q e^{ia(j-r)q} = 0 \tag{4.2.33}$$

But using (4.2.28) in (4.2.33) reduces the double sum to a single sum over q, and we have

$$\sum_q \left[\frac{d^2 \eta_q}{dt^2} + \omega^2(q) \eta_q \right] = 0 \tag{4.2.34}$$

The η_q are independent coordinates and each term in (4.2.34) is independent. This means that the sum in (4.2.34) can be zero only if each individual term is zero. That is,

$$\frac{d^2 \eta_q}{dt^2} + \omega^2(q) \eta_q = 0 \tag{4.2.35}$$

This result could have been obtained directly by differentiating (4.2.31), but the somewhat longer method given here shows that a coordinate transformation defined by (4.2.30) reduces the dynamical problem to a set of independent harmonic oscillators. A similar procedure applied to three-dimensional crystals yields a similar result. If the potential energy is quadratic in the displacements, a set of coordinates can always be found that separates the equation of motion into independent equations of motion for independent harmonic oscillators. This is an important result because it tells us how to get the energy levels for a vibrating solid. They are just the energy levels of a set of harmonic oscillators whose frequencies are determined by (4.2.28) (or its analog in the case of two or three dimensions).

The allowed frequencies of the monatomic chain can be easily computed if it is assumed that the potential energy depends only on the relative distance between nearest neighbors and that the only forces acting in the chain are those between nearest neighbors. The potential energy is taken to be zero when the relative distance between atoms is the mean equilibrium distance, and is nonzero only when the relative displacements are nonzero. then, instead of (4.2.1), we write

$$V(u_0, u_1, \ldots, u_n) = \frac{1}{2} \sum_{i,j} d_{ij}(u_i - u_j)^2$$

the d_{ij} being constants. [These are readily related to the C_{ij} by multiplying out the squares of the relative displacements and comparing the result to equation (4.2.1).]

If we focus on an atom labeled m, we assume it interacts only with the atoms at $(m-1)$ and $(m+1)$, so $d_{m,m-1}$ and $d_{m,m+1}$ exist but all other d_{mj} are zero. Because all the atoms are identical, all the force constants acting between nearest neighbors are the same and can be set equal to a constant K. For this case, equation (4.2.5) reduces to

$$m \frac{d^2 u_r}{dt^2} = K(u_{r+1} + u_{r-1} - 2u_r) \tag{4.2.36}$$

It has already been shown that (4.2.25) is a solution of (4.2.5) and therefore of (4.2.36), so

$$u_r = v(0) e^{-i\omega t} e^{iarq} \tag{4.2.37}$$

$$u_{r+1} = v(0) e^{-i\omega t} e^{ia(r+1)q} \tag{4.2.38}$$

$$u_{r-1} = v(0) e^{-i\omega t} e^{ia(r-1)q} \tag{4.2.39}$$

The second derivative of (4.2.37) is

$$\frac{d^2 u_r}{dt^2} = -v(0) \omega^2 e^{-i\omega t} e^{iarq} \tag{4.2.40}$$

Putting (4.2.37)–(4.2.40) into (4.2.36) gives the relation between the frequency and the wave number as

$$m\omega^2 = -2K(\cos qa - 1) \tag{4.2.41}$$

Using the trigonometric identity $\cos x = 1 - 2\sin^2(x/2)$ in (4.2.41) gives

$$\omega = 2\sqrt{\frac{K}{m}} \left| \sin\left(\frac{qa}{2}\right) \right| \tag{4.2.42}$$

Equation (4.2.42) shows how the frequency is related to the wave number and is called the *frequency dispersion relation* for our nearest-neighbor monatomic chain. Its form is shown in figure 4.2.

The first point to notice from equation (4.2.42) is that there is a maximum frequency given by the fact that the highest value the sine can have is unity. This occurs when $qa/2 = \pi/2$ or $q = \pi/a$. The maximum frequency is given by (4.2.42) as

$$\omega_M = 2\sqrt{\frac{K}{m}} \tag{4.2.43}$$

Frequencies higher than this cannot propagate through the lattice. The shortest wavelength, corresponding to this maximum frequency, is obtained by remembering that q is 2π times the wave number, so $q_m = 2\pi/\lambda_M = \pi/a$ and therefore the minimum wavelength is $\lambda_M = 2a$. The physical interpretation of this is that waves with a length shorter than twice the spacing between atoms cannot

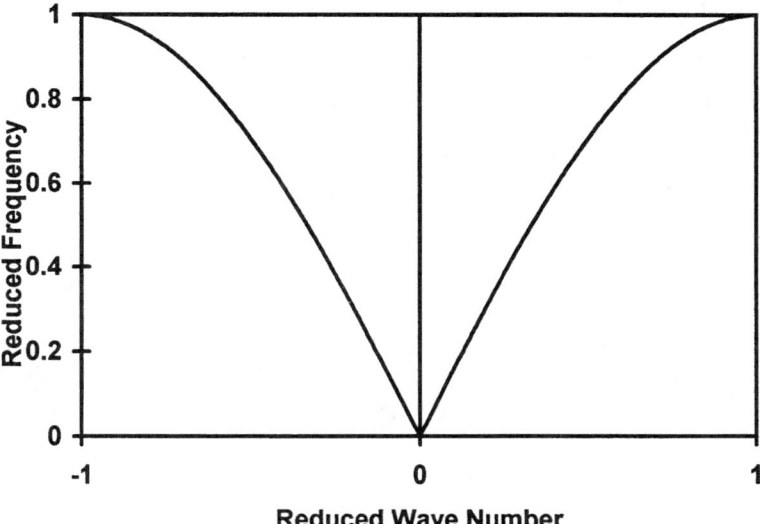

Figure 4.2. Dispersion relation for vibrations of a monatomic chain.

exist because there is nothing to take part in the vibration in the space between the atoms. Wavelengths less than the order of the atomic spacing have no meaning in a discrete structure.

The second important point to note is that the wave velocity is not constant. Since the velocity of a plane wave is defined by $v = \lambda\omega/2\pi = \omega/q$,

$$v = \frac{\omega}{q} = \frac{2}{q}\sqrt{\frac{K}{m}}\sin\left(\frac{qa}{2}\right) \quad (4.2.44)$$

so waves of different frequencies have different velocities. This is why the $\omega = \omega(q)$ function is called the *frequency dispersion relation*.

It is instructive to compare these results to those for the vibrations in a continuous string. This can be done by recognizing that (4.2.36) is a difference equation that approaches a differential equation as the atomic spacing approaches zero. Thus, if we make the transition from the discrete chain to the continuous string, then

$$\frac{u_{j+i} - u_j}{a} \to \frac{\partial u}{\partial x} \quad (4.2.45)$$

$$\frac{1}{a}\left(\frac{u_{j+1} - u_j}{a} - \frac{u_j - u_{j-1}}{a}\right) \to \frac{\partial^2 u}{\partial x^2} \quad (4.2.46)$$

This treats u as a continuous displacement in a continuous string and in the continuum limit (4.2.36) becomes

$$m\frac{\partial^2 u}{\partial t^2} = Ka^2 \frac{\partial^2 u}{\partial x^2} \quad (4.2.47)$$

where the mass m now becomes a linear mass density. But (4.2.47) is just the equation of wave motion in one dimension; that is,

$$\frac{\partial^2 u}{\partial t^2} = v_0^2 \frac{\partial^2 u}{\partial x^2} \qquad (4.2.48)$$

with the wave velocity given by

$$v_0 = \sqrt{\frac{Ka^2}{m}} \qquad (4.2.49)$$

Solutions of (4.2.48) exist that have the form

$$u(x,t) = A e^{i[(2\pi/\lambda_0)x - \omega_0 t]} \qquad (4.2.50)$$

where A is a constant, as can be verified by direct substitution. This is the equation of a one-dimensional wave that is periodic in x and t with a frequency $\nu_0 = \omega_0/2\pi$ and wavelength λ_0 where

$$\nu_0 \lambda_0 = v_0 \qquad (4.2.51)$$

Since v_0 is a constant, (4.2.51) is just the ordinary frequency–wavelength relation for a nondispersive medium. The continuum limit is a representation of one-dimensional sound waves.

The correspondence between the continuum and the discrete cases is apparent from a comparison of equations (4.2.25) and (4.2.50). The quantity am plays the role of x, and q corresponds to $2\pi/\lambda_0$. In the continuum case, however, there is no upper limit on the frequency and the velocity is not a function of wavelength.

When the wavelength is very large, q is very small and the sine in (4.2.42) can be expanded in a power series. If q is small enough, it is sufficient to keep only the first term of the expansion, with the result that

$$\omega = qa\sqrt{\frac{K}{m}} \qquad (q \to 0) \qquad (4.2.52)$$

But $\omega = 2\pi\nu$ and $q = 2\pi/\lambda$, so (4.2.52) gives $\nu = a/\lambda \sqrt{K/m}$, $(q \to 0)$ or, using (4.2.49),

$$\nu\lambda = v_0 \qquad (q \to 0) \qquad (4.2.53)$$

which is the same as (4.2.51).

This analysis shows that in the limit of very large wavelength, the chain behaves as a continuous string carrying a sound wave. This conclusion carries over to the more general case of two or three dimensions and with interactions among atoms that are not restricted to nearest neighbors. All that is required is that the potential be quadratic in the displacements. For long waves, a crystal can be treated as a continuous elastic solid.

The treatment given here is restricted to one dimension but it is readily generalized to three-dimensional crystals. The wave number q then becomes a wave number vector $\mathbf{q}$ and instead of the condition (4.2.26) we have

$$\mathbf{q} = \frac{2\pi}{N_c^{1/3}}(n_1 \mathbf{b}_1 + n_2 \mathbf{b}_2 + n_3 \mathbf{b}_3) \qquad (4.2.54)$$

where n_1, n_2, n_3 are integers ranging from 0 to N_c and $\mathbf{b}_1, \mathbf{b}_2, \mathbf{b}_3$ are reciprocal lattice vectors defined by

$$\mathbf{a}_i \cdot \mathbf{b}_j = \delta_{ij} \qquad i, j = 1, 2, 3 \qquad (4.2.55)$$

The $\mathbf{a}_i$ are the basic lattice vectors defined by the size and shape of the unit cell, N_c is the number of unit cells, and δ_{ij} is the Kronecker delta. The displacement vector $\mathbf{u}_r$ of the rth atom in a three-dimensional crystal is related to a set of independent, harmonic oscillator–type coordinates by equations similar to equation (4.2.30) but generalized to three dimensions.

It is a general result of mechanics that whenever the potential energy of a set of interacting particles is a quadratic function of the particle displacements, a set of coordinates can be found whose equations of motion are those of a set of independent simple harmonic oscillators. This can be proven directly without using the trial solutions we have resorted to here. Such a set of coordinates are said to represent the normal modes of the motion, and what we have done for the linear chain is to analyze its motion into normal modes.

The solution of the equations of motion of real crystals requires that the force constants acting among the atoms be known. A specific calculation must be made for each different material from a detailed knowledge of the crystal energy as a function of atomic displacements. The end result that is important for the thermodynamic properties is the set of frequencies of the normal mode vibrations.

4.3 Partition function and free energy of the harmonic crystal

For the harmonic model, the energy levels of a crystal containing N atoms are the same as those for a set of $3N$ independent simple harmonic oscillators, the jth oscillator having a frequency ν_j. The possible energy levels associated with this frequency are given by the quantum theory of the simple harmonic oscillator as

$$E(n_j) = \left(n_j + \frac{1}{2}\right) h\nu_j \qquad (4.3.1)$$

where n_j is a positive integer that can take on any value from zero to infinity. The energy of the harmonic crystal is the potential energy when all atoms are at their equilibrium positions plus a sum of terms like (4.3.1), one for each of the $3N$ normal modes. The state of the crystal is therefore determined by a set of integers and all possible states are determined by the possible sets of integers n_j. If U_0 is the potential energy when all atoms are at their equilibrium positions, the energy of the crystal when it is in a particular state defined by the integers $\{n_j\} = \{n_1, n_2, n_3, \ldots, n_{3N}\}$ is therefore given by

$$E\{n_j\} = U_0 + \sum_{j=1}^{3N} \left(n_j + \frac{1}{2}\right) h\nu_j \qquad (4.3.2)$$

The partition function is a sum over all states and is given by

$$Z = \sum_{\{n_j\}} e^{-E\{n_j\}/kT} \qquad (4.3.3)$$

But summing over all states means that the sum is to be performed over all possible sets of integers. That is, for each normal mode of frequency ν_j, the sum is carried out over all integers from zero to infinity. Thus, substituting (4.3.2) into (4.3.3) gives

$$Z = \sum_{n_1}\sum_{n_2}\cdots\sum_{n_{3N}} e^{-U_0/kT} e^{-\sum_j [n_j + (1/2)] h\nu_j/kT} \qquad (4.3.4)$$

Now let us define E_0 by

$$E_0 = U_0 + \frac{1}{2}\sum_{j=1}^{3N} h\nu_j \qquad (4.3.5)$$

This is the zero point energy. The sum is the vibrational zero point energy, which derives its name from the fact that at zero temperature none of the higher vibrational quantum states are excited and each normal mode is in its lowest state with energy $h\nu_j$. Using (4.3.5), we rewrite (4.3.4) as

$$Z = e^{-E_0/kT}\sum_{n_1}\sum_{n_2}\cdots\sum_{n_{3N}} e^{-\sum_j n_j h\nu_j/kT} \qquad (4.3.6)$$

Equation (4.3.6) looks complicated but it is easily simplified. Because of the convenient properties of the exponential function, the exponential of a sum is the product of the exponentials of the factors. Therefore, the multiple sum reduces to a product of sums, all of which are alike except for the subscript j on the frequencies, and (4.3.6) becomes

$$Z = e^{-E_0/kT}\sum_{n_1} e^{-n_1 h\nu_1/kT}\sum_{n_2} e^{-n_2 h\nu_2/kT}\cdots\sum_{n_{3N}} e^{-n_{3N} h\nu_{3N}/kT} \qquad (4.3.7)$$

Since all the sums have the same form, the subscript on the ns can be thrown away and (4.3.7) can be written as a product of similar sums with one factor for each frequency:

$$Z = e^{-E_0/kT}\prod_{j=1}^{3N}\sum_{n}^{\infty} e^{-nh\nu_j/kT} \qquad (4.3.8)$$

The infinite sum over n can be performed: if x_j is defined by

$$x_j = e^{-h\nu_j/kT} \qquad (4.3.9)$$

then

$$\sum_{n}^{\infty} e^{-nh\nu_j/kT} = \sum_{n}^{\infty} x_j^n \qquad (4.3.10)$$

But this is just a geometric series whose sum is $1/(1 - x_j)$, so

$$\sum_{n}^{\infty} e^{-nh\nu_j/kT} = \frac{1}{1 - e^{-h\nu_j/kT}} \qquad (4.3.11)$$

Putting this into (4.3.8) gives the partition function as

$$Z = e^{-E_0/kT}\prod_{j=1}^{3N}\frac{1}{1 - e^{-h\nu_j/kT}} \qquad (4.3.12)$$

and the Helmholtz free energy $A = -kT \ln Z$ is

$$A = E_0 + kT \sum_j \ln(1 - e^{-h\nu_j/kT}) \tag{4.3.13}$$

Now a complete theory of the thermodynamic properties of a harmonic crystal, complete with numerical results, can be constructed, provided we know the values of the frequencies. This, of course, is the fundamental problem of the theory of the harmonic crystal.

It was pointed out in section 4.2 that the frequencies can indeed be found from a solution of the equations of motion, and this has actually been done for a number of crystals. There are two disadvantages with such an approach. The first is that the force constants are not generally known to a high degree of accuracy for most crystals. This is an especially serious defect for those many crystals for which the nearest neighbor approximation is not very accurate. The second disadvantage is that a separate numerical calculation is needed for each case. The simplicity and coherence of being able to express statistical mechanical results in a single set of equations are thereby lost.

Historically, a different approach was taken, based on simplified approximations to the frequency spectrum in solids. Such approximations were suggested at about the same time (Einstein in 1907, Debye in 1911) that the dynamical equation of motion methods were being developed (1912) and were so useful that the more rigorous equation of motion methods were investigated vigorously only in more recent times.

It is convenient to replace the sum in (4.3.13) by an integral. This can be done by defining a frequency distribution function $g(\nu)$ such that $Ng(\nu)d\nu$ is the number of modes with frequencies between ν and $\nu + d\nu$. Since there are $3N$ vibrational modes altogether, the normalization condition for $g(\nu)$ is

$$\int_0^\infty g(\nu)d\nu = 3 \tag{4.3.14}$$

The Helmholtz free energy is put into integral form by multiplying the summand in (4.3.13) by $Ng(\nu)d\nu$, replacing the sum by an integral from zero to infinity, and dropping the subscript on the frequency. The result is

$$A = E_0 + NkT \int_0^\infty g(\nu) \ln(1 - e^{-h\nu_j/kT}) d\nu \tag{4.3.15}$$

Similarly, the partition function is expressed in the continuum notation as

$$\ln Z = -\frac{E_0}{kT} - N \int_0^\infty g(\nu) \ln(1 - e^{-h\nu_j/kT}) d\nu \tag{4.3.16}$$

If $g(\nu)$ were known, the problem would be solved. The approximate methods mentioned above are actually approximations to $g(\nu)$, and we will examine these shortly.

At this point it is instructive to establish the relation of the frequency distribution function to dynamical theory by showing that $g(\nu)$ can be obtained from a knowledge of the frequency as a function of wave number. For the one-dimensional case, we start with equation (4.2.26), which, treating q and n as quasi-continuous variables, gives

$$dn = \frac{Na}{2\pi} dq \tag{4.3.17}$$

We can now restrict the n and dn to positive integers by taking the number of vibrations in the frequency range dv to be twice the number of integers in the corresponding range dn, and therefore

$$2dn = Ng(v)dv \tag{4.3.18}$$

Combining these last two equations and solving for $g(v)$ shows that

$$g(v) = \frac{a}{\pi}\frac{dq}{dv} = 2a\frac{dq}{d\omega} \tag{4.3.19}$$

so if the frequency is known as a function of wave number from a solution of the equations of motion, $g(v)$ can be calculated. Similar considerations hold for the three-dimensional case with the general result that the frequency distribution function can be computed from the dispersion relation.

4.4 General heat capacity equations

The heat capacity at constant volume can be obtained from the thermodynamic relation which gives the heat capacity in terms of the second derivative of the Helmholtz free energy. It can also be more conveniently obtained by directly differentiating the energy, which will be done here. From the definition of statistical mechanical averages in the canonical ensemble, the energy of any system is given by

$$U = \frac{1}{Z}\sum_j E_j e^{-E_j/kT} \tag{4.4.1}$$

where the partition function, as usual, is

$$Z = \sum_j e^{-E_j/kT} \tag{4.4.2}$$

the sums in (4.4.1) and (4.4.2) being taken over all possible states.

If Z is differentiated with respect to the variable $1/kT$, the result is

$$\frac{\partial Z}{\partial(1/kT)} = -\sum_j E_j e^{-E_j/kT} \tag{4.4.3}$$

Comparing (4.4.3) with (4.4.1) shows that

$$U = -\frac{\partial \ln Z}{\partial(1/kT)} \tag{4.4.4}$$

The energy levels are kept unchanged during this differentiation so that the derivative in (4.4.4) is at constant volume.

Equation (4.4.4) is perfectly general and is readily applied to the harmonic crystal. Therefore, differentiating (4.3.16) in accord with (4.4.4) gives the energy of a harmonic crystal as

$$U = E_0 + N\int_0^\infty \frac{g(v)hv}{e^{hv/kT}-1}dv \tag{4.4.5}$$

and the derivative of this with respect to the temperature gives the heat capacity at constant volume as

$$C_V = Nk \int_0^\infty \left(\frac{h\nu}{kT}\right)^2 \frac{g(\nu)d\nu}{e^{h\nu/kT} + e^{-h\nu/kT} - 2} \qquad (4.4.6)$$

The procedure leading to (4.4.6) could, of course, just as easily have been carried out in the discrete notation, with the result that

$$C_V = k \sum_j \left(\frac{h\nu_j}{kT}\right)^2 \frac{1}{e^{h\nu_j/kT} + e^{-h\nu_j/kT} - 2} \qquad (4.4.7)$$

From now on, we will use either the discrete or the continuum notation as convenient without further comment.

Some interesting results follow from (4.4.6) even in the absence of any detailed information about $g(\nu)$. First, as shown for the case of the monatomic chain, there must be an upper limit on the possible frequencies in the crystal. Because of the atomistic, discrete structure of crystals, wavelengths shorter than some value of the order of the lattice spacing have no meaning, so frequencies cannot be arbitrarily large and some maximum frequency must exist. As a result, a temperature T must exist that is high enough to ensure that $h\nu/kT \ll 1$, so it is reasonable to seek a high-temperature limiting form of (4.4.6) based on a series expansion of the exponentials. Let us therefore take a look at the function $E(x)$ defined by

$$E(x) = \frac{x^2}{e^x + e^{-x} - 2} \qquad (4.4.8)$$

$E(x)$ is called the *Einstein function*.

A series expansion for the Einstein function can be found by using the series representation of the exponential functions, which are

$$e^x = \sum_{n=0}^\infty \frac{x^n}{n!} \qquad (4.4.9)$$

$$e^{-x} = \sum_{n=0}^\infty \frac{x^n}{n!}(-1)^n \qquad (4.4.10)$$

Putting these expansions into (4.4.8) gives

$$E(x) = \frac{x^2}{2}\left(\frac{x^2}{2!} + \frac{x^4}{4!} + \frac{x^6}{6!} + \frac{x^8}{8!} + \ldots\right)^{-1} \qquad (4.4.11)$$

Inverting the series in the parentheses gives

$$E(x) = 1 - \frac{x^2}{12} + \frac{x^4}{240} - \ldots \qquad (4.4.12)$$

If we identify $h\nu/kT$ with x, then (4.4.6) is

$$C_V = Nk \int_0^\infty g(\nu) E(h\nu/kT) d\nu \qquad (4.4.13)$$

Therefore, replacing x by $h\nu/kT$ in (4.4.12) and substituting the result in (4.4.13) gives the high-temperature result as

$$C_V = Nk \int_0^\infty g(\nu) \left[1 - \frac{1}{12}\left(\frac{h\nu}{kT}\right)^2 + \frac{1}{240}\left(\frac{h\nu}{kT}\right)^4\right] d\nu \qquad (4.4.14)$$

The dots have been dropped and only the first three terms have been retained because this expansion is only useful for high temperatures. If $g(\nu)$ is known, (4.4.14) can be integrated term by term. Actually, the first term does not depend on the form of the distribution function, and it can be evaluated immediately because of the normalization condition (3.3.14). Also, the second and third terms are proportional to the second and fourth moments of the distribution function, so we write (4.4.14) as

$$C_V = 3Nk - \frac{Nk}{4}\left(\frac{h}{kT}\right)^2 \overline{\nu^2} + \frac{Nk}{80}\left(\frac{h}{kT}\right)^4 \overline{\nu^4} \qquad (4.4.15)$$

If the temperature is high enough, all terms but the first can be ignored and to a good approximation $C_V = 3Nk$. This is in fact the Dulong–Petit law, which was first enunciated on the basis of experimental observations in 1819, in which it was noticed that, for many solids, the heat capacity was close to six calories per gram atom. In many cases, room temperature is high enough for the approximate validity of the high-temperature limit. Of course, it was found that some elements did not follow the Dulong–Petit law and that at low temperatures the heat capacity was much less than six calories per gram atom. These facts were important in the early development of the theory of solids.

The value of $3Nk$ for the heat capacity is just what is obtained from classical statistical mechanics in which the partition function would be computed from the energy of classical oscillators. The deviation of the heat capacity from $3Nk$ is a direct quantum mechanical effect. Since the heat capacity is the derivative of the energy, the high-temperature limiting value of the energy is $U = 3NkT$. There are $3N$ oscillators, so the energy per oscillator is kT. This is called the *classical equipartition* of vibrational energy because the energy is equally divided among all the vibrational modes.

What happens at low temperatures? In this case $h\nu/kT$ is very large and the heat capacity approaches zero as the temperature goes to zero. This is evident from taking the limit of $E(x)$:

$$\lim_{x \to \infty} E(x) = \lim_{x \to \infty} \frac{x^2}{e^x + e^{-x} - 2} = 0 \qquad (4.4.16)$$

$E(x)$ starts out at zero when T is zero and increases monotonically, approaching unity asymptotically as T goes to infinity.

4.5 The Einstein model

The simplest choice to make for the vibration spectrum of a harmonic crystal is to assume that all vibrational modes have the same frequency. Einstein made this assumption and in 1907 worked out a theory of heat capacity on this basis that accounts for the experimental data rather well and clarified the quantum mechanical origin of the low-temperature deviations from the Dulong–Petit law.

If all modes have the same frequency ν_E, then the heat capacity (4.4.7) becomes

$$C_V = 3Nk\left(\frac{h\nu_E}{kT}\right)^2 \frac{1}{e^{h\nu_E/kT} + e^{-h\nu_E/kT} - 2} \qquad (4.5.1)$$

This is the Einstein formula for the heat capacity. The fact that it gives a pretty fair representation of the heat capacity is a remarkable success considering that it rests on the assumption that all atoms vibrate independently with the same frequency. It is worthwhile noting that the Einstein theory is quite accurate for the vibrational contribution of the specific heat of diatomic gases. In this case, the interactions among molecules are weak and each molecule is an independent oscillator.

It is convenient to define a parameter with the dimensions of temperature, Θ_E, by

$$\Theta_E = \frac{h\nu_E}{k} \qquad (4.5.2)$$

and rewrite (4.5.1) as

$$C_V = 3Nk\left(\frac{\Theta_E}{T}\right)^2 \frac{1}{e^{\Theta_E/T} + e^{-\Theta_E/T} - 2} \qquad (4.5.3)$$

From (4.4.8), we see that (4.5.3) can be abbreviated by writing it in terms of the Einstein function as

$$C_V = 3NkE\left(\frac{\Theta_E}{T}\right) \qquad (4.5.4)$$

Θ_E is called the *Einstein characteristic temperature* and is obtained by finding the best fit of (4.5.3) to experiment. An advantage of defining a characteristic temperature is that the heat capacity then becomes a universal function of the ratio of the temperature to the characteristic temperature for all crystals. The form of the Einstein heat capacity is shown in figure 4.3, where C_V is plotted against T/Θ_E. Above about $T = \Theta_E/2$ the theory is fairly accurate. Typical values of the Einstein characteristic temperature are in the neighborhood of 300K. Using (4.5.2) gives an estimate of the magnitude of crystal vibration frequencies to be $\nu_E \approx 10^{13}$/sec.

4.6 Superposition of Einstein oscillators

Equation (4.4.6) shows that the most general form of the heat capacity is a superposition of Einstein functions, the contribution of Einstein oscillators of each frequency being determined by the frequency distribution function $g(\nu)$. The heat capacity equation is therefore the sum of many curves similar to that in figure 4.3. All these curves start out with a value of zero at the origin, and as shown from the analysis of the high-temperature limit [equation (4.4.15)], they all approach the same value at high temperatures. This means that the differences among theories with different assumptions about $g(\nu)$, and between these theories and experiment, will occur at low and intermediate temperatures. At higher temperatures all theories of $g(\nu)$ will give about the same results. The reason for this is that the Einstein function $E(x)$ is relatively insensitive to changes in x at low x values.

An important consequence of the Einstein function is that not all frequencies are equally significant at all temperatures. $E(x)$ is small when x is large,

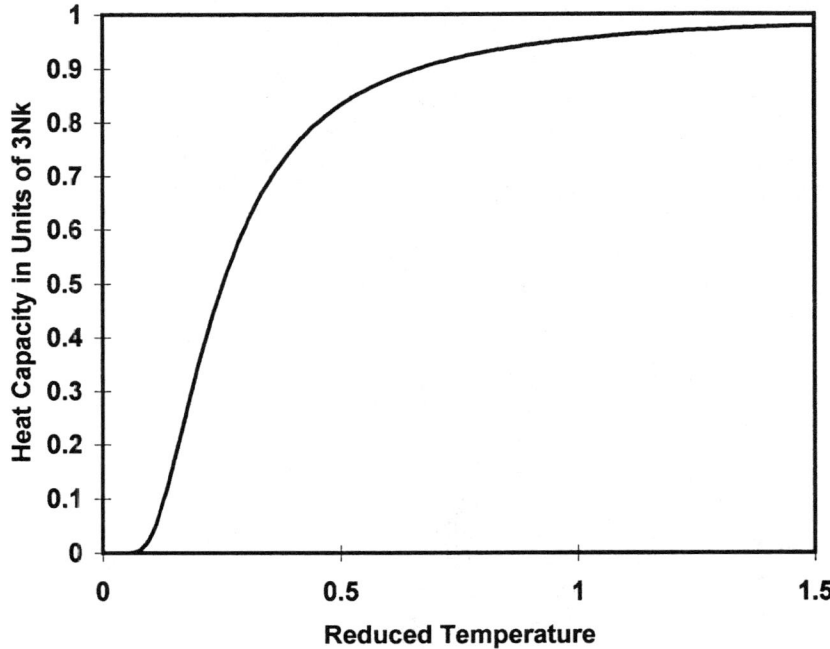

Figure 4.3. The Einstein heat capacity function.

and x is essentially the ratio of frequency to temperature. If the temperature is very low, then $E(x)$ is small unless the frequency is also low. That is, at low-temperatures only-low frequency, or long-wavelength, vibrations contribute significantly to the heat capacity. The high-frequency vibrations are simply choked off by the form of the Einstein function. As the temperature is raised, the frequency values that can contribute to the heat capacity increase. This can be summarized by saying that at low temperatures only the low-frequency modes are excited, and as the temperature increases more and more of the higher frequencies become excited. As a consequence of this we can anticipate that at low temperatures the heat capacity can be treated as arising from low-frequency sound waves. This is consistent with the result in section 4.2 that the equation of motion of a discrete crystal structure reduces to that of an elastic continuum in the long-wavelength limit.

4.7 The Debye model

Debye assumed that the frequencies in a crystal are distributed as though the solid were an isotropic elastic continuum rather than an aggregate of particles. The distribution function $g(v)$ can then be obtained by just counting the number of sound waves that can exist in a given frequency range.

This procedure should work well for long wavelengths because, as shown in section 4.2, the long wavelength limit of the vibrations in a discrete structure gives the same result as a continuum. The reason for this is that long waves consist of the cooperative motion of large numbers of atoms and in such cases the discrete structure is not important. This is the physical reason that the dispersion relation at low frequencies is linear, corresponding to elastic waves being propagated at a constant velocity. But the Debye assumption cannot be right at high frequencies because the short waves are sensitive to the discrete nature of the medium and the dispersion relation is not linear. The discrete structure of the crystal introduces the additional complication that waves shorter than the lattice spacing have no meaning, so there is an upper limit on the frequencies. This difficulty is easily met by introducing a cutoff frequency above which $g(\nu)$ is zero and no vibrations exist. The dispersion problem, however, is simply ignored in the Debye model.

The Debye model works quite well, and at first sight, this is surprising considering the physical assumptions it contains. The reasons for its success can be understood by referring to the results of preceding sections. At low temperatures the form of the Einstein function is such that low frequencies make the major contribution to the heat capacity, and the low frequencies are well described as vibrations in a continuum. At high temperatures, it was shown that the heat capacity is not very sensitive to the form of $g(\nu)$, and if the temperature is high enough it is independent of any information about the vibrations at all. The Debye model should therefore work well for low and for high temperatures. Deviations between the Debye model and experiment should manifest themselves at intermediate temperatures. It turns out that this is indeed the case, but even at intermediate temperatures the troubles are not serious.

To derive an expression for $g(\nu)$ in the Debye model, suppose that the specimen is a cube of side aL (where L is a positive integer and a is a lattice spacing) and attach a coordinate system to one corner such that the Cartesian coordinates of any point in the specimen lie in the range

$$0 \leq x \leq aL, \quad 0 \leq y \leq aL, \quad 0 \leq z \leq aL \tag{4.7.1}$$

Any frequency can be impressed on an elastic continuum but only standing waves concern us here. All other waves would die out rapidly and cannot correspond to the thermal vibrations that exist in a crystal. The possible wavelengths are therefore determined by the dimensions of the specimen.

Consider a particular standing wave with wavelength λ_j traveling in a direction determined by a unit vector **n** that has direction cosines α_1, α_2, α_3 relative to the x, y, and z axes, respectively. For a standing wave traveling in the direction of the x-axis the half-wavelength must be an integral submultiple of the length of the box. For the wave traveling in the direction **n**, this condition must be fulfilled by the x–component of the wave number vector $\mathbf{n}/\lambda_j$. That is,

$$\frac{2}{\lambda_j}\alpha_1 = \frac{j_1}{L} \tag{4.7.2}$$

where j_1 is any positive integer from 1 to L.

Similarly, for the projections of $\mathbf{n}/\lambda_j$ in the y and z directions,

$$\frac{2}{\lambda_j}\alpha_2 = \frac{j_2}{L} \tag{4.7.3}$$

$$\frac{2}{\lambda_j}\alpha_3 = \frac{j_3}{L} \tag{4.7.4}$$

j_2 and j_3 also being any positive integers from 1 to L. Each possible wavelength is associated with a set of three positive integers (j_1, j_2, j_3), which form a cubic lattice in a *j*-space, and we have a problem completely analogous to that of determining the density of states of a gas of independent particles, which we solved in connection with the development of particle statistics in section 3.4. If we define *j* in the integer space by

$$j^2 = j_1^2 + j_2^2 + j_3^2 \qquad (4.7.5)$$

then, in quasi-continuum language, the number of possible vibrations in a given range of wavelength, $d\lambda$, corresponding to a range dj, is

$$\frac{\pi}{2} j^2 dj \qquad (4.7.6)$$

From this, we want to compute the number of vibrations in a given wavelength range. From the theory of sound vibrations in an isotropic elastic solid, we know that three distinct waves are possible for each wavelength. One of these is longitudinal, and two are transverse, so the number of vibrations in a given frequency range should be written as

$$3Ng(\nu)d(\nu) = N_l g_l(\nu)d\nu + 2Ng_t(\nu)d\nu \qquad (4.7.7)$$

$g_l(\nu)$ and $g_t(\nu)$ being the frequency distribution functions for the longitudinal and transverse waves, respectively. Since there is only one longitudinal wave per wavelength, (4.7.6) gives the number of longitudinal waves directly:

$$Ng_l(\nu)d\nu = \frac{\pi}{2} j^2 dj \qquad (4.7.8)$$

and all that is needed is a relation between ν and *j*. Summing the squares of (4.7.2)–(4.7.4) gives

$$\frac{4}{\lambda_j^2}(\alpha_1^2 + \alpha_2^2 + \alpha_3^2) = \frac{1}{L^2}(j_1^2 + j_2^2 + j_3^2) \qquad (4.7.9)$$

or, since the sum of the squares of the direction cosines is unity, using (4.7.5) gives

$$\frac{4}{\lambda^2} = \frac{j^2}{L^2} \qquad (4.7.10)$$

The wavelength is related to the frequency by $\lambda\nu = C$, where C is the wave velocity. Therefore, for longitudinal waves with a velocity C_l, we have

$$\frac{1}{\lambda} = \frac{\nu}{C_l} \qquad (4.7.11)$$

Combining this with equation (4.7.10) gives

$$\frac{4\nu^2}{C_l^2} = \frac{j^2}{L^2} \qquad (4.7.12)$$

and therefore

$$j^2 = \frac{4L^2}{C_l^2} \nu^2 \qquad (4.7.13)$$

from which we get

$$dj = \frac{2L}{C_l}dv \qquad (4.7.14)$$

Now we can use (4.7.13) for j^2 and (4.7.14) for dj in equation (4.7.8) to get the distribution function for longitudinal waves as

$$Ng_l(v) = \frac{4\pi V}{C_l^3}v^2 \qquad (4.7.15)$$

where L^3 has been replaced by the volume V. The calculation for the transverse modes is just the same except for a factor of two because there are two transverse modes for each frequency. The result is

$$Ng_t(v) = \frac{8\pi V}{C_t^3}v^2 \qquad (4.7.16)$$

C_t being the transverse velocity of sound. Adding (4.7.15) and (4.7.16) gives the frequency distribution for the isotropic elastic solid:

$$g(v) = 4\pi \frac{V}{N}\left(\frac{1}{C_l^3} + \frac{2}{C_t^3}\right)v^2 \qquad (4.7.17)$$

It is convenient to lump the constants in (4.7.17) together into a single symbol B defined by

$$B = 4\pi \frac{V}{N}\left(\frac{1}{C_l^3} + \frac{2}{C_t^3}\right) \qquad (4.7.18)$$

and write (4.7.17) as

$$g(v) = Bv^2 \qquad (4.7.19)$$

Notice that B contains constants that can be measured independently of heat capacity data. The frequencies will integrate out in the statistical mechanical equations, so the values of B obtained from the density and velocities of sound in the crystal can be compared to those obtained by fitting heat capacity data. This will give a measure of the accuracy of the Debye model.

The constant B can be related to the cutoff frequency required by the discrete structure of the crystal by using the normalization condition (4.3.14). Calling the cutoff frequency v_D, this gives

$$\int_0^\infty g(v)dv = B\int_0^{v_D} v^2 dv \qquad (4.7.20)$$

Performing the integration we get

$$B = \frac{9}{v_D^3} \qquad (4.7.21)$$

Later, we will need a characteristic Debye temperature, Θ_D, which is defined by

112 STATISTICAL MECHANICS OF SOLIDS

$$\frac{h\nu_D}{k} = \Theta_D \tag{4.7.22}$$

In terms of Θ_D, B is

$$B = 9\left(\frac{h}{k\Theta_D}\right)^3 \tag{4.7.23}$$

An estimate of the values of ν_D and Θ_D can be made from data on the density and velocity of sound in solids using the definition of B given by (4.7.18) and equations (4.7.21) and (4.7.23). If the velocity of sound is of the order of 10^5 cm/sec (we ignore the difference between longitudinal and transverse velocities for the purposes of this calculation) and we take the density to be about 10^{23} atoms/cm^3, then ν_D is of the order of 10^{13} vibrations per second and the Debye characteristic temperature Θ_D is in the neighborhood of 400K.

4.8 Debye energy and heat capacity

The energy and the heat capacity for the Debye model are obtained by substitution of the Debye frequency distribution function (4.7.19) into equations (4.4.5) and (4.4.6). The results are

$$U - E_0 = 9NkT\left(\frac{h}{k\Theta_D}\right)^3 \int_0^{\nu_D} \frac{h\nu}{kT} \frac{\nu^2}{e^{h\nu/kT} - 1} d\nu \tag{4.8.1}$$

$$C_V = 9Nk\left(\frac{h}{k\Theta_D}\right)^3 \int_0^{\nu_D} \left(\frac{h\nu}{kT}\right)^2 \frac{\nu^2}{e^{h\nu/kT} + e^{-h\nu/kT} - 2} d\nu \tag{4.8.2}$$

The constant B in (4.7.19) has been replaced by its value in terms of the Debye characteristic temperature according to (4.7.23), and the cutoff frequency of Debye has been introduced as the upper limit on the integrals. $(U - E_0)$ is the thermal energy and plays an important role not only in the heat capacity but also in the theory of the equation of state.

To simplify the appearance of these equations, define a parameter x by

$$x = \frac{h\nu}{kT} \tag{4.8.3}$$

When $T = \Theta_D$ this parameter has the value x_D given by

$$x_D = \frac{h\nu_D}{kT} = \frac{\Theta_D}{T} \tag{4.8.4}$$

Now define two functions $D_E(x_D)$ and $D(x_D)$ by

$$D_E(x_D) = \frac{3}{x_D^4} \int_0^{x_D} \frac{x^3}{e^x - 1} dx \tag{4.8.5}$$

$$D(x_D) = \frac{3}{x_D^3} \int_0^{x_D} \frac{x^4}{e^x + e^{-x} - 2} dx \tag{4.8.6}$$

Using (4.8.5) and (4.8.6), the energy (4.8.1) and the heat capacity (4.8.2) can now be written in the following compact forms:

$$U - E_0 = 3Nk\Theta_D D_E(x_D) \qquad (4.8.7)$$

$$C_V = 3NkD(x_D) \qquad (4.8.8)$$

$D_E(x_D)$ and $D(x_D)$ are called the *Debye energy function* and the *Debye heat capacity function*, respectively. They cannot be evaluated in closed form, but their properties have been well studied. Tables of these functions are available for use in numerical analyses, but with the widespread use of microcomputers, the tables are no longer really necessary. [Note that $D_E(x_D)$ is the thermal energy per oscillator in units of $k\Theta_D$ and $D(x_D)$ is the vibrational heat capacity per oscillator in units of k.]

Analytic approximations to the Debye energy and heat capacity can be obtained in the limits of low and high temperatures that are useful for the study of crystal properties. For high temperatures $\Theta_D/T = x_D$ is small, so the variables of integration in the integrals of equations (4.8.5) and (4.8.6) are small and it is legitimate to use Taylor series expansions. To do this for the thermal energy, define a function $f(x)$ by

$$f(x) = \frac{x}{e^x - 1} \qquad (4.8.9)$$

and expand it in a Taylor series to the second order in x:

$$f(x) = f(0) + f'(0)x + \frac{1}{2}f''(0)x^2 \qquad (4.8.10)$$

The values of $f(x)$ and its derivatives at $x = 0$ can be computed by using the series expansion for the exponential

$$e^x = 1 + x + \frac{1}{2}x^2 + \frac{1}{6}x^3 \ldots \qquad (4.8.11)$$

in equation (4.8.9) and then taking the limit as x goes to zero. The result is

$$f(0) = 1$$
$$f'(0) = -\frac{1}{2} \qquad (4.8.12)$$
$$f''(0) = \frac{1}{6}$$

Using these values, (4.8.10) becomes

$$f(x) = 1 - \frac{1}{2}x + \frac{1}{12}x^2 \qquad (4.8.13)$$

Multiplying this by x^2 gives us the integrand we were after, which can be substituted into (4.8.5) to yield

$$D_E(x_D) = \frac{3}{x_D^4} \int_0^{x_D} \left(x^2 - \frac{x^3}{2} + \frac{x^4}{12} \right) dx \qquad (4.8.14)$$

Now perform the integration to get

$$D_E(x_D) = \frac{1}{x_D} - \frac{3}{8} + \frac{x_D}{20} \qquad (4.8.15)$$

The Debye heat capacity function in the high-temperature limit is obtained even more simply because (4.8.6) is just x^2 times the Einstein function, and we have already derived the high-temperature limiting formula for the Einstein function. Therefore, multiplying (4.4.12) by x^2, the high-temperature limit of the integrand in (4.8.7) is

$$x^2\left(1 - \frac{x^2}{12} + \frac{x^4}{240}\right) \qquad (4.8.16)$$

Putting this in (4.8.6) and performing the integration gives

$$D(x_D) = 1 - \frac{x_D^2}{20} \qquad (4.8.17)$$

The high-temperature approximations for the Debye energy and heat capacity are obtained by substituting (4.8.15) and (4.8.17) into equations (4.8.7) and (4.8.8), respectively, to get

$$U - E_0 = 3NkT\left[1 - \frac{3}{8}\frac{\Theta_D}{T} + \frac{1}{20}\left(\frac{\Theta_D}{T}\right)\right]^2 \qquad (T \gg \Theta_D) \qquad (4.8.18)$$

$$C_V = 3Nk\left[1 - \frac{1}{20}\left(\frac{\Theta_D}{T}\right)^2\right] \qquad (4.8.19)$$

As required by the general theory of section 4.4, the energy and the heat capacity approach the classical limit in which each oscillator has an energy kT and a heat capacity k.

The low-temperature limit is obtained by first writing the integral in the definition of the Debye energy (4.8.5) as the difference of two integrals:

$$\int_0^{x_D} \frac{x^3}{e^x - 1}dx = \int_0^{\infty} \frac{x^3}{e^x - 1}dx - \int_{x_D}^{\infty} \frac{x^3}{e^x - 1}dx \qquad (4.8.20)$$

The first term on the right-hand side in this equation is a definite integral whose value is known to be $\pi^4/15$. To approximate the second term, remember that x_D is large when T is small. This means that unity can be neglected relative to the exponential term and the last integral in equation (4.8.20) can therefore be approximated by[1]

$$\int_{x_D}^{\infty} x^3 e^{-x}dx = e^{-x_D}(x_D^3 + 3x_D^2 + 6x_D + 6) \qquad (4.8.21)$$

Putting these results in (4.8.20) and multiplying by $3/x_D^4$ gives the low-temperature approximation to (4.8.5) as

$$D_E(x_D) = \frac{3}{x_D^4}\left[\frac{\pi^4}{15} - e^{-x_D}(x_D^3 + 3x_D^2 + 6x_D + 6)\right] \qquad (4.8.22)$$

Actually, an approximation in which terms beyond the first two are retained is not useful, so we take the low-temperature approximation to be

$$D_E(x_D) = \frac{3}{x_D^4}\left(\frac{\pi^4}{15} - e^{-x_D}x_D^3\right) \qquad (4.8.23)$$

The low-temperature approximation to the heat capacity function is derived in a similar manner by writing the integral in (4.8.6) as a sum of two integrals as follows:

$$\int_0^{x_D} \frac{x^4 dx}{e^x + e^{-x} - 2} = \int_0^{\infty} \frac{x^4 dx}{e^x + e^{-x} - 2} - \int_{x_D}^{\infty} \frac{x^4 dx}{e^x + e^{-x} - 2}$$

The definite integral on the right-hand side has the value $4\pi^4/15$, and in the second integral the exponential e^x needs to be retained because x is large. We therefore have[2]

$$\int_0^{x_D} \frac{x^4}{e^x + e^{-x} - 2} dx = \frac{4\pi^4}{15} - x_D^4 e^{-x_D} \tag{4.8.24}$$

where we neglect powers of x_D lower than the fourth. Multiplying (4.8.24) by $3/x_D^3$ gives the low-temperature Debye heat capacity function as

$$D(x_D) = \frac{4\pi^4}{5x_D^3} - 3x_D e^{-x_D} \tag{4.8.25}$$

Equations (4.8.23) and (4.8.25) can now be put into (4.8.7) and (4.8.8) to give the low-temperature approximations to the Debye energy and heat capacity, respectively:

$$U - E_0 = 3NkT \left[\frac{\pi^4}{5} \left(\frac{T}{\Theta_D} \right)^3 - 3e^{-\Theta_D/T} \right] \tag{4.8.26}$$

$$C_V = 3Nk \left[\frac{4\pi^4}{5} \left(\frac{T}{\Theta_D} \right)^3 - 3\frac{\Theta_D}{T} e^{-\Theta_D/T} \right] \quad (T \ll \Theta_D) \tag{4.8.27}$$

The Debye model predicts that at very low temperatures the heat capacity varies as the cube of the temperature. Because only long waves are important at low temperatures and because the long waves are well represented by an elastic continuum model regardless of the choice of frequency distribution function, we would expect the T^3 law to be valid if the temperature is low enough. This is verified by experiment.

The theory presented here is valid only if lattice vibrations are the only cause of the variation of energy with temperature. When other factors are present, such as phase transformations or "free" electrons in metals, additional contributions to the heat capacity must be considered.

4.9 Relation between Einstein and Debye characteristic temperatures

The Einstein model can be thought of as describing crystal vibrations by replacing all of the vibrational frequencies by a single average frequency. Given the Debye model, an equivalent Einstein frequency can be defined as the average frequency for the Debye frequency distribution. This average is

$$\bar{v} = \frac{\int_0^{\infty} v g(v) dv}{\int_0^{\infty} g(v) dv} = \frac{3}{v_D^3} \int_0^{v_D} v^3 dv \tag{4.9.1}$$

or, performing the integral,

$$\bar{\nu} = \frac{3}{4}\nu_D \quad (4.9.2)$$

so if the average of the Debye spectrum is identified with the Einstein frequency, we have

$$\nu_E = 0.75\nu_D$$
$$\Theta_E = 0.75\Theta_D \quad (4.9.3)$$

An alternative method of relating the Einstein and Debye models is to require that the two theories match closely in some temperature range. At low temperatures, analysis shows that the Einstein heat capacity varies as $(\Theta_E/T)^2 e^{-\Theta_E/T}$. Since the Debye heat capacity varies as $(T/\Theta_D)^3$, no sensible comparison can be made. At high temperatures, both theories approach the classical limit of $3Nk$ and the first term in the high-temperature expansion that contains any information about the frequency spectrum is quadratic in the frequencies, as is evident from equation (4.4.15).

For the Debye model, the mean square frequency is obtained from the Debye distribution function (4.7.19) as

$$\overline{\nu^2} = \frac{3}{\nu_D^3}\int_0^{\nu_D} \nu^4 d\nu = \frac{3}{5}\nu_D^2 \quad (4.9.4)$$

Now if the Einstein frequency is identified with the root mean square of the Debye distribution, we get

$$\nu_E = 0.775\nu_D$$
$$\Theta_E = 0.775\Theta_D \quad (4.9.5)$$

The connection between the Einstein and the Debye model is approximately the same for both the mean and the root mean square relations. Use of (4.9.5) gives a rather good match between the Einstein and the Debye heat capacity data at high temperatures, while the overall fit is better when the average given by (4.9.4) is used.

Figure 4.4 shows a comparison of the heat capacity per oscillator in the Einstein and Debye theories when we take $\Theta_E = 0.75\Theta_D$. The two theories give significantly different results only at temperatures lower than about half the Debye temperature.

4.10 Comparison of Debye theory with experiment

The Debye theory has been tested for a large number of solids by fitting the theory to experimental heat capacity data, using Θ_D as a disposable parameter to be computed from the data. The result is that a Θ_D can always be found such that the deviation of the data from that predicted by the Debye theory is small. The Debye theory therefore gives an excellent theoretical description of the heat capacity. Values of the Debye temperature obtained by fitting the theory to experiment are given for a number of crystalline solids in table 4.1. But comparison with heat capacity data is not the most sensitive way to test the Debye theory. Section 4.4 showed that the theory is not very sensitive to the form of

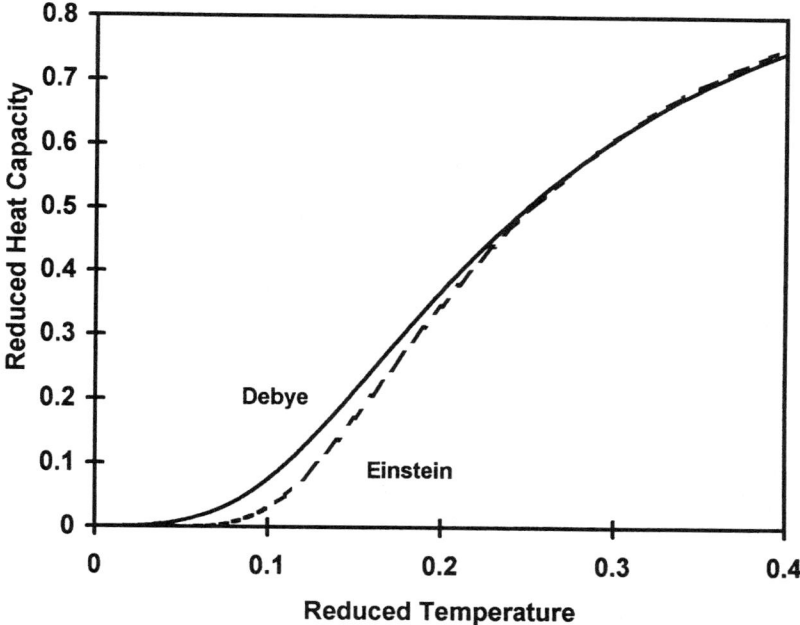

Figure 4.4. Comparison of Debye and Einstein heat capacity functions.

the frequency distribution function, especially at higher temperatures. The Einstein and Debye models therefore give quite similar results. This is comforting in that the Debye theory can be used with an assurance that is not really justified by the fundamental assumptions, but it is not of much help if we want to investigate the validity of these assumptions.

Another way of testing the theory is to start with the definition of the frequency distribution function as being that for an elastic solid continuum. From equations (4.7.18) and (4.7.23), the Debye temperature is related to the velocities of sound by

$$\Theta_D = \frac{h}{k}\left(\frac{9N}{4\pi V}\right)^{1/3}\left(\frac{1}{C_l^3}+\frac{2}{C_t^3}\right)^{-1/3} \tag{4.10.1}$$

The velocity of sound can be measured directly or computed from elastic constants, so (4.10.1) can be used to compute Θ_D from information that is directly connected to the core of the model. The Debye temperatures obtained in this way can then be compared with those obtained from heat capacity data. Such a comparison is given in table 4.2, which shows that the Debye temperature computed from the velocity of sound is not very different from that obtained from heat capacity data. In many cases, the two results differ by less than 5%. Some solids, however, show a considerable difference between the two methods. This is particularly true for anisotropic crystals, as would be expected because the theory is based on that of an isotropic elastic continuum.

A third test of the theory is to see if the data actually follow the T^3 law at low temperatures. If the Debye theory is correct, then a plot of C_V data against T^3 should be linear. The evidence is very good that the T^3 law holds for simple

Table 4.1: Debye Temperatures of Some Elements from Heat Capacity Data

Element	Θ_D (°K)	Element	Θ_D (°K)
A	85	Mg	318
Ag	215	Mn	400
Al	394	Mo	380
As	285	Na	150
Au	170	Ne	63
B	1250	Ni	375
Be	1000	Pb	88
C (diamond)	1860	Pd	275
C (graphite)	420	Pr	74
Ca	230	Pt	230
Cd	120	Sb	200
Co	385	Si	625
Cr	460	Sn (gray)	260
Cu	315	Sn (white)	170
Fe	420	Ta	225
Ga	240	Tb	100
Ge	360	Ti	380
Gd	152	Tl	96
Hg	100	V	390
In	129	W	310
K	100	Zn	234
Li	132	Zr	250

From a compilation by De Launay (1956, p. 233). The *American Institute of Physics Handbook* (1957) contains another compilation based on low-temperature heat capacity data.

Table 4.2: Comparison of Debye Temperatures Based on Heat Capacity and Velocities of Sound

Element	Θ_D (sound velocity; °K)	Θ_D (heat capacity; °K)
Al	399	394
Fe	467	420
Cu	329	315
Ag	212	215
Cd	168	120
Pt	226	230
Pb	72	88

Taken from a compilation by Slater, J.C.; 1939; *Introduction to Chemical Physics*; McGraw-Hill, New York, p. 237.

solids if the temperature is low enough. It was previously thought that the T^3 law is valid if the temperature is below $\Theta_D/12$. More recent work, however, shows that the temperature must be below $\Theta_D/50$ for the T^3 law to be accurate. Of course, this is just a question of the degree of accuracy that can be obtained in experiments.

Again, however, neither the computation of Θ_D from elastic constants nor the verification of the T^3 law is a very sensitive test because the long-wavelength limit is expected to give fairly good results at low temperatures and at high temperatures the theory is well described using a single root mean square frequency. A more sensitive test is to use precision heat capacity data to compute Θ_D from equation (4.8.8) for each temperature at which data are taken. This is easily done by using tables of the Debye function. If the Debye theory is correct, then the Debye temperature computed in this way should be the same for all temperatures. Any deviation from a constant in a Θ_D versus T plot reflects an inaccuracy in the Debye theory. This is a sensitive method and shows that for many crystals Θ_D is not constant with respect to temperature. However, deviations from constancy are generally smaller than 20% and often less than 10%. Crystals with the diamond or hexagonal structures exhibit greater deviations than those with face-centered or body-centered cubic structures.

The general conclusion to be drawn from the comparison with experiment is that the Debye theory is very useful and quite accurate considering its doubtful axiomatic ancestry. Detailed comparisons show that the theory contains some defects that can be removed only by going to an actual lattice dynamics calculation of the frequency distribution function.

4.11 The phonon gas

In the harmonic approximation, the statistical thermodynamic functions are written in terms of simple harmonic oscillator frequencies. This is possible because a coordinate transformation always exists that transforms a quadratic form in the atomic coordinates to normal mode coordinates, thereby converting the dynamical equations of motion, which involve all the coordinates, to a set of equations each containing only one normal mode coordinate. The many-body problem of the crystal is then reduced to a set of one-body problems. But a set of independent one-body equations of motion is just what we get for a system of independent particles, so normal mode analysis reduces the vibrating crystal to an analog of a set of independent particles.

The equivalence of the harmonic approximation to an independent particle model can be shown by considering a system of independent particles with the following properties:

1. Each particle can exist in any of a set of energy levels $\varepsilon_j = h\nu_j$.
2. There is no restriction on the number of particles that can occupy any given level.
3. There is no restriction on the total number of particles in the system; that is, particles are not conserved.

If the ν_j are the possible frequencies of electromagnetic radiation, then the particles are photons and the theory of black body radiation can be developed from these three properties. If the ν_j are the frequencies of a vibrating solid, then the "particles" are called *phonons* and, as shown below, the statistical thermodynamics worked out in the preceding sections is equivalent to that for a phonon gas.

The second property listed above requires that phonons be described by Bose–Einstein statistics. Because of the third property, however, the probability distribution function is not quite the same as for material bosons. Phonons can be created and destroyed, so in solving the variational problem for the statistical number of complexions, the restriction that $dN = 0$ does not apply in this case. The corresponding Lagrangian therefore does not appear in the distribution function. For phonons, $\mu = 0$. Instead of the Bose–Einstein distribution for material particles worked out above, we then have, for the number of phonons in an energy range ε_j to $\varepsilon_j + d\varepsilon_j$,

$$N_j = \frac{\omega_j}{e^{\varepsilon_j/kT} - 1} \tag{4.11.1}$$

or, in continuum language,

$$N(\varepsilon)d\varepsilon = \frac{\omega(\varepsilon)d\varepsilon}{e^{\varepsilon/kT} - 1} \tag{4.11.2}$$

To write this in terms of the frequencies, recall that $Ng(\nu)d\nu$ is the number of modes in a frequency range ν to $(\nu + d\nu)$, which we now identify with the number of phonon states in the same frequency range. This amounts to choosing a density of states $\omega(\varepsilon)$ that agrees with the frequency distribution in crystals. Equation (4.11.2) now becomes

$$N(\nu)d\nu = \frac{N_g(\nu)d\nu}{e^{h\nu/kT} - 1} \tag{4.11.3}$$

where $N(\nu)d\nu$ is the number of phonons with frequency range ν to $(\nu + d\nu)$.

The energy of the system of phonons is now readily obtained by multiplying (4.11.3) by the energy $h\nu$ of a phonon of frequency ν and integrating over all frequencies. The result is

$$U_p = \int_0^\infty h\nu N(\nu)d\nu = \int_0^\infty \frac{g(\nu)h\nu d\nu}{e^{h\nu/kT} - 1} \tag{4.11.4}$$

The heat capacity at constant volume of the phonons is just the temperature derivative of (4.11.4), which is

$$C_V = Nk\int_0^\infty g(\nu)\left(\frac{h\nu}{kT}\right)^2 \frac{d\nu}{e^{h\nu/kT} + e^{-h\nu/kT} - 2} \tag{4.11.5}$$

We want to get the Helmholtz free energy by using the relation $A = U - TS$, so we need the entropy. From thermodynamics, this is given by an integral of the heat capacity as

$$S = \int_0^T \frac{C_V}{T}dT \tag{4.11.6}$$

Using (4.11.5), this gives, for the phonon gas,

$$S_p = Nk\int_0^\infty g(\nu)\int_0^T \frac{1}{T}\left(\frac{h\nu}{kT}\right)^2 \frac{dTd\nu}{e^{h\nu/kT} + e^{-h\nu/kT} - 2} \tag{4.11.7}$$

The inner integral can be evaluated by means of the transformation $y = e^{h\nu/kT}$, so that the inner integral becomes

$$-\int_\infty^{e^{h\nu/kT}} \frac{\ln y}{(y-1)^2} dy \qquad (4.11.8)$$

Now let us integrate (4.11.8) by parts to get

$$-\int_\infty^{e^{h\nu/kT}} \frac{\ln y}{(y-1)^2} dy = \frac{\ln y}{y-1}\bigg|_\infty^{e^{h\nu/kT}} - \int_\infty^{e^{h\nu/kT}} \frac{dy}{y(y-1)}$$

$$= \frac{h\nu}{kT(e^{h\nu/kT}-1)} - \ln(1-e^{-h\nu/kT}) \qquad (4.11.9)$$

Substituting this for the inner integral in (4.11.7) gives us the phonon entropy as

$$S_p = \frac{N}{T}\int_0^\infty \frac{g(\nu)h\nu d\nu}{e^{h\nu/kT}-1} - Nk\int_0^\infty g(\nu)\ln(1-e^{-h\nu/kT})d\nu \qquad (4.11.10)$$

Multiplying (4.11.10) by the temperature and subtracting the result from the energy equation (4.11.4) gives the Helmholtz free energy for the phonon gas $A_p = U_p - TS_p$ as

$$A_p = NkT\int_0^\infty g(\nu)\ln(1-e^{-h\nu/kT})d\nu \qquad (4.11.11)$$

Now compare equations (4.11.4), (4.11.5), and (4.11.11), to the equations for the energy, heat capacity, and Helmholtz free energy from the theory of harmonic vibrations given by equations (4.4.5), (4.4.6), and (4.3.15). The two sets of equations are identical except for the presence of the zero point energy term E_0 in the earlier set of equations. This term could have been included in those for the phonon gas also, if the phonons had been defined as moving in a constant potential U_0 and if $h\nu_j/2$ were added to the definition of the phonon energy ε_j.

This analysis shows that, as far as the vibrations are concerned, the harmonic crystal can be treated as a phonon gas. This concept is widely used in the theory of solids. Note in passing that the frequency distribution function for photons is given by

$$g_{ph}(\nu) = \frac{8\pi V\nu^2}{c^3} \qquad (4.11.12)$$

where c is the velocity of light. The theory of black body radiation is the theory of bosons with the distribution function given by (4.11.12). The similarity to the Debye solid is apparent through the dependence of the frequency distribution on the square of the frequency. The major difference is that a black body is a true continuum with respect to radiation in that there is no cutoff frequency.

Exercises

4.1 Derive the frequency distribution function from the dispersion relation for a monatomic chain with nearest neighbor forces undergoing longitudinal vibrations.

4.2 The third law of thermodynamics states that the entropy of a perfect crystal is zero at absolute zero of temperature. Verify this in two ways:

A. by recognizing that for a perfect crystal there is only one ground state (i.e., wave function; neglect possible quantum degeneracy) and that the ground state is realized at absolute zero, and
B. by using the Debye theory for the harmonic crystal and the thermodynamic formula for the entropy as an integral containing the heat capacity.

4.3 A crystal contains N_I impurity interstitial atoms, each of which vibrates with three localized modes (three-dimensional Einstein oscillator with Einstein temperature $= \Theta_E$). The crystal has N identical atoms of the major component and a Debye temperature of Θ_D, which is not changed by the presence of the interstitials. Find the equation for the total heat capacity (at constant volume) as a function of atomic fraction $n_I = N_I/N$ of interstitials in terms of the Einstein and Debye temperatures. Write this equation for the case when the Einstein characteristic temperature is twice the Debye temperature for a temperature that is equal to the Debye temperature. What is the fractional contribution of the interstitials to the heat capacity for N_I/N of 0.5%? Use high-temperature approximations (to order T^{-2}) for the heat capacity equations.

4.4 Derive the formula for the vibrational zero point energy for a Debye solid and compute the zero point energy in electron volts if the Debye temperature is 300 K.

4.5 Show that, in the limit of high temperatures, the entropy of a harmonic crystal in terms of the normal mode frequencies is

$$S = 3Nk - k\sum_{j=1}^{3N}\ln\left(\frac{h\nu_j}{kT}\right)$$

4.6 Show that, in the Debye theory, the high-temperature limit of the entropy (derived in problem 4.5) becomes $S = 3Nk[(4/3) - \ln(\Theta_D/T)]$.

4.7 Find the low-temperature limit of the heat capacity of a harmonic crystal in the Einstein approximation and compare it to the low-temperature limit in the Debye theory. Why does the Einstein model fail at low temperatures?

4.8 Work out the heat capacity theory for a two-dimensional Debye solid. Show that at very low temperatures, the heat capacity (constant volume) varies as the square of the absolute temperature. You need not evaluate the proportionality constant.

4.9 Given a crystal consisting of diatomic molecules, assume that the vibrational energy can be separated into two parts: a crystal vibrational energy and a molecular vibrational energy. Let the crystal vibrations have a Debye frequency ν_D corresponding to a Debye temperature Θ_D, and let the molecules all have the same intravibrational frequency ν_M corresponding to a characteristic temperature Θ_M.

A. How would you compute Θ_D and Θ_M from low-temperature heat capacity data?
B. If the crystal lattice and molecular vibrational contributions are equal at a temperature corresponding to $T/\Theta_M = 0.1$, what is the value of T/Θ_D at that temperature? What is the value of v_M/v_D?

Use low temperature approximations throughout.

4.10 Compute the values of the heat capacity per atom of an Einstein crystal, in units of k, when the temperature is equal to the Einstein characteristic temperature, and for a Debye crystal when the temperature is equal to the Debye characteristic temperature. What is the ratio of these two heat capacities?

Notes

1. This integral is readily evaluated in terms of simple forms by a successive integration by parts.
2. This requires only one integration by parts to be evaluated to the fourth order.

5
Anharmonic Properties and the Equation of State

5.1 The crystal potential energy

The equation of state is the pressure as a function of temperature and volume. It is contained in the thermodynamic relation

$$P = -\left(\frac{\partial A}{\partial V}\right)_T \tag{5.1.1}$$

All that is needed is a statistical mechanical expression for the Helmholtz free energy, which is then differentiated according to (5.1.1) to give the P–V–T relation. For the harmonic solid, the Helmholtz free energy is given by equation (4.3.13).

If practical use is to be made of (5.1.1), the static crystal energy U_0 as a function of volume must be known. In fact, the pressure term arising from the variation of U_0 with volume is the largest contribution, so a major task of equation of state theory is to determine $U_0 = U_0(V)$. The general form of this function is clear from a consideration of some simple observations on the condensation of gases and on the compressibility of solids. Since cooling a gas reduces the kinetic energy of its atoms, this means that when they are far apart the atoms attract one another. At high temperatures the attractive forces are not strong enough to overcome the kinetic energy, but as the temperature is lowered the kinetic energy decreases, the attractive forces take over, and condensation occurs. But at very large distances from each other, the attractive forces must be small. Thus, we infer an attractive force between atoms that increases in strength as the atoms come closer together.

But if only an attractive force existed, all matter would condense to a vanishingly small volume at very low temperatures. In fact, once a condensed phase is formed, further decrease in temperature decrease the volume only by small amounts, and compressibility experiments show that large external forces are needed to decrease the volume of a solid or a liquid appreciably. We therefore conclude that when atoms are very close together they repel each other and that the repulsive force increases to very high values with decreasing distance between the atoms. Also, the fact that solids have a definite volume at low temperatures shows that there is a most stable configuration corresponding to a minimum in the potential energy.

These observations can be summarized by the following description of the interatomic interactions: atoms exert both attractive and repulsive forces on each other. When they are very far apart their interactions are negligible, but as they approach each other their mutual attraction increases with decreasing interatomic distance due to the attractive forces. The attraction dominates until the atoms get close enough to a point at which the repulsive forces become large. These repulsions increase rapidly with decreasing distance. The equilibrium volume of a crystal is determined by the balance of attractive and repulsive forces when the potential energy of the crystal is a minimum.

A great deal of effort has been expended toward finding the theoretical form for $U_0 = U_0(V)$. For rare gas crystals, the energy has been successfully computed by assuming that the atoms interact according to a Lennard–Jones potential of the form

$$\phi_{ij} = -\frac{A}{r_{ij}^6} + \frac{B}{r_{ij}^{12}} \qquad (5.1.2)$$

where ϕ_{ij} is the potential energy of interaction of two atoms i and j when they are a distance r_{ij} apart, and A and B are constants. The first term in (5.1.2) represents the attractive potential, while the second term represents repulsion. The static energy of a rare gas crystal is obtained by assuming that it is a pairwise sum of Lennard–Jones potentials summed over the crystal lattice.

For simple metals, theoretical crystal energies have been computed from the nearly free electron model, which assumes that the metal consists of ion cores immersed in a free electron gas formed from the valence electrons. The form of the crystal energy is then found to be

$$U_0 = Nz\left[E_{ce}(r_s) + \frac{2.21}{r_s^2} + \frac{1.2}{r_s} - \frac{0.916}{r_s} + 0.031\ln r_s - 0.115\right] \qquad (5.1.3)$$

N is the number of atoms, and z is the valence of the metal; r_s is the Wigner–Seitz radius, defined as the radius of a sphere whose volume is equal to the atomic volume. That is,

$$\frac{4}{3}\pi r_s^3 = \frac{V}{N} \qquad (5.1.4)$$

In equation (5.1.3) the energy is expressed in rydbergs and the Wigner–Seitz radius is in units of the Bohr radius. The first term on the right-hand side is the interaction of the electrons with the ion cores. A simple approximate expression for this term can be obtained from a quantum mechanical treatment of a single electron near the edge of a unit cell in the field of the ion core at the center. The result is

$$E_{ce} = -\frac{3}{r_s} + \frac{r_0^2}{r_s^3} \qquad (5.1.5)$$

where r_0 is a constant. The second term in (5.1.3) arises from the kinetic energy of the electrons in the electron gas, the third term is the Coulomb energy of the mutual repulsion of the electrons, and the third term is the exchange energy, which arises from the fact that, because of the exclusion principle, electrons of like spin tend to avoid each other. The last two terms constitute the correlation energy, which results from cooperative interactions among the electrons that are not accounted for by assuming that they are independent particles.

The Lennard–Jones and the nearly free electron theories given above are only two of many approaches that have been developed in efforts to describe the

126 STATISTICAL MECHANICS OF SOLIDS

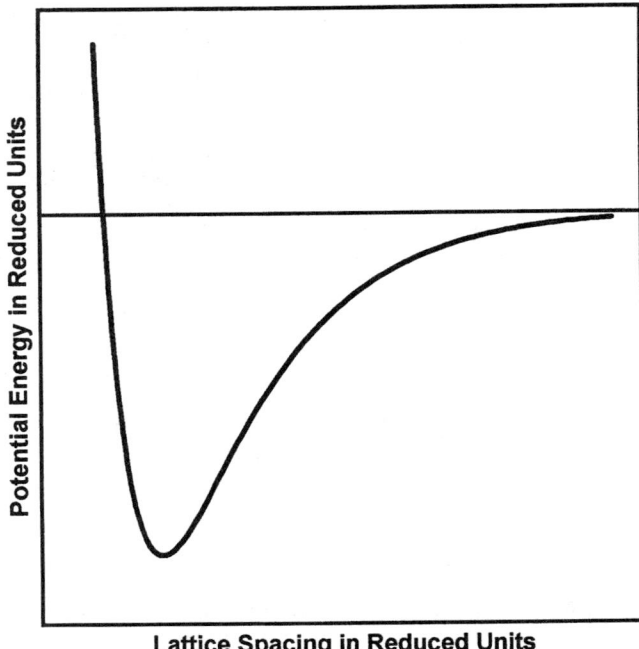

Figure 5.1. Crystal energy as a function of lattice spacing.

crystal potential energy. The most rigorous work consists of large-scale computer calculations based on accurate approximations to the Schrodinger equation for crystals. Despite the fact that these are individual numerical calculations, they can be organized in a remarkably simple way. In fact, a simple form for the static potential energy function exists that is valid for a large number of metals, given by Rose et al.[1]

The starting point is the set of first-principles quantum mechanical calculations of the crystal energy as a function of lattice parameter that have been performed for a large number of elements. Each of these calculations gave a curve similar to that shown in figure 5.1, which is a typical cohesive energy curve showing that the energy of a crystal is a minimum at the equilibrium lattice spacing (taken as unity in the figure). The energy scale is defined relative to the minimum energy. At very close distances the energy is very high because of atomic repulsions resulting from electron overlaps, while at very large distances the energy goes to a constant value corresponding to the energy of atoms at infinite separation.

Rose and colleagues in 1984 found the remarkable result that the energy–lattice parameter curves for elements for which the first-principles calculations were performed could all be reduced to a single universal curve by proper scaling. They did this by first defining a function g as the ratio of the crystal energy to the magnitude of its value at the minimum (zero pressure):

$$g(r_s) = \frac{U_0(r_s)}{|U_0(r_s^0)|} \qquad (5.1.6)$$

(The energy is written as a function of the Wigner–Seitz radius r_s rather than the lattice parameter.) This brings the scaled minimum energy values for all crystals to the same value, namely $g(r_s^0) = -1$, but still does not bring all the energy curves into coincidence.

The scaling of the distance is somewhat more complex. If all distances are translated by the minimum Wigner–Seitz radius, the different energy curves are brought closer together and their minima are all at the same point, but they still do not coincide because the curves all have different curvatures. Let us look at the curvature by expanding $g(r_s)$ to the second order in r_s:

$$g(r_s) = g(r_s^0) + \left(\frac{dg}{dr_s}\right)_0 (r_s - r_s^0) + \frac{1}{2}\left(\frac{d^2g}{dr_s^2}\right)_0 (r_s - r_s^0)^2$$

or, since $g(r_s^0) = -1$ and the first derivative is zero at the energy minimum,

$$g(r_s) = -1 + \frac{1}{2}\left(\frac{d^2g}{dr_s^2}\right)_0 (r_s - r_s^0)^2 \qquad (5.1.7)$$

This is the equation of a parabola, which crosses the r_s axis ($g = 0$) for values of $r_s = r_s'$, given by

$$\frac{1}{2}\left(\frac{d^2g}{dr_s^2}\right)_0 (r_s' - r_s^0)^2 = 1 \qquad (5.1.8)$$

or

$$r_s' = r_s^0 \pm \sqrt{\frac{2}{g_0''}} \qquad (5.1.9)$$

g_0'' being the second derivative in (5.1.8). The two values of r_s' given by (5.1.9) define a length given by

$$L = 2\sqrt{\frac{2}{g_0''}} \qquad (5.1.10)$$

which is the distance intercepted on the r_s axis by the parabola (5.1.8). This leads us to try using (5.1.10) as a scaling factor for the distance. The numerical factor in (5.1.10) is irrelevant, so we choose a scaling distance L as

$$L = \sqrt{\frac{1}{g_0''}} = \sqrt{\frac{|U_0(r_s^0)|}{U_0(r_s^0)''}} \qquad (5.1.11)$$

$U_0(r_s^0)''$ is the second derivative of the crystal energy with respect to the Wigner–Seitz radius, evaluated at the equilibrium volume (zero pressure). It is related to the bulk modulus by

$$B_0^0 = \frac{1}{\kappa_0^0} = V\left(\frac{d^2U_0}{dV^2}\right)_0 \qquad (5.1.12)$$

The superscripts on the bulk modulus and the compressibility indicate that they refer to zero pressure while the subscripts refer to the fact that (5.1.12) includes only the static contribution and all vibrational contributions are neglected. Using the definition of the Wigner–Seitz radius, we get

$$\left(\frac{d^2U_0}{dV^2}\right)_0 \equiv U_0'' r_s^2 = 12\pi r_s^0 B_0^0 N \tag{5.1.13}$$

so the scaling length (5.1.11) can be written as

$$L = \sqrt{\frac{|u_0^0|}{12\pi r_s^0 B_0^0}} \tag{5.1.14}$$

where u_0^0 is the potential energy per atom of the crystal, evaluated at the zero pressure, zero temperature, equilibrium value of the Wigner–Seitz radius.

Note that the scaling length contains only the energy and bulk modulus at the Wigner–Seitz radius for which the energy is a minimum (for the static crystal). These quantities are obtained either from experiment or from the first principles calculation.

Now we define a reduced distance a by

$$a = \frac{r_s - r_s^0}{L} \tag{5.1.15}$$

r_s^0 being the Wigner–Seitz radius at the static energy minimum (zero pressure).

When the results of first principles calculations for the potential energy as a function of volume are used to plot g against a for many metals, it is found that for metallic elements all points fall on the same curve to a high degree of accuracy. This curve is well represented by the equation

$$\frac{U_0}{|U_0^0|} = g(a) = -(1 + a + 0.05a^3)e^{-a} \tag{5.1.16}$$

The static contribution to the equation of state is readily obtained by differentiating the static crystal energy in (5.1.16):

$$P_0 = \frac{dU_0}{dV} = -\frac{1}{4\pi r_s^2 N}\frac{dU_0}{dr_s} = -\frac{|U_0^0|}{4\pi r_s^2 N}\frac{dg}{dr_s} = -\frac{|u_0^0|}{4\pi r_s^2 L}\frac{dg}{da}$$

or

$$P_0 = -\frac{|u_0^0|}{4\pi r_s^2 L}e^{-a}(a - 0.15a^2 + 0.05a^3) \tag{5.1.17}$$

From the definition of the scaling distance L and the Wigner–Seitz radius, (5.1.17) is readily converted to

$$P_0 = B_0^0 \frac{3(\tilde{V}^{1/3} - 1)}{\tilde{V}^{2/3}} e^{-a}(1 - 0.15a + 0.05a^2) \tag{5.1.18}$$

where $\tilde{V} = V/V_0$ is the ratio of the volume to the volume at zero pressure. This is the static part of the equation of state.

Note that we have used the fact that the parameter a can be written in terms of $\tilde{V} = V/V_0$, since from (5.1.15) we have

$$a = \frac{r_s^0}{L}(\tilde{V}^{1/3} - 1) \tag{5.1.19}$$

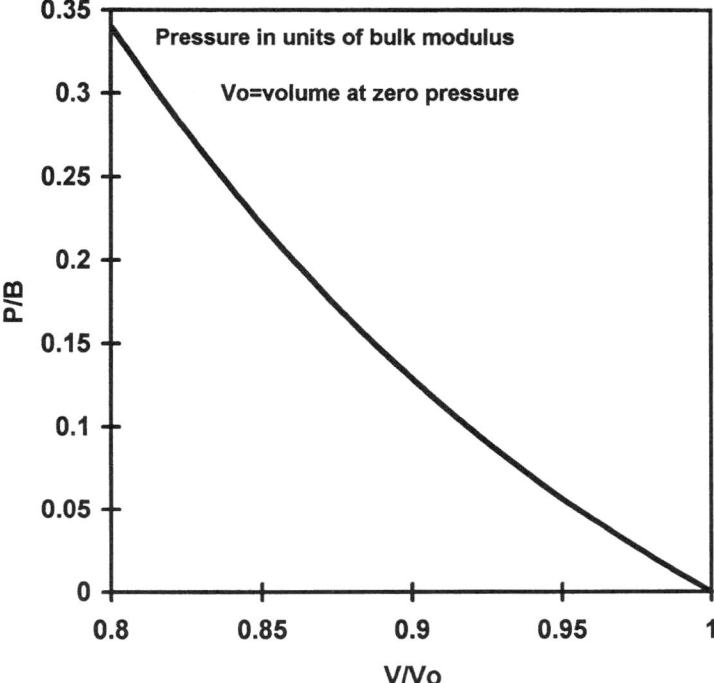

Figure 5.2. Static equation of state of copper.

Equation (5.1.18) therefore states that the static contribution to the pressure–volume relation for all pressures can be obtained from the cohesive energy, the bulk modulus, and Wigner–Seitz radius at zero pressure.

The theory can be tested by computing an equation of state from (5.1.18) and comparing it with experiment. When this was done, it was found to work very well for a great many metals out to very high pressures. Of course, the experimental data are at a finite temperature while the derivation of the theory neglected all atomic vibrations. But at room temperature or lower the contribution of the vibrations to the equation of state is small.

A calculation of the equation of state for copper is shown in figure 5.2. The pressure is plotted in units of the bulk modulus, and the volume is in units of its relative deviation from the value at zero pressure.

5.2 Anharmonic properties and the Gruneisen assumption

The full equation of state, including the vibrational contributions, is obtained by differentiating equation (4.3.13) according to equation (5.1.1). In doing this, it must be recognized that the frequencies are functions of volume. When the volume changes, the interaction forces among the atoms change because of the altered distances between them. In general, we can expect that, as the atoms get closer together, the frequencies increase because of the increased force of atomic interactions. This is clearly evident in the theory of the linear chain worked out in section 4.2. The force constants in harmonic theory are second derivatives of interatomic potential functions, which makes them the first derivatives of the interatomic force functions, evaluated at the mean atomic position. Since the slope of the force function increases with decreasing interatomic distance, the force constants, and therefore the frequencies, increase. Conversely, the frequencies will decrease with increasing volume. Differentiating (4.3.13) gives

$$P = -\left(\frac{\partial E_0}{\partial V}\right)_T - kT\sum_j \frac{1}{e^{h\nu_j/kT}-1}\left(\frac{h}{kT}\right)\frac{\partial \nu_j}{\partial V} \tag{5.2.1}$$

The derivatives of the frequencies are an essential part of the equation of state, and there is no easy way of getting them from first principles. To deal with the variation of frequency with volume, Gruneisen assumed that

$$\frac{d\ln \nu_j}{d\ln V} = -\gamma \tag{5.2.2}$$

where γ is a positive constant that is the same for all frequencies in a given solid. Equation (5.2.2) is just the result of assuming that the frequency varies inversely as some power of the volume as follows:

$$\frac{\nu_j}{\nu_j^0} = \left(\frac{V_0}{V}\right)^{\gamma} \tag{5.2.3}$$

ν_j being the frequency of the jth mode when the volume is V, and ν_j^0 being the frequency when the volume has some reference value V_0, normally taken as the volume at zero pressure. The constant γ is taken to be the same for all frequencies because it is then much easier to develop a theory from which calculations can be made. γ is a disposable parameter to be determined from experimental data, and the validity of this approach is measured by the utility and applicability of its results.

Adopting the Gruneisen assumption, equation (5.2.2) gives

$$\frac{d\nu_j}{dV} = -\frac{\gamma \nu_j}{V} \tag{5.2.4}$$

$$\frac{d^2\nu_j}{dV^2} = (\gamma^2 + \gamma)\frac{\nu_j}{V^2} \tag{5.2.5}$$

Using these relations, it is possible to write statistical mechanical formulas for volume-dependent properties and make several interesting connections among the thermodynamic properties of a harmonic crystal. To do this, we need the energy and heat capacity equations, which in the discrete notation are given by equations (4.4.5) and (4.4.6) as

$$U - E_0 = kT\sum_j \frac{h\nu_j}{kT}\frac{1}{e^{h\nu_j/kT}-1} \tag{5.2.6}$$

$$C_V = k\sum_j \left(\frac{h\nu_j}{kT}\right)^2 \frac{1}{e^{h\nu_j/kT}+e^{-h\nu_j/kT}-2} \tag{5.2.7}$$

Now use (5.2.4) and rewrite (5.2.1) as

$$P = -\left(\frac{\partial E_0}{\partial V}\right)_T + \frac{\gamma k T}{V} \sum_j \frac{1}{e^{h\nu_j/kT} - 1}\left(\frac{h\nu_j}{kT}\right) \quad (5.2.8)$$

But the sum in (5.2.8) is the same as that in (5.2.6), and therefore we can write

$$P = -\left(\frac{\partial E_0}{\partial V}\right)_T + \frac{\gamma}{V}(U - E_0) \quad (5.2.9)$$

This is the Mie–Gruneisen equation of state. The pressure is displayed as a sum of two terms. The first term does not contain the temperature explicitly and arises from the variation of the potential energy of the nonvibrating crystal with volume plus the variation of the zero point vibrational energy with volume. The second term is a pressure that is proportional to the thermal energy $(U - E_0)$.

The volume derivatives of the frequencies are required for both the compressibility κ and the thermal expansion coefficient α since they are determined by the thermodynamic equations

$$\frac{1}{\kappa} = -V\left(\frac{\partial^2 A}{\partial V^2}\right)_T = -V\left(\frac{\partial P}{\partial V}\right)_T \quad (5.2.10)$$

$$\frac{\alpha}{\kappa} = -\left[\frac{\partial}{\partial V}\left(\frac{\partial A}{\partial T}\right)_V\right]_T \quad (5.2.11)$$

Differentiating equation (5.2.8) for the pressure with respect to volume and using equations (5.2.4) and (5.2.5) for the volume derivatives of the frequencies then gives the bulk modulus according to (5.2.10) as

$$B = \frac{1}{\kappa} = V\left(\frac{\partial^2 E_0}{\partial V^2}\right)_T + \frac{\gamma^2 + \gamma}{V}(U - E_0) - \frac{\gamma^2 C_V T}{V} \quad (5.2.12)$$

In arriving at (5.2.12), the differentiation yielded sums over the normal modes that were the same as those for the thermal energy and the heat capacity as given by equations (5.2.6) and (5.2.7).

Differentiating the free energy equation (4.3.13) first with respect to volume and then with respect to temperature and using the definition of the Gruneisen constant (5.2.4), we get, from (5.2.11),

$$\frac{\alpha}{\kappa} = \frac{\gamma C_V}{V} \quad (5.2.13)$$

This is known as Gruneisen's equation and can be used to compute the Gruneisen constant since all other quantities are measurable and can be obtained from experiment.

Equations (5.2.9), (5.2.12), and (5.2.13) show that the introduction of the Gruneisen parameter allows the equation of state and its accompanying parameters (κ and α) to be expressed in terms of thermal properties, that is, the heat capacity and the thermal energy. The theoretical development of the thermal properties can therefore be carried over bodily into equation of state theory. In particular, the Debye theory, which was originally designed to describe the heat capacity, can now be used for a theory of the equation of state just by substituting the Debye expressions (4.8.7) and (4.8.8) for the thermal energy and the heat capacity into equations (5.2.9), (5.2.12), and (5.2.13). The results are

$$P = -\left(\frac{\partial E_0}{\partial V}\right)_T + \frac{3Nk\Theta_D\gamma}{V}D_E(x_D) \qquad (5.2.14)$$

$$\frac{1}{\kappa} = V\left(\frac{\partial^2 E_0}{\partial V^2}\right)_T + \frac{\gamma^2 + \gamma}{V}3Nk\Theta_D D_E(x_D) - \frac{3NkT\gamma^2}{V}D(x_D) \qquad (5.2.15)$$

$$\frac{\alpha}{\kappa} = \frac{3Nk\gamma}{V}D(x_D) \qquad (5.2.16)$$

The low- and high-temperature limits of equations (5.2.14)–(5.2.16) are obtained by substituting the low- and high-temperature approximations for the Debye energy and heat capacity functions as given by equations (4.8.15), (4.8.17), (4.8.23), and (4.8.25). For $T \gg \Theta_D$, the results are

$$P = -\left(\frac{\partial E_0}{\partial V}\right)_T + \frac{3NkT\gamma}{V}\left[1 - \frac{3\Theta_D}{8T} + \frac{1}{20}\left(\frac{\Theta_D}{T}\right)^2\right] \qquad (5.2.17)$$

$$\frac{1}{\kappa} = V\left(\frac{\partial^2 E_0}{\partial V^2}\right)_T + \frac{3NkT\gamma}{V}\left[1 - 3\frac{(\gamma+1)}{8}\frac{\Theta_D}{T} + \frac{(2\gamma+1)}{20}\left(\frac{\Theta_D}{T}\right)^2\right] \qquad (5.2.18)$$

$$\frac{\alpha}{\kappa} = \frac{3Nk\gamma}{V}\left[1 - \frac{1}{20}\left(\frac{\Theta_D}{T}\right)^2\right] \qquad (5.2.19)$$

while for $T \ll \Theta_D$,

$$P = -\left(\frac{\partial E_0}{\partial V}\right)_T + \frac{3Nk\Theta_D\gamma\pi^4}{5V}\left(\frac{T}{\Theta_D}\right)^4 \qquad (5.2.20)$$

$$\frac{1}{\kappa} = V\left(\frac{\partial^2 E_0}{\partial V^2}\right)_T - \frac{3Nk\Theta_D\gamma}{V}\frac{(3\gamma-1)\pi^4}{5}\left(\frac{T}{\Theta_D}\right)^4 \qquad (5.2.21)$$

$$\frac{\alpha}{\kappa} = \frac{12Nk\gamma\pi^4}{5V}\left(\frac{T}{\Theta_D}\right)^3 \qquad (5.2.22)$$

where we have retained only the dominant term in the low-temperature expansion.

From experimental data we know that the equations of state and compressibility of metallic, covalent, and ionic solids do not vary much with temperature when the temperature is low ($T \ll \Theta_D$) This means that the thermal contribution to equation of state properties is small at low temperatures. At high temperatures, the thermal contribution is larger but still does not exceed the zero temperature term. Note that although we refer to the terms containing the derivatives of E_0 as zero point or zero temperature terms, they are not really temperature independent. Because of thermal expansion, the volume changes with temperature, and since the "zero point" terms are volume dependent, they are indirect functions of temperature. These terms are truly temperature independent in heat capacity theory because that theory deals with constant volume systems. Of course, if we want to write a theory for the heat capacity at constant pressure, then the indirect temperature dependence of the zero point energy must be taken into account.

Let us recall that the zero point energy is given by equation (4.3.5) as

$$E_0 = U_0 + \sum_{j=1}^{3N} \frac{h\nu_j}{2} \tag{5.2.23}$$

The second term is the zero point vibrational energy and is easily evaluated in the Debye theory by converting the sum to an integral and using the Debye distribution function:

$$\sum_{j=1}^{3N} \frac{h\nu_j}{2} \rightarrow \frac{9N}{2\nu_D^3} \int_0^{\nu_D} \nu^3 d\nu = \frac{9}{8} Nk\Theta_D \tag{5.2.24}$$

In the Debye theory, the zero point energy of the crystal is therefore

$$E_0 = U_0 + \frac{9}{8} Nk\Theta_D \tag{5.2.25}$$

The derivative of the zero point vibrational energy is readily obtained within the Gruneisen assumption because it applies to ν_D as well as to all other frequencies, and since ν_D is proportional to the Debye temperature, (5.2.4) and (5.2.5) can be rewritten as

$$\frac{d\Theta_D}{dV} = -\frac{\gamma \Theta_D}{V} \tag{5.2.26}$$

$$\frac{d^2\Theta_D}{dV^2} = (\gamma^2 + \gamma)\frac{\Theta_D}{V^2} \tag{5.2.27}$$

Therefore, the first and second derivatives of the zero point energy (2.25) are

$$\frac{\partial E_0}{\partial V} = \frac{\partial U_0}{\partial V} - \frac{9Nk\Theta_D}{8V}\gamma \tag{5.2.28}$$

$$\frac{\partial^2 E_0}{\partial V^2} = \frac{\partial^2 U_0}{\partial V^2} + \frac{9Nk\Theta_D}{8V^2}(\gamma^2 + \gamma) \tag{5.2.29}$$

These equations show how the zero point contributions to the pressure and the bulk modulus vary with volume in the Debye theory. The first derivative of U_0 is negative, and its magnitude decreases with the increasing interatomic separation resulting from a volume increase, so the U_0 derivative contributes a term to the pressure, (5.2.20), that decreases with temperature. Θ_D varies as $V^{-\gamma}$, so the last term in (5.2.28) varies as $V^{-\gamma-1}$, and since the volume increases with temperature, the magnitude of the last term in (5.2.28) decreases with temperature, so the contribution of the zero point vibrational energy to the pressure (5.2.20) decreases with temperature.

The total zero point energy therefore contributes a term to the pressure that decreases with temperature. However, this variation is small because, if we start at absolute zero, the potential energy is a minimum and, as we see from (5.2.22), the thermal expansion coefficient goes to zero as T goes to zero. This means that the dependence of volume at low temperatures is very small, so U_0 and its derivatives are only weakly dependent on temperature at low temperatures. The temperature dependence of the zero point contributions to both the pressure and the bulk modulus can therefore be neglected at low temperatures. According to (5.2.20) and (5.2.21), they then both vary as T^4 for $T \ll \Theta_D$.

At high temperatures, however, the thermal expansion is not negligible, so U_0 and its derivatives vary with temperature and there is a contribution from the zero point energy to both the pressure and the bulk modulus. According to equations (5.2.17) and (5.2.18), the magnitude of the thermal energy contribution to both the pressure and the bulk modulus increases with temperature. The static energy terms, on the other hand, decrease with increasing temperature. That is, as the temperature goes up, the thermal vibrations make the crystal harder while the spreading apart of the atoms makes it softer. The latter effect dominates, and the bulk modulus always decreases with increasing temperature.

The low-temperature T^4 dependence of the bulk modulus is actually observed in solids with an accuracy that increases with decreasing temperature.

5.3 Heat capacity at constant pressure

The canonical ensemble leads to the Helmholtz free energy as the thermodynamic potential most naturally related to the energy levels of the physical system. This means that the theory gives the heat capacity at constant volume rather than at constant pressure. Experiments, however, are most conveniently done at constant pressure, and it is the constant pressure heat capacity that is measured.

From thermodynamics, the relation between constant volume and constant pressure heat capacities is

$$C_P = C_V + TV \frac{\alpha^2}{\kappa} \qquad (5.3.1)$$

The presence of the compressibility and thermal expansion in this relation shows that a theory of the constant pressure heat capacity must somehow include the variation of the energy levels with volume. For crystals, the easiest way to do this is to adopt the Gruneisen assumption by using equation (5.2.13), which when combined with (5.3.1) gives

$$C_P = C_V(1 + \gamma \alpha T) \qquad (5.3.2)$$

Section 5.2 showed that the thermal expansion coefficient approaches zero as the temperature decreases and increases with increasing temperature. At low temperatures, therefore, C_P and C_V are nearly equal. Even at high temperatures the difference is not large. Taking $\gamma = 2$ and $\alpha = 10^{-5}$ as typical values for metals, then at $T = 1000\,\mathrm{K}$, we get $\gamma\alpha = 2 \times 10^{-2}$. The two heat capacities therefore differ by only a few percent.

5.4 Debye theory and the Gruneisen assumption

In the Debye theory, the vibrational frequencies v_j are the possible sound frequencies. The longitudinal frequencies depend on the volume according to equation (4.7.13), which, when solved for the frequencies and written in terms of the volume $V = L^3$, gives

$$v_j = \frac{jC_l}{2V^{1/3}} \qquad \text{(longitudinal waves)} \qquad (5.4.1)$$

ANHARMONIC PROPERTIES AND THE EQUATION OF STATE

An exactly, analogous relation exists for the transverse waves, for which the volume dependence is just like (5.4.1) except that it contains the transverse velocity of sound rather than the longitudinal velocity:

$$v_j = \frac{jC_t}{2V^{1/3}} \quad \text{(transverse waves)} \tag{5.4.2}$$

Taking logarithms of these two equations and then differentiating gives

$$\frac{d\ln v_j}{d\ln V} = -\frac{1}{3} + \frac{d\ln C}{d\ln V} \tag{5.4.3}$$

where we have written only one of the equations with the understanding that C can be either the transverse or the longitudinal velocity of sound.

From the theory of elasticity, the transverse and longitudinal velocities in an isotropic body are related to its elastic properties by

$$C_t = \sqrt{\frac{\mu}{\rho}} \tag{5.4.4}$$

$$C_l = \sqrt{\frac{\lambda + 3\mu/2}{\rho}} \tag{5.4.5}$$

where ρ is the density and μ and λ are the Lame constants, which are related to the compressibility κ and Poisson's ratio σ by the equations

$$\frac{1}{\kappa} = \lambda + \frac{2\mu}{3} \tag{5.4.6}$$

$$\sigma = \frac{\lambda}{2(\lambda + \mu)} \tag{5.4.7}$$

The left-hand side of (5.4.3) is just the definition of the Gruneisen constant, and it is clear from the existence of two velocities of sound that if the Gruneisen assumption is applied to the Debye theory, then two Gruneisen constants should be used. From (5.4.3), these are given by

$$\gamma_t = \frac{1}{3} - \frac{d\ln C_t}{d\ln V} \tag{5.4.8}$$

$$\gamma_l = \frac{1}{3} - \frac{d\ln C_l}{d\ln V} \tag{5.4.9}$$

Using (5.4.4) and (5.4.5), and the fact that the density is inversely proportional to the volume, (5.4.8) and (5.4.9) give the Gruneisen constants in terms of the Lame constants as

$$\gamma_t = -\frac{1}{6} - \frac{1}{2}\frac{d\ln\mu}{d\ln V} \tag{5.4.10}$$

$$\gamma_l = -\frac{1}{6} - \frac{1}{2}\frac{d\ln\left(\lambda + \frac{2\mu}{3}\right)}{d\ln V} \tag{5.4.11}$$

To be completely consistent with its assumptions, the Debye theory requires two Gruneisen constants to describe the variation of the frequencies, and therefore of the Debye cutoff frequency ν_D and the Debye temperature Θ_D, with volume. The Debye theory contains only one cutoff frequency, but this is just a result of assuming that the cutoff frequency is the same for both transverse and longitudinal waves. The introduction of two cutoff frequencies and two Gruneisen constants would greatly complicate the theory. Not only would the algebra be a lot more cumbersome, but two Debye temperatures and two Gruneisen constants would then have to be determined from experiment. The accuracy of the underlying assumptions of the theory does not justify adding these complications, and a single Debye temperature and a single Gruneisen constant are usually adequate for a good description of heat capacity and a reasonably fair description of equations of state.

5.5 Vibrational anharmonicity

From equation (5.2.13), we see that the thermal expansion coefficient vanishes if $\gamma = 0$. But if the thermal expansion is zero, then the derivatives of the zero point energy in equations (5.2.14) and (5.2.15) are temperature independent and the other terms vanish, so neither the equation of state nor the compressibility depend on temperature. Also, equation (5.3.2) shows that the heat capacities at constant pressure and at constant volume are the same if the Gruneisen constant vanishes. The interpretation of a zero value for the Gruneisen constant is that the crystal vibrations are precisely harmonic. This can be seen by looking at equation (4.2.28), which gives the normal mode frequencies of a linear chain in terms of the force constants C_{ij}. For a harmonic crystal, the potential energy is truly a quadratic function of the atomic displacements; the force constants are independent of atomic positions and therefore independent of volume. Equation (4.2.28) shows that in this case the frequencies are also independent of volume and therefore Gruneisen's constant is zero.

Adopting a nonzero Gruneisen constant is a replacement of the harmonic assumption that introduces anharmonic effects in an arbitrary way. The relation between the Gruneisen method and anharmonicity can be seen if we start with an anharmonic potential. To the fourth order in the atomic displacements, the potential energy of the crystal Φ is

$$\Phi = U_0 + \frac{1}{2}\sum_{i,j} C_{ij}u_i u_j + \frac{1}{3!}\sum_{i,j,k} C_{ijk}u_i u_j u_k + \frac{1}{4!}\sum_{i,j,k,l} c_{ijkl}u_i u_j u_k u_l \qquad (5.5.1)$$

where the us are the atomic displacements and the Cs are constants defined as derivatives of the potential Φ with respect to displacements, evaluated at zero displacements. (To avoid having too many symbols, the subscripts are meant to label the atoms as well as the components of the displacements.) Equation (5.5.1) is just a Taylor expansion of the potential energy to fourth order. (Both third- and fourth-order terms should be retained when going beyond the harmonic approximation because they both contribute comparable amounts to physical properties. The reason for this is that, when the higher order contributions to physical properties are treated by perturbation theory, the third-order terms appear in the second-order perturbation while the fourth-order terms appear in the first-order perturbation. Thus, although the fourth-order force constants are roughly an order of magnitude smaller, they are still important.)

Equation (5.5.1) holds for arbitrary displacements of the atoms and therefore holds for the case when the displacements are the result of a change in volume of the static crystal. In this case, the displacements are the changes in the position of atoms when the crystal volume is changed and the potential energy Φ becomes the energy of the static crystal as a function of volume. The constant U_0 is now the energy of the static crystal at zero pressure, which we have called U_0^0. For the case of a static crystal under pressure, equation (5.5.1) is written as

$$U_0(V) = U_0^0 + \frac{1}{2}\sum_{i,j} C_{ij}\lambda_i\lambda_j + \frac{1}{3!}\sum_{i,j,k} C_{ijk}\lambda_i\lambda_j\lambda_k + \frac{1}{4!}\sum_{i,j,k,l} C_{ijkl}\lambda_i\lambda_j\lambda_k\lambda_l \quad (5.5.2)$$

The λ_j are the changes in atomic position when the static crystal changes its volume.

In the harmonic approximation, only second-order terms are used and the force constants are the second derivatives of the potential. But if the crystal is anharmonic, these second-order derivatives are not independent of volume, as can be seen by taking the second derivative of (5.5.2):

$$\frac{\partial^2 U_0}{\partial \lambda_i \partial \lambda_j} = C_{ij} + \sum_k C_{ijk}\lambda_k + \text{(higher order terms)} \quad (5.5.3)$$

So when anharmonicity is included, the second derivatives depend on the atomic displacements and therefore on the volume of the crystal. This shows that the frequencies are volume dependent.

What if we want to retain the simplicity of the harmonic assumption and write the potential as a quadratic form even if we know it is not entirely correct? Equation (5.5.3) shows that it would be more accurate to take the force "constants" to be functions of volume than to treat them as being completely constant. This is precisely what is done with the Gruneisen method; it amounts to shifting the minimum of the potential as the volume changes.

This discussion clarifies the meaning of the Gruneisen method and also shows that it only accounts for a part of the anharmonic effect because it does not truly reflect the existence of the anharmonic terms but tries to make the volume dependence of the second-order derivatives wholly responsible for the higher order terms. It is therefore not surprising that the Gruneisen method gives only a semiquantitative description of the volume-dependent properties of crystals.

5.6 Theory of the Gruneisen parameter

The Gruneisen assumption is quite useful and shows that the thermal expansion, the variation of compressibility with temperature, the temperature dependence of the equation of state, and the difference between constant pressure and constant volume heat capacities are anharmonic properties in that their existence is the result of vibrational anharmonicity.

But the Gruneisen assumption suffers from some defects even aside from the fact that it takes all vibrational frequencies to have the same dependence on volume. First, it assumes that γ is truly a constant that is independent of crystal volume. Second, it is completely empirical and treats γ as a disposable constant to be determined from the values of the anharmonic properties. Considerable progress can be made on removing these defects.[2]

The discussion in section 5.5 shows that if the crystal potential is known as a function of volume, we should be able to calculate the Gruneisen constant. The way to do this is to remember that the vibration frequencies are determined by the second derivatives of the potential energy. If we assume that only the nearest neighbor interactions are important, then the vibration frequencies are proportional to the square root of the second derivatives of the potential with respect to changes in the nearest neighbor distance. All we need then is the potential energy as a function of nearest neighbor distance. For this, we choose the "universal" potential energy of Rose et al. (1984) as given by equation (5.1.16), which can be used to get the Gruneisen parameter.

First, we write γ in terms of the Wigner–Seitz radius as follows:

$$\gamma = -\frac{1}{3}\frac{d\ln v_j}{d\ln r_s} \tag{5.6.1}$$

Assuming that the frequencies are all proportional to the square root of the second derivative of the energy, using (5.1.16), this becomes

$$\gamma = -\frac{1}{6}\frac{d\ln}{d\ln r_s}\left(\frac{d^2 g}{dr_s^2}\right) = -\frac{1}{6}\frac{d\ln g''}{d\ln r_s} = -\frac{r_s}{6g''}\frac{dg''}{dr_s}$$

or

$$\gamma = -\frac{r_s}{6}\frac{g'''}{g''} \tag{5.6.2}$$

where g'' and g''' are the second and third derivatives, respectively, of the reduced energy $g(a)$ with respect to r_s. Since $dr_s = L da$, this becomes

$$\gamma = -\frac{r_s}{6L}\frac{g^{(3)}}{g^{(2)}} \tag{5.6.3}$$

where $g^{(2)}$ and $g^{(3)}$ are the second and third derivatives of $g(a)$ with respect to a. The derivatives of $g(a)$ are

$$g^{(1)} \equiv \frac{dg}{da} = e^{-a}(a - 0.15a^2 + 0.05a^3) \tag{5.6.4}$$

$$g^{(2)} \equiv \frac{d^2 g}{da^2} = e^{-a}(1 - 1.3a + 0.3a^2 - 0.05a^3) \tag{5.6.5}$$

$$g^{(3)} \equiv \frac{d^3 g}{da^3} = e^{-a}(-2.3 + 1.9a - 0.45a^2 + 0.05a^3) \tag{5.6.6}$$

The crystal anharmonicity depends on the ratio of the third to the second derivatives of the potential energy. This ratio is obviously a measure of anharmonicity since the second derivative determines the harmonic forces. It would appear that only the third derivative enters into the anharmonicity as measured by (5.6.3), but this is true only for a crystal volume corresponding to the minimum of potential energy. At other volumes, the entire form of the potential function is needed to get the derivatives, and this implicitly brings in the potential to all orders of the volume change from that at the minimum of energy.

Substituting (5.6.5) and (5.6.6) into (5.6.3), the Gruneisen parameter becomes

$$\gamma = 0.383\eta F(a) \tag{5.6.7}$$

where $F(a)$ is a function of the reduced distance a given by

$$F(a) = \frac{1 - 0.826a + 0.1957a^2 - 0.0217a^3}{1 - 1.3a + 0.3a^2 - 0.05a^3} \tag{5.6.8}$$

and η is defined by

$$\eta = \frac{r_s}{L} \tag{5.6.9}$$

At zero pressure, $r_s = r_s^0$, $a = 0$ and $F(a) = 1$, so

$$\gamma_0 = 0.383\eta_0 \tag{5.6.10}$$

where the subscript indicates zero pressure.

Note that the Gruneisen "constant" is not really a constant. It depends on the volume through the function $F(a)$, which is why we call it the Gruneisen parameter rather than the Gruneisen constant. Note also that the Gruneisen parameter is proportional to the parameter η, so the ratio of the Wigner–Seitz radius to the scaling distance L is a measure of anharmonicity. That this must be so is evident from the definition of the scaling distance as the distance marked off on the zero energy axis by a parabola whose curvature is that of the potential energy function. Clearly, the narrower this parabola relative to the interatomic distance, the greater the departure of the potential energy from harmonicity.

Equation (5.6.7) gives us a way of calculating the Gruneisen parameter as a function of volume directly from the crystal potential energy. The theory can be tested by comparing calculations from the crystal potential energy with those computed from experiment. This was done for a number of elements by putting values of η_0 tabulated by Rose et al. into (5.6.10) and using experimental data in equation (5.2.13) to get γ at zero pressure and room temperature. The thermal expansion coefficients and densities were taken from standard tables in the *American Institute of Physics Handbook*, and the bulk moduli were taken from the compilation of Rose et al. (1984). The heat capacities at 300 K were computed from Debye theory using the Debye temperatures listed in chapter 4. The Debye theory is sufficiently accurate for this purpose. The results are shown in table 5.1. The agreement between the two methods of calculating the Gruneisen parameter is quite good for most of the simple metals, but for silicon, germanium, and a few others the agreement is poor.

There are two possible sources of error in the theoretically computed values. The first is that the "universal" energy curve of Rose et al. (1984) may not give a sufficiently accurate representation of the energy, and the second is that the lattice vibrations are not described with sufficient accuracy by nearest neighbor interactions alone. If we keep in mind that the calculation requires the second and third derivatives of the static crystal energy with respect to volume, it is clear that small errors in the energy curve can lead to significant errors in the calculated Gruneisen parameter. Also, the interatomic forces in metals are not so short-ranged that the vibration frequencies are completely determined by nearest neighbor interactions. Given these considerations, the agreement between the calculated and experimental Gruneisen parameters is remarkable.

The variation of the Gruneisen parameter with volume can be computed from (5.6.7) and (5.6.8). The result of such a calculation for copper is shown in figure 5.3. The volume axis in figure 5.3 extends to a 20% compression of the solid, which corresponds to extremely high pressures. The rate of variation of the Gruneisen parameter with volume is therefore not large. At pressures

Table 5.1: Values of the Gruneisen Parameter at Zero Pressure

Element	γ_0 (experiment)	γ_0 (theory)	Element	γ_0 (experiment)	γ_0 (theory)
Li	1.06	1.33	Pd	2.47	2.49
Be	1.28	1.56	Ag	2.65	2.31
Na	1.49	1.44	Cd	3.09	3.14
Mg	1.71	2.17	In	2.75	1.98
Al	2.41	1.83	Cs	—	1.62
Si	0.44	1.89	Ba	—	1.71
K	1.65	1.53	Hf	—	1.81
Ca	1.09	1.76	Ta	1.66	1.91
V	1.36	1.87	W	1.71	2.21
Cr	0.90	2.17	Re	—	2.39
Fe	1.88	2.00	Ir	—	2.53
Co	2.37	2.06	Pt	2.89	2.51
Ni	2.11	1.98	Au	3.17	2.62
Cu	2.14	2.02	Tl	2.36	2.23
Zn	2.75	2.78	Pb	3.06	2.47
Ge	0.77	1.96	Ce	—	1.21
Rb	—	1.62	Eu	—	1.84
Y	—	1.64	Gd	—	1.66
Zr	0.92	1.74	Dy	—	1.88
Nb	—	1.88	Er	—	1.92
Mo	1.63	2.27	Yb	—	1.53
Ru	—	2.35			

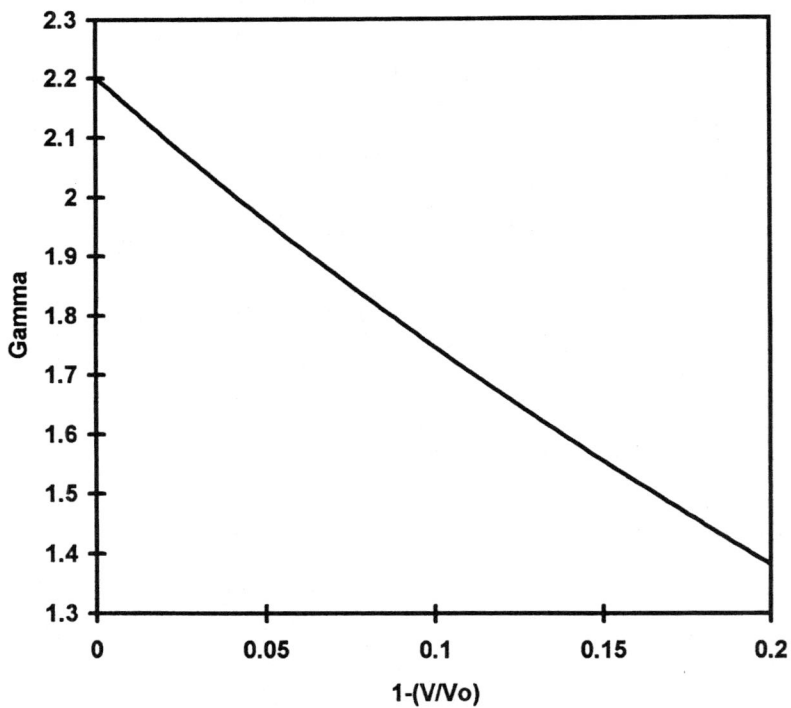

Figure 5.3. Variation of the Gruneisen parameter with volume for copper.

corresponding to compressions of 4%, the Gruneisen parameter changes by 10%. For pressures that are not too high, the assumption of a constant Gruneisen parameter is not bad. However, for pressures that are now attainable in modern high-pressure apparatus and in shock-wave experiments, the variation of γ with volume should be taken into account.

The equations in sections 5.2 and 5.3 for the anharmonic properties were derived assuming that γ was independent of volume and must therefore be modified when the volume dependence is taken into account. This is easily done by going through exactly the same steps that led to equations (5.2.9), (5.2.12), (5.2.13), and (5.3.2) but letting $\gamma = \gamma(V)$ so that the derivatives of the Gruneisen parameter do not vanish. It turns out that the only one of these equations to be affected is (5.2.12) for the compressibility, because this is the only anharmonic property that depends on the second derivative of the vibration frequencies with respect to volume. The bulk modulus must now be modified by appending a term given by

$$-\frac{d\gamma}{dV}\left[(U-E_0)+\frac{9Nk\Theta_D}{8V}\frac{d\gamma}{dV}\right] \tag{5.6.11}$$

The second term in the bracket arises from the zero point vibrational energy as computed from Debye theory. Of course, in addition to adding this term to the equation for the bulk modulus, the volume-dependent Gruneisen parameter must be used in all the equations for the anharmonic properties.

Exercises

5.1 The Lennard–Jones potential for the energy of interaction between two rare gas atoms is given by $\varphi(r) = -A/r^6 + B/r^{12}$, where A and B are constants and r is the distance between the two atoms. From the requirement that the energy be a minimum at the equilibrium distance, show that the Lennard–Jones potential can be written as

$$\varphi(r) = A\left(\frac{1}{2}\frac{r_0^6}{r^{12}} - \frac{1}{r^6}\right)$$

where r_0 is the equilibrium distance between atoms.

5.2 Assume that the energy of a rare gas crystal is a sum of pairwise Lennard–Jones potentials and that only nearest neighbor interactions are important. Assume that the z nearest neighbors are all at the same distance r_0 at static equilibrium.

Write the expression for the crystal energy in terms of these potentials and compute the numerical value of the Gruneisen constant at the static equilibrium (zero pressure) volume. (Hint: equation (5.6.2) in obviously correct when g is any crystal energy as a function of a distance.)

5.3 Show that at high temperatures and zero pressure, the ratio of the thermal energy to the zero pressure, zero temperature potential energy of the crystal depends approximately only on the reduced parameter a in the universal energy equation. That is, prove that, to a good approximation, $kT/|u_0^0| = f(a)$.

5.4 In terms of the function derived in exercise 5.3, derive a formula for the linear thermal expansion coefficient at high temperatures.

Notes

1. Rose, J.H., J.R. Smith, F. Guinea, and J. Ferrante; (1984), *Physics Review B; vol. 29*, p. 2963.
2. Girifalco, L.A.; 1995; "Extended Mie-Gruneisen Theory Applied to C-60 in the Disordered FCC Phase"; *Physics Review B; vol. 52*, p. 9910; Girifalco, L.A., and K. Kniaz; 1997; "Anharmonicity in Metals from the Universal Equation of State"; *Journal of Materials Research; vol. 12*, no. 2, p. 311.

6

Free Electron Theory in Metals and Semiconductors

6.1 Free electrons in metals

One of the earliest successful approximations in the theory of metals was the free electron theory. This approximation starts with the notion that the metal contains a collection of ion cores, each core consisting of a nucleus and shells of tightly bound electrons, and a number of electrons that are not bound to any particular nucleus. The electrons are assumed to be essentially independent and are treated as being completely free, the metal simply acting as a box to contain them. The statistical counterpart of this assumption is that electrons in metals are described by particle statistics.

An essential parameter in free electron theory is the number of electrons per atom that are to be considered free rather than being tied up in ion cores. Usually, the number of free electrons is taken to be the number of valence electrons of the atoms. When applied to real solids, the free electron concentration can be treated as a disposable parameter to be determined by experiment, thereby providing a check on our choice of the number of free electrons per atom.

A much more basic question is involved in trying to understand how it is that the free electron theory of metals is successful at all. We know that none of the electrons in a metal can really be independent; they interact with each other and with the ion cores through strong Coulomb forces. However, there are several factors that allow the electrons to be treated as if they were free. First, on a large-scale average, the ionic charges just cancel the electronic charges. The electrons move in a background of positive charge that tends to neutralize them, so in the free electron model we regard the periodic potential of the ion cores to be smoothed out to some average value. Second, if we solve the problem of a charged electron in a "sea" of mobile charged particles, we find that the electrical potential of an electron is screened by the other electrons, thereby greatly reducing the range over which the interaction of two electrons is appreciable. To a first approximation, then, the electron-electron interactions can be neglected.

Also, there is a remarkable result from quantum theory that states that a perfectly periodic potential offers no resistance to electron motion. The electrical resistivity observed in metals arises from imperfections in the periodic structure and not from the ideal periodic potential itself. These imperfections include atomic vibrations, vacant sites, dislocations, impurity atoms, grain boundaries, and anything else that interrupts the regular periodic array of the ion cores. This does not mean that the regular ion-core structure has no influence on the electron motion. In fact it does, but this influence manifests itself in the inertial properties of the electron. It is often sufficient to assign some fictitious "effective mass" m^* to the electron instead of the true electron mass m. The electron then moves through the crystal as if it were a Fermi–Dirac particle of mass m^*.

There is thus some justification from theory, and certainly in practice, for applying free electron theory to metals. The theory works best for alkali and alkaline earth metals and reasonably well for the noble metals. It is interesting to note that for the alkali metals the electrons-in-a-box model is not far from the truth. A calculation based on experimental values of ion core and metallic radii shows that the ion cores in sodium take up only about one tenth the volume of the metal. Nine tenths of it is empty except for the valence electrons. Also, quantum theoretic calculations show that the wave function in sodium is very much like that for a free electron over the 90% of the volume not occupied by the ion cores.

6.2 Statistics for the electron gas

Our first job is to show that semiclassical particle statistics are not adequate for electrons and that quantum statistics must be used. This can be done by recalling that, from section 3.3, the quantum statistics reduce to semiclassical statistics only if

$$e^{-\mu/kT} \gg e^{-\varepsilon_j/kT} \tag{6.2.1}$$

for all values of ε_j. Since ε_j is always positive, the condition (2.1) for the applicability of semiclassical statistics can only be strengthened if we write

$$e^{-\mu/kT} \gg 1 \tag{6.2.2}$$

The theory of the ideal gas has already been worked out on the assumption that (6.2.2) is satisfied, and we obtained a formula for the chemical potential in terms of particle mass, particle density, and temperature:

$$e^{-\mu/kT} = \left(\frac{2\pi mkT}{h^2}\right)^{3/2} \frac{V}{N} \tag{6.2.3}$$

This allows us to test the validity of (6.2.2) for various kinds of particles. If (6.2.2) is fulfilled when we compute $e^{-\mu/kT}$ from (6.2.3), then the use of semiclassical statistics is justified. Otherwise, quantum statistics is necessary.

Equation (6.2.3) shows that high particle mass, high temperature, and low particle density contribute to the accuracy of the semiclassical statistics. Let us evaluate (6.2.3) for a collection of hydrogen atoms. Inserting the values of h, k, and the mass of a hydrogen atom gives

$$e^{-\mu/kT} = 1.89 \times 10^{20} T^{3/2} \frac{V}{N} \tag{6.2.4}$$

Let us assume that the particle density is $N/V = 1.89 \times 10^{20}$ atoms/cm³. This corresponds to a pressure of about seven atmospheres at room temperature. Then (6.2.4) becomes $e^{-\mu/kT} = T^{3/2}$, and for temperatures as low as 10K we get $e^{-\mu/kT} = 31.6$. This means that even at quite low temperatures a collection of hydrogen atoms is described rather accurately by semiclassical statistics. Quantum statistics is not needed for collections of atoms or molecules except at very low temperatures and high densities.

But a similar calculation for electrons gives the opposite result. If we use the electronic mass in (6.2.3) we get

$$e^{-\mu/kT} = 2.42 \times 10^{15} T^{3/2} \frac{V}{N} \tag{6.2.5}$$

Let us take $N/V = 2.42 \times 10^{22}$ electrons/cm^3. This is roughly the magnitude of the density of free electrons in metals. Then we get

$$e^{-\mu/kT} = 10^{-7} T^{3/2} \tag{6.2.6}$$

and we see that $e^{-\mu/kT} \ll 1$ at all but the highest temperatures. Even at $T = 10,000$ K $e^{-\mu/kT} = 0.1$, so at all temperatures of practical interest it is necessary to use the Fermi–Dirac quantum statistics for electrons in metals.

In semiconductors, the density of free electrons ranges from 10^9 to 10^{17} electrons/cm^3, so the semiclassical statistics are sometimes valid depending on the particular density and temperature.

6.3 The distribution of free electrons

Since electrons are fermions, in the continuum notation the results of section 3.3 give

$$N(\varepsilon) = \frac{\omega(\varepsilon)}{e^{(\varepsilon-\mu)/kT} + 1} \tag{6.3.1}$$

where $N(\varepsilon)$ is a distribution function such that $N(\varepsilon)d\varepsilon$ is the number of electrons in an energy range between ε and $\varepsilon + d\varepsilon$, and $\omega(\varepsilon)$ is the density of states in energy. The energy levels are just those for a particle-in-a-box given by (3.4.7) or (3.4.10). The density of states is therefore just like that for the ideal gas except for a minor modification. Since every electron can exist in two spin states, each with the same energy, there are two electrons that can have an energy determined by a set of integers (j_1, j_2, j_3) rather than just one. For electrons, therefore, the density of states is just twice that for an ideal gas, and we have

$$\omega(\varepsilon) = \frac{8\pi V \sqrt{2}}{h^3} m^{3/2} \sqrt{\varepsilon} \tag{6.3.2}$$

or, using the constant C defined in equation (3.4.16),

$$\omega(\varepsilon) = CV\sqrt{\varepsilon} \tag{6.3.3}$$

The reason for defining the constant C with the factor 1/2 in chapter 3 is now evident: we did not want to have factors of two in all the equations that arise from electron statistics.

We now define the Fermi function by

$$f(\varepsilon) = \frac{1}{e^{(\varepsilon-\mu)/kT} + 1} \tag{6.3.4}$$

and write the distribution function (6.3.1) as

$$N(\varepsilon) = CV\sqrt{\varepsilon}f(\varepsilon) \qquad (6.3.5)$$

The density of states in terms of wave number, momentum, and velocity are just like those for an ideal gas in equations (3.6.12)–(3.6.14) except for the factor of two that takes into account the electron spin:

$$\omega(k) = V\frac{k^2}{\pi^2} \qquad (6.3.6)$$

$$\omega(p) = V\frac{p^2}{\pi^2\hbar^3} \qquad (6.3.7)$$

$$\omega(v) = V\frac{m^3v^2}{\hbar^2} \qquad (6.3.8)$$

The distribution functions in the wave number, velocity and momentum representations are obtained just by multiplying (6.3.6), (6.3.7), and (6.3.8), respectively, by the Fermi function.

The properties of the Fermi function are of basic importance and should be thoroughly understood. The Fermi function is the probability that a particular state of energy ε is occupied by an electron, as can be seen from the fact that when it is multiplied by the number of states with given energy, it yields the number of electrons with that energy.

From the calculations in section 6.2, we know that μ is positive and much greater than kT for ordinary temperatures. Also, $f(\varepsilon) < 1$ since the exponential is always positive. The Fermi function therefore approaches unity for low energies, but it must decrease to zero for high enough energy. How wide is the energy range over which $f(\varepsilon)$ differs from either unity or zero? Note that for $\varepsilon = \mu$, $f(\varepsilon) = f(\mu) = 1/2$; if $\varepsilon < \mu$, then the greater the difference $(\mu - \varepsilon)$, the closer $f(\varepsilon)$ is to unity. For an energy difference $(\mu - \varepsilon) = 3kT$, (6.3.4) tells us that $f(\varepsilon)$ differs from unity by only 5%. If $\varepsilon > \mu$, then the greater the difference $(\mu - \varepsilon)$, the closer $f(\varepsilon)$ is to zero. Again a difference of $3kT$ yields a $f(\varepsilon)$ that differs from zero by about 5%. We therefore conclude that $f(\varepsilon)$ differs significantly from unity or zero over an energy range, centered on μ, of about five or six times the thermal energy kT. The lower the temperature, the more rapid the drop of the Fermi function from unity to zero in the vicinity of μ. In the limit of zero temperature, $f(\varepsilon)$ is a step function whose value is unity for energies up to μ and zero for energies greater than μ. The chemical potential μ obviously plays a critical role in electron statistics and is called the *Fermi energy*. The Fermi function at zero and at a finite temperature is shown in figure 6.1.

For solids, the Fermi energy is much greater than the thermal energy at ordinary temperatures. For metals, μ is about 2–6 eV, while at room temperature kT is about 1/40 eV, so $\mu/kT = 80$–240. This means that the width of the transition region from $f(\varepsilon) = 1$ to $f(\varepsilon) = 0$ is very narrow, and the Fermi function is almost a step function at temperatures that are not too high. In fact, in metals room temperature is a low-temperature for electrons and low-temperature approximations are adequate.

In this analysis, we have incidentally arrived at the physical interpretation of the Fermi energy. It is the energy of that state for which the probability of being occupied by an electron is 0.5.

In the limit as $T \to 0$, $f(\varepsilon) \to 1$ for $\varepsilon < \mu$ and $f(\varepsilon) \to 0$ for $\varepsilon > \mu$, so at $T = 0$ the distribution function (6.3.5) becomes

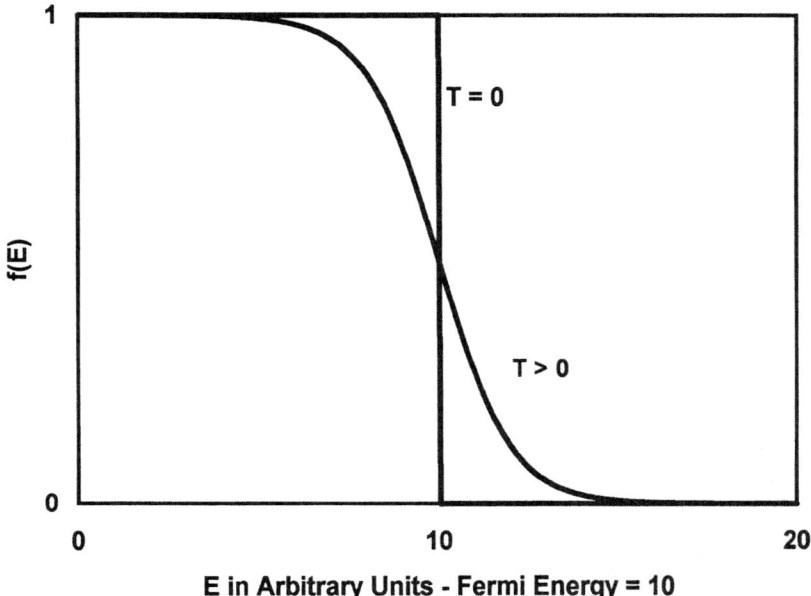

Figure 6.1. The Fermi function.

$$N(\varepsilon) = CV\sqrt{\varepsilon} \quad \varepsilon < \mu_0$$
$$N(\varepsilon) = 0 \quad \varepsilon > \mu_0 \quad (T = 0) \quad (6.3.9)$$

where μ_0 is the value of the Fermi energy at absolute zero.

Since μ_0 is the maximum energy that an electron can have at $T = 0$, an expression for μ_0 can be obtained by integrating (6.3.9).

$$\int_0^{\mu_0} N(\varepsilon)d\varepsilon = CV \int_0^{\mu_0} \sqrt{\varepsilon}d\varepsilon \quad (6.3.10)$$

But the left-hand side of this equation is just the total number of electrons N, so performing the integral on the right and solving for μ_0 gives

$$\mu_0 = \left(\frac{3N}{2CV}\right)^{2/3} \quad (6.3.11)$$

or, using the definition of C,

$$\mu_0 = \frac{h^2}{8m}\left(\frac{3N}{\pi V}\right)^{2/3} \quad (6.3.12)$$

So the Fermi energy at absolute zero is completely determined by the electron density N/V. The magnitude of the Fermi energy to be expected in metals can be estimated from this equation, and it turns out to be several electron volts.

It is also easy to get the average energy of the electrons. The definition of the average is

$$\bar{\varepsilon} = \frac{1}{N}\int_0^\infty \varepsilon N(\varepsilon)d\varepsilon \qquad (6.3.13)$$

and since at absolute zero $N(\varepsilon)$ is given by (6.3.9), the average energy at absolute zero is given by

$$\bar{\varepsilon}_0 = \frac{CV}{N}\int_0^\infty \varepsilon^{3/2}d\varepsilon \qquad (6.3.14)$$

which upon integration yields

$$\bar{\varepsilon}_0 = \frac{2CV}{5N}\mu_0^{5/2} \qquad (6.3.15)$$

If (6.3.11) is solved for CV/N and the result substituted into (6.3.16), the result is

$$\bar{\varepsilon}_0 = \frac{3}{5}\mu_0 \qquad (6.3.16)$$

so at $T = 0$ the average electron energy is three fifths of the Fermi energy.

These results at absolute zero provide estimates that are reasonably good even at ordinary finite temperatures because the thermal effect on electrons at the densities found in metals is small. Also, the zero temperature results are a starting point for the analysis of the temperature dependence of the thermodynamic properties of free electrons.

6.4 Thermodynamic properties of the free electron gas

All the statistical thermodynamic formulas for a free electron gas can be derived if the Fermi energy and the internal energy are known as functions of temperature. The equation for the Fermi energy is obtained from the requirement that the total number of electrons is the integral of the distribution function over all energies:

$$N = \int_0^\infty N(\varepsilon)d\varepsilon \qquad (6.4.1)$$

Since $N(\varepsilon)d\varepsilon$ is the number of electrons with energy in the range $d\varepsilon$, $\varepsilon N(\varepsilon)d\varepsilon$ is their energy and the total energy is therefore

$$U = \int_0^\infty \varepsilon N(\varepsilon)d\varepsilon \qquad (6.4.2)$$

Using (6.3.5) for the distribution function in terms of the Fermi function, equations (6.4.1) and (6.4.2) become

$$N = CV\int_0^\infty \varepsilon^{1/2}f(\varepsilon)d\varepsilon \qquad (6.4.3)$$

$$U = CV\int_0^\infty \varepsilon^{3/2}f(\varepsilon)d\varepsilon \qquad (6.4.4)$$

The dependence of μ and U on temperature can be obtained by evaluating the integrals in (6.4.3) and (6.4.4). These integrals cannot be evaluated in closed form, but a method exists to compute integrals of this type as rapidly convergent series. This method depends on the fact that the derivative of the Fermi function with respect to energy is practically zero except in the vicinity of the Fermi energy, where it is large, as can be seen by inspection of figure 6.1.

The integrals in (6.4.3) and (6.4.4) are examples of the Fermi integral, which is defined by the general form

$$I = \int_0^\infty g(\varepsilon)f(\varepsilon)d\varepsilon \qquad (6.4.5)$$

where $f(\varepsilon)$ is the Fermi function and $g(\varepsilon)$ is any monotonically increasing function of the energy whose value is zero when $\varepsilon = 0$. The method of getting the series solution of (6.4.5) depends on expressing it in terms of the derivative of the Fermi function. This can be done if (6.4.5) is integrated by parts to give

$$I = F(\varepsilon)f(\varepsilon)\Big|_0^\infty - \int_0^\infty F(\varepsilon)\frac{df}{d\varepsilon}d\varepsilon \qquad (6.4.6)$$

where the function $F(\varepsilon)$ is defined by

$$F(\varepsilon) = \int_0^\varepsilon g(x)dx \qquad (6.4.7)$$

But, from its definition, $F(\varepsilon)$ is zero when $\varepsilon = 0$, and the Fermi function vanishes as $\varepsilon \to \infty$, so the first term on the right of (6.4.6) is zero and we have

$$I = -\int_0^\infty F(\varepsilon)\frac{df}{d\varepsilon}d\varepsilon \qquad (6.4.8)$$

A rapidly converging series for the integral, based on the fact that the derivative in (6.4.8) is small everywhere except when ε is near μ, is derived in appendix 5 (A.5.20). For the temperatures and Fermi energies encountered in the study of solids, an excellent approximation is obtained by retaining only the first two terms of the series expansion. That is,

$$I = F_0(\mu) + \frac{(\pi kT)^2}{6} F_2(\mu) \qquad (6.4.9)$$

where

$$F_0(\mu) = F(\mu) \qquad (6.4.10)$$

and

$$F_2(\mu) = \left(\frac{d^2F}{d\varepsilon^2}\right)_{\varepsilon=\mu} \qquad (6.4.11)$$

Now equations (6.4.3) and (6.4.4) can be written in terms of Fermi integrals as

$$\frac{N}{CV} = \int_0^\infty \varepsilon^{1/2} f(\varepsilon)d\varepsilon \qquad (6.4.12)$$

$$\frac{U}{CV} = \int_0^\infty \varepsilon^{3/2} f(\varepsilon)d\varepsilon \qquad (6.4.13)$$

and these are the starting points for getting the Fermi energy and the total energy of the electrons as functions of temperature.

Let us work on the Fermi energy by solving for N/CV from equation (6.3.11) and putting the result in (6.4.12) to get

$$\frac{2}{3}\mu_0^{3/2} = \int_0^\infty \varepsilon^{1/2} f(\varepsilon) d\varepsilon \tag{6.4.14}$$

The right-hand side of (6.4.14) is just a Fermi integral of the form of (6.4.5), with $g(\varepsilon)$ defined by

$$g(\varepsilon) = \varepsilon^{1/2} \tag{6.4.15}$$

so using (6.4.7),

$$F(\varepsilon) = \int_0^\varepsilon \varepsilon^{1/2} d\varepsilon = \frac{2}{3}\varepsilon^{3/2} \tag{6.4.16}$$

The value of $F_0(\mu) = F(\mu)$ is therefore

$$F_0(\mu) = \frac{2}{3}\mu^{3/2} \tag{6.4.17}$$

and the second derivative evaluated at $\varepsilon = \mu$ is

$$F_2(\mu) = \frac{1}{2}\mu^{-1/2} \tag{6.4.18}$$

According to equation (6.4.9), the temperature dependence of the Fermi integral (6.4.14) is given by

$$\frac{2}{3}\mu_0^{3/2} = F_0(\mu) + \frac{(\pi kT)^2}{6} F_2(\mu) \tag{6.4.19}$$

Using (6.4.17) and (6.4.18), this becomes

$$\frac{2}{3}\mu_0^{3/2} = \frac{2}{3}\mu^{3/2} + \frac{(2\pi kT)^2}{48}\mu^{-1/2} \tag{6.4.20}$$

or, raising both sides of (6.4.20) to the 2/3 power,

$$\mu_0 = \mu\left[1 + \frac{1}{8}\left(\frac{\pi kT}{\mu}\right)^2\right]^{2/3} \tag{6.4.21}$$

This is a truncated version of the temperature dependence of the Fermi energy, but an idea of its accuracy can be obtained by putting in reasonable values of the temperature and Fermi energy. The higher the temperature and the lower the Fermi level, the more slowly the series converges and the more inaccurate the approximation (6.4.21). Let us therefore be conservative and choose kT to be about one tenth of an electron volt and the Fermi energy to be just one electron volt. This corresponds to a metal with a low Fermi energy at a temperature of 1200K. Even so, the second term in (6.4.21) amounts to only 10^{-2}, and if the next term in the series were to be included, it would have a value of about 2×10^{-7}. Keeping only the first temperature-dependent term is therefore sufficiently accurate for electrons in metals. The Fermi energy varies with temperature so slowly that it can be replaced by μ_0 in the second term of (6.4.21). This allows us to solve for μ as a function of T, which is what we were after:

$$\mu = \mu_0 \left[1 + \frac{1}{8}\left(\frac{\pi kT}{\mu_0}\right)^2\right]^{-2/3}$$

Since the second term in the bracket is quite small, there is no loss of accuracy if the right-hand side is expanded in a Taylor series and only the first two terms are retained. The result is a somewhat simpler formula

$$\mu = \mu_0 \left[1 - \frac{1}{12}\left(\frac{\pi kT}{\mu_0}\right)^2\right] \tag{6.4.22}$$

which shows that the Fermi level decreases with temperature.

A similar calculation can be performed to get a formula for the energy. The application of (6.4.9) to (6.4.13) gives

$$\frac{U}{CV} = \frac{2}{5}\mu^{5/2} + \frac{(\pi kT)^2}{4}\mu^{1/2} \tag{6.4.23}$$

The energy can be written as a function of μ_0 by substituting (6.4.22) into (6.4.23) to get

$$\frac{U}{CV} = \frac{2}{5}\mu_0^{5/2}\left[1 - \frac{1}{12}\left(\frac{\pi kT}{\mu_0}\right)^2\right]^{5/2} + \frac{(\pi kT)^2}{4}\mu_0^{1/2}\left[1 - \frac{1}{12}\left(\frac{\pi kT}{\mu_0}\right)^2\right]^{1/2}$$

The powers in the brackets can be expanded by the binomial theorem or the Taylor series. Doing this and retaining terms only up to $(kT/\mu_0)^2$ gives

$$\frac{U}{N} = \frac{3}{5}\mu_0\left[1 + \frac{5}{12}\left(\frac{\pi kT}{\mu_0}\right)^2\right] \tag{6.4.25}$$

and thus the energy is a slowly increasing function of temperature. [In getting (6.4.25), we used equation (6.3.11).]

The heat capacity at constant volume is obtained by differentiating (6.4.25) with respect to temperature, with the result that

$$C_V^e = \frac{\pi^2 Nk}{2}\left(\frac{kT}{\mu_0}\right) \tag{6.4.26}$$

so the heat capacity of the electrons is linear with temperature. Its numerical value is small relative to the vibrational heat capacity, except at very low temperatures, where the electronic contribution becomes comparable to that of the T^3 term in the low-temperature limit of the vibrational heat capacity.

The entropy is easily computed in terms of the integral of the heat capacity.

$$S = \int_0^T \frac{C_V^e}{T} dT \tag{6.4.27}$$

which becomes, on using (6.4.26),

$$S = \frac{\pi^2 Nk}{2}\left(\frac{kT}{\mu_0}\right) \qquad (6.4.28)$$

so the entropy of a free electron gas is equal to its heat capacity.

From the thermodynamic formula $A = U - TS$, (6.4.25) and (6.4.28) give the Helmholtz free energy as

$$A = \frac{3}{5}N\mu_0\left[1 - \frac{5}{12}\left(\frac{\pi kT}{\mu_0}\right)^2\right] \qquad (6.4.29)$$

This completes the collection of basic equations for the thermodynamic properties of the free electron gas in terms of the zero-temperature Fermi level.

6.5 Electronic heat capacity in metals

Before the development of quantum theory and Fermi–Dirac statistics, there was a great inconsistency in the interpretation of experimental data for metals. The high values of the observed electrical and thermal conductivities could only be understood by postulating the existence of large numbers of free electrons. If electrons were present, however, classical statistical mechanics predicted that they should contribute an amount to the heat capacity of $3k$ per electron. As a result, the heat capacity per atom should be about two or three times that of insulators. In fact, however, it is found that metals and insulators have about the same heat capacity per atom, which is close to the classical Dulong–Petit value at room temperature and above. The resolution of this inconsistency by the Fermi–Dirac statistics is evident in equation (6.4.26).

Most metals have Fermi levels between 2eV and 10eV, so kT/μ_0 is between about 0.05 and 0.01 for temperatures in the neighborhood of 1200K. For these ranges, equation (6.4.26) shows that the electronic heat capacity is about 0.25–$0.05k$ per particle. This means that the free electrons can contribute only something like 2–8% of the Dulong–Petit value to the heat capacity at high temperatures. This relative contribution decreases with increasing temperature until the temperature approaches the Debye temperature.

Despite its small value, the electronic heat capacity can be separated from that arising from crystal vibrations. The electronic heat capacity decreases linearly with decreasing temperature, but at low temperatures the lattice heat capacity decreases as T^3 in accord with the Debye law. At low enough temperatures the electronic contribution is a large fraction of the lattice contribution. At low temperatures, the total heat capacity of a metal is

$$C_V(\text{Total}) = A_1 T + A_2 T^3 \qquad (6.5.1)$$

where the first term on the right is the electron contribution and the second term is the lattice contribution. Both constants in (6.5.1) can be determined from experimental values of the heat capacity as a function of temperature by plotting C_V/T against T^2. The intercept gives A_1 while the slope gives A_2.

The theoretical value of the constant A_1 is given by the coefficient of T in (6.4.26):

$$A_1 = \frac{\pi^2 Nk^2}{2\mu_0} \qquad (6.5.2)$$

or, using (6.3.13) for μ_0,

Table 6.1: Effective Mass and Fermi Level for Some Metals

Metal	Effective Mass	Fermi Level (e.v.)
Li	1.19	4.72
Na	1.0	3.12
K	0.99	2.14
Rb	0.97	1.82
Cs	0.98	1.53
Cu	0.99	7.04
Ag	1.01	5.51
Au	1.01	5.51

The Fermi levels were computed from free electron theory. The effective masses were computed from pseudopotential theory. (See Cohen, M.L., and V. Heine; 19••; *Solid State Physics* (1970) *vol. 24*; F. Seitz and D. Turnbull, Eds.; Academic Press, New York. The effective masses of the noble metals were computed by Kambe, K.; 1955; *Physics Review*; vol. 99, p. 419.)

$$A_1 = \frac{4m\pi^2 k^2 N}{h^2} \left(\frac{\pi V}{3N}\right)^{2/3} \quad (6.5.3)$$

while A_2 is given as the coefficient of the T^3 law in the Debye theory as $A_2 = 12\pi^4 kN/5\Theta_D^3$. If the number of free electrons per unit volume is known, A_1 can be computed from (6.5.3).

It is quite reasonable to take the number of free electrons to equal the number of valence electrons. The computed value of A_1 is not always in good agreement with experiment when the actual electronic mass is used in the calculation. The reason for this is to be found in the quantum theory of electrons moving in a three-dimensional periodic potential. The effect of the periodic potential is to alter inertial properties of the electrons so that, while electrons in metals can still be treated by free electron theory with reasonable accuracy, the electronic mass must be replaced by the effective mass m^*. The ratio m^*/m is a measure of the influence of the crystal potential on the motion of the quasi-free electrons. This ratio, along with values of the Fermi energy, is shown for a number of metals in table 6.1.

6.6 Equation of state of the free electron gas

The equation of state can be derived from the Helmholtz free energy by using the thermodynamic relation

$$P = -\left(\frac{\partial A}{\partial V}\right)_T \quad (6.6.1)$$

We have an expression for the Helmholtz free energy, but it has to be written as a function of volume. To do this, we just substitute equation (6.3.11) into (6.4.29) to get

$$A = \frac{3N}{5}\left(\frac{3N}{2C}\right)^{2/3} V^{-2/3} - \frac{N(\pi kT)^2}{4}\left(\frac{2C}{3N}\right)^{2/3} V^{2/3} \quad (6.6.2)$$

Now differentiate (6.6.2) with respect to volume and take its negative. This gives the pressure as

$$P = \frac{2N}{5}\left(\frac{3N}{2C}\right)^{2/3} V^{-5/3} + \frac{N(\pi kT)^2}{6}\left(\frac{2C}{3N}\right)^{2/3} V^{-1/3} \quad (6.6.3)$$

This is the equation of state for the free electron gas. It can be written in terms of the Fermi energy μ_0 by using (6.3.11) in (6.6.3). The result is

$$P = \frac{2N\mu_0}{5V}\left[1 + \frac{5}{12}\left(\frac{\pi kT}{\mu_0}\right)^2\right] \quad (6.6.4)$$

An estimate of the pressure of the electron gas in metals can be made from equation (6.6.4). For an electron density of 10^{22} electrons/cm^3 and a Fermi energy of 5 eV, the leading term in (6.6.4) gives a pressure of about 3×10^4 atmospheres. This means that, as far as the free electrons are concerned, a metal should fly apart into atoms. The metal is held together, of course, by the attractive interaction between the electrons and the ions. It is clear that any theory of the cohesion of metals must take the free electrons into account.

If (6.6.4) is compared to equation (6.4.25), we see that the PV product is just two thirds of the energy:

$$PV = \frac{2}{3}U \quad (6.6.5)$$

This relation is also correct for the ideal gas, as can be seen from our results in section 3.4 [see equation (3.4.33)].

The bulk modulus, $B = 1/\kappa$, is defined in terms of the derivative of the pressure by

$$B = \frac{1}{\kappa} = -V\left(\frac{\partial P}{\partial V}\right)_T \quad (6.6.6)$$

so taking the volume derivative of (6.6.3) and putting it in (6.6.6) gives the compressibility as

$$\frac{1}{\kappa} = \frac{2N}{3V}\left(\frac{3N}{2CV}\right)^{2/3} + \frac{N(\pi kT)^2}{18V}\left(\frac{2CV}{3N}\right)^{2/3} \quad (6.6.7)$$

or, using (6.3.11),

$$\frac{1}{\kappa} = \frac{2N\mu_0}{3V}\left[1 + \frac{1}{12}\left(\frac{\pi kT}{\mu_0}\right)^2\right] \quad (6.6.8)$$

Again, the leading term is sufficient to get an estimate of the bulk modulus for free electrons in metals. Taking $N/V = 10^{22}$ and $\mu_0 = 5$ eV gives $1/\kappa = 5 \times 10^{10}$ dynes/cm^2, which is of the order of the experimental results. The free electrons therefore contribute an appreciable amount to the bulk modulus of metals.

The equation of state properties of real metals depend on the electron-ion interactions as well as on the free electrons. This effect is rather insensitive to temperature, so we could ignore it in heat capacity theory and a satisfactory model could be built on the basis of a combination of crystal vibrations and free electrons. But the electron-ion interactions are functions of volume and must be included in any realistic equations of state for metals. If the volume is decreased this interaction becomes stronger, so the electron-ion interaction decreases both the pressure and the bulk modulus. That this is a large effect is shown by the fact that it more than offsets the tendency of the free electrons to make the crystal blow apart.

6.7 Thomas–Fermi theory

The free electron theory presumes that the potential energy and the electron density are constants. This is not so in a metal, and some simplified method of estimating the spatial dependence is desirable. The Thomas–Fermi theory provides such a method. It starts with the electron distribution function in terms of momentum, which is just the product of the Fermi function and the density of states given by equation (6.3.7):

$$N(p) = \frac{8\pi V p^2}{h^3} f(\varepsilon) \tag{6.7.1}$$

All thermal effects will be neglected by taking $T = 0$. Thus, if we let p_F be the momentum of an electron when its energy is the Fermi energy, then (6.7.1) becomes

$$N(p) = \frac{8\pi V p^2}{h^3} \quad p < p_F$$
$$N(p) = 0 \quad p > p_F \tag{6.7.2}$$

If this equation is integrated over all momenta, the result is just the total number of electrons:

$$N = \int_0^{p_F} N(p)dp = \int_0^{p_F} \frac{8\pi V p^2}{h^3} dp \tag{6.7.3}$$

so the electron density $n = N/V$ is given by

$$n = \frac{8\pi}{3h^3} p_F^3 \tag{6.7.4}$$

The basic assumption of Thomas–Fermi theory is that (6.7.4) is valid even when the electron density depends on position. That is, for an electron density that is space dependent, we assume that

$$n(\mathbf{r}) = \frac{8\pi}{3h^3} [p_F(\mathbf{r})]^3 \tag{6.7.5}$$

where $n(\mathbf{r})$ and $p_F(\mathbf{r})$ are the electron density and Fermi momentum at the position $\mathbf{r}$. The variations in electron density are the result of a space-dependent potential acting on the electrons. The connection between this potential Φ and the electron density is given from electrostatics by the Poisson equation as

$$\nabla^2 \Phi = 4\pi e n(\mathbf{r}) \tag{6.7.6}$$

In this equation, e is the magnitude of the charge on an electron, $-en(\mathbf{r})$ is the charge density at $\mathbf{r}$, and Φ is the electrostatic potential, so $-e\Phi$ is the potential energy of an electron at **r**. Since the system is in self-equilibrium, the chemical potential must be a constant throughout the system. Also, μ_0 is the total energy of an electron with momentum p_F, so the law of conservation of energy requires that

$$\mu_0 = \frac{p_F^2(\mathbf{r})}{2m} - e\Phi \tag{6.7.7}$$

where μ_0 is independent of position.

Solving (6.7.7) for $p_F^3(\mathbf{r})$ and substituting into (6.7.5) gives

$$n(\mathbf{r}) = \frac{8\pi}{3h^3}[2m(\mu_0 + e\Phi)]^{3/2} \tag{6.7.8}$$

This equation allows Poisson's equation to be written in terms of the potential only by putting it into (6.7.6) to get

$$\nabla^2 \Phi = \frac{32\pi^2 e}{3h^3}[2m(\mu_0 + e\Phi)]^{3/2} \tag{6.7.9}$$

Once the boundary conditions are given, (6.7.9) can be solved for $\Phi(\mathbf{r})$, which can then be substituted into (6.7.8) to get $n(\mathbf{r})$. The Thomas–Fermi method therefore gives the spatial variation of both the electron density and the potential in which the electrons move.

The solution of (6.7.9) generally requires numerical methods, but an approximation exists for which analytic solutions can be found. If we restrict ourselves to systems for which the Fermi energy is much greater than the electrostatic potential energy, then we have $e\Phi \ll \mu_0$. Equation (6.7.8) can then be linearized by first writing it in the form

$$n(\mathbf{r}) = \frac{8\pi}{3h^3}(2m\mu_0)^{3/2}\left(1 + \frac{e\Phi}{\mu_0}\right)^{3/2} \tag{6.7.10}$$

and then treating $e\Phi/\mu_0$ as a small quantity so that to a good approximation

$$\left(1 + \frac{e\Phi}{\mu_0}\right)^{3/2} = 1 + \frac{3e\Phi}{2\mu_0} \tag{6.7.11}$$

Therefore, (6.7.10) becomes

$$n(\mathbf{r}) = \frac{8\pi}{3h^3}(2m\mu_0)^{3/2}\left(1 + \frac{3e\Phi}{2\mu_0}\right) \tag{6.7.12}$$

This equation can be made to look a little simpler by using (6.3.12), which relates μ_0 to the electron density that would exist in the absence of a potential. This electron density is N/V, which we will call n_0. Solving (6.3.13) for $N/V = n_0$ and putting the result in (6.7.12) gives

$$n(r) = n_0\left(1 + \frac{3e\Phi}{2\mu_0}\right) \tag{6.7.13}$$

Substituting (6.7.13) into the Poisson equation (6.7.6) gives

$$\nabla^2 \Phi = 4\pi e n_0 \left(1 + \frac{3e\Phi}{2\mu_0}\right) \qquad (6.7.14)$$

This is the linearized Thomas–Fermi equation.

The Thomas–Fermi theory developed so far holds only in regions where there is no positive charge, because only the negative charge density was included on the right-hand side of the Poisson equation. But it should contain the total charge density, so in regions where a positive charge density exists the appropriate term must be added. If the positive charge exists as point charges, then it is represented by delta functions centered on the location of the point charges. In this case, Poisson's equation is solved in the region outside these points and the positive charge need not be included explicitly. Its existence, however, is taken into account in the boundary conditions.

In the application of free electron theory to metals, the positive charges are assumed to be smeared out into a uniform density that just cancels the charge of the electrons. The total charge is then zero and Poisson's equation reduces to Laplace's equation $\nabla^2 \Phi = 0$. This is called the *jellium model* of a metal. Let us investigate what happens if we put a point charge q into this uniform system. This can be done, for example, by replacing one of the atoms in the metal by another of a different valence, or by removing an atom to form a vacancy. In the first case, the point charge is the difference between the valencies of the host and the impurity atoms. In the second case, the electrical equivalent of removing an atom is to introduce a negative charge of the same magnitude as the positive charge that was removed. To be specific, a cadmium impurity (Cd^{2+}) in silver (Ag^+) would introduce a unit positive charge, $q = 1$, while a vacancy in silver introduces a unit negative charge $q = -1$. Analysis of this problem is clearly useful for the theory of impurities and point defects.

The introduction of a point charge into the jellium model gives rise to a nonuniform potential and induces a charge redistribution of the electron gas in the vicinity of the charge. The uniform positive background charge is assumed to remain unchanged at its value of en_0 everywhere except at the excess point charge because it is anchored in place by the ions on the lattice. In the linearized approximation, the redistributed electron density is related to the potential by (6.7.13). Subtracting this from the positive charge background density en_0 then gives the total charge density at all points except at the excess point charge as

$$en_0 - en_0\left(1 + \frac{3e\Phi}{2\mu_0}\right) = -n_0\frac{3e^2\Phi}{2\mu_0} \qquad (6.7.15)$$

Multiplying this by -4π and putting it in the Poisson equation gives the linearized Thomas–Fermi equation for the potential around an excess point charge:

$$\nabla^2\Phi = \frac{6\pi e^2 n_0 \Phi}{\mu_0} \qquad (6.7.16)$$

The field around a point charge embedded in an initially uniform electron gas must be spherically symmetric, so the Laplacian in (6.7.16) can be written in terms of spherical coordinates centered on the point charge as

$$\nabla^2\Phi = \frac{d^2\Phi}{dr^2} + \frac{2}{r}\frac{d\Phi}{dr} \qquad (6.7.17)$$

r being the distance from the point charge. Using (6.7.17), equation (6.7.16) becomes

$$\frac{d^2\Phi}{dr^2} + \frac{2}{r}\frac{d\Phi}{dr} = \frac{\Phi}{\lambda^2} \qquad (6.7.18)$$

where λ is defined as

$$\lambda^2 = \frac{\mu_0}{6\pi e^2 n_0} \qquad (6.7.19)$$

Define a function $f(r)$ by

$$f(r) = r\Phi \qquad (6.7.20)$$

from which we get, by direct differentiation,

$$\frac{d^2\Phi}{dr^2} + \frac{2}{r}\frac{d\Phi}{dr} = \frac{1}{r}\frac{d^2 f}{dr^2} \qquad (6.7.21)$$

so equating the right-hand sides of (6.7.18) and (6.7.21) gives

$$\frac{d^2 f}{dr^2} - \lambda^{-2} f = 0 \qquad (6.7.22)$$

But this is just a simple second-order differential equation with constant coefficients whose auxiliary equation is $m^2 - \lambda^{-2} = 0$ and whose general solution is therefore

$$f = C_1 e^{r/\lambda} + C_2 e^{-r/\lambda} \qquad (6.7.23)$$

Using the definition of f given in (6.7.20), the general solution for the potential is therefore

$$\Phi(r) = \frac{C_1}{r} e^{r/\lambda} + \frac{C_2}{r} e^{-r/\lambda} \qquad (6.7.24)$$

To get the solution specific to our problem, we need to know the boundary conditions. Clearly, the potential must approach the Coulomb potential of the point charge at very small distances because the effects of the electron gas must be negligible very close to the origin. At very large distances, the effect of the point charge must vanish, so the potential approaches zero. This gives us the two boundary conditions we need as

$$\lim_{r \to 0} \Phi(r) = \frac{q}{r}$$
$$\lim_{r \to \infty} \Phi(r) = 0 \qquad (6.7.25)$$

From the second of these boundary conditions the constant C_1 must be zero, while from the first boundary condition we must have $C_2 = q$. Equation (6.7.24) therefore becomes

$$\Phi(r) = \frac{q}{r} e^{-r/\lambda} \qquad (6.7.26)$$

The potential field around an excess charge embedded in a free electron gas therefore has the form of a Coulomb field modified by a decreasing exponential function. The physical interpretation of this is that the mobile electrons screen the interaction between an electron and the excess point charge by piling up around the origin (if q is positive) or leaving a hole around it (if q is negative). The resulting spatial distribution of the electrons is obtained by substituting equation (6.7.26) into (6.7.13):

$$n(r) = n_0\left(1 + \frac{3eq}{2\mu_0 r} e^{-r/\lambda}\right) \qquad (6.7.27)$$

From equations (6.7.26) and (6.7.27), we see that for distances much greater than λ, the effect of the excess point charge is small; λ is a measure of the effectiveness of the screening of the point charge by the mobile electrons and is called the *screening distance*. The value of the screening distance can be calculated from the Fermi energy and the electron density using its defining equation (6.7.19). For typical metals, it turns out to be of the order of one angstrom, so an excess point charge is effectively screened out over distances comparable to the lattice spacing.

As is evident from the nature of its approximations, the linearized Thomas–Fermi theory gives only a rough approximation to the actual potential and electron density around impurities and point defects in real metals. However, it has been used profitably in both alloy and defect theory to give qualitative and semiquantitative results. The physical significance of the above development goes beyond its applicability to such problems because it provides at least a partial justification of the use of free electron theory in metals. The screening distance is short. This means that the interaction between any two charges is negligible unless the charges are close together. Ignoring all electron-electron interactions in the metal is therefore not as bad an approximation as it sounds. The theory of electron-electron interactions is more complex than can be presented here, but more sophisticated investigations bear out these conclusions.

6.8 Review of results of band theory

For metals, the analysis of the statistical mechanics of electrons could be made just by assuming that they contain free electrons. For semiconductors, however, a few simple results of the band theory of solids are also required. Accordingly, these results are reviewed here. Let us start by recalling that each atomic level of a free atom gives rise to a band of energy levels when a large number of such atoms are condensed to form a solid. These levels are very close together in energy and form a quasi continuum. Because there are infinite numbers of atomic levels, there are infinite numbers of bands. Also, because each atomic energy level can hold two electrons, the number of levels in each band is twice the number of atoms in the solid. The gap width between bands, the width of the band itself, and the presence or absence of overlap among the bands depend on the details of the interactions among the electrons and ions of the solid.

The most obvious success of the band theory is that it accounts for the existence of metals, insulators, and semiconductors. Consider a monovalent metal. A free atom of this metal would have one electron in an outer level, and the energy band corresponding to this level would be only half full because it can accommodate two electrons per atom. With this in mind, assume that an electric field is applied to the metal. In order for electrons to move under the influence of the field and thereby contribute to an electric current, they must absorb some energy, thereby converting potential energy from the field into kinetic energy of moving electrons. Electrons at the bottom of the band cannot do this because if they were to absorb energy, this would put them in a level slightly above their normal level. But this upper level is already occupied by two electrons, and according to the Pauli exclusion principle it can accept no more, so the electron must stay where it is. Electrons just at the top of the occupied portion of the band, however, can contribute to electrical conductivity because they can absorb energy from the field by just jumping into the empty levels above them. Furthermore, when an electron jumps into an empty level in the top part of the band, it leaves an empty level behind it. This empty level is called a "hole," and it too can contribute to conductivity since electrons near it in energy can jump into it under the influence of the applied field. This situation is characteristic of metals. The high mobility of the free electrons in metals is the result of the fact that they exist in partially filled bands. Such partially filled bands are called *conduction bands*.

If the free atoms from which the solid is made contain two electrons each in their outer levels, then the corresponding band in the crystal contains two electrons per atom and is completely filled. Such a filled band is called a valence band. Assume that there is no overlapping of bands so that the band above the filled one is completely empty and the two bands are separated by a gap. Also assume that the energy gap is large compared to the energy that an applied field can give to an electron. No electrons can exist with energies in this gap, so it is clear from the above discussion that the electrons in the filled band cannot absorb energy from an applied electric field because there are no levels near them to jump into. If the two bands overlap, then a set of $4N$ contiguous states exist containing $2N$ electrons and the material is a metal.

All this is at absolute zero. The effect of temperature is different for metals, insulators, and semiconductors. For a metal, the thermal energy excites some electrons near the top of the occupied part of the conduction band into nearby levels that would be empty at absolute zero. At normal temperatures, the fraction of such electrons is not large since the thermal energy is of the order of kT and metallic bands are much wider than this. If we raise the temperature of an insulator, there is a finite probability that a thermal fluctuation will occur that supplies enough energy to an electron at the top of the filled band to kick it into the upper, empty band. If this happens, a current can flow because the electron has empty states available to it. The greater the energy gap, the larger the energy needed for the kick and the more improbable the required thermal fluctuation. If the gap is large relative to kT, very few electrons will be excited to the conduction band and the material remains an insulator. But if the energy gap is not too large relative to kT, an appreciable number of electrons reach the conduction band and the material is a semiconductor. The designation "semiconductor" is a recognition of the fact that the concentration of electrons excited to the conduction band is much less than that in a metal, so the electrical conductivity is much less than that in metals. Energy band diagrams for metals, insulators, and semiconductors are shown in figure 6.2.

One of the most important results of band theory is that in many cases the energy of an electron near the bottom of a band is given, at least approximately, by

$$\varepsilon = E_B + \gamma \mathbf{k}^2 \tag{6.8.1}$$

where E_B is the energy at the bottom of the band, $\mathbf{k}$ is the wave number vector, and γ is a constant. Note that in the theory of the particle statistics of free electrons, the energy was a quadratic function of the wave number vector.

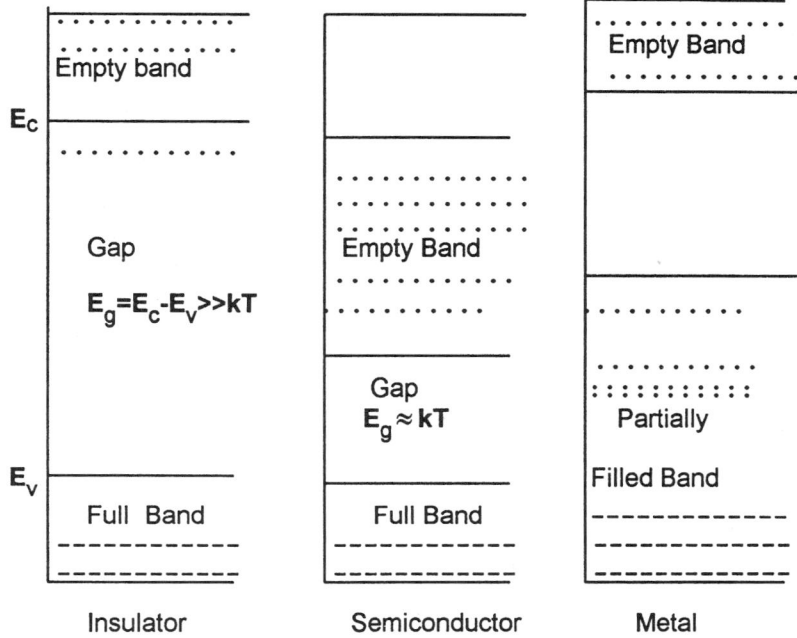

Figure 6.2. Energy bands in insulators, semiconductors, and metals.

Equation (6.8.1) can be put into a form just like that for a free electron by defining the effective mass as

$$m^* = \frac{\hbar^2}{2\gamma} \tag{6.8.2}$$

Then (6.8.1) can be written as

$$\varepsilon = E_B + \frac{\hbar^2 \mathbf{k}^2}{2m^*} \tag{6.8.3}$$

This equation is just like that for a free particle with an effective mass m^* moving in a constant potential. The constant γ, and therefore the effective mass, depends on the interaction of the electron with the rest of the crystal. If the scale of energy is defined such that $E_B = 0$ and if the effective mass is used rather than the actual electronic mass, free electron theory can be used whenever the energy is a quadratic function of the wave number vector.

For electrons near the top of a band, the energy is also found to be approximately a quadratic function of a wave number vector, but in this case, the wave number is measured relative to the top of the band. That is,

$$\varepsilon = E_T - \frac{\hbar^2}{2m^*}(\mathbf{k} - \mathbf{k}_T)^2 \tag{6.8.4}$$

where E_T is the energy of an electron at the top of the band and $\mathbf{k}_T$ is its wave number vector. Note that the effective masses at the top and at the bottom of a band are not necessarily the same, nor are they necessarily the same for different bands in the same material. Equation (6.8.4) can be written as

$$(E_T - \varepsilon) = \frac{\hbar^2 \tilde{k}^2}{2m^*} \tag{6.8.5}$$

where $\tilde{k}^2 = (\mathbf{k} - \mathbf{k}_T)^2$. This shows that the electron can be treated as a free particle with wave number vector $\tilde{\mathbf{k}}$, mass m^*, and energy $(E_T - \varepsilon)$. These results are particularly useful for semiconductors because the electrons that contribute to the conductivity in these materials are generally near the top and the bottom of bands.

For the statistical mechanics of electrons in semiconductors, it is convenient to treat the holes left behind in a valence band when electrons are excited to the conduction band as "particles." This is a useful concept because the holes act as positive charge carriers, which can be shown as follows: in a filled band no current can flow, so the sum of all the electron velocities must vanish. This means that if there are N electrons in a band and $\mathbf{v}_i$ is the velocity of the ith electron, then

$$\sum_{i=1}^{N} \mathbf{v}_i = 0 \tag{6.8.6}$$

Now pick out a particular electron, label it j, and rewrite (6.8.6) as

$$-\mathbf{v}_j = \sum_{i \neq j}^{N} \mathbf{v}_i \tag{6.8.7}$$

where the sum is over all electrons except the jth. The right-hand side of (6.8.7) is therefore the sum of all electron velocities for a band that is missing one electron. Equation (6.8.7) shows that, for such a band, the effect is that of an electron moving with the negative of the velocity it would have in the band. The hole therefore acts as if it were a positive charge carrier.

6.9 Impurity levels in semiconductors

Impurities have profound effects on the properties of semiconductors, and the operation of semiconductor devices depends on these effects. It is therefore important to have some knowledge of how the distribution of energy levels is related to these impurities.

Let us replace a tetravalent atom in an elemental semiconductor (e.g., silicon or germanium) by a pentavalent atom such as phosphorus. Four of the five valence electrons take part in the tetrahedral bonding in the diamond structure of the semiconductor. The fifth electron, however, is not needed for covalent bonding and, at absolute zero of temperature, stays in the vicinity of the impurity ion since it is attracted by its positive charge. To a first approximation, the electron-impurity ion system behaves like a hydrogen atom, and the binding energy of the electron to the impurity ion can be computed from quantum theory analogous to the calculation of the ionization energy of hydrogen. The fact that this system is embedded in a crystal can be accounted for by recalling that in a dielectric medium the force of attraction between two charges is just the Coulomb force divided by the dielectric constant κ of the medium. Working out the hydrogen atom problem with a potential $e^2/\kappa r$ gives the ionization energy in electron volts as

$$E_I = \frac{13.6}{\kappa^2} \tag{6.9.1}$$

Using typical values of κ for semiconductors (11.9 for silicon, 16.1 for germanium), the ionization energy is then about 0.05–0.10 eV. When the impurity ionizes, it must go into the conduction band. This means that the energy level of an electron attached to an impurity is about 0.05–0.10 eV below the lowest level of the conduction band.

Note that the ionization energy is in the range of thermal energies at room temperature. Thermal excitation is therefore sufficient to ionize a significant fraction of the impurity atoms, thereby releasing electrons to the empty conduction band where they can take part in electrical conduction. Impurities of the type just discussed are called *donors* because they donate electrons to the conduction band. The semiconductor is then an extrinsic n-type conductor since conduction is by negative electrons and the conduction is not an intrinsic property of the pure crystal.

Trivalent atoms also introduce impurity energy levels into the crystal. In this case, the impurity needs an extra electron to complete the tetrahedral bonding, and it can get this electron from the full valence band. The impurity-electron system is then a negative ion but the binding energy of the electron to the impurity can still be estimated by analogy to a hydrogen atom in a dielectric medium. The impurity level is now, therefore, about 0.05–0.10 eV above the highest energy in the valence band. When it ionizes, the impurity creates a hole in the valence band that contributes to conduction as if it were a positive particle. The material is then an extrinsic p-type semiconductor.

6.10 Electron distribution in intrinsic semiconductors

The statistics of electrons in pure semiconductors can be understood by starting with the simple two-band model shown in figure 6.3. This model states that only electrons in two bands contribute to electronic conductivity and that these bands are separated by a forbidden range of energies that constitute the energy gap.

At absolute zero, all the states in the lower (valence) band are filled with electrons while all states in the upper band are empty. The zero of energy is taken to be at the bottom of the valence band. E_V is the energy at the top of the valence band, while E_C and E_C^T are the energies at the bottom and top of the conduction band, respectively. At nonzero temperatures, a fraction of the electrons in the valence band are excited to the conduction band, leaving an equivalent number of holes in the valence band.

We will assume that E_V, E_C, and E_C^T are all considerably greater than kT, and that the energy gap $(E_C - E_V)$ is large enough that the conduction band contains only a small fraction of the electrons available from the valence band. For real materials of technological interest, these conditions are usually satisfied.

The purpose of the present analysis is to determine the energy distribution function for the electrons in the conduction band and the holes in the valence band, and to determine the position of the Fermi energy relative to energies of the bands. The importance of the electron and hole distributions and of the Fermi energy is that these determine the conductivity of semiconductors. The electrons we are interested in are obviously those near the bottom of the conduction band and near the top of the valence band. We can then adopt the approximate results of band theory that the electrons near the bottom of a band are just like free electrons with kinetic energies $(\varepsilon - E_C)$ and those near the top

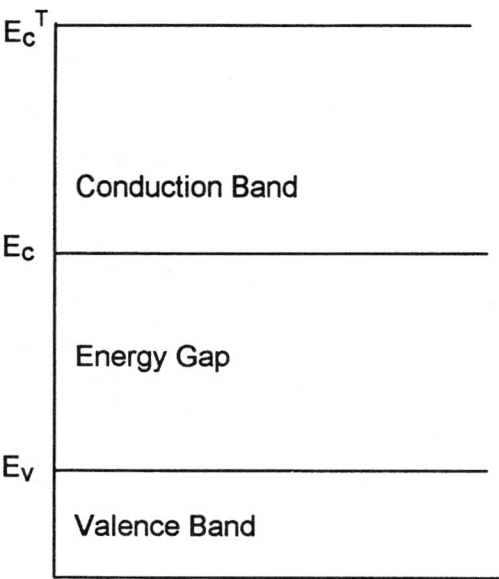

Figure 6.3. Two-band model for intrinsic semiconductors.

of the valence band are just like free electrons with energies $(E_V - \varepsilon)$ with, of course, effective masses appropriate to each band. Then the distribution functions of electrons in the two bands are

$$N_C(\varepsilon) = \frac{\omega_C(\varepsilon)}{e^{(\varepsilon-\mu)/kT} + 1} \tag{6.10.1}$$

$$N_V(\varepsilon) = \frac{\omega_V(\varepsilon)}{e^{(\varepsilon-\mu)/kT} + 1} \tag{6.10.2}$$

$\omega_C(\varepsilon)$ and $\omega_V(\varepsilon)$ are the density of states near the bottom of the conduction band and near the top of the valence band, respectively. They are given by expressions just like those for free electrons except that the energy is replaced by $(\varepsilon - E_C)$ in ω_C and by $(E_V - \varepsilon)$ in ω_V. That is,

$$\omega_C = \frac{8\pi\sqrt{2}}{h^3} m_C^{3/2} V \sqrt{\varepsilon - E_C} \tag{6.10.3}$$

$$\omega_V = \frac{8\pi\sqrt{2}}{h^3} m_V^{3/2} V \sqrt{E_V - \varepsilon} \tag{6.10.4}$$

The possibility that the effective masses in the two bands might be different has been indicated by the subscripts C and V.

In the valence band, we are more interested in the distribution of holes than of electrons. This is easily obtained from the fact that the Fermi function f is the probability that a particular state contains an electron. The probability that the state is empty is therefore $(1 - f)$, and this is just f_h, the probability of the existence of a hole. Therefore, the Fermi function for holes is

$$f_h = 1 - f = 1 - \frac{1}{e^{(\varepsilon-\mu)/kT}+1} = \frac{1}{e^{(\mu-\varepsilon)/kT}+1} \quad (6.10.5)$$

Multiplying this by the density of states for the valence band gives the energy distribution for the holes:

$$N_h(\varepsilon) = \frac{\omega_V(\varepsilon)}{e^{(\mu-\varepsilon)/kT}+1} \quad (6.10.6)$$

In intrinsic semiconductors the band gap is several electrons volts and the Fermi level is near the middle of the gap. For a semiconductor with a band gap of several electron volts and a Fermi level right at the middle of the gap, equations (6.10.1) and (6.10.6) give an electron or hole concentration of about 10^{16} per cubic centimeter at room temperature. Both $N_C/\omega_C(\varepsilon)$ and $N_h/\omega_h(\varepsilon)$ are therefore small relative to unity. But (6.10.1) and (6.10.6) show that then the exponentials in these equations must be large compared to unity, which can therefore be ignored. Equations (6.10.1) and (6.10.6) then reduce to

$$N_C(\varepsilon) = \omega_C(\varepsilon) e^{(\mu-\varepsilon)/kT} \quad (6.10.7)$$

$$N_h(\varepsilon) = \omega_V(\varepsilon) e^{(\varepsilon-\mu)/kT} \quad (6.10.8)$$

Equation (6.10.7) can be true only if $(\mu - \varepsilon) < 0$ for all values of the energy of conduction electrons. This means that the Fermi energy must be less than the energy of the bottom of the conduction band. Also, (6.10.8) can be true only if $(\varepsilon - \mu) < 0$ for all values of the energy of the valence electrons, and this means that the Fermi energy must be greater than that of the top of the valence band. The Fermi energy must therefore be somewhere in the energy gap.

The actual position of the Fermi energy is obtained from the fact that the number of electrons in the conduction band must be equal to the number of holes in the valence band. The total number of electrons per unit volume of crystal is just the integral of (6.10.7) from the bottom to the top of the conduction band divided by the volume. Using the density of states expression (6.10.3), the integral of (6.10.7) per unit volume is

$$n = \frac{8\pi\sqrt{2}}{h^3} m_c^{3/2} \int_{E_c}^{E_c^T} \sqrt{(\varepsilon - E_c)} e^{(\mu-\varepsilon)/kT} d\varepsilon \quad (6.10.9)$$

Change variables in the integral to $x = (\varepsilon - E_c)/kT$ to get

$$n = \frac{4\pi(2m_ckT)^{3/2}}{h^3} e^{(\mu-E_c)/kT} \int_0^{(E_c^T - E_c)/kT} \sqrt{x} e^{-x} dx \quad (6.10.10)$$

The upper limit of this integral is the ratio of the width of the conduction band to kT, and this ratio is large, so practically no accuracy is lost if the upper limit is replaced by infinity. The value of the integral is then $\sqrt{\pi}/2$ and (6.10.10) becomes

$$n = 2\left(\frac{2\pi m_c kT}{h^2}\right)^{3/2} e^{(\mu-E_c)/kT} \quad (6.10.11)$$

A completely analogous calculation for the number of holes per unit volume of crystal gives the result

$$p = 2\left(\frac{2\pi m_V kT}{h^2}\right)^{3/2} e^{(E_V - \mu)/kT} \qquad (6.10.12)$$

The preexponential factor in (6.10.11) has an interesting interpretation as can be seen by considering the quantity ρ_c defined by

$$\rho_c = \left(\frac{2\pi m_c kT}{h^2}\right)^{3/2} \qquad (6.10.13)$$

The Boltzmann factor in (6.10.11) multiplied by two (because of electron spin) is the probability that there is an electron in the conduction band, and when this is multiplied by ρ_c we get the number of electrons in the conduction band. ρ_c can then be thought of as a density of states for the conduction band in the sense that the number of electrons in the conduction levels is the product of a probability and a density of states. A precisely analogous argument leads us to interpret ρ_V, which is defined by

$$\rho_V = \left(\frac{2\pi m_V kT}{h^2}\right)^{3/2} \qquad (6.10.14)$$

as a density of states for the valence band with respect to the number of holes it contains.

The Fermi level is easily obtained by equating n to p and solving for μ to get

$$\mu = \frac{E_c + E_V}{2} + \frac{3kT}{4} \ln \frac{m_V}{m_c} \qquad (6.10.15)$$

This shows that if the two effective masses are equal, the Fermi level is directly in the middle of the gap. In real materials the effective masses are not usually very different, so μ is generally near the center of the gap.

The concentrations of electrons and of holes can be expressed in terms of the value of the energy gap $E_g = E_c - E_V$ as follows. Multiplying (6.10.11) by (6.10.12) eliminates the Fermi level from the equations with the result

$$np = 4\left(\frac{2\pi kT}{h^2}\right)^3 (m_c m_V)^{3/2} e^{-E_g/kT} \qquad (6.10.16)$$

But for an intrinsic semiconductor $n = p$, so taking the square root of both sides of (6.10.16) gives

$$n = p = 2\left(\frac{2\pi kT}{h^2}\right)^{3/2} (m_c m_V)^{3/4} e^{-E_g/2kT} \qquad (6.10.17)$$

The electron and hole concentrations in an intrinsic semiconductor are independent of the Fermi level and depend only on the energy gap. This is an intuitively reasonable result since it is in accord with the fact that the process of transferring an electron from the valence band to the conduction band depends on thermal excitation, which must give the electron an energy at least equal to the gap energy.

Note that the only step in which it was assumed that the semiconductor is intrinsic was that in which the square root of (6.10.16) was taken to give equation (6.10.17) for the equal concentrations of electrons and holes. In extrinsic semiconductors, the equilibrium distribution function for electrons in the conduction and valence bands is the same as for intrinsic semiconductors. This means that equations (6.10.11) and (6.10.12) for the electron and hole concentrations, and equation (6.10.16) for their product, will also be valid for extrinsic semiconductors.

Equation (6.10.17) shows that the charge carrier concentration is an increasing function of temperature. Since the electrical conductivity is proportional to the charge concentration, it will therefore increase with temperature in a manner controlled by the Boltzmann factor in (6.10.17) containing E_g. The thermal vibrations tend to decrease the electrical conductivity as more of them are excited and as their amplitudes increase with increasing temperature. But this effect is overshadowed by the more rapid increase in carrier concentration. In metals, the electron concentration changes very slowly with temperature and is nearly constant for the usual temperature ranges encountered in practice. The temperature derivative of the conductivity is controlled largely by the electron-phonon scattering process. The conductivity in metals therefore exhibits an approximately linear decrease with temperature, while semiconductors show a rapid increase with temperature.

6.11 Electron statistics in extrinsic semiconductors

In intrinsic semiconductors, the energy levels in each band form a quasi continuum and each level can hold two electrons of opposite spin. The Fermi particle statistics can therefore be applied directly. For semiconductors containing donor or acceptor impurities, however, some modifications are necessary. To show the origin of these modifications, we must go back to the statistical counting procedure used to derive the distribution in energy.

Consider a semiconductor containing both donor and acceptor impurities whose energy band diagram is shown in figure 6.4. The levels in the conduction and in the valence bands are treated in the same way as for the particle statistics, and the number of complexions is given by equations that are completely analogous to those for free fermions. For the valence band, the number of complexions is

$$W_V = \prod_j \frac{\omega_V^j!}{N_V^j!(\omega_V^j - N_V^j)!} \qquad (6.11.1)$$

where N_V^j is the number of electrons in the ω_V^j levels whose energies are centered on a level ε_j in the valence band. For the conduction band, the number of complexions is

$$W_C = \prod_i \frac{\omega_c^i!}{N_c^i!(\omega_c^i - N_c^i)!} \qquad (6.11.2)$$

where N_V^j is the number of electrons in the ω_V^j levels whose energies are centered on a level ε_j in the conduction band.

Now consider the number of ways in which a given number of electrons can be distributed among the acceptor levels. When an electron moves from the valence band to an acceptor level, the impurity atom becomes negatively charged. Any attempt to place another electron in that level requires a very high energy to overcome the electrostatic repulsion. This means that each level can hold only one electron. If N_A is the number of acceptor impurity atoms and N_A^- is the number of electrons in these levels (i.e., the number of ionized acceptors), then a straightforward application of the

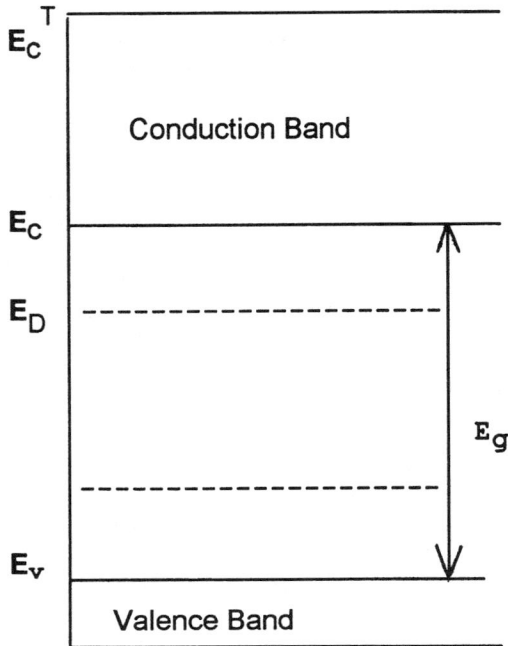

Figure 6.4. Two-band model for extrinsic semiconductors.

Fermi–Dirac counting method (see appendix 1) would give the number of complexions as

$$\frac{N_A!}{N_A^-!(N_A - N_A^-)!} \tag{6.11.3}$$

However, because of the existence of electron spin, this is not quite correct. A neutral trivalent impurity has an unpaired electron, and this electron has two possible spin states. If the spin of the unpaired electron is reversed, a new state is produced. Since the number of neutral acceptors is just the number of empty acceptor levels $(N_A - N_A^-)$, the number of states that can be produced just by reversing spins of the unpaired electrons is $2^{(N_A - N_A^-)}$. To get the number of complexions for acceptors, (6.11.3) must be multiplied by this factor. Therefore, the number of ways of distributing N_A^- electrons among N_A acceptor levels is

$$w_A = \frac{2^{(N_A - N_A^-)} N_A!}{N_A^-!(N_A - N_A^-)!} \tag{6.11.4}$$

Similar considerations hold for the donor levels. When a donor impurity gives up an electron to the conduction band, it becomes ionized, and the energy to ionize it further is prohibitive because the second electron must overcome the electrostatic attraction of the ion. The neutral pentavalent atom has an unpaired electron that can take on two spin values. The number of neutral atoms is just the number of electrons in donor levels. Each time the spin of an electron in a donor level is reversed, a new state is produced, and the total number of such possible permutations is $2^{N_D^0}$ where N_D^0 is the number of electrons in donor levels. For donors, therefore, the total number of complexions is

$$w_D = \frac{2^{N_D^0} N_D!}{N_D^0!(N_D - N_D^0)!} \tag{6.11.5}$$

where N_D is the total number of donor atoms. Equation (6.11.5) is just the product of the normal Fermi–Dirac statistical count and the spin factor $2^{N_D^0}$.

The total number of complexions is the product of the separate counts, so the number of ways of arranging electrons among the available states such that N_V^j are in the jth valence band levels, N_C^i are in the ith conduction band levels, N_A^- are in acceptor levels, and N_D^0 are in donor levels is

$$w = w_V w_C w_A w_D \tag{6.11.6}$$

The distribution in energy can now be derived in the usual way by maximizing $\ln w$ subject to the conditions of constant total number of electrons N and constant total energy U. The maximization must be carried out with respect to N_V^j, N_C^i, N_A^-, and N_D^0, so the variational problem to be solved is

$$\delta \ln w = \sum_j \frac{\partial \ln w_V}{\partial N_V^j} \delta N_V^j + \sum_i \frac{\partial \ln w_C}{\partial N_C^i} \delta N_C^i$$

$$+ \frac{\partial \ln w_A}{\partial N_A^-} \delta N_A^- + \frac{\partial \ln w_D}{\partial N_D^0} \delta N_D^0 = 0 \tag{6.11.7}$$

$$\delta N = \sum_j \delta N_V^j + \sum_j \delta N_C^i + \delta N_A^- + \delta N_D^0 = 0 \tag{6.11.8}$$

$$\delta U = \sum_j \varepsilon_j \delta N_V^j + \sum_i \varepsilon_i \delta N_C^i + E_A \delta N_A^- + E_D \delta N_D^0 = 0 \tag{6.11.9}$$

where E_A and E_D are the energies of the acceptor and donor levels, respectively. Multiplying (6.11.8) and (6.11.9) by the Lagrangian multipliers $-a$ and $-b$, respectively, adding the results to (6.11.7), and equating the coefficients of the variations to zero gives

$$\frac{\partial \ln w_V}{\partial N_V^j} = a + b\varepsilon_j \tag{6.11.10}$$

$$\frac{\partial \ln w_C}{\partial N_C^i} = a + b\varepsilon_i \tag{6.11.11}$$

$$\frac{\partial \ln w_A}{\partial N_A^-} = a + bE_A \tag{6.11.12}$$

$$\frac{\partial \ln w_D}{\partial N_D^0} = a + bE_D \tag{6.11.13}$$

The first two of these equations lead directly to the usual Fermi–Dirac distribution for the bands. Performing the differentiations in the last two and using

Stirling's approximation, however, gives the following results for the distribution in the donor and acceptor levels:

$$N_D^0 = \frac{N_D}{(1/2)e^{(a+bE_D)} + 1} \tag{6.11.14}$$

$$N_A^- = \frac{N_A}{2e^{(a+bE_A)} + 1} \tag{6.11.15}$$

The Lagrangian multipliers have the same connection with thermodynamics as derived previously, so these equations may be written as

$$N_D^0 = \frac{N_D}{(1/2)e^{(E_D-\mu)/kT} + 1} \tag{6.11.16}$$

$$N_A^- = \frac{N_D}{2e^{(E_A-\mu)/kT} + 1} \tag{6.11.17}$$

The probability that an acceptor level contains an electron is N_A^-/N_A, and the probability that a donor level contains an electron is N_D^0/N_D. We therefore define the Fermi functions for donor and acceptor levels by

$$f_D = \frac{1}{(1/2)e^{(E_D-\mu)/kT} + 1} \tag{6.11.18}$$

$$f_A = \frac{1}{2e^{(E_A-\mu)/kT} + 1} \tag{6.11.19}$$

Equation (6.11.17) gives the number of ionized acceptors and equation (6.11.19) gives the probability that an acceptor is ionized. To get the number of ionized donors, N_D^+, write

$$N_D^+ = N_D - N_D^0 = N_D(1 - f_D) \tag{6.11.20}$$

$(1 - f_D)$ is just the probability that a donor level is empty and is given by

$$f_D^h \equiv (1 - f_D) = \frac{1}{2e^{(\mu-E_D)/kT} + 1} \tag{6.11.21}$$

and the number of ionized donors is

$$N_D^+ = \frac{N_D}{2e^{(\mu-E_D)/kT} + 1} \tag{6.11.22}$$

Similarly, the probability that an acceptor level is neutral (empty) is

$$f_A^h = 1 - f_A = \frac{1}{(1/2)e^{(\mu-E_A)/kT} + 1} \tag{6.11.23}$$

and the number of empty acceptor levels is

$$N_A^0 = N_A - N_A^- = \frac{N_A}{(1/2)e^{(\mu-E_A)/kT} + 1} \tag{6.11.24}$$

This completes the list of the statistical formulas needed for a description of extrinsic semiconductors.

6.12 Mass action laws for extrinsic semiconductors

Whenever $(\varepsilon - \mu) \gg kT$ for electrons, or $(\mu - \varepsilon) \gg kT$ for holes, the Fermi function reduces to a Boltzmann form. Section 6.10 showed that this is true for intrinsic semiconductors with a large energy gap and that this implies that the electron and hole concentrations in the conduction and valence bands, respectively, are small relative to the number of available states. The Fermi level then lies near the center of the energy gap. If small amounts of donor or acceptor dopants are added to such a semiconductor, then the number of electrons and holes in the conduction bands may be increased, but if the impurity concentration is low enough, the number of electrons and holes will still be small relative to the number of available energy states.

For an extrinsic semiconductor of the type just described, it is convenient to describe the transfer of electrons among the energy levels by a set of chemical ionization reactions. Four such reactions are possible, (1) the ionization of a donor in which an electron goes to the conduction band:

$$D^0 \leftrightarrow D^+ + e^-(\text{CB}) \tag{6.12.1}$$

(2) the ionization of an acceptor in which an electron goes from the valence band to an acceptor level, leaving behind a hole:

$$A^0 \leftrightarrow A^- + h^+(\text{VB}) \tag{6.12.2}$$

(3) the transfer of an electron from a neutral donor to a neutral acceptor:

$$A^0 + D^0 \leftrightarrow A^- + D^+ \tag{6.12.3}$$

and (4) the transfer of an electron from the valence band to the conduction band:

$$e^-(\text{VB}) + h^+(\text{CB}) \leftrightarrow e^-(\text{CB}) + h^+(\text{VB}) \tag{6.12.4}$$

In these equations, e^- represents an electron, or a filled state, while h^+ represents a hole, or an empty state. CB and VB stand for conduction band and valence band, respectively.

Now define four equilibrium constants, corresponding to each of the four reactions as follows:

$$K_D = \frac{nN_D^+}{N_D^0} \tag{6.12.5}$$

$$K_A = \frac{pN_A^-}{N_A^0} \tag{6.12.6}$$

$$K_{AD} = \frac{N_A^- N_D^+}{N_A^0 N_D^0} \tag{6.12.7}$$

$$K_{np} = \frac{np}{\left(\frac{2N}{V} - p\right)\left(\frac{2N}{V} - n\right)} \tag{6.12.8}$$

In the last of these equations, N is the number of atoms in the crystal and $2N/V$ is the number of available states per unit volume in each band. But n, the number of electrons per unit volume, and p, the number of holes per unit volume, are both much smaller than the concentration of states in the bands, so n and p can be neglected in the denominator of (6.12.8), which can then be written as

$$K = np \tag{6.12.9}$$

where K is a new equilibrium constant defined by $K = K_{np}4N^2/V^2$. The equilibrium constants can be written in terms of the energy parameters of the system by substituting the appropriate expressions for n, N_D^+, N_D^0, N_D^0, p, and N_A^0. For the semiconductor we are considering here (large gap, low impurity concentration), equations (6.10.11) and (6.10.12) are valid. Using these equations for n and p, and (6.11.16), (6.11.17), (6.11.22), and (6.11.24) in the right-hand sides of (6.12.5)–(6.12.7) and (6.12.9), and remembering that $f_D + f_D^h = 1$, we get

$$\frac{nN_D^+}{N_D^0} = \rho_C e^{-E_{CD}/kT} \tag{6.12.10}$$

$$\frac{pN_A^-}{N_A^0} = \rho_V e^{-E_{AV}/kT} \tag{6.12.11}$$

$$\frac{N_A^- N_D^+}{N_A^0 N_D^0} = \frac{1}{4} e^{E_{DA}/kT} \tag{6.12.12}$$

$$np = 4\rho_C \rho_V e^{-E_g/kT} \tag{6.12.13}$$

The energies in these expressions are defined by

$$\begin{aligned} E_{CD} &= E_C - E_D \\ E_{AV} &= E_A - E_V \\ E_{DA} &= E_D - E_A \\ E_g &= E_C - E_V \end{aligned} \tag{6.12.14}$$

and ρ_C and ρ_V are defined by equations (6.10.13) and (6.10.14).

Equations (6.12.10)–(6.12.13) must be satisfied in our semiconducting system. Their value lies in the fact that the right-hand sides are functions of temperature that depend only on the nature of the crystal and its impurities, and not on concentrations. To illustrate the use of these equations, consider the case in which the energy gap is so large that a negligible number of electrons are excited from the valence band to the conduction band. The number of conduction electrons is then equal to the number of ionized donors,

$$N_D^+ = Vn \tag{6.12.15}$$

and the number of holes is equal to the number of ionized acceptors,

$$N_A^- = Vp \tag{6.12.16}$$

Substituting these equations into (6.12.10) and (6.12.11), and remembering that

$$N_D^0 = N_D - N_D^+$$

$$N_A^0 = N_A - N_A^-$$

(6.12.17)

we get

$$\frac{n^2}{n_D - n} = \rho_C e^{-E_{CD}/kT}$$

(6.12.18)

$$\frac{p^2}{n_A - p} = \rho_V e^{-E_{AV}/kT}$$

(6.12.19)

where $n_D = N_D/V$ and $n_A = N_A/V$, respectively. These equations show how an increase in the impurity concentration increases the concentration of charge carriers. Note that (6.12.13) is still valid. This means that an increase in concentration of donor atoms increases the electron concentration because of (6.12.18), but it decreases the hole concentration because of (6.12.13). Likewise, an increase in acceptor impurity concentration increases the concentration of holes but decreases the electron concentration. It follows that increasing the donor concentration suppresses the ionization of acceptors and vice versa.

6.13 Relation between Fermi level and impurity concentration

The concentrations of electrons and holes are given by equation (6.10.11) and (6.10.12). Since n and p can vary greatly with different impurity content, it is clear that the Fermi level is a sensitive function of the impurity concentrations and ionization energies. To get explicit values of concentrations in terms of the parameters of the system, it is necessary to know how the Fermi level depends on impurity concentrations.

In metals, where all electrons are treated by particle statistics, the Fermi level is easily found from the condition that the number of electrons is a constant and is therefore determined by the electron density. In semiconductors, the situation is somewhat more complex. The number of electrons in the bands depends on the values of the various energy levels and on the concentration of donor and acceptor impurities. For an intrinsic semiconductor the Fermi level is determined by the requirement that the numbers of electrons and holes are equal, thus preserving electrical neutrality, and found to be about midway between the top of the valence band and the bottom of the conduction band. In extrinsic semiconductors, the Fermi level is also determined by the requirement of electrical neutrality. The number of electrons plus the number of ionized acceptors must equal the number of holes plus the number of ionized donors. That is,

$$n + n_A^- = p + n_D^+$$

(6.13.1)

where $n_A^- = N_A^-/V$ and $n_D^+ = N_D^+/V$. Putting (6.10.11), (6.10.12), (6.11.17), and (6.11.22) into (6.13.1) gives

$$2\rho_C e^{(\mu - E_C)/kT} + \frac{n_A}{2e^{(E_A - \mu)/kT} + 1} = 2\rho_V e^{(E_V - \mu)/kT} + \frac{n_D}{2e^{(\mu - E_D)/kT} + 1}$$

(6.13.2)

where we have used the definitions of valence band and conduction band density of states given by equations (6.10.13) and (6.10.14).

Equation (6.13.2) determines μ. If the impurity concentrations and energy levels are known, μ can be found by numerical or graphical procedures. Simplified solutions can be found for some important special cases.

Case 1. Weak ionization

If $(E_A - \mu)$ and $(\mu - E_D)$ are large enough, or the temperature is low enough, unity in the denominators of (6.13.2) can be neglected relative to the exponentials. Then some straightforward algebra gives

$$\mu = \frac{E_D + E_A}{2} + \frac{kT}{2}\ln\left[\frac{2\rho_v e^{(E_V - E_D)/kT} + n_D/2}{2\rho_c e^{(E_A - E_C)/kT} + n_A/2}\right] \tag{6.13.3}$$

The energies $(E_V - E_D)$ and $(E_A - E_C)$ differ from the gap energy by the ionization energies for donor and acceptor impurities, respectively and section 6.9 showed that these are of the order of 0.05–0.1 eV. We can therefore get a rough estimate of the exponential terms in (6.13.3) by assuming that the effective mass is close to the actual electron mass and replacing the energies in the exponentials by the gap energy. Doing this, and assuming a gap energy of 1 eV shows that, unless the impurity content is extremely small, the impurity concentration completely dominates the exponential term at room temperature and we can write (6.13.3) as

$$\mu = \frac{E_D + E_A}{2} + \frac{kT}{2}\ln\frac{n_D}{n_A} \tag{6.13.4}$$

For the range of impurities encountered in practice, the position of the Fermi level therefore depends on the relative concentration of donors and acceptors. Increasing the donor concentration pushes the Fermi level closer to the conduction band while increasing the acceptor concentration pushes it towards the valence band.

Case 2. Donor impurities only

If only donor impurities are present, then $n_A^- = 0$ in (6.13.1). Furthermore, the holes in the valence band arise only from thermal excitation from the valence band to the conduction band and, provided the gap width is not too small, their number will be much smaller than the number of electrons in the conduction band. This is readily seen from the mass action law (6.12.13). The hole concentration can then be neglected in (6.13.1) and the charge neutrality condition becomes

$$n = n_D^+ \tag{6.13.5}$$

Substituting (6.10.11) and (6.11.22) in this equality gives

$$2\rho_c e^{(\mu - E_C)/kT} = \frac{n_D}{2e^{(\mu - E_D)/kT} + 1} \tag{6.13.6}$$

This is readily transformed into a quadratic equation for $e^{\mu/kT}$ whose solution is

$$e^{\mu/kT} = \frac{1}{4} e^{E_D/kT} \left[-1 + \sqrt{1 + \frac{4n_D}{\rho_C} e^{(E_C-E_D)/kT}} \right] \qquad (6.13.7)$$

The positive root is chosen because the left-hand side of (6.13.7) is greater than zero.

From our earlier calculation of ρ_C, it follows that

$$\frac{4n_D}{\rho_C} e^{(E_C-E_D)/kT} \ll 1$$

for most practical cases. Therefore, the square root can be expanded, retaining only terms to the first order, to get

$$e^{\mu/kT} = \frac{n_D}{2\rho_C} e^{E_C/kT} \qquad (6.13.8)$$

or

$$\mu = E_C + kT \ln \frac{n_D}{2\rho_C} \qquad (6.13.9)$$

For n-type semiconductors, therefore, the Fermi level is close to the bottom of the conduction band, provided the temperature is not extremely high.

Case 3. Acceptor impurities only

If only acceptors are present, $n_D^+ = 0$ and $n \ll p$. Then (6.13.1) becomes

$$p = n_A^- \qquad (6.13.10)$$

and from (6.10.12) and (6.11.17) we get

$$2\rho_V e^{(E_V-\mu)/kT} = \frac{n_D}{2e^{(E_A-\mu)/kT} + 1} \qquad (6.13.11)$$

Proceeding as we did for the n-type semiconductor give

$$\mu = E_V - kT \ln \frac{n_A}{2\rho_V} \qquad (6.13.12)$$

and for p-type conductors, we see that the Fermi level is near the top of the valence band.

Exercises

6.1 For a two-dimensional Fermi gas, show that the average energy $\bar{\varepsilon}$ of a particle at absolute zero is related to the chemical potential μ_0 at absolute zero by $\bar{\varepsilon} = \mu_0/2$.

6.2 What is the percentage change of the Fermi level in copper when the temperature changes from 0 to 500K?

6.3 Over what energy range does the Fermi function differ from either zero or unity by 1% at 300K? What fraction of the Fermi energy is this range? Do the same calculations for $T = 1000$K.

6.4 For sodium, calculate the ratio of the electronic to the vibrational specific heat at temperatures of 0.1, 1, and 10K.

6.5 A white dwarf star is so dense that all electrons are stripped from nuclei and can be treated by free electron theory. A typical white dwarf has an electron density of 10^{30} electrons/cm^3. Calculate the temperature at which the zero temperature approximation for the Fermi level fails by 1% and compare this to the internal temperature of 10 million degrees at the center of the star. What is the velocity of an electron at the Fermi level in the white dwarf and what is the ratio of the Fermi velocity to the velocity of light?

6.6 ^{79}Au has a nuclear radius of 5×10^{-13} cm. Assuming the nucleons can be treated as an ideal Fermi gas, compute the Fermi energy. At what temperature is the thermal energy equal to the Fermi energy?

6.7 From the linearized Thomas–Fermi theory for a point charge in jellium, compute the total amount of excess charge that is beyond the screening distance λ if the point charge has the same magnitude as the charge on an electron.

6.8 For silver, compute the Thomas–Fermi screening distance and the ratio of this distance to the nearest neighbor distance, assuming one free electron per atom.

6.9 Show that, in Thomas–Fermi theory, the screening distance λ for the potential of a point charge q in jellium is $\lambda = K\sqrt{r_s}$, where r_S is the Wigner–Seitz radius. What is the numerical value of the proportionality constant K?

6.10 Given that pure germanium has a band gap of 0.67 eV and assuming that the effective masses of electrons and holes are each equal to the electron mass, compute the electron and hole concentrations at 300K.

6.11 Germanium has a band gap of 0.67 eV and a Wigner–Seitz radius of 1.67 angstroms. Assume it contains 10^{-4} atomic percent of a donor impurity whose energy level is 0.01 eV below the conduction band energy.

  A. What percentage of donor impurities are ionized at 300K?
  B. What is the ratio of the electron concentration of the material with 10^{-4} atomic percent impurity to that of pure germanium at 300K?

7

Statistical-Kinetic Theory of Electron Transport

7.1 Free electrons in external fields and temperature gradients

The free electrons in metals and semiconductors carry charge and are responsible for their electrical conductivity and for some important magnetic effects. They also carry energy and therefore contribute to thermal conductivity. In metals, their concentration is so high that electrons are the major contributors to thermal conductivity. It is therefore important to study the response of free electrons to external electric and magnetic fields and to temperature gradients.

When a system of free electrons is at equilibrium, the number of electrons in any volume element is statistically constant and there is no net flow. This means that macroscopic measurements will find the electron concentration to be independent of time. Of course, the electrons are constantly moving and there will be microscopic fluctuations in the number of electrons in a volume element, but on the average, just as many will be entering as leaving the volume element per second. If an electric or magnetic field acts on the system, however, the electrons will accelerate in a direction determined by the field. A drift motion will thereby be superimposed on the random movements of the electrons, giving rise to a directional net flow.

Consider an electron with velocity **v** in a state labeled by the wave number vector **k**. Its energy and momentum are given by

$$\varepsilon_k = \frac{mv^2}{2} = \frac{\hbar^2 k^2}{2m} \tag{7.1.1}$$

$$\mathbf{p} = m\mathbf{v} = \hbar\mathbf{k} \tag{7.1.2}$$

At a time $t = 0$, switch on an electric field **E**. The field exerts a force $-e\mathbf{E}$ on the electron, and because the force is the time rate of change of the momentum (Newton's second law of motion), we have

$$\frac{d\mathbf{p}}{dt} = \hbar\frac{d\mathbf{k}}{dt} = -e\mathbf{E} \tag{7.1.3}$$

For a constant field, this equation is easily integrated to give **k** as a function of time. Then, if the initial time is taken to be zero, (7.1.3) gives

$$\mathbf{k} = \mathbf{k}_0 - \frac{e\mathbf{E}}{\hbar}t \qquad (7.1.4)$$

where t is the time after switching and $\mathbf{k}_0$ is the wave number vector before the field was switched on. Equation (7.1.4) is valid for all the free electrons, and it is clear that the effect of the field is to shift the entire Fermi distribution opposite to the field direction by an amount that is the same for all electrons and proportional to the time. Because $\mathbf{k} = m\mathbf{v}/\hbar$ [see equation (7.1.2)], we can rewrite (7.1.4) as

$$\mathbf{v} = \mathbf{v}_0 - \frac{e\mathbf{E}}{m}t \qquad (7.1.5)$$

which states that the field imposes a constantly increasing velocity, opposite to the direction of the field, on the initial velocity $\mathbf{v}_0$ of the electron.

An analogous situation exists if the electron is acted on by a constant magnetic field $\mathbf{H}$. In this case, the force, in c.g.s./e.s.u. units, acting on an electron of velocity $\mathbf{v}$ is given by electrodynamics as $-e/c\,\mathbf{v} \times \mathbf{H}$, where c is the velocity of light. From Newton's second law, the acceleration of the electron in the magnetic field is

$$\frac{d\mathbf{v}}{dt} = -\frac{e}{mc}\mathbf{v} \times \mathbf{H} \qquad (7.1.6)$$

If the average velocity at a time t after turning on the magnetic field is defined by

$$\mathbf{v}_{av} = \frac{1}{t}\int_0^t \mathbf{v}\,dt \qquad (7.1.7)$$

then (7.1.6) integrates to

$$\mathbf{v} = \mathbf{v}_0 - \frac{e}{mc}\mathbf{v}_{av} \times \mathbf{H}t \qquad (7.1.8)$$

Therefore, the effect of the magnetic field is to superimpose a velocity on the electron that continually increases with time and is perpendicular to both the magnetic field and the average electron velocity. Since the wave number vector is proportional to the velocity, equation (7.1.8) can also be written as

$$\mathbf{k} = \mathbf{k}_0 - \frac{e}{mc}\mathbf{k}_{av} \times \mathbf{H}t \qquad (7.1.9)$$

The average wave number vector, $\mathbf{k}_{av}$, is not the same for all electrons since it depends on the initial value $\mathbf{k}_0$, so unlike the electric field, a magnetic field does not affect all electrons equally and the overall effect is more complex than a simple shift of all electrons in the Fermi distribution. This is a result of the fact that magnetic forces are velocity dependent.

Equations (7.1.5) and (7.1.8) show that the electrons are constantly accelerated, and the velocities would reach very high values if some mechanism did not exist that opposes the field effects. In real metals and semiconductors, a variety of mechanisms operate that continually undo the accelerating effects of the external fields.

According to quantum theory, there is no resistance whatever to the motion of an electron in a perfectly periodic potential. Real crystals, however, are not perfect. Anything that disturbs the periodicity may scatter electrons and contribute to the resistance they meet in their flight. Electrons are scattered by atomic vibrations, impurity atoms, vacant lattice sites, interstitials, dislocations, grain boundaries, precipitates, surfaces, or anything else that upsets the ideal crystal structure. External fields cannot accelerate electrons in solids indefinitely. The electrons eventually interact with some imperfection, transfer energy and momentum to it, and move on again in some other direction. Each such interaction can be thought of as a collision that erases the electron's memory of its previous response to the external field. After the collision, the field has to start afresh to exert its influence on the electron. The result is a balance between the field, which tries to bring the electrons away from equilibrium, and the collisions, which try to restore equilibrium.

Let $\bar{\tau}$ be the average time between collisions of an electron with crystal imperfections. This is the average time between scattering events, so the average change in wave number vectors caused by the external field is obtained by replacing t in equations (7.1.4) and (7.1.9) by $\bar{\tau}$. That is, if we define $d\mathbf{k} = \mathbf{k} - \mathbf{k}_0$, then

$$d\mathbf{k} = -\frac{e\bar{\tau}}{\hbar}\mathbf{E} \quad \text{(electric field)} \tag{7.1.10}$$

$$d\mathbf{k} = -\frac{e\bar{\tau}}{mc}\mathbf{k} \times \mathbf{H} \quad \text{(magnetic field)} \tag{7.1.11}$$

In getting (7.1.11), it is assumed that the average time between collisions is so small that the average wave number vector has its instantaneous value.

The shift in **k**-vectors means that there will be a shift in the Fermi distribution. For all practically realizable fields, however, this shift will be small because the amount of energy that the electrons can absorb from the field is small compared to their original kinetic energy. This means that if $f(\mathbf{k})$ is the probability that an electron is in a state **k**, then we can write

$$f(\mathbf{k}) = f_0(\mathbf{k}) + \Delta f(\mathbf{k}) \tag{7.1.12}$$

where $f_0(\mathbf{k})$ is the equilibrium Fermi function and $\Delta f \ll f_0$.

Now consider a free electron gas in a temperature gradient. Because temperature is a measure of kinetic energy, the average velocity in hot regions will be greater than in cold regions. These electrons carry energy with them, so there is a corresponding heat flow.

The shift in the Fermi distribution resulting from a temperature gradient has two origins. First, the distribution function depends on temperature explicitly, so if a temperature gradient exists, the distribution will vary with position. Second, the distribution depends on the Fermi level and, to the extent that this varies with temperature, it will vary with position. Of these two effects, the first is the more important one in metals because the Fermi level varies only slightly with temperature. The second effect is more important in semiconductors than in metals because the Fermi level can have a greater variation with temperature.

Note that electrons carry both kinetic energy and charge. Therefore, the heat flow caused by a temperature gradient is accompanied by electrical effects. Conversely, the current caused by electric and magnetic fields is accompanied by thermal effects. The theory to be developed here describes these effects as well as the ordinary electrical and thermal conductivity.

7.2 The statistical-kinetic method

The theory of electron transport in metals and semiconductors can be worked out by kinetic methods in which particle and energy fluxes are expressed in terms of particle velocities and the particle distribution function based on Fermi–Dirac statistics. The distribution function is then related to external fields and temperature gradients. The methods of equilibrium statistical mechanics are not adequate for this task and they must be supplemented in two ways. First, because the system is not in equilibrium, the distribution function for the electrons must be modified. For fields and gradients of experimental interest, the perturbed distributions are not too far from the equilibrium results given by Fermi functions, and it is possible to find reasonably accurate approximations for the gradients of the distribution.

Second, a more fundamental difficulty is that equilibrium statistical mechanics has nothing to say about the details of transitions of electrons from one state to another. But this is just what is needed for a theory of transport. When an electron in a particular energy state takes part in a collision, it is scattered into another state, and the rate at which this happens is central to transport theory. The theory must therefore be supplemented by introducing transition probabilities that give the rate at which electrons move from one state to another. This enables equations for the flux of electrons to be derived.

The flux equations can be obtained by a kinetic argument as follows. Choose a plane in the system across which the flux is to be calculated, and consider a small surface area Δs in this plane. In a small time interval dt, an electron of velocity $\mathbf{v}$ will reach the flux plane if it is anywhere within a distance $\Delta x = \mathbf{v} \cdot \mathbf{i}_n dt$, where $\mathbf{i}_n$ is a unit vector normal to the flux plane and Δx is measured in the $\mathbf{i}_n$ direction. Therefore, all the electrons in a volume element $\mathbf{v} \cdot \mathbf{i}_n dt \Delta s$ will cross the area element Δs in time dt. If $N(\mathbf{r}, \mathbf{v}, t)d\mathbf{r}d\mathbf{v}$ is the number of electrons at time t in a volume element $d\mathbf{r}$ with velocities in the range $\mathbf{v}$ to $\mathbf{v} + d\mathbf{v}$, then the number that cross Δs in time dt with velocity $\mathbf{v}$ is

$$N(\mathbf{r}, \mathbf{v}, t)\mathbf{v} \cdot \mathbf{i}_n dt \Delta s d\mathbf{v} \qquad (7.2.1)$$

The total number of electrons crossing Δs is obtained by integrating over all velocities.

$$\Delta s dt \int N(\mathbf{r}, \mathbf{v}, t)\mathbf{v} \cdot \mathbf{i}_n d\mathbf{v} \qquad (7.2.2)$$

The flux of electrons in the $\mathbf{i}_n$ direction is the number of electrons crossing a unit area per unit time, which is just (7.2.2) divided by $\Delta s dt$. Doing this to get the flux, and multiplying by the charge per electron $-e$, we get the component of the flux of charge, or electric current, that is perpendicular to our flux plane, as

$$I_n = -e\int \mathbf{v} \cdot \mathbf{i}_n N(\mathbf{r}, \mathbf{v}, t)d\mathbf{v} \qquad (7.2.3)$$

The component I_n is related to the flux vector $\mathbf{I}$ by $I_n = \mathbf{I} \cdot \mathbf{i}_n$, and therefore (7.2.3) becomes

$$\mathbf{I} = -e\int \mathbf{v} N(\mathbf{r}, \mathbf{v}, t)d\mathbf{v} \qquad (7.2.4)$$

The flux of energy is obtained in a similar way. Equation (7.2.1) is multiplied by the kinetic energy $mv^2/2$, integrated over the velocity $\mathbf{v}$ and divided by $\Delta s dt$ to get the energy flux $\mathbf{J}$ as

STATISTICAL-KINETIC THEORY OF ELECTRON TRANSPORT 181

$$\mathbf{J} = \frac{m}{2} \int v^2 \mathbf{v} N(\mathbf{r}, \mathbf{v}, t) d\mathbf{v} \qquad (7.2.5)$$

For an equilibrium system, $N(\mathbf{r}, \mathbf{v}, t)$ reduces to the equilibrium velocity distribution given by Fermi particle statistics. But our system is not at equilibrium, and for a nonequilibrium, system, $N(\mathbf{r}, \mathbf{v}, t)$ is the distribution function resulting from a balance between the effects of the fields and the collisions. To describe the way that collisions restore equilibrium, a conditional transition probability function $\Lambda(\mathbf{v}, \mathbf{v}')$ is defined such that, if an electron has an initial velocity $\mathbf{v}$, then the probability that its velocity changes to a value in the range $\mathbf{v}'$ to $\mathbf{v}' + d\mathbf{v}'$ in a time dt is $\Lambda(\mathbf{v}, \mathbf{v}')d\mathbf{v}'dt$. The physically reasonable assumption is made that $\Lambda(\mathbf{v}, \mathbf{v}')$ depends only on the initial and final states and is independent of the external fields, the time, or the electron's position. The transition probabilities are called *conditional* because they give the probability of going to a final state $\mathbf{v}'$, given the condition that they are in a specific initial state $\mathbf{v}$. (Note that these transition probabilities are special cases of the transition probabilities defined in section 2.13 to describe the time evolution of the entropy.)

The rate at which the distribution function changes because of collisions can now be expressed in terms of the conditional transition probabilities. We do this by choosing a volume element $d\mathbf{r}$ and considering electrons with a velocity $\mathbf{v}$ in the range $d\mathbf{v}$. Then we count the number of these electrons that leave the volume element and subtract this from the number that enter the volume element. This leads us to the rate of change of the distribution function.

The decrease, during time dt, in the number of electrons in volume element $d\mathbf{r}$ with velocity in the range $d\mathbf{v}$ is the number of such electrons, given by $N(\mathbf{r}, \mathbf{v}, t)d\mathbf{r}d\mathbf{v}$, times the probability that one of them will jump to some other state with velocity $\mathbf{v}'$ in the range $d\mathbf{v}'$, given by $\Lambda(\mathbf{v}, \mathbf{v}')d\mathbf{v}'dt$, integrated over all possible states with final velocity $\mathbf{v}'$. This is

$$d\mathbf{r}d\mathbf{v}dt \int N(\mathbf{r}, \mathbf{v}, t)\Lambda(\mathbf{v}, \mathbf{v}')d\mathbf{v}' \qquad (7.2.6)$$

Similarly, the increase in the number of electrons in the volume element $d\mathbf{r}$ with velocity in the range $d\mathbf{v}$ is the number of electrons with velocity $\mathbf{v}'$ in the range $d\mathbf{v}'$ times the probability that such an electron will go into a state with velocity $\mathbf{v}$ (in the range $d\mathbf{v}$) and integrating over all possible $\mathbf{v}'$. This is

$$d\mathbf{r}d\mathbf{v}dt \int N(\mathbf{r}, \mathbf{v}', t)\Lambda(\mathbf{v}', \mathbf{v})d\mathbf{v}' \qquad (7.2.7)$$

If (7.2.6) is subtracted from (7.2.7) and the result divided by $d\mathbf{r}d\mathbf{v}dt$, we get the rate of change with time of $N(\mathbf{r}, \mathbf{v}, t)$.

$$\left(\frac{\partial N(\mathbf{r}, \mathbf{v}, t)}{\partial t}\right)_c = \int N(\mathbf{r}, \mathbf{v}', t)\Lambda(\mathbf{v}', \mathbf{v})d\mathbf{v}' - N(\mathbf{r}, \mathbf{v}, t)\int \Lambda(\mathbf{v}, \mathbf{v}')d\mathbf{v}' \qquad (7.2.8)$$

The subscript c is put on the derivative to emphasize that this is the rate of change resulting from collisions. Equation (7.2.8) is useful for the development of transport theory. It contains the physical collision mechanisms in $\Lambda(\mathbf{v}, \mathbf{v}')$.

7.3 The Boltzmann transport equation

If our system were left to itself, the collision derivative would eventually go to zero and the particle distribution function would approach its equilibrium value. However, the actions of external fields and temperature gradients impart a drift velocity to the random motion of the particles that keeps them from reaching a state of equilibrium. The total rate of change of the distribution function is therefore the sum of two terms: the collision derivative discussed in section 7.2 and a drift derivative arising from fields and temperature gradients. That is,

$$\frac{\partial N(\mathbf{r}, \mathbf{v}, t)}{\partial t} = \left(\frac{\partial N}{\partial t}\right)_c + \left(\frac{\partial N}{\partial t}\right)_d \qquad (7.3.1)$$

The first term on the right is the collision derivative discussed above, while the second term is the drift derivative.

The drift derivative can be related to the external influences by following the motion of a given group of electrons. At time t, the number of electrons with positions in $d\mathbf{r}$ and velocities in $d\mathbf{v}$ is

$$N(\mathbf{r}, \mathbf{v}, t) d\mathbf{r} d\mathbf{v} \qquad (7.3.2)$$

At a later time $t + dt$, the positions and velocities of all electrons in this group have changed to

$$\mathbf{r} + \Delta \mathbf{r} = \mathbf{r} + \mathbf{v}\Delta t$$
$$\mathbf{v} + \Delta \mathbf{v} = \mathbf{v} + \mathbf{a}\Delta t \qquad (7.3.3)$$

where $\mathbf{a} = d\mathbf{v}/dt$ is the particle acceleration. But the number of electrons in the group has not changed, so

$$N(\mathbf{r} + \mathbf{v}\Delta t, \mathbf{v} + \mathbf{a}\Delta t, t + \Delta t) = N(\mathbf{r}, \mathbf{v}, t) \qquad (7.3.4)$$

Expand the left-hand side in a Taylor series and, since Δt can be arbitrarily small, retain only the first-order terms. The result is

$$\Delta t \left(\mathbf{v} \cdot \frac{\partial N}{\partial \mathbf{r}} + \mathbf{a} \cdot \frac{\partial N}{\partial \mathbf{v}} + \frac{\partial N}{\partial t} \right) = 0 \qquad (7.3.5)$$

The vector derivatives are defined by

$$\frac{\partial N}{\partial \mathbf{r}} = \frac{\partial N}{\partial x}\mathbf{i}_1 + \frac{\partial N}{\partial y}\mathbf{i}_2 + \frac{\partial N}{\partial z}\mathbf{i}_3$$

$$\frac{\partial N}{\partial \mathbf{v}} = \frac{\partial N}{\partial v_x}\mathbf{i}_1 + \frac{\partial N}{\partial v_y}\mathbf{i}_2 + \frac{\partial N}{\partial v_z}\mathbf{i}_3 \qquad (7.3.6)$$

$\mathbf{i}_1, \mathbf{i}_2, \mathbf{i}_3$ being unit vectors of a Cartesian coordinate system.

The time derivative in (7.3.5) is just the drift derivative, so

$$\left(\frac{\partial N}{\partial t}\right)_d = -\left(\mathbf{v} \cdot \frac{\partial N}{\partial \mathbf{r}} + \mathbf{a} \cdot \frac{\partial N}{\partial \mathbf{v}}\right) \qquad (7.3.7)$$

This equation connects the transport properties to the external fields through the acceleration $\mathbf{a}$, and to the temperature gradient through the derivative $\partial N/\partial \mathbf{r}$.

Now the time derivative of the distribution function can be written explicitly in terms of collision events and external fields by substituting (7.2.8) for the collision derivative and (7.3.7) for the drift derivative into (7.3.1):

$$\frac{\partial N(\mathbf{r}, \mathbf{v}, t)}{\partial t} = -\left(\mathbf{v} \cdot \frac{\partial N}{\partial \mathbf{r}} + \mathbf{a} \cdot \frac{\partial N}{\partial \mathbf{v}}\right) + \int N(\mathbf{r}, \mathbf{v}', t)\Lambda(\mathbf{v}', \mathbf{v})d\mathbf{v}' - N(\mathbf{r}, \mathbf{v}, t)\int \Lambda(\mathbf{v}, \mathbf{v}')d\mathbf{v}' \qquad (7.3.8)$$

This is the Boltzmann transport equation.

If the derivative on the left of (7.3.8) is zero, the system is said to be at steady state. In this case, all gradients and fields are constant in time because the time dependence of the distribution function is the result of the time dependence of the fields and gradients. The system then has a distribution function that is independent of time, although it may have a form different than that at equilibrium, and (7.3.8) reduces to

$$\mathbf{v} \cdot \frac{\partial N}{\partial \mathbf{r}} + \mathbf{a} \cdot \frac{\partial N}{\partial \mathbf{v}} = \int N(\mathbf{r}, \mathbf{v}', t)\Lambda(\mathbf{v}', \mathbf{v})d\mathbf{v}' - N(\mathbf{r}, \mathbf{v}, t)\int \Lambda(\mathbf{v}, \mathbf{v}')d\mathbf{v}' \qquad (7.3.9)$$

This is the steady state Boltzmann transport equation. Note that steady state only means that the overall dependence of the distribution on time vanishes and that this is the result of an equal balance between the drift and the collision terms. Between collisions the electrons are accelerated, so the distribution function is changing. During a collision the past history of collisions is erased, so the distribution changes again. The continuous description in terms of derivatives is an approximation to the rapidly varying and sometimes discontinuous changes taking place as the result of collisions.

For constant electric and magnetic fields $\mathbf{E}$ and $\mathbf{H}$, the acceleration of an electron with velocity $\mathbf{v}$ is given in terms of the Lorentz force (in c.g.s./e.s.u. units) by

$$m\mathbf{a} = -e\left(\mathbf{E} + \frac{\mathbf{v} \times \mathbf{H}}{c}\right) \qquad (7.3.10)$$

Also, since $\mathbf{E}$ and $\mathbf{H}$ do not vary with position, the spatial derivative of the distribution function must arise from the temperature gradient and therefore

$$\frac{\partial N}{\partial \mathbf{r}} = \frac{\partial N}{\partial T}\frac{\partial T}{\partial \mathbf{r}} = \frac{\partial N}{\partial T}\nabla T \qquad (7.3.11)$$

Equations (7.3.10) and (7.3.11) give us the left-hand side of (7.3.9) in terms of the fields and temperature gradient. That is,

$$\mathbf{v} \cdot \frac{\partial N}{\partial \mathbf{r}} + \mathbf{a} \cdot \frac{\partial N}{\partial \mathbf{v}} = \nabla T \cdot \mathbf{v}\frac{\partial N}{\partial T} - \frac{e}{m}\mathbf{E} \cdot \frac{\partial N}{\partial \mathbf{v}} - \frac{e}{mc}(\mathbf{v} \times \mathbf{H}) \cdot \frac{\partial N}{\partial \mathbf{v}} \qquad (7.3.12)$$

The right-hand side of (7.3.9), which is just the collision derivative, is more difficult to handle because we do not have a general expression for the transition probabilities. A workable procedure is to introduce a quantity $\tau(\mathbf{v})$ that is defined by

$$\left(\frac{\partial N}{\partial t}\right)_c = -\frac{N(\mathbf{r}, \mathbf{v}, t) - N^0(\mathbf{v})}{\tau(\mathbf{v})} \qquad (7.3.13)$$

where $N^0(\mathbf{v})$ is the equilibrium distribution function in the absence of fields or gradients.

$\tau(\mathbf{v})$ is assumed to be a function only of the electron's velocity. It has the dimensions of time and is a measure of the rate at which the particle distribution function regresses to its equilibrium value. To see this, assume that we suddenly switch off all fields and gradients (so that the drift derivative is zero) at a time we call $t = 0$. Then integration of (7.3.13) gives

$$N(t) - N^0 = \Delta N e^{-t/\tau(\mathbf{v})} \qquad (7.3.14)$$

where we have written ΔN for the initial deviation from equilibrium $[N(0) - N^0]$. The interpretation of $\tau(\mathbf{v})$ is now obvious. It is the characteristic relaxation time for electrons of velocity $\mathbf{v}$ governing the approach to equilibrium of a perturbed distribution. Its value depends on the collision mechanism, so τ should be different for different mechanisms.

If several collision mechanisms are operating in the system, then each mechanism contributes to the collision derivative and (7.3.13) should be written as

$$\left(\frac{\partial N}{\partial t}\right)_c = -\frac{N - N^0(\mathbf{v})}{\tau_1(\mathbf{v})} - \frac{N - N^0(\mathbf{v})}{\tau_2(\mathbf{v})} - \ldots \qquad (7.3.15)$$

where there are as many terms on the right-hand side, and as many relaxation times, as there are collision mechanisms. Equation (7.3.15) can be put into the simpler form of (7.3.13) by defining a combined relaxation time by

$$\frac{1}{\tau(\mathbf{v})} = \frac{1}{\tau_1(\mathbf{v})} + \frac{1}{\tau_2(\mathbf{v})} + \ldots \qquad (7.3.16)$$

This definition allows us to derive general equations of transport without worrying about the collision mechanisms beforehand. However, it should be kept in mind that an overall relaxation time may be the result of a number of different processes working simultaneously. Numerical calculation of transport coefficients from theory would require that the theory of each of the mechanisms be worked out. The relaxation times are usually taken to be functions only of the magnitude of the velocity, an assumption that is reasonable for the scattering mechanisms encountered in solids.

Using the relaxation time concept, we now equate the right-hand side of (7.3.12), which is just the negative of the drift derivative, to (7.3.13) and solve for $N(\mathbf{r}, \mathbf{v}, t)$ to get

$$N = N^0 - \tau \frac{\partial N}{\partial T} \mathbf{v} \cdot \nabla T + \frac{e}{m} \tau \mathbf{E} \cdot \frac{\partial N}{\partial \mathbf{v}} + \frac{e}{mc} \tau (\mathbf{v} \times \mathbf{H}) \cdot \frac{\partial N}{\partial \mathbf{v}} \qquad (7.3.17)$$

The arguments in the functions N and τ have been dropped for simplicity.

Equation (7.3.17) is a formal solution to the problem of determining $N(\mathbf{r}, \mathbf{v}, t)$. If $\tau(\mathbf{v})$ is computed, the solution becomes explicit, but for the time being we will go as far as we can without a detailed theory for the relaxation times.

The last three terms in (7.3.17) describe the departure of the electron distribution from equilibrium. This description is facilitated by a formal device in which a vector $\mathbf{h}$ is defined by

$$N = N^0 + \mathbf{v} \cdot \mathbf{h} \qquad (7.3.18)$$

or, from (7.3.17),

$$\mathbf{v}\cdot\mathbf{h} = -\tau \frac{\partial N}{\partial T}\mathbf{v}\cdot\nabla T + \frac{e}{m}\tau\mathbf{E}\cdot\frac{\partial N}{\partial \mathbf{v}} + \frac{e}{mc}\tau(\mathbf{v}\times\mathbf{H})\cdot\frac{\partial N}{\partial \mathbf{v}} \quad (7.3.19)$$

Now we express the derivatives on the right-hand side of (7.3.19) in terms of $\mathbf{h}$ by using (7.3.18):

$$\mathbf{v}\cdot\mathbf{h} = -\tau\mathbf{v}\cdot\nabla T\left[\frac{\partial N^0}{\partial T} + \frac{\partial(\mathbf{v}\cdot\mathbf{h})}{\partial T}\right] + \frac{e}{m}\tau\mathbf{E}\cdot\left[\frac{\partial N^0}{\partial \mathbf{v}} + \frac{\partial(\mathbf{v}\cdot\mathbf{h})}{\partial \mathbf{v}}\right]$$
$$+ \frac{e\tau}{mc}(\mathbf{v}\times\mathbf{H})\cdot\left[\frac{\partial N^0}{\partial \mathbf{v}} + \frac{\partial(\mathbf{v}\cdot\mathbf{h})}{\partial \mathbf{v}}\right] \quad (7.3.20)$$

Because we are dealing with systems in which the deviation from equilibrium is small, we have $\partial(\mathbf{v}\cdot\mathbf{h})/\partial T \ll \partial N^0/\partial T$ and $\partial(\mathbf{v}\cdot\mathbf{h})/\partial \mathbf{v} \ll \partial N^0/\partial \mathbf{v}$ and, to a first approximation, it would seem that the departures from equilibrium can always be neglected in evaluating derivatives of the distribution function. This is true for the first two terms in (7.3.20). In the third term, which is the effect of magnetic fields, however, retention of the derivative $\partial(\mathbf{v}\cdot\mathbf{h})/\partial \mathbf{v}$ is essential. This can be seen by working out the triple product in the last term:

$$(\mathbf{v}\times\mathbf{H})\cdot\left[\frac{\partial N^0}{\partial \mathbf{v}} + \frac{\partial(\mathbf{v}\cdot\mathbf{h})}{\partial \mathbf{v}}\right] = (\mathbf{v}\times\mathbf{H})\cdot\left[m\mathbf{v}\frac{\partial N^0}{\partial \varepsilon} + \frac{\partial(\mathbf{v}\cdot\mathbf{h})}{\partial \mathbf{v}}\right] \quad (7.3.21)$$

In the first derivative on the right, we have used the fact that the kinetic energy of an electron is $\varepsilon = m(v_x^2 + v_y^2 + v_z^2)/2$ to convert the velocity gradient to an energy derivative.

From vector algebra, $\mathbf{v}\times\mathbf{H}\cdot\mathbf{v} = 0$, so the first term on the right of (7.3.21) vanishes. Now we can write (7.3.20) to a first-order approximation by neglecting the derivatives containing $\mathbf{v}\cdot\mathbf{h}$ in the first two brackets on the right, but using the last term of (7.3.21) for the third bracket. The result is

$$\mathbf{v}\cdot\mathbf{h} = -\tau\mathbf{v}\cdot\nabla T\frac{\partial N^0}{\partial T} + \frac{e\tau}{m}\mathbf{E}\cdot\frac{\partial N^0}{\partial \mathbf{v}} + \frac{e\tau}{mc}(\mathbf{v}\times\mathbf{H})\cdot\frac{\partial(\mathbf{v}\cdot\mathbf{h})}{\partial \mathbf{v}} \quad (7.3.22)$$

As long as no magnetic fields are present, derivatives of the distribution function can be treated as if the system were at equilibrium. In the presence of a magnetic field, however, this procedure would give a zero result for magnetic effects. In all magnetic terms, the departure from equilibrium must be explicitly included in evaluating derivatives of the distribution.

7.4 Formal flux equations

The flux equation for the electric current can now be obtained in terms of the fields and temperature gradient by substituting (7.3.17) into (7.2.4):

$$\mathbf{I} = -e\int \mathbf{v}\left(N^0 - \tau\frac{\partial N}{\partial T}\mathbf{v}\cdot\nabla T + \frac{e\tau}{m}\frac{\partial N}{\partial \mathbf{v}}\cdot\mathbf{E} + \frac{e\tau}{mc}\mathbf{v}\times\mathbf{H}\cdot\frac{\partial N}{\partial \mathbf{v}}\right)d\mathbf{v} \quad (7.4.1)$$

The function N^0 is just the equilibrium particle distribution in velocity for free electrons, given by particle statistics as

$$N^0 = \frac{2m^3}{h^3} \frac{1}{e^{m(v^2-v_F^2)/2kT}+1}$$

$$= \frac{2m^3}{h^3} \frac{1}{e^{(\epsilon-\mu)/2kT}+1} \qquad (7.4.2)$$

The first term in the integral in (7.4.1) refers to the equilibrium distribution, and we would expect that this does not contribute to an electron flow. This is actually the case because N^0 is symmetric in the velocity, so $\mathbf{v}N^0$ is antisymmetric, thereby making the integral of $\mathbf{v}N^0$ over the velocity zero. Therefore, (7.4.1) becomes

$$\mathbf{I} = e\int \tau \mathbf{v} \frac{\partial N}{\partial T}(\mathbf{v}\cdot\nabla T)d\mathbf{v} - \frac{e^2}{m}\int \tau \mathbf{v}\left(\frac{\partial N}{\partial \mathbf{v}}\cdot\mathbf{E}\right)d\mathbf{v} - \frac{e^2}{mc}\int \tau \mathbf{v}(\mathbf{v}\times\mathbf{H})\cdot\frac{\partial N}{\partial \mathbf{v}}d\mathbf{v} \qquad (7.4.3)$$

In the same way, the heat current vector (flux of kinetic energy) is obtained by inserting (7.3.17) into (7.2.5):

$$\mathbf{J} = -\frac{m}{2}\int \tau \mathbf{v} v^2 \frac{\partial N}{\partial T}(\mathbf{v}\cdot\nabla T)d\mathbf{v} + \frac{e}{2}\int \tau \mathbf{v} v^2 \left(\frac{\partial N}{\partial \mathbf{v}}\cdot\mathbf{E}\right)d\mathbf{v} + \frac{e}{2c}\int \tau \mathbf{v} v^2(\mathbf{v}\times\mathbf{H})\cdot\frac{\partial N}{\partial \mathbf{v}}d\mathbf{v}$$

$$(7.4.4)$$

The entire theory of steady state thermal and electrical conduction is contained in these two flux equations. Note that all the external influences contribute to both the electric and thermal fluxes. Thus, there is an electric current arising not only from the electric field, but also from the temperature gradient and the magnetic field. Similarly, the electric and magnetic fields, as well as the temperature gradient, contribute to the heat flow.

All that is needed to extract the transport coefficients from the flux equations is a computation of the various integrals. In computing these integrals, it will be assumed that the derivatives of the distribution function can be replaced by the values they would have for the equilibrium distribution function unless magnetic fields are present. In the latter case, it turns out that an approximation can be developed that relates the derivatives to those for the equilibrium distribution. Since we are dealing with small perturbations from equilibrium, these procedures are sufficiently accurate.

7.5 The electrical conductivity of metals

To calculate the electrical conductivity of a metal, consider a system for which there is no magnetic field and the temperature is uniform so there is no temperature gradient. Also, let the electric field have a nonzero component only in the x-direction. Equation (7.4.3) then reduces to

$$I_x = -\frac{e^2 E_x}{m}\int \tau v_x \frac{\partial N}{\partial v_x} d\mathbf{v} \qquad (7.5.1)$$

The components of the current flow I_y and I_z in the y- and z-directions vanish because the corresponding integrals vanish as a result of their being antisymmetric in the velocity components v_y and v_z, respectively. That is, the flux of electric charge is in the same direction as the electric field.

Since the electrical conductivity σ is defined by $I_x = \sigma E_x$, it is given by (7.5.1) as

$$\sigma = -\frac{e^2}{m}\int \tau v_x \frac{\partial N}{\partial v_x} d\mathbf{v} \tag{7.5.2}$$

The first step in evaluating this integral is to use the relation

$$\varepsilon = \frac{mv^2}{2} = \frac{m(v_x^2 + v_y^2 + v_z^2)}{2} \tag{7.5.3}$$

to get

$$\frac{\partial N}{\partial v_x} = \frac{\partial N}{\partial \varepsilon}\frac{\partial \varepsilon}{\partial v_x} = mv_x \frac{\partial N}{\partial \varepsilon} \tag{7.5.4}$$

Using this, (7.5.2) becomes

$$\sigma = -e^2 \int \tau v_x^2 \frac{\partial N}{\partial \varepsilon} d\mathbf{v} \tag{7.5.5}$$

If we assume that the relaxation time is a function only of the magnitude of the velocity and not of its direction, then the integral remains unchanged if v_x is replaced by v_y or v_z. Therefore, since $v^2 = v_x^2 + v_y^2 + v_z^2$, equation (7.5.5) can be written as

$$\sigma = -\frac{e^2}{3} \int \tau v^2 \frac{\partial N}{\partial \varepsilon} d\mathbf{v} \tag{7.5.6}$$

The integrand in (7.5.6) is spherically symmetric with respect to velocity, so $d\mathbf{v}$ can be expressed in spherical coordinate form as

$$d\mathbf{v} = 4\pi v^2 dv \tag{7.5.7}$$

or, using (7.5.3),

$$d\mathbf{v} = \frac{4\pi v}{m} d\varepsilon \tag{7.5.8}$$

Equation (7.5.6) therefore becomes

$$\sigma = -\frac{8\pi m^2 e^2}{3h^3} \int_0^\infty \tau v^3 \frac{\partial f}{\partial \varepsilon} d\varepsilon \tag{7.5.9}$$

where N in the derivative has been replaced by its equilibrium value $2m^3 f(\varepsilon)/h^3$, in keeping with our restriction to small deviations from equilibrium. $f(\varepsilon)$ is the Fermi function from particle statistics, so the conductivity is now expressed in terms of a Fermi integral. If only the first term in the expression for the Fermi integral is retained, we have

$$\int_0^\infty \tau v^3 \frac{\partial f}{\partial \varepsilon} d\varepsilon = -\tau(\mu)v(\mu)^3 \tag{7.5.10}$$

where $v(\mu)$ is the speed of an electron at the Fermi level. Since $\mu = mv(\mu)^2/2$, equation (7.5.10) can be written in terms of the Fermi energy as

$$\int_0^\infty \tau v^3 \frac{\partial f}{\partial \varepsilon} d\varepsilon = -\tau(\mu)\left(\frac{2\mu}{m}\right)^{3/2} \qquad (7.5.11)$$

Now put this into (7.5.9) and neglect the variation of μ with temperature so that the zero temperature relation between the electron density and the Fermi level for free electrons can be used. The result is

$$\sigma = \frac{ne^2\tau(\mu)}{m} \qquad (7.5.12)$$

The electrical conductivity is now reduced to a very simple form. Note that only the relaxation time of electrons with energy equal to the Fermi energy appears in this equation. This is a result of using the zero temperature approximation for a Fermi gas. However, the temperature dependence of the Fermi level is so weak that this is quite a good approximation. The most important temperature dependence arises from the scattering of the electrons by lattice vibrations. Approximate quantum mechanical analysis shows that for this mechanism the relaxation time is inversely proportional to temperature. Since the electron density n is insensitive to temperature, this means that the resistivity of metals increases linearly with temperature. The mass in (7.5.12) should be interpreted as the effective mass of electrons in the conduction band of the metal. This is also a very weak function of temperature.

For an electron density of $10^{22}/cm^3$ and a specific conductivity of 10^5/ohm·cm, which is characteristic of most metals at room temperature, (7.5.12) gives a value of 10^{-14} seconds for τ. This is of the order of the mean time between atomic vibrations. The relaxation time is a measure of the average time between collisions, so multiplying it by the velocity of an electron with energy μ gives a measure of the mean free path between collisions. For a Fermi energy of 4 eV, the room temperature mean free path between scattering events turns out to be about 120 angstroms.

7.6 Thermal conductivity and the Wiedemann–Franz law

If the only external influence on our system is a temperature gradient, then electrons will move from hot to cold regions because their velocities are higher in the regions of higher temperature. As a result of this, the electron concentration will increase in the direction of decreasing temperature and an electric field will be established in the system. This internal field induces an electron flow in the opposite direction to that resulting from the temperature gradient. An analysis of thermal conduction, therefore, requires the use of both the charge flow and the heat flow equations. If the temperature gradient is along the x-axis, the electrical potential induced by the flow of electrons is also along the x-axis and the heat flow equation (7.4.4) is

$$J_x = -\frac{m}{2}\int \tau v_x^2 v^2 \frac{\partial N}{\partial T} d\mathbf{v} \frac{dT}{dx} + \frac{e}{2}\int \tau v_x v^2 \frac{\partial N}{\partial v_x} d\mathbf{v} E_x^T \qquad (7.6.1)$$

where E_x^T is the electric field induced by the temperature gradient. For a constant gradient, this induced field is also constant.

Thermal conductivity measurements are carried out under open electrical circuit conditions. The electric current vector in (7.4.3) therefore vanishes. Since, in the present case, $\nabla T \to dT/dx$, $\mathbf{E} \to E_x^T$, and $\mathbf{H} = 0$, we can set $\mathbf{I} = 0$ in (7.4.3) and solve for E_x^T, with the result that

$$E_x^T = \frac{m \int \tau v_x^2 \frac{\partial N}{\partial T} d\mathbf{v}}{e \int \tau v_x \frac{\partial N}{\partial v_x} d\mathbf{v}} \frac{dT}{dx} \qquad (7.6.2)$$

Now the induced electric field E_x^T can be replaced in (7.6.1) by (7.6.2) to give the heat flow in terms of the temperature gradient alone. The result is

$$J_x = -\frac{m}{2}\left(\frac{I_2 I_3 - I_1 I_4}{I_2}\right)\frac{dT}{dx} \qquad (7.6.3)$$

The Is are integrals defined by

$$I_1 = \int \tau v_x^2 \frac{\partial N}{\partial T} d\mathbf{v} \qquad (7.6.4)$$

$$I_2 = \int \tau v_x \frac{\partial N}{\partial v_x} d\mathbf{v} \qquad (7.6.5)$$

$$I_3 = \int \tau v_x^2 v^2 \frac{\partial N}{\partial T} d\mathbf{v} \qquad (7.6.6)$$

$$I_4 = \int \tau v_x v^2 \frac{\partial N}{\partial v_x} d\mathbf{v} \qquad (7.6.7)$$

To evaluate these integrals, it is convenient to express them in terms of energy rather than velocity so that the method of evaluating Fermi integrals can be used. To do this, use (7.5.4) and (7.5.8), replace v_x^2 by $v^2/3$ and remember that $N = 2m^3 f(\varepsilon)/h^3$ and $v = \sqrt{2\varepsilon/m}$. The result is

$$I_1 = \frac{16\pi\sqrt{2m}}{3h^3} \int \tau \varepsilon^{3/2} \frac{\partial f}{\partial T} d\varepsilon \qquad (7.6.8)$$

$$I_2 = \frac{16\pi m^{3/2}\sqrt{2}}{3h^3} \int \tau \varepsilon^{3/2} \frac{\partial f}{\partial \varepsilon} d\varepsilon \qquad (7.6.9)$$

$$I_3 = \frac{32\pi}{3h^3}\sqrt{\frac{2}{m}} \int \tau \varepsilon^{5/2} \frac{\partial f}{\partial T} d\varepsilon \qquad (7.6.10)$$

$$I_4 = \frac{32\pi\sqrt{2m}}{3h^3} \int \tau \varepsilon^{5/2} \frac{\partial f}{\partial \varepsilon} d\varepsilon \qquad (7.6.11)$$

Equations (7.6.8) and (7.6.10) contain temperature derivatives and these must be converted to derivatives with respect to energy so that the series expansion method of evaluating Fermi integrals can be used. It is easy to show that

$$\frac{\partial f}{\partial T} = f^2 e^{(\varepsilon-\mu)/kT}\left(\frac{(\varepsilon-\mu)}{kT^2} + \frac{1}{kT}\frac{d\mu}{dT}\right) \qquad (7.6.12)$$

Also, since

$$\frac{\partial f}{\partial \varepsilon} = \frac{f^2}{kT} e^{(\varepsilon-\mu)/kT} \qquad (7.6.13)$$

equation (7.6.12) can be written as

$$\frac{\partial f}{\partial T} = g(\varepsilon)\frac{\partial f}{\partial \varepsilon} \qquad (7.6.14)$$

where the function $g(\varepsilon)$ is defined by

$$g(\varepsilon) = -\frac{d\mu}{dT} - \frac{\varepsilon-\mu}{T} = \frac{\pi^2 k^2 T}{6\mu_0} - \frac{\varepsilon-\mu}{T} \qquad (7.6.15)$$

The last equality is obtained by evaluating $d\mu/dT$ from

$$\mu = \mu_0\left[1 - \frac{1}{12}\left(\frac{\pi kT}{\mu_0}\right)^2\right]$$

which is the free electron result for the variation of the Fermi level with temperature given by equation (6.4.22).

Now substitute (7.6.14) into (7.6.8) and (7.6.10) to get

$$I_1 = \frac{16\pi\sqrt{2m}}{3h^3}\int \tau g(\varepsilon)\varepsilon^{3/2}\frac{\partial f}{\partial \varepsilon}d\varepsilon \qquad (7.6.16)$$

$$I_3 = \frac{32\pi}{3h^3}\sqrt{\frac{2}{m}}\int \tau g(\varepsilon)\varepsilon^{5/2}\frac{\partial f}{\partial \varepsilon}d\varepsilon \qquad (7.6.17)$$

Now all the integrals I_1 to I_4 can be treated by the series expansion formula for Fermi integrals. When this formula was applied to the electrical conductivity calculation, only the first term in the series was retained since all other terms are negligibly small. But the thermal conductivity would vanish if only the first term were retained. To get a nonzero result, it is necessary to retain the first two terms of the expansion in evaluating the integrals I_1 to I_4. A straightforward application of the series expansion method, up to the first two terms, to equations (7.6.9), (7.6.11), (7.6.16), and (7.6.17) gives [see equations (6.4.8) and (6.4.9)]

$$I_1 = -\frac{16\pi\sqrt{2m}}{3h^3}\left[\tau(\mu)g(\mu)\mu^{3/2} - \frac{(\pi kT)^2}{6}(\tau g \varepsilon^{3/2})''\right] \qquad (7.6.18)$$

$$I_2 = \frac{16\pi m^{3/2}\sqrt{2}}{3h^3}\left[\tau(\mu)\mu^{3/2} - \frac{(\pi kT)^2}{6}(\tau \varepsilon^{3/2})''\right] \qquad (7.6.19)$$

$$I_3 = -\frac{32\pi}{3h^3}\sqrt{\frac{2}{m}}\left[\tau(\mu)g(\mu)\mu^{5/2} - \frac{(\pi kT)^2}{6}(\tau g \varepsilon^{5/2})''\right] \qquad (7.6.20)$$

STATISTICAL-KINETIC THEORY OF ELECTRON TRANSPORT 191

$$I_4 = -\frac{32\pi\sqrt{2m}}{3h^3}\left[\tau(\mu)\mu^{5/2} - \frac{(\pi kT)^2}{6}(\tau\varepsilon^{5/2})''\right] \quad (7.6.21)$$

In these equations, the double primes indicate that second differentiations with respect to energy of the quantities in the parentheses have been performed, the second derivatives being evaluated at $\varepsilon = \mu$. That is,

$$(\tau g\varepsilon^{3/2})'' = \left[\frac{d^2}{d\varepsilon^2}(\tau g\varepsilon^{3/2})\right]_{\varepsilon=\mu}$$

and so on. Working out the derivatives according to this definition gives

$$(\tau\varepsilon^{3/2})'' = \left[\frac{3}{4}\tau\varepsilon^{-1/2} + 3\varepsilon^{1/2}\frac{d\tau}{d\varepsilon} + \varepsilon^{3/2}\frac{d^2\tau}{d\varepsilon^2}\right]_{\varepsilon=\mu} \quad (7.6.22)$$

$$(\tau g\varepsilon^{3/2})'' = \left[\frac{3}{4}\tau g\varepsilon^{-1/2} + 3\varepsilon^{1/2}\frac{d(\tau g)}{d\varepsilon} + \varepsilon^{3/2}\frac{d^2(\tau g)}{d\varepsilon^2}\right]_{\varepsilon=\mu} \quad (7.6.23)$$

$$(\tau g\varepsilon^{5/2})'' = \left[\frac{15}{4}\tau g\varepsilon^{1/2} + 5\varepsilon^{3/2}\frac{d(\tau g)}{d\varepsilon} + \varepsilon^{5/2}\frac{d^2(\tau g)}{d\varepsilon^2}\right]_{\varepsilon=\mu} \quad (7.6.24)$$

$$(\tau\varepsilon^{5/2})'' = \left[\frac{15}{4}\tau\varepsilon^{1/2} + 5\varepsilon^{3/2}\frac{d\tau}{d\varepsilon} + \varepsilon^{5/2}\frac{d^2\tau}{d\varepsilon^2}\right]_{\varepsilon=\mu} \quad (7.6.25)$$

We now have all the formulas we need to get the heat flow equation. First substitute equations (7.6.18)–(7.6.21) into (7.6.3), retaining only terms to second order in the temperature, to get

$$J_x = -\frac{8\pi^3\sqrt{2m}}{9h^3}(kT)^2\left\{g(\mu)(\tau\varepsilon^{5/2})'' - (\tau g\varepsilon^{5/2})'' - \mu\left[g(\mu)(\tau\varepsilon^{3/2})'' - (\tau g\varepsilon^{3/2})''\right]\right\}\frac{dT}{dx} \quad (7.6.26)$$

Now use equations (7.6.22)–(7.6.25) in (7.6.26). The result is

$$J_x = -\frac{16\pi^3\sqrt{2m}}{9h^3}(kT)^2\mu^{3/2}\left[g\left(\frac{d\tau}{d\varepsilon}\right)_\mu - \left(\frac{d(\tau g)}{d\varepsilon}\right)_\mu\right]\frac{dT}{dx}$$

or

$$J_x = \frac{16\pi^3\sqrt{2m}}{9h^3}(kT)^2\mu^{3/2}\tau(\mu)\left(\frac{dg}{d\varepsilon}\right)_\mu\frac{dT}{dx} \quad (7.6.27)$$

From (7.6.15), we get

$$\left(\frac{dg}{d\varepsilon}\right)_\mu = -\frac{1}{T} \quad (7.6.28)$$

Putting this in (7.6.27), and using the first-order expression μ_0 from free electron theory for μ [equation (6.3.12)], $\mu_0 = h^2/8m(3N/\pi V)^{2/3}$, equation (7.6.27) becomes

Table 7.1: Lorenz Number for Metals at 100°C

Metal	$L \times 10^{13}$	Metal	$L \times 10^{13}$
Mg	2.57	Cd	2.70
Al	2.47	Sn	2.76
Ni	2.53	W	3.55
Cu	2.60	Ir	2.76
Zn	2.60	Pt	2.88
Mo	3.10	Au	2.88
Rh	2.82	Rb	2.85
Pd	2.63	Ag	2.63

From a compilation by Wilson (1965).

$$J_x = -\frac{\pi^2 k^2 T \tau(\mu) n}{3m} \frac{dT}{dx} \quad (7.6.29)$$

This is what we were after because, by definition, the negative of the coefficient of the temperature gradient is the thermal conductivity, K:

$$K = \frac{\pi^2 k^2 T \tau(\mu) n}{3m} \quad (7.6.30)$$

Just as in the case for the electrical conductivity, the relaxation time enters only through its value for $\varepsilon = \mu$.

If (7.6.30) is compared to equation (7.5.12) for the electrical conductivity, we see that the number L, defined by

$$L = \frac{K}{\sigma T} = \frac{\pi^2 k^2}{3 e^2} \quad (7.6.31)$$

contains only fundamental constants. This is the Wiedemann–Franz law, and the quantity L is called the Lorenz number. In the free electron theory it is a universal constant whose value is 2.71×10^{-13} (in c.g.s./e.s.u. units). Experimental values of the Lorenz number for some metals are given in table 7.1. The agreement between theory and experiment is quite good for a number of metals, and this is evidence for the applicability of the free electron theory.

7.7 The isothermal Hall effect

If a magnetic field is imposed on a metal or semiconductor carrying a current, it modifies both the current and the applied electric field. This is the basis of the Hall effect. Its importance lies in the fact that it gives a direct experimental measure of the electron concentration and the sign of the charge carrier. The Hall experiment is carried out by imposing an electric field in the x-direction and a magnetic field in the z-direction on a specimen as shown in figure 7.1.

An electron whose initial motion is in the x-direction is deflected by the magnetic field, so its velocity has a component in the y-direction. But in the experiment, the electric circuit is closed only in the x-direction, so there is no current flow in the y-direction. This means that a constant field E_y is

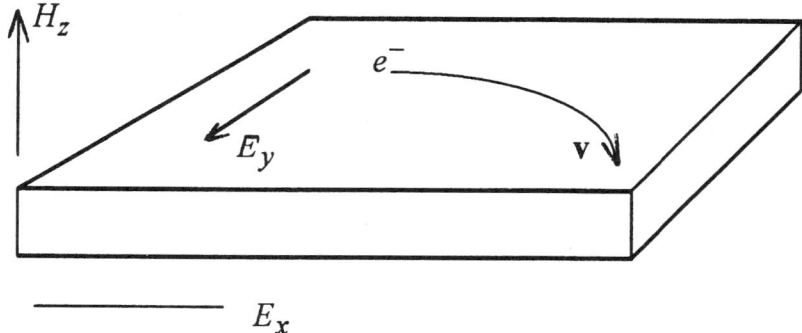

Figure 7.1. The isothermal Hall effect.

established in the y-direction. This is called the Hall voltage. A measure of the effect is given by the Hall coefficient, which is defined by

$$R_H = \frac{E_y}{I_x H_z} \tag{7.7.1}$$

where I_x is the current in the x-direction and H_z is the applied magnetic field.

The derivation of equations for the Hall coefficient is somewhat different for metals and for semiconductors. In metals, electrons constitute the only type of charge carrier and the electrons are degenerate so they obey Fermi–Dirac statistics. In semiconductors, both electrons and holes can carry charge, and because the electron density is usually much smaller than in metals, the electron statistics are well represented by the semiclassical approximation. However, the methodology of the derivations is the same in both cases. In this section, we will assume we have a metal that can be treated by the one-band effective mass model. After becoming familiar with this treatment, it is a simple matter to extend and modify it for semiconductors.

Switching on the magnetic field has the effect of converting the electric field vector from $E_x \mathbf{i}_1$ to $(E_x \mathbf{i}_1 + E_y \mathbf{i}_2)$, which is equivalent to rotating the electric field through an angle ϕ given by

$$\tan \phi = \frac{E_y}{E_x} \tag{7.7.2}$$

ϕ is called the *Hall angle*.

For the isothermal Hall effect, the temperature gradients are zero and the flux equation (7.4.3) reduces to

$$\mathbf{I} = -\frac{e^2}{m} \int \tau \mathbf{v} \left(\frac{\partial N}{\partial \mathbf{v}} \cdot \mathbf{E} \right) d\mathbf{v} - \frac{e^2}{mc} \int \tau \mathbf{v} (\mathbf{v} \times \mathbf{H}) \cdot \frac{\partial N}{\partial \mathbf{v}} d\mathbf{v} \tag{7.7.3}$$

The first term is just the electrical conduction $\sigma \mathbf{E}$ as shown in section 7.5. The second term requires a little more attention. As discussed in section 7.3, the equilibrium part of the derivative $\partial N / \partial \mathbf{v}$ contributes nothing to magnetic effects, and an explicit calculation of the departure from equilibrium must be made. From the definition of $\mathbf{h}$ given by equation (7.3.18), the derivative of N in the second term can be replaced by the derivative of $\mathbf{v} \cdot \mathbf{h}$ (since there is no contribution to magnetic effects from the derivative of N^0). That is, in the second term of (7.7.3),

$$\frac{\partial N}{\partial \mathbf{v}} \to \frac{\partial (\mathbf{v} \cdot \mathbf{h})}{\partial \mathbf{v}} \tag{7.7.4}$$

so

$$\int \tau \mathbf{v}(\mathbf{v} \times \mathbf{H}) \cdot \frac{\partial N}{\partial \mathbf{v}} d\mathbf{v} = \int \tau \mathbf{v}(\mathbf{v} \times \mathbf{H}) \cdot \frac{\partial (\mathbf{v} \cdot \mathbf{h})}{\partial \mathbf{v}} d\mathbf{v} \tag{7.7.5}$$

An estimate of the derivative in the integral on the right-hand side of equation (7.7.5) can be obtained by starting with equation (7.3.22). Since the temperature gradient is zero for the present case, this equation is

$$\mathbf{v} \cdot \mathbf{h} = \frac{e\tau}{m} \mathbf{E} \cdot \frac{\partial N^0}{\partial \mathbf{v}} + \frac{e\tau}{mc} (\mathbf{v} \times \mathbf{H}) \cdot \frac{\partial (\mathbf{v} \cdot \mathbf{h})}{\partial \mathbf{v}} \tag{7.7.6}$$

A series solution of (7.7.6) can be obtained by first treating the second term as if the magnetic field is small enough that the term in (7.7.6) containing the derivative of $\mathbf{v} \cdot \mathbf{h}$ can be neglected to get a first approximation. Thus, if we differentiate (7.7.6) and neglect the second term, we get

$$\frac{\partial (\mathbf{v} \cdot \mathbf{h})}{\partial \mathbf{v}} = \frac{e\tau}{m} \frac{\partial}{\partial \mathbf{v}} \left(\mathbf{E} \cdot \frac{\partial N^0}{\partial \mathbf{v}} \right) \quad \text{(1st approx.)} \tag{7.7.7}$$

But, since $\varepsilon = mv^2/2$,

$$\frac{\partial N^0}{\partial \mathbf{v}} = m\mathbf{v} \frac{\partial N^0}{\partial \varepsilon} \tag{7.7.8}$$

so (7.7.7) becomes

$$\frac{\partial (\mathbf{v} \cdot \mathbf{h})}{\partial \mathbf{v}} = \frac{e\tau}{m} \frac{\partial}{\partial \mathbf{v}} \left(\mathbf{E} \cdot m\mathbf{v} \frac{\partial N^0}{\partial \varepsilon} \right) = e\tau \frac{\partial N^0}{\partial \varepsilon} \mathbf{E} \tag{7.7.9}$$

Substitution of (7.7.9) into (7.7.6) gives the first approximation to $\mathbf{v} \cdot \mathbf{h}$ as

$$\mathbf{v} \cdot \mathbf{h} = e\tau \mathbf{E} \cdot \mathbf{v} \frac{\partial N^0}{\partial \varepsilon} + \frac{e^2 \tau^2}{mc} \mathbf{E} \cdot (\mathbf{v} \times \mathbf{H}) \frac{\partial N^0}{\partial \varepsilon} \tag{7.7.10}$$

A second approximation can be obtained by differentiating equation (7.7.10) and treating everything but $\mathbf{v}$ as a constant. The result is

$$\frac{\partial (\mathbf{v} \cdot \mathbf{h})}{\partial \mathbf{v}} = e\tau \frac{\partial N^0}{\partial \varepsilon} \mathbf{E} - \frac{e^2 \tau^2}{mc} \frac{\partial N^0}{\partial \varepsilon} (\mathbf{E} \times \mathbf{H}) \quad \text{(2nd approx.)} \tag{7.7.11}$$

This approximation procedure can be continued indefinitely to get higher approximations, but for our purposes, the second approximation given by (7.7.11) is adequate. Actually, the first approximation is usually sufficient because the experiment can always be done with small magnetic fields. But we carry the second approximation through to the end of the calculation to show that the second term in (7.7.11) turns out to be quadratic in the magnetic field and to show how higher approximations are obtained. Substituting (7.7.11) into (7.7.5) gives

$$\int \tau \mathbf{v}(\mathbf{v} \times \mathbf{H}) \cdot \frac{\partial N}{\partial \mathbf{v}} d\mathbf{v} = e \int \tau^2 \mathbf{v} \frac{\partial N^0}{\partial \varepsilon} (\mathbf{v} \times \mathbf{H}) \cdot \mathbf{E} d\mathbf{v} + \mathbf{a}(H^2) \qquad (7.7.12)$$

where $\mathbf{a}(H^2)$ is the integral arising from the second term in (7.7.11) and is easily shown to be quadratic in the magnetic field. Since $\mathbf{H} = H_z \mathbf{i}_3$ and $\mathbf{E} = E_x \mathbf{i}_1 + E_y \mathbf{i}_2$, working out the product $(\mathbf{v} \times \mathbf{H}) \cdot \mathbf{E}$ reduces (7.7.12) to

$$\int \tau \mathbf{v}(\mathbf{v} \times \mathbf{H}) \cdot \frac{\partial N}{\partial \mathbf{v}} d\mathbf{v} = eH_z \int \tau^2 \mathbf{v} \frac{\partial N^0}{\partial \varepsilon} (v_y E_x - v_x E_y) d\mathbf{v} + \mathbf{a}(H^2) \qquad (7.7.13)$$

The first integral on the right can be evaluated by using the series expansion formula for the Fermi integral. First, write the velocity vector in terms of its components and remember that the integral is zero for an antisymmetric integrand, so only the terms quadratic in the velocity components survive. The integral on the right of (7.7.13) therefore becomes

$$\int \tau^2 \frac{\partial N^0}{\partial \varepsilon} (v_y^2 E_x \mathbf{i}_2 - v_x^2 E_y \mathbf{i}_1) d\mathbf{v}$$

Also, because of spherical symmetry, v_y^2 and v_x^2 can be replaced by $v^2/3$, so this integral can be written as

$$\frac{1}{3} \int \tau^2 \frac{\partial N^0}{\partial \varepsilon} v^2 d\mathbf{v} (E_x \mathbf{i}_2 - E_y \mathbf{i}_1)$$

Now use the relationsn $N^0 = (2m^3/h^3) f(\varepsilon)$, $v^2 = (2\varepsilon/m)$, and $d\mathbf{v} = 4\pi v^2 dv$ to reduce this integral to $(8\pi m^2/3h^3) \int \tau^2 (\partial f/\partial \varepsilon) v^3 d\varepsilon (E_x \mathbf{i}_2 - E_y \mathbf{i}_1)$. Using the first term of the Fermi integral expansion, this becomes $(8\pi m^2/3h^3)[-\tau^2(\mu) v^3(\mu_0)][E_x \mathbf{i}_2 - E_y \mathbf{i}_1]$.

Now use the relation between velocity and energy to replace velocity by Fermi level and (6.3.12) to express the Fermi level in terms of electron density. Put the result into the right-hand side of (7.7.13) and use the definition of the electrical conductivity given by equation (7.5.2) to get

$$\int \tau \mathbf{v}(\mathbf{v} \times \mathbf{H}) \cdot \frac{\partial N}{\partial \mathbf{v}} d\mathbf{v} = \frac{\tau(\mu)}{e} \sigma H_z (E_y \mathbf{i}_1 - E_x \mathbf{i}_2) + \mathbf{a}(H^2) \qquad (7.7.14)$$

Finally, we substitute (7.7.14) into (7.7.3) (again using the definition of the electrical conductivity for the first term), with the result that

$$\mathbf{I} = \sigma \mathbf{E} - \frac{eH_z \tau(\mu) \sigma}{mc} (E_y \mathbf{i}_1 - E_x \mathbf{i}_2) + \mathbf{a}(H^2) \qquad (7.7.15)$$

Writing this in component form, remembering that $I_y = I_z = 0$, we get

$$I_x = \sigma E_x - \frac{eH_z \tau(\mu) \sigma}{mc} E_y + a_x(H^2) \qquad (7.7.16)$$

$$E_y = -\frac{eH_z \tau(\mu)}{mc} E_x + \frac{a_y(H^2)}{\sigma} \qquad (7.7.17)$$

The Hall voltage E_y contains a term linear in the magnetic field. The current, however, does not; the lowest power of H_z it contains is 2, as can be verified by substituting (7.7.17) for E_y into (7.7.16). Since we have assumed the magnetic field to be small, quadratic terms in the magnetic field will be neglected. Equations (7.7.16) and (7.7.17) then reduce to

$$I_x = \sigma E_x \tag{7.7.18}$$

$$E_y = -\frac{eH_z \tau(\mu)}{mc} E_x \tag{7.7.19}$$

The Hall coefficient (7.7.1) then becomes

$$R_H = -\frac{e\tau(\mu)}{mc\sigma} \tag{7.7.20}$$

or, substituting for σ from (7.5.12),

$$R_H = -\frac{1}{nec} \tag{7.7.21}$$

The Hall coefficient contains only the electron density, the velocity of light, and the electronic charge. Furthermore, if the current were carried by holes instead of electrons, the only effect would be to convert e to $-e$. Hall experiments are therefore useful in determining the sign of the charge carriers as well as their concentration.

7.8 Electrical conductivity in semiconductors

The theory developed so far in this chapter presumes that only one type of charge carrier, namely, the electron, is present in our system. But it is obvious that the theory is easily generalized to treat a system containing both electrons and holes if the holes are treated as particles and we adopt the two-band effective mass model of a semiconductor. This is done by working out the flux equations for holes in the same way as for electrons. The total flux is then the sum of the electron flux and the hole flux. Doing this for the flux of charge, the conductivity of a semiconductor is given by a formula containing two terms similar to the right-hand side of (7.5.2). The electron and hole conductivity can therefore be treated separately and then combined. For the contribution of electrons to the conductivity, equation (7.5.2) is

$$\sigma_e = -\frac{e^2}{m} \int \tau v_x \frac{\partial N_c}{\partial v_x} d\mathbf{v} \tag{7.8.1}$$

N_c is the distribution function for electrons in the conduction band, which is given by equation (6.10.1) with a density of states given by equation (6.10.3). If the energy of the electron is measured relative to the energy E_c at the bottom of the conduction band so that an energy scale ε' is defined by

$$\varepsilon' = \varepsilon - E_c \tag{7.8.2}$$

then (6.10.1) and (6.10.3) become

$$N_c(\varepsilon') = \frac{\omega_c(\varepsilon')}{e^{(\varepsilon'-\mu')/kT}+1} = \omega_c(\varepsilon')f(\varepsilon') \tag{7.8.3}$$

where

$$\mu' = \mu - E_c \tag{7.8.4}$$

and

$$\omega_c(\varepsilon') = \frac{8\pi\sqrt{2}}{h^3}m_c^{3/2}\sqrt{\varepsilon'} \tag{7.8.5}$$

Written in this way, the distribution function has precisely the same form as in the preceding sections of this chapter. This means that equation (7.5.9) is still valid if ε is replaced by ε' so the conductivity from the electrons is

$$\sigma_e = -\frac{8\pi m_c^2 e^2}{3h^3}\int_0^\infty \tau v^3 \frac{\partial f(\varepsilon')}{\partial \varepsilon'}d\varepsilon' \tag{7.8.6}$$

The kinetic energy is $m_c v^2/2 = \varepsilon'$, so (7.8.6) becomes

$$\sigma_e = -\frac{16\pi e^2 \sqrt{2m_c}}{3h^3}\int_0^\infty \tau\varepsilon'^{3/2} \frac{\partial f(\varepsilon')}{\partial \varepsilon'}d\varepsilon' \tag{7.8.7}$$

It is not necessary to use the series expansion formula to evaluate the integral in (7.8.7) because we can adopt the semiclassical approximation, in which the exponential in (7.8.3) is much greater than unity. That is,

$$f(\varepsilon') = e^{(\mu'-\varepsilon')/kT} \tag{7.8.8}$$

so

$$\frac{\partial f(\varepsilon')}{\partial \varepsilon'} = -\frac{1}{kT}e^{(\mu'-\varepsilon')/kT} \tag{7.8.9}$$

Furthermore, we assume that τ is a constant that we will call τ_c. This is not too serious an approximation since the integrand in (7.8.7) is largest near the bottom of the band and decreases rapidly as the energy increases because of the exponential in (7.8.9). With this assumption, and using (7.8.9), the integral in (7.8.7) becomes

$$\int_0^\infty \tau\varepsilon'^{3/2}\frac{\partial f(\varepsilon')}{\partial \varepsilon'}d\varepsilon' = -\frac{\tau_c}{kT}e^{\mu'/kT}\int_0^\infty \varepsilon'^{3/2}e^{-\varepsilon'/kT}d\varepsilon'$$

$$= -\tau_c(kT)^{3/2}e^{\mu'/kT}\int_0^\infty x^{3/2}e^{-x}dx$$

$$= -\frac{3\sqrt{\pi}}{4}\tau_c(kT)^{3/2}e^{\mu'/kT}$$

Equation (7.8.7) therefore reduces to

$$\sigma_e = \frac{4e^2\tau_c\sqrt{2\pi^3 m_c}}{h^3}(kT)^{3/2}e^{(\mu-E_c)/kT} \qquad (7.8.10)$$

This is considerably simplified if we use equation (6.10.11) for the electron concentration to get

$$\sigma_e = \frac{ne^2\tau_c}{m_c} \qquad (7.8.11)$$

Note that this has precisely the same form as equation (7.5.12) for the conductivity of metals.

The conductivity due to holes, σ_h, is calculated in exactly the same way except that, instead of defining an energy scale relative to the bottom of the conduction band, we measure energies relative to the top of the valence band so that, instead of equations (7.8.2)–(7.8.5), we use

$$\varepsilon'' = E_V - \varepsilon \qquad (7.8.12)$$

$$\mu'' = E_V - \mu \qquad (7.8.13)$$

$$N_V(\varepsilon'') = \omega_V(\varepsilon'')f(\varepsilon'') \qquad (7.8.14)$$

$$f(\varepsilon'') = \frac{1}{e^{(\varepsilon''-\mu'')/kT}+1} \qquad (7.8.15)$$

$$\omega_V(\varepsilon'') = \frac{8\pi\sqrt{2}}{h^3}m_V^{3/2}\sqrt{\varepsilon''} \qquad (7.8.16)$$

Now go through the same steps as for the calculation of the electron conductivity. Make the semiclassical approximation, assume τ is a constant called τ_V, and use equation (6.10.12) for the concentration of holes p. This results in the hole conductivity given by

$$\sigma_h = \frac{pe^2\tau_V}{m_V} \qquad (7.8.17)$$

The total conductivity of an intrinsic semiconductor is the sum of the electron and hole conductivity:

$$\sigma = e^2\left(\frac{n\tau_c}{m_c} + \frac{p\tau_V}{m_V}\right) \qquad (7.8.18)$$

The electron and hole concentrations are temperature dependent. For an intrinsic semiconductor, the electron and hole concentrations are equal and their temperature dependence is given by equation (6.10.17). Using this temperature dependence, equation (7.8.18) becomes

$$\sigma = A(T)e^{-E_g/2kT} \qquad (7.8.19)$$

where $A(T)$ is defined by

$$A(T) = 2e^2\left(\frac{2\pi kT}{h^2}\right)^{3/2}(m_c m_V)^{3/4}\left(\frac{\tau_c}{m_c} + \frac{\tau_V}{m_V}\right) \qquad (7.8.20)$$

Equation (7.8.19) displays the exponential dependence of the conductivity of an intrinsic semiconductor on temperature. The factor $A(T)$ is a slowly varying function of temperature and depends on the relaxation times and effective masses. In all cases the temperature variation of the exponential predominates and the energy gap E_g can be determined with satisfactory accuracy from measured conductivity values by plotting $\ln\sigma$ versus $1/T$.

An essential difference between the theory developed here and that worked out for metals is that, in metals, only the relaxation time at the Fermi level enters into the final equations. The energy dependence of the relaxation time is therefore not needed for metals. In semiconductors, however, a knowledge of the relaxation time as a function of energy is needed if we are to evaluate the integrals accurately. We have chosen the relaxation times for electrons and holes to be constants. But this is not the best choice that could be made. The energy dependence of the relaxation times depends on the particular collision processes that are operating, but one process that is always present, and that often predominates, is the scattering of electrons by lattice vibrations. A quantum theoretic treatment shows that the relaxation time is inversely proportional to the square root of the energy and therefore to the velocity of the electron. Therefore, $v\tau_c$ is a constant and we can define a distance parameter λ_e by

$$\lambda_e = v\tau_c \quad (7.8.21)$$

that is independent of the energy. Since the relaxation time measures the mean time between collisions, λ_e has the interpretation of a mean free path.

Now let us calculate the electron conductivity using (7.8.21) instead of the assumption that τ_c is a constant. The integral in (7.8.7) then becomes

$$\int_0^\infty \tau\varepsilon'^{3/2} \frac{\partial f(\varepsilon')}{\partial \varepsilon'} d\varepsilon' = -\frac{\lambda_e}{kT} e^{\mu/kT} \int_0^\infty \frac{\varepsilon'^{3/2}}{v} e^{-\varepsilon'/kT} d\varepsilon'$$

$$= -\frac{\lambda_e}{kT} e^{\mu/kT} \sqrt{\frac{m_c}{2}} \int_0^\infty \varepsilon' e^{-\varepsilon'/kT} d\varepsilon'$$

$$= -\lambda_e kT \sqrt{\frac{m_c}{2}} e^{\mu/kT} \quad (7.8.22)$$

where we have replaced τ_c by λ_e/v. Putting (7.8.22) into (7.8.7) gives the electron conductivity as

$$\sigma_e = \frac{4e^2 \lambda_e n}{3\sqrt{2\pi m_c kT}} \quad (7.8.23)$$

In getting (7.8.23), we have used equation (6.10.11) for the electron concentration in an intrinsic semiconductor.

If we define a mean free path for holes by $\lambda_h = v\tau_V$, then an identical procedure can be carried out for the hole conductivity to give

$$\sigma_h = \frac{4e^2 \lambda_h p}{3\sqrt{2\pi m_V kT}} \quad (7.8.24)$$

The total conductivity is the sum of (7.8.23) and (7.8.24):

$$\sigma = \frac{4e^2}{3\sqrt{2\pi kT}} \left(\frac{\lambda_c}{\sqrt{m_c}} n + \frac{\lambda_h}{\sqrt{m_V}} p \right) \quad (7.8.25)$$

This has the same form as (7.8.18) if we formally identify the relaxation times as follows:

$$\tau_c \to \frac{4\lambda_e m_c}{3\sqrt{2\pi m_c kT}} \tag{7.8.26}$$

$$\tau_h \to \frac{4\lambda_h m_V}{3\sqrt{2\pi m_V kT}} \tag{7.8.27}$$

Again using equation (6.10.17), the temperature dependence of the conductivity for intrinsic semiconductors is given by

$$\sigma = A'(T) e^{-E_g/2kT} \tag{7.8.28}$$

where $A'(T)$ is defined by

$$A'(T) = \left(\frac{16\pi kTe^2}{h^3}\right)(m_c m_V)^{3/4}\left(\frac{\lambda_e}{\sqrt{m_c}} + \frac{\lambda_h}{\sqrt{m_V}}\right) \tag{7.8.29}$$

The Hall coefficient in semiconductors can be computed in a straightforward manner using the methods of section 7.7, the semiclassical approximation of the distribution function, and the assumption of constant mean free path. We leave it to the reader to show that, for electronic conduction in semiconductors, the Hall coefficient is

$$R_H = -\frac{3\pi}{8}\frac{1}{nec} \tag{7.8.30}$$

while for conduction by holes

$$R_H = \frac{3\pi}{8}\frac{1}{pec} \tag{7.8.31}$$

Note that in arriving at equation (7.8.28), or (7.8.19), the only step in which the semiconductor was assumed to be intrinsic was that in which we took $n = p$ and used equation (6.10.17). This means that equations (7.8.18) and (7.8.25) are valid for extrinsic as well as intrinsic semiconductors. Explicit expressions for the conductivity in terms of impurity concentrations and ionization energies can be obtained by an application of the formulas for the concentrations of electrons and holes in extrinsic semiconductors given in chapter 6. The temperature dependence of the conductivity can then be rather complex, and the conductivity itself can vary over orders of magnitude. In general, however, the conductivity can be thought of as arising from two effects: an intrinsic effect due to the excitation of valence electrons across the energy gap, and an extrinsic effect due to the ionization of donor or acceptor impurities. Normally, the energy gap is considerably larger than the ionization energies, so for the concentrations of impurities found in doped semiconductors the extrinsic contribution dominates at ordinary temperatures. For small-enough impurity concentrations, the major contribution is intrinsic. Note, however, that even nominally pure semiconductor crystals contain some trace amounts of impurities. This means that as the temperature is lowered, there is an extrinsic contribution from these impurities, and if the temperature is low enough the crystal becomes an extrinsic conductor.

A technical point of interest is that the use of the semiclassical approximation for the Fermi function in semiconductors is the result of the fact that the charge carriers have energies that place them in the "Boltzmann tail" of the distribution. The expansion of the Fermi integrals used for electron transport theory in metals therefore cannot be used since this expansion is valid for energies close to the Fermi energy.

Exercises

7.1 For copper at 300 K, take the electrical conductivity to be 6.25×10^5 ohm·cm and the density of conduction electrons to be $n = 8.5 \times 10^{22}/\text{cm}^3$. The conversion factor for the conductivity from practical units to e.s.u. is $\sigma(\text{e.s.u.}) = 9 \times 10^{11}\, \sigma(\text{practical})$. Assume that the transport properties can be described by free electron theory with an effective mass equal to the electron mass. Use this information to do the following:

A. Compute the relaxation time for electrons at the Fermi level.
B. Compute the thermal conductivity at 300 K.
C. Compute the Hall coefficient for an applied magnetic field of 2 tesla (20,000 oersteds) applied in the z-direction in copper. What is the Hall angle in degrees?
D. What is the increase in velocity of an electron at the Fermi energy during one of its flight times (between collisions) in a field of 100 volts/cm? What is the increase in energy?

7.2 The mobility of electrons μ_e in a conductor is defined by the relation $\sigma = n\mu_e$, n being the electron density. Compute the mobility for copper. (Do not confuse the mobility with the Fermi energy.)

7.3 When an electric field acts in the x-direction on a piece of copper in a closed circuit, the net flux of electrons across a unit area normal to the x-axis is $I_x = nev_d$, where n is the electron concentration and v_d is the drift velocity of the electrons in the x-direction.

A. Derive this expression and compute the drift velocity for an applied electric field of 10 volts/cm. [Note: 1 volt(practical)=10(statvolts; e.s.u.).]
B. Find an expression for the mobility in terms of the drift velocity and applied potential.

7.4 Given that pure germanium has a band gap of 0.67 eV and assuming that the mean free paths for electrons and holes are equal and that the effective masses of electrons and holes are each equal to the electron mass, solve the following problems:

A. If the conductivity at 300 K is 3.8/ohm·cm, compute the mean free path.
B. Compute the ratio of the conductivity at 400 K to that at 300 K.
C. Compute the electron mobility at 300 K.
D. For an electrical potential of 10 volts/cm, compute the increase in velocity and the increase in energy of an electron in the conduction band when it travels a distance equal to the mean free path. Assume that the kinetic energy is zero immediately after a collision.

Order-Disorder Alloys

8.1 Order-Disorder Structures

A binary order-disorder structure is defined as a two-component crystal with the following properties:

1. At absolute zero of temperature, atoms of each component separately occupy a sublattice of the crystal. Each of the two interpenetrating sublattices contains only one type of atom. The structure is then said to be completely ordered.
2. At sufficiently high temperatures, both types of atoms are distributed throughout both sublattices at random. The structure is then said to be completely disordered.

Two examples of order-disorder structures are shown in figure 8.1, in which A represents the structure of β-brass (CuZn). If we ignore the identity of the atoms, the structure is body-centered cubic (BCC). The BBC lattice contains two interpenetrating simple cubic (SC) sublattices, one sublattice consisting of the cube corners, and the other sublattice consisting of the cube centers, as shown in figure 8.1A. In the completely ordered state, the body centers of the unit cubes are all occupied by atoms of one type (say, Cu) while the cube corners are all occupied by atoms of the other type (say, Zn). The two sublattices are completely equivalent, and it is immaterial which type of atom is assigned to a given sublattice. In the completely disordered state, any site can be occupied by an atom of either type, the probability that a given site contains an atom of a given type being one half.

Figure 8.1B shows the structure of Cu_3Au. The complete lattice is face-centered cubic (when the identity of the atoms is ignored) and can be resolved into two sublattices, one formed from the cube corner sites and the other formed from the cube face centers. The cube corners form an SC sublattice while the cube face centers form a body-centered tetragonal (BCT) sublattice. The BCT sublattice contains three times as many sites as the SC sublattice. In contrast with the β-brass structure, the two sublattices in this case are not equivalent. In the completely ordered state, the SC sublattice sites are all occupied by Au atoms and the BCT sublattice sites are all occupied by Cu atoms. In the completely disordered state, the probability that a given site contains an Au atom is 1/4 while the probability that a site contains a Cu atom is 3/4.

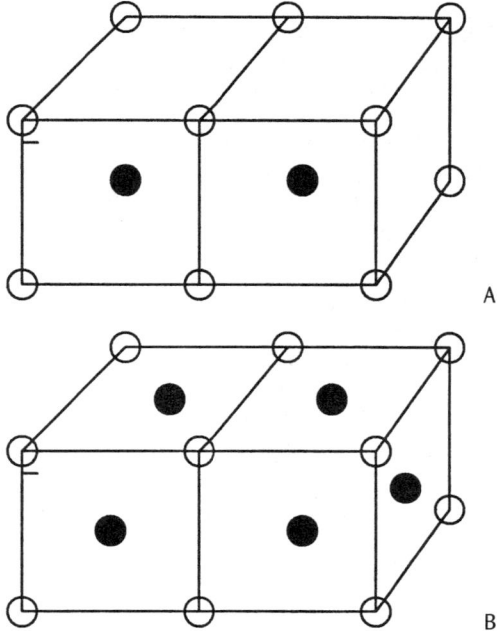

Figure 8.1. Order-disorder lattice structures. A, Body-centered cubic stimustince of β-brass (CuZu). B, Face-centered cubic structure of Cu₃ Au.

8.2 The order-disorder transition

Consider an AB alloy of the β-brass structure that is completely ordered at absolute zero. The energy of the ordered state must then be lower than that of the disordered state. This means that AB contacts are energetically favored over AA or BB contacts. As the temperature is raised, the thermal vibrations disrupt the perfect order of the crystal, so some AA and BB bonds will form. This is just a result of the fact that the entropy is larger for disordered than for ordered systems. At a given temperature, the free energy is a minimum and a balance between the energy and entropy requirements is established, so, in general, the crystal is partially ordered. The degree of order is bound to decrease with increasing temperature, and a temperature must exist above which the degree of order is negligible. However, changes in the degree of order can be observed only if the energy of the ordered state is not too much lower than that of the disordered state. If the energy difference is too large, the crystal will remain ordered right up to the melting point. A system that illustrates this to an extreme degree is sodium chloride, which may be regarded as an ordered alloy (Na⁺Cl⁻) with an ordering energy much greater than kT_m (T_m = melting point). On the other hand, if the energy difference is very small, the crystal will be disordered at all but very low temperatures. In general, alloys exhibiting order-disorder behavior have an energy in the disordered state that exceeds the energy in the ordered state by something less than the thermal energy at the melting point. The crystal will then be disordered at all temperatures for which the energy difference is small relative to kT.

The order-disorder transition can be observed directly in a number of alloys by X-ray or neutron diffraction techniques. At low temperatures, the ordered arrangement of the alloy constituents gives rise to superlattice lines that are not observed at high temperatures. A number of other changes in physical properties accompany the transition from the ordered to the disordered state. These include the electrical resistivity (which is generally increased by disorder), mechanical properties (elastic constants are generally decreased by disorder), and atomic diffusion (in β-brass, the activation energies for diffusion are decreased by increasing disorder). In theories of the order-disorder transition, it is usually assumed that the lattice parameter remains constant, but this is not always the case (in some alloys, the density increases by up to 5% as a result of disordering).

A particularly important effect is that of the heat capacity. For an order-disorder alloy, heat capacity versus temperature has a typical Debye-type curve at low temperatures. As the temperature is increased, however, a temperature is reached where the heat capacity increases above the Debye curve, and at a reasonably well-defined critical temperature the heat capacity very quickly drops to a value characteristic of a normal crystal. The importance of the heat capacity is that it measures the absorption of energy and it is a measurable thermodynamic property for which theories can be developed. Direct comparisons can therefore be made between theoretical concepts and experimental data.

The order-disorder transition is a typical cooperative phenomenon in which the ease of a transition increases rapidly with the extent to which it has already occurred. A qualitative understanding of this cooperative action can be based on the fact that an AB contact is favored over an AA or a BB contact. In a perfectly ordered AB alloy with two sublattices labeled α and β, all A atoms are on α sites and all B atoms are on β sites. Now replace a B atom on a β site by an A atom. This "wrong" A atom is surrounded by α sites, each containing an A atom; z AA contacts have been created, where z is the coordination number. But AA contacts are energetically less favorable than AB contacts, so it is now easier for one of the A atoms on the z nearest neighbors to be replaced by a B atom. Replacing a "right" A atom on one of the z sites by a "wrong" B atom requires less energy than would be the case if the original β site were occupied by a B atom. The existence of some disorder in the crystal thus promotes further disorder. The conversion of unlike to like pairs absorbs heat, so equal increases in temperature produce larger and larger increases in disorder. The heat capacity increases ever more rapidly with temperature until a critical temperature is reached at which the crystal becomes completely disordered. The order-disorder contribution to the heat capacity then drops rapidly to zero because, after the crystal is completely disordered, further temperature increases can have no more effect on the degree of disorder.

The statistical mechanical problem for order-disorder systems consists of establishing a quantitative measure for the degree of order, determining the degree of order as a function of temperature, and computing the effect of order on thermodynamic properties.

8.3 Description of the degree of order

Consider a two-component order-disorder alloy with composition A_nB_m having two sublattices labeled α and β. In the completely ordered state all A atoms are on the α sublattice and all B atoms are on the β sublattice. In the partially ordered state, however, this is not true and the number of atoms of each type on each sublattice must be specified. Let

$$N_{A\alpha} = \text{number of A atoms on } \alpha \text{ sites}$$
$$N_{A\beta} = \text{number of A atoms on } \beta \text{ sites}$$

ORDER-DISORDER ALLOYS   205

$N_{B\beta}$ = number of B atoms on β sites
$N_{B\alpha}$ = number of B atoms on α sites
N_A = total number of A atoms in the crystal
N_B = total number of B atoms in the crystal
N = total number of atoms in the crystal
γ_A = fraction of A atoms in the crystal
γ_B = fraction of B atoms in the crystal

The following relations are obvious:

$$\gamma_A = \frac{n}{m+n} \tag{8.3.1}$$

$$\gamma_B = \frac{m}{m+n} \tag{8.3.2}$$

$$\gamma_A + \gamma_B = 1 \tag{8.3.3}$$

$$N_A + N_B = N \tag{8.3.4}$$

$$N_{A\alpha} + N_{B\alpha} = N_{A\alpha} + N_{A\beta} = N_A = \gamma_A N \tag{8.3.5}$$

$$N_{A\beta} + N_{B\beta} = N_{B\alpha} + N_{B\beta} = N_B = \gamma_B N \tag{8.3.6}$$

Also, from (8.3.5) or (8.3.6),

$$N_{A\beta} = N_{B\alpha} \tag{8.3.7}$$

The fractions of sites of a given sublattice occupied by atoms of a given type are defined by

$$f_{A\alpha} = \frac{N_{A\alpha}}{\gamma_A N} \tag{8.3.8}$$

$$f_{A\beta} = \frac{N_{A\beta}}{\gamma_B N} \tag{8.3.9}$$

$$f_{B\beta} = \frac{N_{B\beta}}{\gamma_B N} \tag{8.3.10}$$

$$f_{B\alpha} = \frac{N_{B\alpha}}{\gamma_A N} \tag{8.3.11}$$

In a perfectly ordered crystal, all α sites are occupied by A atoms and all β sites are occupied by B atoms, so the occupation fractions for "right" occupancy on each sublattice are unity. That is, $f_{A\alpha} = f_{B\beta} = 1$. In a completely disordered crystal, $f_{A\alpha} = \gamma_A$ and $f_{B\beta} = \gamma_B$ since the atoms are distributed at random. We need a parameter to describe the crystal when the degree of order is between these two values. Such a parameter is defined by

$$R = \frac{f_{A\alpha} - \gamma_A}{1 - \gamma_A} \tag{8.3.12}$$

Clearly, $R = 1$ for perfect order and $R = 0$ for complete disorder. Also, it should not matter which sublattice we focus on to define our parameter. That this is the case can be seen by starting from the relation

$$\gamma_A(1 - f_{A\alpha}) = \gamma_B(1 - f_{B\beta}) \tag{8.3.13}$$

which follows from equations (8.3.5)–(8.3.7). Using (8.3.13), it is easy to convert (8.3.12) to

$$R = \frac{f_{B\beta} - \gamma_B}{1 - \gamma_B} \tag{8.3.14}$$

The parameter R is therefore invariant with respect to the interchange $(A\alpha) \leftrightarrow (B\beta)$. R is a measure of the degree of order defined by the occupation of atom types on the sublattices and is called the *long-range order parameter* since it describes the extent to which the sublattices are filled with "right" kinds of atoms and says nothing about local configurations.

It is convenient to express the site occupation probabilities in terms of the long-range order parameter. From (8.3.12) and (8.3.14),

$$f_{A\alpha} = \gamma_A + \gamma_B R \tag{8.3.15}$$

$$f_{B\beta} = \gamma_B + \gamma_A R \tag{8.3.16}$$

and using (8.3.5) and (8.3.6), equations (8.3.8)–(8.3.11) give

$$f_{B\alpha} = \frac{\gamma_B}{\gamma_A}(1 - f_{B\beta}) \tag{8.3.17}$$

$$f_{A\beta} = \frac{\gamma_A}{\gamma_B}(1 - f_{A\alpha}) \tag{8.3.18}$$

which, when combined with (8.3.15) and (8.3.16), become

$$f_{A\beta} = \gamma_A(1 - R) \tag{8.3.19}$$

$$f_{B\alpha} = \gamma_B(1 - R) \tag{8.3.20}$$

An alternate method of describing the degree of order is to count the number of AB pairs. In a completely ordered crystal, the number of unlike pairs will be a maximum, while in a completely disordered crystal, the number of unlike pairs will be a minimum. Such a description is local since it is not concerned with the occupancy of the entire sublattices, but with nearest neighbor configurations. Let Q_{AA}, Q_{BB}, and Q^t_{BB} be the number of AA, BB, and AB pairs, respectively. Then, if Q is the total number of pairs and q is the fraction of pairs of unlike atoms, we have

$$Q = Q_{AA} + Q_{BB} + Q^t_{AB} = \frac{zN}{2} \tag{8.3.21}$$

$$q = \frac{Q^t_{AB}}{Q} \tag{8.3.22}$$

(The coordination number z is 8 for β-brass and 12 for Cu₃Au.) We have put a superscript t on the number of unlike pairs because it is sometimes necessary to identify the sublattice that is occupied by each type of atom. In a 50-50 AB alloy with α and β sublattices, for example, we will want to reserve the notation Q_{AB} for the number of pairs such that an A atom is on an α site and a B atom is on an adjacent β site, while Q_{BA} is the number of pairs such that a B atom is on an α site and an A atom is on an adjacent β site. In this case, $Q^t_{AB} = Q_{AB} + Q_{BA}$. This notation is important when the crystal configuration is explicitly described in terms of pairs because an (A on α, B on β) pair defines a different state than a (B on α, A on β) pair. This will be the case for the quasi-chemical method.

In a completely ordered state, q has a maximum value which we will call q_m, and in the completely disordered state, it has a minimum value q_0. The short-range order parameter σ is defined by

$$\sigma \equiv \frac{q - q_0}{q_m - q_0} \quad (8.3.23)$$

For an AB alloy (e.g., β-brass), all pairs are unlike pairs when the alloy is completely ordered, and there are just as many like as unlike pairs for complete disorder, so $q_m = 1$ and $q_0 = 1/2$. Equation (8.3.23) then gives

$$\sigma = 2q - 1 \quad \text{(AB alloy)} \quad (8.3.24)$$

Note that in this case, the fraction of unlike pairs is $q = (1 + \sigma)/2$ and the fraction of like pairs is $(1 - q) = (1 - \sigma)/2$. The short-range order parameter σ is therefore the difference between the fractions of unlike and like pairs.

The short-range order parameter is zero for complete disorder and unity for complete order, but it is only at these two points that it equals the long-range order parameter. The configurational state of the system is defined by numbering the lattice sites and specifying the type of atom on each site. For a given set of occupation numbers [$N_{A\alpha}$, $N_{A\beta}$, $N_{B\alpha}$, $N_{B\beta}$] there are a great many distinguishable configurational states, and many sets [Q_{AA}, Q_{BB}, Q_{AB}, Q_{BA}] of the number of pairs are consistent with a given set of sublattice occupation numbers. The long-range order parameter must therefore be a function of the average short-range order parameter, the average being taken over all configurational states consistent with a given set of occupation numbers.

The relation between the short- and long-range order parameters is equivalent to the relation between the number of pairs and the probability of single-site occupancy by atom type. It is important to have this relation because, first, the site occupancy is essential for counting the number of crystal configurations, and second, the energy of the crystal depends on the interactions of pairs of atoms.

Let us first compute the average number of AA pairs in terms of the long-range order parameter. In general, AA pairs can be formed in two ways: (1), we can have an A atom on an α site and another A atom on an adjacent β site, or (2) we can have an A atom on a β site and another A atom on an adjacent β site. Note that in β-brass–type alloys, the second type of pair does not exist because all neighbors to a β site are α sites. In Cu₃Au, however, a β site has 12 neighbors, only four of which are α sites since the β sublattice has three times as many sites as the α sublattice. Note that we are adopting the convention that each α site has z neighbors, all of which are β sites, while each β site has z neighbors, of which $z\gamma_A/\gamma_B$ are α sites and $z(\gamma_B - \gamma_A)/\gamma_B$ are β sites.

To get the number of pairs of type 1, choose an atom on an α site. This atom is surrounded by z β sites, each of which has a probability $f_{A\beta}$ of containing an

A atom. The number of such AA pairs around our chosen site is therefore $zf_{A\beta}$, and multiplying this by the total number of A atoms on α sites gives the total number of AA pairs of type 1 as

$$N_{A\alpha} z f_{A\beta} = \gamma_A N(\gamma_A + \gamma_B R) z \gamma_A (1-R)$$
$$= zN\gamma_A^2 (1-R)(\gamma_A + \gamma_B R) \qquad (8.3.25)$$

where (8.3.20) has been used to replace $f_{A\beta}$, and (8.3.8) along with (8.3.15) was used to replace $N_{A\alpha}$

To get the number of AA pairs of type 2, choose an A atom on a β site, which has $z(\gamma_B - \gamma_A)/\gamma_B$ neighbors that are β sites. Of these, the probability that one is occupied by an A atom is $f_{A\beta}$. Since the number of β sites containing A atoms is $N_{A\beta}$, the number of pairs of type 2 is

$$zN_{A\beta} \frac{(\gamma_B - \gamma_A)}{2\gamma_B} f_{A\beta} = z\gamma_B \frac{N(\gamma_B - \gamma_A)}{2\gamma_B} \gamma_A^2 (1-R)^2$$
$$= \frac{1}{2} zN\gamma_A^2 (\gamma_B - \gamma_A)(1-R)^2 \qquad (8.3.26)$$

The factor of 1/2 is included because, since the A atoms in the pair are both on the same sublattice, the number of pairs would otherwise be counted twice.

Adding (8.3.25) and (8.3.26) gives the total average number of AA pairs

$$Q_{AA} = \frac{zN}{2} \gamma_A^2 (1-R^2) \qquad (8.3.27)$$

Proceeding in the same way, we get the average number of BB and AB pairs as

$$Q_{BB} = \frac{zN}{2} (\gamma_B^2 - \gamma_A^2 R^2) \qquad (8.3.28)$$

$$Q_{AB}^t = \frac{zN}{2} (2\gamma_A \gamma_B - 2\gamma_A^2 R^2) \qquad (8.3.29)$$

Dividing (8.3.29) by $Q = zN/2$ gives the average fraction of unlike pairs

$$\bar{q} = 2(\gamma_A \gamma_B - \gamma_A^2 R^2) \qquad (8.3.30)$$

From (8.3.23), the average short-range order parameter is

$$\bar{\sigma} = \frac{\bar{q} - q_0}{q_m - q_0} \qquad (8.3.31)$$

and substitution of (8.3.30) into (8.3.31) gives the relation between the short-range and long-range order parameters. For an AB alloy, $\gamma_A = \gamma_B$ and (8.3.30) becomes

$$\bar{q} = \frac{1 + R^2}{2} \quad \text{(AB alloy)} \qquad (8.3.32)$$

also, from (8.3.24),

$$\bar{\sigma} = 2\bar{q} - 1 \quad \text{(AB alloy)} \qquad (8.3.33)$$

so that

$$\bar{\sigma} = R^2 \quad \text{(AB alloy)} \tag{8.3.34}$$

Equation (8.3.30) can be used to compute q_m and q_0 because when $R = 1$, σ can be unity only if $\bar{q} = q_m$, and when $R = 0$, σ can be zero only if $\bar{q} = q_0$. Thus,

$$q_m = 2\gamma_A\gamma_B + 2\gamma_A^2 \tag{8.3.35}$$

$$q_0 = 2\gamma_A\gamma_B \tag{8.3.36}$$

For an AB alloy, this gives $q_m = 1$ and $q_0 = 1/2$ as expected. For a Cu$_3$Au-type alloy, this gives $q_m = 1/2$ and $q_0 = 3/8$.

It must be emphasized that the averages considered here are not statistical mechanical averages over states. They are simple arithmetic averages consistent with a given value of R. The relations given between the short- and long-range order parameters are not true in general for the statistical thermodynamic values of these quantities, except within the framework of the Bragg–Williams approximation, which will be developed subsequently.

The long- and short-range order parameters do not provide a complete specification of the nature of the ordering because they neglect the occupation of next neighbors, next nearest neighbors, triplets of sites, and so on. Descriptions of order that go beyond the pair description described here have been used, but they lead to theories that are quite complex and pose serious calculational difficulties. The simpler approach given here is a reasonably satisfactory basis for the description of order-disorder systems because it is normally the first and most important term in the more complex theories. The important physical insights are contained in the simple description based on the long- and short-range order parameters, and it yields good qualitative and even semiquantitative results. Our next task is to incorporate this description into a statistical mechanical framework from which the equations for the thermodynamic properties as a function of order can be obtained.

8.4 The order-disorder partition function

The central problem of the statistical mechanics of order-disorder systems is the evaluation of the partition function in terms of the order parameters. In the theory of monatomic crystals there was no need to consider the occupancy of lattice sites since the interchange of two identical atoms did not alter the quantum state of the crystal. But for alloys it is necessary to take the distribution of atoms into account because the exchange of unlike atoms among sites does give a new quantum state. Each such distribution constitutes a different configurational state of the system. We therefore construct a canonical ensemble in which R is identical for every member of the ensemble, but the members have different configurations, consistent with a given value of R. The partition function for this ensemble can then be evaluated provided certain approximations are made. This gives the Helmholtz free energy as a function of the degree of long-range order. The free energy is then minimized with respect to R to get the equilibrium degree of order as a function of temperature.

Let $Z(R)$ be the partition function for a given value of the long-range order parameter:

$$Z(R) = \sum_{r,v} e^{-(W_r + E_v^c)/kT} \tag{8.4.1}$$

where W_r is the potential energy of the rth configurational state (for a given R) and E_v^r is the vibrational energy of the crystal when its atoms are arranged in the rth configuration. The sum is taken over all configurational and vibrational states of the system.

The vibrational energy of the rth configuration is given by

$$E_v^r = \sum_j \left(n_j + \frac{1}{2}\right) h v_j^r \qquad (8.4.2)$$

where n_j are the integers that specify the quantum levels of the normal modes of vibration and v_j^r is the frequency of the jth normal mode of the crystal with a configuration r. The vibrational partition function is given by (see section 4.3)

$$Z_v^r(R) = e^{-\varepsilon_0^r/kT} \prod_{j=1}^{3N} \frac{1}{1-e^{-hv_j^r/kT}} \qquad (8.4.3)$$

where ε_0^r is the zero point energy of the rth configuration. Equation (8.4.1) can be written as

$$Z(R) = \sum_r Z_v^r(R) e^{-W_r/kT} \qquad (8.4.4)$$

and it is clear that a complete evaluation of the partition function requires an analysis of the dependence of the normal mode frequencies on the configurational state. Such an analysis would be very difficult to carry out, and no complete theory in which the vibrations are treated along with the configurational states has been constructed. Instead, it is assumed that the vibrational and configurational parts of the problem can be separated. This is done by working with the configurational partition function defined by

$$Z_c(R) = \sum_r e^{-W_r/kT} \qquad (8.4.5)$$

It is then assumed that the thermodynamic functions computed from Z_c can be added to those obtained from the vibrational states. In practice, this means that when experimental data are examined some estimate of the vibrational contribution must be made. This estimate is subtracted from the data, and the difference is compared to that computed from theory based on Z_c. Such an approach can be successful only if the vibrational partition function is much less sensitive to the degree of order than is the configurational partition function. That this is approximately correct for many alloys arises from the fact that the order-disorder transition usually occurs at temperatures above the Debye temperature. For β-brass, for example, the effect of order on the heat capacity is most pronounced at temperatures from 550 to 750 K. In this temperature range the heat capacity is well represented by the high-temperature approximation (see chapter 4), the leading term of which is independent of the frequency spectrum. Furthermore, it was shown in chapter 4 that the thermodynamic properties, particularly the heat capacity, are insensitive to the precise form of the frequency distribution function. Therefore, an analysis based on the separation of configurational and vibrational partition functions will give results that are at least semiquantitatively correct.

To get the degree of order at equilibrium, we just minimize the Helmholtz free energy obtained from Z_c with respect to R. That is, we solve the equation

$$\frac{\partial A_c}{\partial R} = 0 \tag{8.4.6}$$

where

$$A_c = -kT \ln Z_c \tag{8.4.7}$$

The equilibrium value of R is then inserted back into the free energy function to give the equilibrium Helmholtz free energy.

To evaluate the configurational partition function, the potential energy W_r must be known. This means that we need to know how the energy depends on the configurations. The basic assumption made in order-disorder theory is that the energy consists of nearest-neighbor pairwise contributions. In metals, the conduction electrons are not localized, and the crystal energy is volume dependent, so the pairwise assumption seems like a rash one that is justified only by the fact that it is too hard to do anything else. However, pseudopotential theory shows that it is often possible to express the energy of a metal as arising from pairwise interactions, providing the crystal volume is constant. Volume changes accompanying the order-disorder transition are usually quite small, so this condition is met in many alloys of interest. The remaining problem is that interactions among atoms that are not nearest neighbors are ignored. But the interaction energy falls off rapidly with distance. For these reasons, representing the energy by sums of nearest neighbor interactions works well enough to yield the important features of the order-disorder transition.

Let $-v_{AA}$, $-v_{BB}$, and $-v_{AB}$ be the nearest-neighbor interaction energies for AA, BB, and AB pairs, respectively. Then the configurational energy is given by

$$W_r = -v_{AA} Q_{AA} - v_{BB} Q_{BB} - v_{AB} Q^t_{AB} \tag{8.4.8}$$

the number of pairs Q_{AA}, Q_{BB}, and Q^t_{AB} being those for the configuration labeled by r. The interaction energies are chosen so that v_{AA}, v_{BB}, and v_{AB} are positive constants since the state of zero energy is taken to be that in which all atoms are very far apart.

To this point, the discussion applies to a binary order-disorder alloy of any stoichiometric composition, but we now particularize the theory to the case of a 50-50 AB alloy. For this case, equation (8.4.8) can be written in several alternate ways that are often useful. Since $q = Q^t_{AB}/Q$ is the fraction of unlike pairs and since the total number of pairs is given by (8.3.21), we have

$$Q_{AA} + Q_{BB} = Q(1 - q) \tag{8.4.9}$$

In a 50-50 alloy, $Q_{AA} = Q_{BB}$, so (8.4.9) gives

$$Q_{AA} = Q_{BB} = \frac{Q}{2}(1 - q) \tag{8.4.10}$$

Using this equation and the definition of the fraction of unlike pairs, equation (8.4.8) becomes

$$W_r = -\frac{Q}{2}(v_{AA} + v_{BB}) - qQ\left(v_{AB} - \frac{1}{2}v_{AA} - \frac{1}{2}v_{BB}\right) \tag{8.4.11}$$

The ordering energy is defined by

$$v \equiv v_{AB} - \frac{1}{2}(v_{AA} + v_{BB}) \tag{8.4.12}$$

This is the energy change accompanying the formation of one AB pair by the interchange of atoms from an AA and a BB pair. With this definition (8.4.11) becomes

$$W_r = w_0 - Qvq \tag{8.4.13}$$

where

$$w_0 = -\frac{Q}{2}(v_{AA} + v_{BB}) \tag{8.4.14}$$

is the average energy of pure A and pure B. From (8.4.13) and (8.3.24), the configurational energy can be written in terms of the short-range order parameter as

$$W_r = w_0 - \frac{Qv}{2}(1+\sigma) \tag{8.4.15}$$

It is understood in this equation that the short-range order parameter is that which is consistent with the configuration r.

Another expression for the energy that is often used is obtained by assigning a parameter S_i to each lattice site such that $S_i = +1$ if the site contains an A atom and $S_i = -1$ if it contains a B atom. For two adjacent sites labeled i and j, the product $S_i S_j$ is always +1 if the sites contain like atoms and -1 if they contain unlike atoms. Therefore,

$$\sum_{<i,j>} S_i S_j = Q_{AA} + Q_{BB} - Q_{AB}^t$$
$$= Q - 2Q_{AB}^t$$
$$= Q(1-2q) \tag{8.4.16}$$

the summation being carried out over all nearest neighbor pairs. From (8.4.16)

$$qQ = \frac{1}{2}\left(Q - \sum_{<i,j>} S_i S_j\right) \tag{8.4.17}$$

and putting this into (8.4.13) gives

$$W_r = w_0 - \frac{v}{2}\left(Q - \sum_{<i,j>} S_i S_j\right) \tag{8.4.18}$$

Finally, the energy can be written in another useful form by defining two parameters α_i and β_j such that $\alpha_i = 1$ if the ith site of the α sublattice contains an A atom and is zero otherwise. Similarly, $\beta_j = 1$ if the jth site of the β sublattice contains an A atom and is zero otherwise. Also let a parameter $\gamma_{ij} = 1$ if two lattice sites i and j are adjacent and be zero otherwise. Then

$$\sum_{i,j} \gamma_{ij} \alpha_i \beta_j = Q_{AA} = \frac{Q}{2}(1-q) \tag{8.4.19}$$

the sum being over all i and j because there is a term +1 for every nearest-neighbor AA pair, all other terms being zero. The sum in (8.4.19) is taken over all lattice sites of both sublattices. Comparing (8.4.19) with (8.4.13) gives

$$W_r = w_0 - vQ + 2v\sum_{i,j}\gamma_{ij}\alpha_i\beta_j \qquad (8.4.20)$$

Accurate evaluation of the configurational partition function is difficult because knowledge is required of the number of configurations that have a given energy, subject to the restriction that the long-range order is fixed. This poses a complex combinatorial problem, and approximation methods must be used. Two general approximation methods will be presented in this chapter: the Kirkwood method and the quasichemical method.

8.5 The Kirkwood method

In 1938 J.G. Kirkwood presented a method by which the configurational free energy A_c can be developed as a series in the long-range order parameter. Within the framework of the assumptions given above, namely, the nearest-neighbor pairwise formula for the configurational energy and the separability of the configurational and vibrational effects, the theory yields an exact series expansion. However, the series does not converge with great rapidity, and mathematical complexities limit the usefulness of including high-order terms. But it is the simplest theory that is at least formally exact, and it illuminates the nature of the approximations that must be made to make the calculations tractable.

Let $\overline{W}$ be the unweighted average of the energy over all configurational states for a given degree of long-range order. That is,

$$\overline{W} = \frac{1}{g(R)}\sum_r W_r \qquad (8.5.1)$$

where $g(R)$ is the total number of configurational states consistent with a given long-range order parameter R. $g(R)$ is easily evaluated since it is just the number of ways of arranging $N_{A\alpha}$ atoms of type A and $N_{B\alpha}$ atoms of type B on the α sublattice and $N_{A\beta}$ A atoms and $N_{B\beta}$ B atoms on the β sublattice. For an A_nB_m alloy, this is

$$g(R) = \frac{(\gamma_A N)!(\gamma_B N)!}{N_{A\alpha}!N_{B\alpha}!N_{A\beta}!N_{B\beta}!} \qquad (8.5.2)$$

From section 8.3, the occupation numbers for A and B atoms on α and β sites are related to the long-range order parameter by

$$N_{A\alpha} = N\gamma_A(\gamma_A + \gamma_B R) \qquad (8.5.3)$$

$$N_{B\beta} = N\gamma_B(\gamma_B + \gamma_A R) \qquad (8.5.4)$$

$$N_{A\beta} = N_{B\alpha} = N\gamma_A\gamma_B(1-R) \qquad (8.5.5)$$

For the case of a 50-50 AB alloy, (8.5.3)–(8.5.5) reduce to

$$N_{A\alpha} = N_{B\beta} = \frac{N}{4}(1+R) \qquad (8.5.6)$$

214   STATISTICAL MECHANICS OF SOLIDS

$$N_{A\beta} = N_{B\alpha} = \frac{N}{4}(1-R) \tag{8.5.7}$$

and (8.5.2) becomes

$$g(R) = \frac{(N/2)!(N/2)!}{\left[\frac{N}{4}(1+R)\right]!\left[\frac{N}{4}(1-R)\right]!\left[\frac{N}{4}(1+R)\right]!\left[\frac{N}{4}(1-R)\right]!} \tag{8.5.8}$$

Now expand the exponential in the partition function in a Taylor series about $\overline{W}$. The result is

$$e^{-W_k/kT} = e^{-\overline{W}/kT}\left[1 - \left(\frac{W_k - \overline{W}}{kT}\right) + \frac{1}{2}\left(\frac{W_k - \overline{W}}{kT}\right)^2 - \cdots\right]$$

$$= e^{-\overline{W}/kT}\sum_{j=0}^{\infty}\left(\frac{-1}{kT}\right)^j \frac{(W_k - \overline{W})^j}{j!} \tag{8.5.9}$$

The configurational partition function, and therefore the configurational free energy, is now obtained by substituting (8.5.9) into (8.4.5) to get

$$Z_c = e^{-A_c/kT} = e^{-\overline{W}/kT}\sum_{k}\sum_{j=0}^{\infty}\left(\frac{-1}{kT}\right)^j \frac{(W_k - \overline{W})^j}{j!} \tag{8.5.10}$$

Equation (8.5.10) is an expansion in terms of the moments of the energy, which are defined by

$$M_j = \frac{1}{g(R)}\sum_k (W_k - \overline{W})^j \tag{8.5.11}$$

In terms of the moments, (8.5.10) is

$$e^{-A_c/kT} = g(R)e^{-\overline{W}/kT}\sum_{j=0}^{\infty}\left(\frac{-1}{kT}\right)^j \frac{M_j}{j!} \tag{8.5.12}$$

Taking the logarithm of (8.5.12) to get the free energy gives a result that contains an inconvenient logarithm of the sum over the jth moments. This can be dealt with by the following device: define a power series in $x = (-1/kT)$ by the relation

$$\sum_{n=1}^{\infty}\frac{B_n}{n!}x^n = \ln\sum_{j=0}^{\infty}\frac{M_j}{j!}x^j \tag{8.5.13}$$

where the B_n are to be determined in terms of the M_j. Differentiate both sides of (8.5.13) with respect to x to give

$$\sum_{n=1}^{\infty}\sum_{j=0}^{\infty}\frac{n}{n!\,j!}B_n M_j x^{n+j-1} = \sum_{j=0}^{\infty}\frac{jM_j}{j!}x^{j-1} \tag{8.5.14}$$

The B_n are readily evaluated by equating equal powers of x in both series. The result is

$$B_1 M_0 = M_1$$
$$B_1 M_1 + B_2 M_0 = M_2$$
$$B_3 M_0 + 2 B_2 M_1 + B_1 M_2 = M_3$$
$$B_4 M_0 + 3 B_3 M_1 + 3 B_2 M_2 + B_1 M_3 = M_4 \qquad (8.5.15)$$
$$\ldots$$

This is considerably simplified by using the fact that from the definition of the moments (8.5.11) we get

$$M_0 = 1, \qquad M_1 = 0 \qquad (8.5.16)$$

so we can solve (8.5.15) for the Bs to get

$$B_1 = 0$$
$$B_2 = M_2$$
$$B_3 = M_3$$
$$B_4 = M_4 - 3 M_2^2 \qquad (8.5.17)$$
$$\ldots$$

Now put these values of the Bs into the left-hand side of (8.5.13) and replace x by its definition $-1/kT$. The result, to the fourth order in $1/kT$, is

$$\ln \sum_{j=0}^{\infty} \frac{M_j}{j!} x^j = \frac{M_2}{2(kT)^2} - \frac{M_3}{3!(kT)^3} + \frac{M_4 - 3 M_3}{4!(kT)^3} - \ldots \qquad (8.5.18)$$

Finally, using (8.5.18), the configurational free energy is obtained from the logarithm of equation (8.5.12):

$$A_c = -kT \ln g(R) + \overline{W} - \frac{M_2}{2kT} + \frac{M_3}{6(kT)^2} - \ldots \qquad (8.5.19)$$

The problem of computing the configurational free energy is thereby reduced to that of computing the moments of the energy. Note that equation (8.5.19) is not restricted to 50-50 alloys since there was no step in the derivation that used such a limitation. The alloy stoichiometry enters in the computation of the moments and the number of configurations $g(R)$. The series has been written only up to the third moment term, and even this is of higher order than is really useful, not only because higher moments are more difficult to compute, but also because the resulting equations get very cumbersome.

We now apply the method to the binary AB alloy by choosing the appropriate expressions for $g(R)$ and the average energy $\overline{W}$. $g(R)$ is then given by (8.5.8). To get rid of the factorials in (8.5.8), use Stirling's approximation so that

$$\ln g(R) = \frac{N}{2}[2 \ln 2 - (1+R) \ln(1+R) - (1-R) \ln(1-R)] \qquad (8.5.20)$$

The average energy $\overline{W}$ is related to the order parameter by taking the average of equation (8.4.15):

$$\overline{W} = w_0 - \frac{Qv}{2}(1+\overline{\sigma}) \qquad (8.5.21)$$

Since the average short-range order parameter is R^2 for a 50-50 alloy [equation (8.3.34)], equation (8.5.21) is

$$\overline{W} = W(0) + \frac{Qv}{2}R^2 \qquad (8.5.22)$$

$W(0)$ is defined by the value of the average energy at zero long-range order so

$$W(0) = -\frac{Q}{4}(v_{AA} + v_{BB} + 2v_{AB}) \qquad (8.5.23)$$

Substitution of (8.5.20) and (8.5.23) into (8.5.19) gives the free energy in terms of the long-range order parameter:

$$A_c = -\frac{NkT}{2}[2\ln 2 - (1+R)\ln(1+R) - (1-R)\ln(1-R)]$$
$$+ W(0) - \frac{Nzv}{4}R^2 - \frac{M_2}{2kT} + \frac{M_3}{6(kT)^2} - \ldots \qquad (8.5.24)$$

where Q has been replaced by $Nz/2$. This gives the configurational free energy for a given R, since the moments are functions of R. If the value of R corresponding to equilibrium is inserted into (8.5.24), the configurational contribution to all of the thermodynamic properties can be obtained.

To get the equilibrium degree of order as a function of temperature, just set the derivative of A_c with respect to R equal to zero. This gives

$$\ln\left(\frac{1+R}{1-R}\right) = \frac{zv}{kT}R + \frac{1}{N(kT)^2}\frac{\partial M_2}{\partial R} - \frac{1}{3N(kT)^3}\frac{\partial M_3}{\partial R} + \ldots \qquad (8.5.25)$$

The configurational energy can be derived from an application of equation (4.4.4), which for the present case is

$$U_c = \frac{\partial(A_c/kT)}{\partial(1/kT)} \qquad (8.5.26)$$

From the derivation of this equation, it is clear that the differentiation is purely formal and that all quantities except $1/kT$ are treated as constants in the differentiation. Then, dividing (8.5.24) by kT and applying (8.5.26) gives

$$U_c = W(0) - \frac{Nzv}{4}R^2 - \frac{M_2}{kT} + \frac{M_3}{2(kT)^2} - \ldots \qquad (8.5.27)$$

for the configurational energy. The configurational entropy is

$$S_c = \frac{U_c - A_c}{T}$$
$$= \frac{Nk}{2}[2\ln 2 - (1+R)\ln(1+R) - (1-R)\ln(1-R)] - \frac{1}{T}\left[\frac{M_2}{2kT} - \frac{M_3}{3(kT)^2} - \ldots\right] \qquad (8.5.28)$$

The heat capacity is the derivative of (8.5.27) with respect to temperature:

$$C_V^c = -\frac{Nzv}{2}R\frac{dR}{dT} - \frac{d}{dT}\left[\frac{M_2}{kT} - \frac{M_3}{2(kT)^2} + \ldots\right] \qquad (8.5.29)$$

8.6 The Bragg–Williams approximation

In the Bragg–Williams approximation, all terms containing second and higher moments are neglected. This is equivalent to assuming that the configurational energy is equal to its arithmetic average, as can be seen by comparing equation (8.5.27) to equation (8.5.22) in which the first two terms are $\overline{W} = W(0) - Nz\nu R^2/2$. At first sight, this is a crude assumption, but the Bragg–Williams approximation succeeds in displaying the main qualitative features of the order-disorder transition and its mathematical development is simple. We will therefore treat the Bragg–Williams method in some detail. Methods of solving the order-disorder problem to a higher degree of accuracy are presented further below.

Neglecting all terms containing M_2 and higher order moments, equations (8.5.24), (8.5.25), (8.5.27), (8.5.28), and (8.5.29) reduce to

$$A_c = W(0) - \frac{Nz\nu}{4}R^2 - \frac{NkT}{2}[2\ln 2 - (1+R)\ln(1+R) - (1-R)\ln(1-R)] \tag{8.6.1}$$

$$\ln\left(\frac{1+R}{1-R}\right) = \frac{z\nu}{kT}R \tag{8.6.2}$$

$$U_c = W(0) - \frac{Nz\nu}{4}R^2 \tag{8.6.3}$$

$$S_c = \frac{Nk}{2}[2\ln 2 - (1+R)\ln(1+R) - (1-R)\ln(1-R)] \tag{8.6.4}$$

$$C_V^c = -\frac{Nz\nu}{2}R\frac{dR}{dT} \tag{8.6.5}$$

The relation between the thermodynamic quantities and the degree of order can be shown more clearly if we remove the constants that clutter up the above equations by defining the following quantities:

$$a = \frac{4[A_c - W(0)]}{Nz\nu} \tag{8.6.6}$$

$$s = \frac{2S_c}{Nk} \tag{8.6.7}$$

$$u = \frac{4[U_c - W(0)]}{Nz\nu} \tag{8.6.8}$$

$$\phi = \frac{2kT}{z\nu} \tag{8.6.9}$$

The quantities a, and u are the dimensionless free energy and energy, respectively, measured from the energy of the completely disordered crystal in units of $Nzv/4$. s is a dimensionless entropy, and ϕ is a dimensionless temperature. Using these definitions, equations (8.6.1)–(8.6.5) give

$$a = u - \phi s \tag{8.6.10}$$

$$\ln\left(\frac{1+R}{1-R}\right) = \frac{2R}{\phi} \tag{8.6.11}$$

$$u = -R^2 \tag{8.6.12}$$

$$s = 2\ln 2 - (1+R)\ln(1+R) - (1-R)\ln(1-R) \tag{8.6.13}$$

$$C_V^c = -NkR\frac{dR}{d\phi} \tag{8.6.14}$$

Figure 8.2 shows u and s as functions of the degree of order. The configurational entropy increases from zero at perfect order to the entropy of mixing for a random solution (2ln 2) at complete disorder. Similarly, the dimensionless configurational energy decreases from zero for the completely disordered crystal to unity at perfect order.

In figure 8.3, the dimensionless free energy is shown as a function of R for several different temperatures ranging from $\phi = 0.8$ to $\phi = 1.05$. This figure shows that for values of $\phi < 1$, there is a minimum in the free energy-order curve for each temperature. This minimum defines the equilibrium value of R at that temperature.

As the temperature is raised, the minimum shifts to lower values of R until, at $\phi = 1$, the minimum coincides with $R = 0$. Also, for $\phi < 1$, as $R \to 0$ the free energy-order curve approaches a zero slope. This means that above $\phi = 1$ there is no nonzero value of R for which the slope of the free energy temperature curve is zero, so $\phi = 1$ defines a critical temperature above which there is

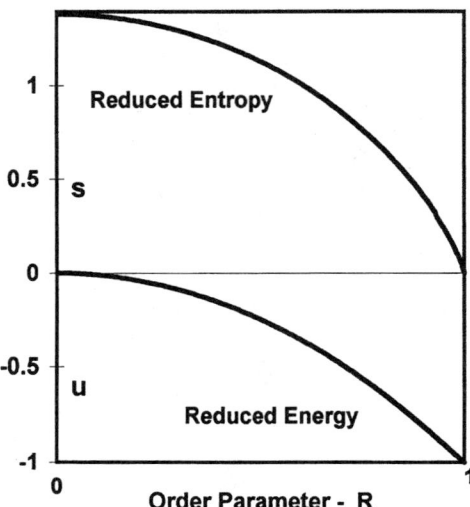

Figure 8.2. Reduced energy and entropy as a function of order in the Bragg–Williams approximation.

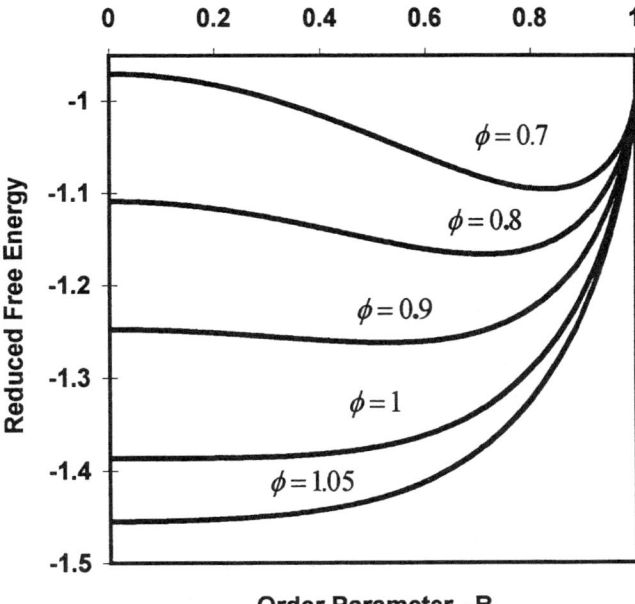

Figure 8.3. Free energy as a function of order in the Bragg–Williams approximation.

no long-range order. From (8.6.9), this critical temperature T_c is defined by $1 = 2kT_c/zv$, or

$$T_c = \frac{zv}{2k} \qquad (8.6.15)$$

At the critical temperature the first and second derivatives of the free energy with respect to the long-range order parameter, as well as the order parameter itself, are zero. Equation (8.6.15) for the critical temperature is easily derived from these conditions.

The equilibrium degree of order as a function of temperature is given by (8.6.11) or, equivalently, by (8.6.2). At high temperatures R is small, so the logarithms can be approximated by the first three terms of their series expansions. That is, $\ln(1+R) = R - R^2/2 + R^3/3$, and $\ln(1-R) = -R - R^2/2 - R^3/3$, so for high enough temperatures (8.6.11) becomes

$$2R + \frac{2}{3}R^3 = \frac{2R}{\phi} \qquad (8.6.16)$$

or

$$R^2 = 3\left(\frac{1-\phi}{\phi}\right) \quad \text{(high } T\text{)} \qquad (8.6.17)$$

When $R = 0$, $\phi = 1$ which is the same result as (8.6.15).

A more accurate high-temperature approximation can be obtained by including more terms in the expansions of the logarithms. Thus, if we take

$$\ln(1+R) = R - \frac{R^2}{2} + \frac{R^3}{3} - \frac{R^4}{4} + \frac{R^5}{5}$$
$$\ln(1-R) = -R - \frac{R^2}{2} - \frac{R^3}{3} - \frac{R^4}{4} - \frac{R^5}{5}$$

(8.6.18)

then (8.6.11) becomes

$$R^2 = \frac{5}{6}\left[\sqrt{1 + \frac{36(1-\phi)}{5\phi}} - 1\right] \quad (R < 0.5) \tag{8.6.19}$$

This equation is a good approximation up to about $R = 1/2$, where it is in error by 6%.

Equation (8.6.11) can be written in another form that is convenient for getting numerical solutions of R versus ϕ. From the definition of the hyperbolic tangent,

$$\tanh\left(\frac{R}{\phi}\right) = \frac{e^{2R/\phi} - 1}{e^{2R/\phi} + 1} \tag{8.6.20}$$

it is easy to see that (8.6.11) can be converted to

$$R = \tanh\left(\frac{R}{\phi}\right) \tag{8.6.21}$$

or, solving for ϕ,

$$\phi = \frac{R}{\tanh^{-1} R} \tag{8.6.22}$$

so that numerical values of ϕ versus R can be calculated from standard tables of the hyperbolic tangent. But with the modern ubiquity of microcomputers, it is now just as easy to use (8.6.11) directly. A graph of R versus ϕ is shown in figure 8.4, which shows that at low temperatures R decreases slowly with temperature, but as the temperature increases, the decrease in R becomes more and more rapid. R changes from 0.8 to zero in a range of ϕ corresponding to only 27% of the range from absolute zero to the critical temperature.

With the help of the numerically computed values of R as a function of ϕ, the energy, entropy, and free energy are readily determined as functions of temperature. These are shown graphically in figure 8.5 in the dimensionless forms defined by equations (8.6.10)–(8.6.13).

The energy and entropy at first rise slowly with increasing temperature, and then for $\phi > 0.4$ they increase very rapidly. The free energy is almost constant until ϕ is about 0.5 and then steadily drops to the value characteristic of a random solution at $\phi = 1$. This behavior is typical of cooperative phenomena.

At low temperatures, R/ϕ is large, so a low-temperature approximation can be obtained by using the following series expansion for the hyperbolic tangent:

$$\tanh\left(\frac{R}{\phi}\right) = 1 - 2e^{-2R/\phi} + 2e^{-4R/\phi} - \ldots \tag{8.6.23}$$

For values of ϕ as high as 0.9 (corresponding to $R = 0.53$), the third term in this series is only 6% of the second term. For $R > 0.5$, therefore, it is sufficiently

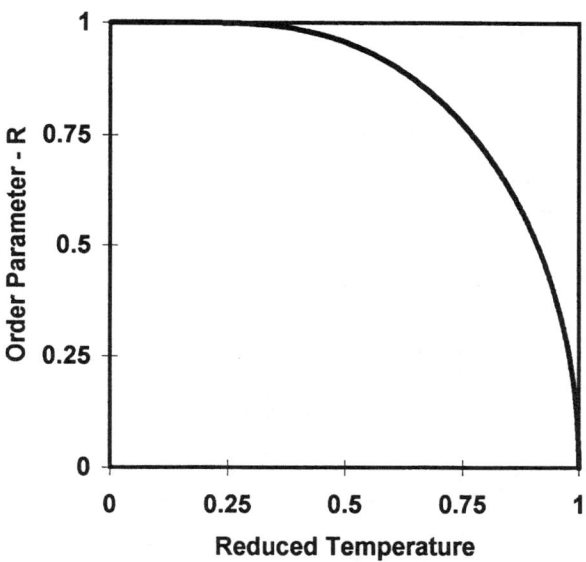

Figure 8.4. Long-range order versus reduced temperature in the Bragg–Williams approximation.

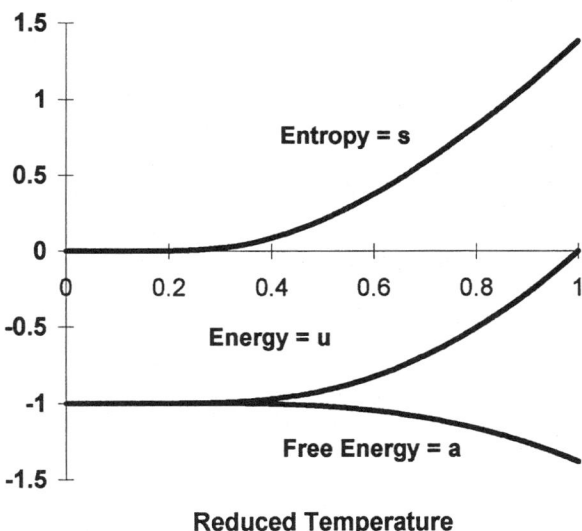

Figure 8.5. Reduced energy, entropy, and free energy as a function of reduced temperature in the Bragg–Williams approximation.

accurate to retain only the first two terms of the expansion, and (8.6.21) then becomes

$$R = 1 - 2e^{-2R/\phi} \tag{8.6.24}$$

or

$$\phi = \frac{2R}{\ln 2 - \ln(1-R)} \quad (R > 0.5) \tag{8.6.25}$$

To get the heat capacity from this, the derivative $dR/d\phi$ must first be evaluated from (8.6.11), which we write as $\ln(1+R) - \ln(1-R) = 2R/\phi$, so that $dR/(1+R) + dR/(1-R) = 2dR/\phi - 2Rd\phi/\phi^2$, from which we get

$$\frac{dR}{d\phi} = \frac{R(1-R^2)}{\phi(1-\phi-R^2)} \tag{8.6.27}$$

Putting this into (8.6.14) gives

$$C_V^c = -Nk\frac{R^2(1-R^2)}{\phi(1-\phi-R^2)} \tag{8.6.28}$$

This is a form that allows the heat capacity to be computed as a function of temperature from the numerical values of R versus ϕ.

Note that the maximum value of the Bragg–Williams configurational heat capacity is $1.5k$ per atom, as can be seen by differentiating the high-temperature expression for the R versus ϕ relation and substituting it into (8.6.14). That is, from (8.6.19),

$$2R\frac{dR}{d\phi} = -\frac{3}{\phi^2\sqrt{1 + \frac{36(1-\phi)}{5\phi}}} \tag{8.6.29}$$

so, as ϕ goes to unity, $dR/d\phi$ goes to 3/2 and the heat capacity goes to $3Nk/2$.

8.7 The second moment approximation

The Bragg–Williams approximation reproduces the general features of order-disorder systems, but its results are, at best, only semiquantitative. Its limitations are a result of neglecting the correct distribution of the configurational energy and replacing all energies of the distribution by the average energy. A more accurate theory is obtained by retaining the second moment in the Kirkwood method. To do this, an explicit calculation of M_2 as a function of order is required. The calculation is a lattice-counting exercise whose result is (see appendix 6)

$$M_2 = \frac{Nzv^2}{8}(1-R^2)^2 \tag{8.7.1}$$

and the configurational free energy, energy, and entropy are then obtained to the second moment approximation by inserting (8.7.1) into (8.5.24), (8.5.27), and (8.5.28) and neglecting all terms containing third and higher order moments.

To get the equilibrium order parameter as a function of temperature from (8.5.25), the derivative of M_2 with respect to R is needed. From (8.7.1) this is

$$\frac{dM_2}{dR} = -\frac{Nzv^2}{2}(1-R^2)R \qquad (8.7.2)$$

Also, the second moment approximation to the heat capacity requires that we put the derivative of M_2/kT with respect to T into equation (8.5.29). From (8.7.1) this derivative is

$$\frac{\partial}{\partial T}\left(\frac{M_2}{kT}\right) = -\frac{Nzv^2}{8kT}\left[\frac{(1-R^2)^2}{T} + 4R(1-R^2)\frac{dR}{dT}\right] \qquad (8.7.3)$$

Substitution of (8.7.2) into (8.5.25) and neglecting higher moments gives the equilibrium relation between long-range order and temperature as

$$\ln\left(\frac{1+R}{1-R}\right) = \frac{zv}{kT}R\left[1 - \frac{v}{2kT}(1-R^2)\right] \qquad (8.7.4)$$

and using (8.7.3) in (8.5.29) gives the second moment approximation to the heat capacity as

$$C_V^c = -\frac{Nzv}{2}R\frac{dR}{dT}\left[1 - \frac{v}{kT}(1-R^2)\right] + \frac{Nzv^2}{8kT^2}(1-R^2)^2 \qquad (8.7.5)$$

The numerical methods of computing the thermodynamic functions are somewhat more cumbersome than in the Bragg–Williams case. The critical temperature, however, is easily obtained from a high-temperature approximation by expanding the logarithms in (8.7.4) to the third order just as in the Bragg–Williams case. Equation (8.7.4) then gives

$$2 + \frac{2R^2}{3} = \frac{zv}{kT} - \frac{zv^2}{2(kT)^2}(1-R^2) \qquad \text{(high temperature)} \qquad (8.7.6)$$

The critical temperature T_c is obtained from (8.7.6) by setting $R = 0$ to get

$$\frac{2kT_c}{zv} = 1 - \frac{v}{2kT_c} \qquad (8.7.7)$$

from which

$$\frac{2kT_c}{zv} = \frac{1}{2}\left(1 \pm \sqrt{1 - \frac{4}{z}}\right) \qquad (8.7.8)$$

The appropriate root to be chosen in (8.7.8) is determined by requiring that it reduce to the Bragg–Williams result when the second term on the right of (8.7.7) is neglected. Without this second term, $4/z$ would not appear under the radical in (8.7.8) and the right-hand side would reduce to $(1 \pm 1)/2$. Choosing the positive sign would then give the Bragg–Williams result for T_c, so we must take (8.7.8) to be

$$\frac{2kT_c}{zv} = \frac{1}{2}\left(1 + \sqrt{1 - \frac{4}{z}}\right) \qquad (8.7.9)$$

From the definition of the critical temperature in the Bragg–Williams approximation, we have

$$\frac{2k}{zv} = \frac{1}{T_c(\text{BW})} \tag{8.7.10}$$

Combining this with (8.7.9) gives a relation between the critical temperatures computed from the second moment and the Bragg–Williams approximations:

$$\frac{T_c}{T_c(\text{BW})} = \frac{1}{2}\left(1 + \sqrt{1 - \frac{4}{z}}\right) \tag{8.7.11}$$

For $z = 8$, this becomes

$$\frac{T_c}{T_c(\text{BW})} = 0.854 \tag{8.7.12}$$

For a given ordering energy, the second moment approximation therefore gives a lower critical temperature than the Bragg–Williams approximation.

The numerical computations for the Bragg–Williams case can be used as a basis for computing the temperature variation of the long-range order parameter from (8.7.4) by treating R as the independent variable and defining a function $f(R)$ by

$$f(R) \equiv 2R \ln\left(\frac{1-R}{1+R}\right) = \frac{R}{\tanh^{-1} R} \tag{8.7.13}$$

This is just the ratio of the temperature to the critical temperature in the Bragg–Williams approximation as given by equation (8.6.22), so $f(R) = \phi(\text{BW})$ is known when Bragg–Williams calculations are made.

Using (8.7.13) in (8.7.4) gives

$$\frac{zv}{2kT}\left[1 - \frac{v}{2kT}(1 - R^2)\right] = \frac{1}{f(R)} \tag{8.7.14}$$

From the definition of the Bragg–Williams critical temperature,

$$\frac{v}{2k} = \frac{T_c(\text{BW})}{z} \tag{8.7.15}$$

and from (8.7.11)

$$T_c(\text{BW}) = \kappa T_c \tag{8.7.16}$$

where κ is determined by (8.7.11) and is 1.172 for $z = 8$. Using (8.7.15) and (8.7.16), we rewrite (8.7.14) as

$$\kappa \frac{T_c}{T}\left[1 - \frac{\kappa T_c}{zT}(1 - R^2)\right] = \frac{1}{f(R)} \tag{8.7.17}$$

Solving this for T/T_c gives

$$\frac{T}{T_c} = \frac{\kappa}{2} f(R)\left[1 + \sqrt{1 - \frac{4(1 - R^2)}{z f(R)}}\right] \tag{8.7.18}$$

where the positive square root was taken to ensure that $T/T_c \to 1$ as $R \to 0$. Equation (8.7.18) provides a convenient way to get numerical values for the order–temperature relation.

An important difference between the Kirkwood and Bragg–Williams approximations for the heat capacity can be seen immediately by examining (8.7.5). Above the critical temperature, $R = 0$ and $dR/dT = 0$, so (8.7.5) gives

$$C_V^c = \frac{Nzv^2}{8kT^2} \quad (T > T_c) \tag{8.7.19}$$

The configurational heat capacity does not immediately vanish above the critical temperature, but falls off as T^{-2}. That is, the ordering phenomenon contributes to the heat capacity even when the long-range order parameter is zero. This is a result of the fact that short-range order persists even after there is no long-range order, and the short-range order vanishes only when the temperature is increased above that needed to get rid of long-range order. The Bragg–Williams approximation completely misses this phenomenon since it predicts a complete vanishing of the heat capacity as soon as the critical temperature is exceeded.

Now let us compute the heat capacity at the critical temperature. From the high-temperature approximation (8.7.6), and using the relation

$$\frac{zv}{2kT_c} = \kappa \tag{8.7.20}$$

it is easy to show that

$$R\frac{dR}{dT}\bigg|_{T=T_c} = \frac{3\kappa}{2T_c}\left(\frac{2\kappa - z}{z - 3\kappa^2}\right) \tag{8.7.21}$$

Put this in (8.7.5), set $R = 0$, and use (8.7.20) to get the maximum configurational heat capacity:

$$C_V^c(T \to T_c) = \frac{3}{2}Nk\kappa^2\left(\frac{z - 2\kappa}{z - 3\kappa^2}\right)\left(1 - \frac{2\kappa}{z}\right) + Nk\frac{\kappa^2}{2z} \tag{8.7.22}$$

for $z = 8$, $\kappa = 1.172$, and

$$C_V^c(T \to T_c) = 2.207Nk \tag{8.7.23}$$

In the second moment approximation, therefore, the peak in the heat capacity curve is higher than in the Bragg–Williams approximation.

8.8 The quasi-chemical approximation

In the quasi-chemical method, it is assumed that the order-disorder crystal can be treated as if it were composed of independent pairs of AA, BB, and AB atoms. That is, in computing the number of ways of distributing the pairs on sublattices, the fact that the existence of one pair places conditions on the distribution of other pairs is ignored. The result is that the order-disorder transition is treated as an exchange of atoms that converts like pairs into unlike pairs. This is analogous to a bimolecular gas phase reaction. Such a procedure works for 50-50 AB alloys, but for more complex structures the "molecules" have to be groups of atoms larger than pairs. We will therefore restrict ourselves to AB alloys with two sublattices.

If the usual course of derivation is followed, we would determine the number of complexions $g(R)$ from these assumptions, and this would permit a calculation of the free energy to be made. In this procedure, it is assumed that $g(R)$ is proportional to the number of ways of putting pairs in the crystal such that the pairs are independent. We will take an alternate course that starts with the energy and turns out to be very convenient.

In a 50-50 AB alloy, the energy for the kth configuration can be written as

$$W_k = 2vQ_{AA} - v_{AB}Q \tag{8.8.1}$$

as can be seen by combining the definition of the ordering energy given by equation (8.4.12) with the expression for the conservation of the number of pairs, which is $Q = Q_{AA} + Q_{BB} + (Q_{AB} + Q_{BA}) = Q_{AA} + Q_{BB} + Q_{AB}^t$, $Q_{AA} = Q_{BB}$, $Q = 2Q_{AA} + Q_{AB}$. Note that we are using the notation in which $Q_{AB}(Q_{BA})$ is the number of pairs such that an A(B) atom is on an α site while a B(A) atom is on a β site.

For a crystal with a given long-range order parameter, the configurational energy is just the statistical mechanical average of (8.8.1):

$$U_c(R) = \overline{W_k} = 2v\overline{Q_{AA}} - v_{AB}Q \tag{8.8.2}$$

From here on, the bars indicate that the average is statistical mechanical, not arithmetic. The average is taken over all $g(R)$ configurations consistent with a given R.

The configurational free energy can be obtained from the energy by integrating (8.5.26) to get

$$A_c = A_c(T \to \infty) + kT \int_0^{1/kT} U_c d(1/kT) \tag{8.8.3}$$

From the relation between the free energy and the partition function, the high-temperature limit of the free energy is

$$A_c(T \to \infty) = -kT \ln g(R) \tag{8.8.4}$$

because at high temperatures the exponential in the partition function goes to unity and the sum of unity over all configurations is just the number of configurations. Equation (8.8.4) is therefore

$$A_c = kT \int_0^{1/kT} U_c d(1/kT) - kT \ln g(R) \tag{8.8.5}$$

To integrate this, U_c is needed as a function of temperature. In the quasi-chemical method, this is obtained by treating the formation of pairs as a chemical reaction of the type

$$(AA) + (BB) \leftrightarrow (AB) + (BA) \tag{8.8.6}$$

That is, a pair such that an A atom is on an α site and another A atom is on an adjacent β site reacts with a pair in which two adjacent sites contain B atoms to give two pairs, one of which has an A on α and a B on β while the other has a B on α and an A on β.

The energy change accompanying the conversion of two like pairs (AA) and (BB) to two unlike pairs (AB) and (BA) is

$$-v_{AB} - v_{AB} + v_{AA} + v_{BB} = -2v_{AB} + v_{AA} + v_{BB} = -2v \tag{8.8.7}$$

Since vibrational effects are ignored, the free energy of the reaction is just the transformation energy $-2v$. Then, from the theory of the equilibrium constant for gas reactions,

$$\frac{\overline{Q}_{AA}\overline{Q}_{BB}}{\overline{Q}_{AB}\overline{Q}_{BA}} = e^{-2v/kT} \tag{8.8.8}$$

This formula treats the pairs as independent, so it certainly cannot be correct. However, the quasi-chemical model leads to more accurate results than does the Bragg–Williams approximation, which completely neglects the statistical mechanical weighting of pair formation. Actually, it turns out that the quasi-chemical model gives results that are similar to the Kirkwood second moment approximation.

Equation (8.8.8) is useful because it can be converted to a form that gives Q_{AA} as a function of R. Therefore, the energy (8.8.2) and finally the free energy (8.8.6) can be obtained as functions of the long-range order parameter. The equilibrium relation between R and T can then be obtained in the usual way by minimizing the free energy, and this solves the problem of getting the thermodynamic properties as functions of temperature.

The number of pairs that have an A atom on an α site is $zN_{A\alpha}$. Therefore,

$$\overline{Q}_{AB} + \overline{Q}_{AA} = zN_{A\alpha} \tag{8.8.9}$$

Similarly,

$$\overline{Q}_{BA} + \overline{Q}_{BB} = zN_{B\alpha} \tag{8.8.10}$$

To forestall any confusion, note that in the present case it is correct to relate the statistical mechanical average to the occupation numbers of atoms on sites, whereas in relating the short- and long-range order in section 8.3 the average number of pairs were *arithmetic* averages. The reason for this is that in section 8.3 the occupation numbers were used to describe the simultaneous occupation of adjacent sites and required a product of probabilities that were not weighted. In the present case, however, we are counting the number of atoms in a crystal with atoms distributed according to the distribution function of an ensemble.

Now let x be defined by

$$x = e^{2v/kT} \tag{8.8.11}$$

Then, if (8.8.9) and (8.8.10) are solved for $\overline{Q}_{AB}$ and $\overline{Q}_{BA}$, and the result substituted into (8.8.9), remembering that for an AB alloy $\overline{Q}_{AA} = \overline{Q}_{BB}$ we get

$$\frac{\overline{Q}_{AA}^2}{(zN_{A\alpha} - \overline{Q}_{AA})(zN_{A\beta} - \overline{Q}_{AA})} = \frac{1}{x} \tag{8.8.12}$$

This is a quadratic equation for $\overline{Q}_{AA}$, whose solution is

$$\overline{Q}_{AA} = \frac{zN}{4(x-1)}\left[\sqrt{1 + \frac{16N_{A\alpha}N_{B\beta}}{N^2}(x-1)} - 1\right] \tag{8.8.13}$$

where we have used equation (8.3.5). The positive root was chosen because the number of AA pairs has to be positive. Recalling equations (8.5.3) and (8.5.4) for the relations between $N_{A\alpha}$, $N_{A\beta}$, and R (for the case of AB alloys), this becomes

$$\overline{Q}_{AA} = \frac{zN}{4(x-1)}\left[\sqrt{1+(1-R^2)(x-1)} - 1\right] \quad (8.8.14)$$

Putting (8.8.14) into (8.8.2) gives the energy in terms of R as

$$U_c = \frac{zvN}{2(x-1)}\left[\sqrt{1+(1-R^2)(x-1)} - 1\right] - v_{AB}Q \quad (8.8.15)$$

This is the integrand of the free energy expression (8.8.5), which now becomes

$$A_c = \frac{zvN}{2}\int_0^{1/kT}\left[\frac{\sqrt{1+(1-R^2)(x-1)}-1}{(x-1)}\right]d\left(\frac{1}{kT}\right) - v_{AB}Q - kT\ln g(R) \quad (8.8.16)$$

The change of variable to α defined by

$$\alpha^2 = 1+(1-R^2)(x-1) = 1+(1-R^2)(e^{2v/kT} - 1) \quad (8.8.17)$$

transforms the integral in (8.8.16) to

$$\frac{1}{v}\int_1^\alpha \frac{(1-R^2)\alpha\, d\alpha}{(\alpha+1)(\alpha^2-R^2)}$$

$$= \frac{(1-R)}{2v}\int_1^\alpha \frac{d\alpha}{(\alpha-R)} + \frac{(1+R)}{2v}\int_1^\alpha \frac{d\alpha}{(\alpha+R)} - \frac{1}{v}\int_1^\alpha \frac{d\alpha}{(\alpha+1)} \quad (8.8.18)$$

The right-hand side is the result of reducing the integrand on the left-hand side to partial fractions.

Performing the integrations and substituting into (8.8.16) gives the free energy as

$$A_c = \frac{zNkT}{4}\left[(1-R)\ln\left(\frac{\alpha-R}{1-R}\right) + (1+R)\ln\left(\frac{\alpha+R}{1+R}\right) - 2\ln\left(\frac{\alpha+1}{2}\right)\right]$$
$$- Qv_{AB} - kT\ln g(R) \quad (8.8.19)$$

To get the equilibrium R versus T relation, set the derivative of (8.8.19) equal to zero:

$$\frac{\partial A_c}{\partial R} = \frac{zNkT}{4}\left[\ln\left(\frac{(\alpha+R)(1-R)}{(\alpha-R)(1+R)}\right) + \frac{\partial \alpha}{\partial R}\left(\frac{1-R}{\alpha-R} + \frac{1+4}{\alpha+R} - \frac{2}{\alpha+1}\right)\right]$$
$$\frac{zNkT}{4}\left[\frac{2R(1-\alpha)}{\alpha^2-R^2}\right] - kT\frac{\partial \ln g}{\partial R} = 0 \quad (8.8.20)$$

The derivatives on the right-hand side are readily evaluated. From (8.8.17),

$$\frac{\partial \alpha}{\partial R} = \frac{R(1-\alpha^2)}{\alpha(1-R^2)} \quad (8.8.21)$$

and from (8.5.20),

$$\frac{\partial \ln g}{\partial R} = -\frac{N}{2}\ln\left(\frac{1+R}{1-R}\right) \qquad (8.8.22)$$

Substitution of (8.8.21) and (8.8.22) into (8.8.20) gives the desired functional relation between the equilibrium value of the long-range order parameter and the temperature:

$$\ln\left(\frac{1+R}{1-R}\right) = \frac{z}{z-2}\ln\left(\frac{\alpha+R}{\alpha-R}\right) \qquad (8.8.23)$$

The critical temperature is determined by expanding the logarithms in series for small R and letting $R \to 0$ as follows:

$$2R + \frac{R^3}{3} + \ldots = \frac{z}{z-2}\left(\frac{2R}{\alpha} + \frac{2R^3}{3\alpha^3} + \ldots\right) \qquad (8.8.24)$$

When $R \to 0$, $\alpha^2 \to e^{2v/kT_c}$, so (8.8.24) gives

$$\frac{v}{kT_c} = \ln\left(\frac{z}{z-2}\right) \qquad (8.8.25)$$

It is of interest to compare the values of the critical temperatures given by the various approximations. Table 8.1 gives the critical temperature in units of v/k for a crystal with $z = 8$. The agreement between the Kirkwood second moment approximation and the quasi-chemical approximation is good. It can be considerably improved by including third moments in the Kirkwood method. In fact, if the free energy given by (8.8.19) is expressed in series form by expanding the logarithms in $(\alpha - 1)$ and then expanding α in v/kT, it can be shown that the result is identical to the Kirkwood series up to terms cubic in v/kT. Thus, although the quasi-chemical method is based on a simplified physical model, its accuracy is comparable to that of the Kirkwood method for practical purposes.

Equation (8.8.23) can be used to compute the long-range order parameter as a function of temperature as follows: solving for α, (8.8.23) gives

$$\alpha = \left[\frac{\left(\frac{1+R}{1-R}\right)^{(z-2)/z} + 1}{\left(\frac{1+R}{1-R}\right)^{(z-2)/z} - 1}\right] R \qquad (8.8.26)$$

so a table of α versus R can be constructed. From (8.8.17),

Table 8.1: Critical Temperatures According to Various Approximations

Approximation	kT_c/v
Bragg–Williams	4
Kirkwood (second moment)	3.414
Quasi chemical	3.476

$$\frac{kT}{v} = 2\left[\ln\left(\frac{\alpha^2 - R^2}{1-R^2}\right)\right]^{-1} \qquad (8.8.27)$$

and from α versus R, the temperature can be computed in units of v/k. Combining (8.8.27) with (8.8.25), we can get the reduced temperature as

$$\frac{T}{T_c} = 2\ln\left(\frac{z}{z-2}\right)\left[\ln\left(\frac{\alpha^2 - R^2}{1-R^2}\right)\right]^{-1} \qquad (8.8.28)$$

Once R is computed as a function of T, the free energy and the other thermodynamic functions can be obtained from (8.8.20).

8.9 Comparison with experiment

Experimental data on both the degree of order and the heat capacity exist for β-brass as a function of temperature. These data allow a sensitive test to be made of the statistical theories of the order-disorder transformation. The most direct test is a comparison of the experimental long-range order parameter[1] with theoretical values and this is shown in figure 8.6. The Bragg–Williams curve (BW) is the same as in figure 8.4. The quasi-chemical curve (QC) was computed from equation (8.8.30), using a table of α versus R computed from equation (8.8.27).

Figure 8.6 shows that both the Bragg–Williams and the quasi-chemical theories reproduce the general features of the order-disorder transition in that the degree of order changes slowly at low temperatures, but as the temperature is increased, the crystal becomes ever more rapidly disordered. Both theories, however, predict a change from the ordered to the disordered state that is spread out over a larger temperature range than is actually the case. The data

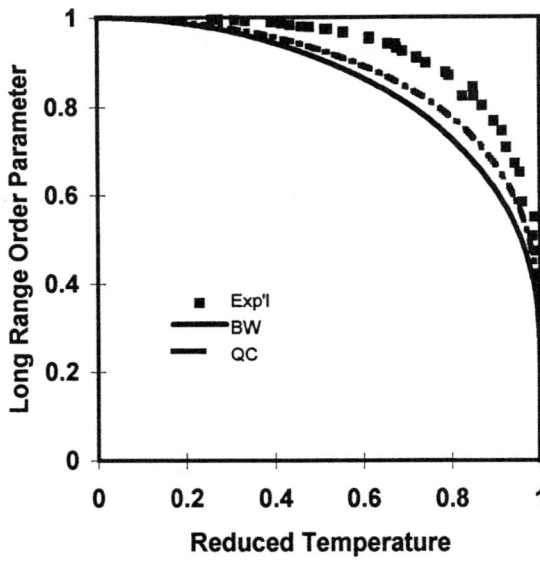

Figure 8.6. Comparison of order-disorder data to theory for CuZn. Exp'l: experimental data; BW: Bragg–Williams approximation; QC: quasi-chemical estimation.

show that the onset of disorder is precipitous, taking place over a relatively narrow range of temperature. The theories capture the overall shape of the transition, but not the rapidity of its appearance with increasing temperature. The quasi-chemical method clearly is an improvement over the Bragg–Williams method, but not sufficiently so to yield the rapid change in order over a short temperature range shown by the data.

Experimental values for the heat capacity are shown in figure 8.7 along with calculations from Bragg–Williams and the quasi-chemical theories. The Bragg–Williams curve was computed from equation (8.6.28). The quasi-chemical curve was obtained by differentiating equation (8.8.15) with respect to temperature to get an expression for the heat capacity. The resulting equation contained both α and its derivative with respect to ϕ, so a table of α versus ϕ was computed from equation (8.8.27) and numerically differentiated to get the needed derivative.

Figure 8.7 shows that the heat capacity resulting from the order-disorder transition goes to a much higher value than that of either the Bragg–Williams or the quasi-chemical theories. Also, the calculated values display a transition that takes place over a larger temperature range than that shown by experiment. This is consistent with the results shown in figure 8.6. The statistical theories underestimate both the sharpness of the transition and its magnitude. An important feature of the data is that the order-disorder contribution to the heat capacity does not instantly vanish immediately above the transition temperature. There is an appreciable tail at high temperatures. The interpretation of this is that, although the long-range order vanishes above the critical temperature, there is still some short-range order, and its decrease with temperature yields a contribution to the heat capacity. The quasi-chemical theory is

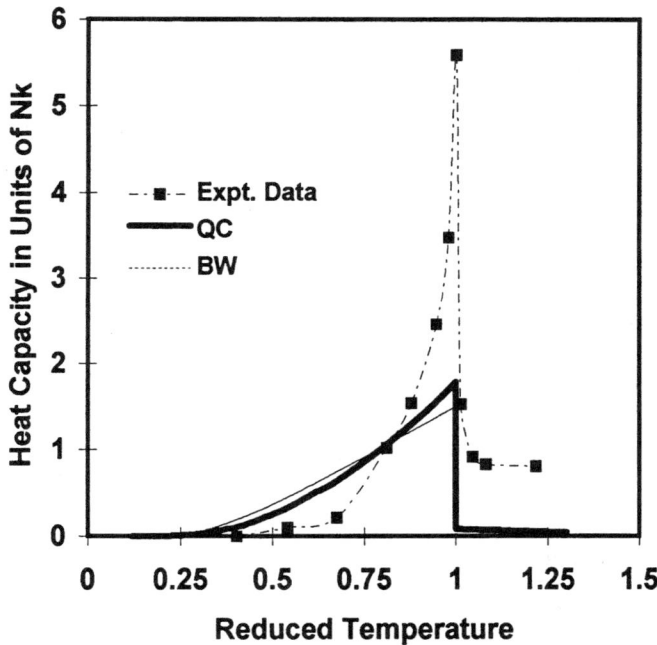

Figure 8.7. Comparison of Bragg–Williams (BW) and quasi-chemical (QC) heat capacity to experiment for CuZn: $T_c = 740K$.

much better than the Bragg–Williams in this regard since it does show a residual short-range order heat capacity above the critical temperature while the Bragg–Williams theory does not.

A few minor points need to be noted about our use of the experimental data.[2] First, the data were for an alloy containing 0.475 atomic percent zinc rather than 0.50. Second, the critical temperature for the heat capacity data was taken to be 740K rather than the 736K used for the order parameter plot of figure 8.6. The reason for this was that the raw heat capacity data fit this value well. Third, the experimental data were at constant pressure rather than at constant volume. But the difference between constant volume and constant pressure values of the heat capacity of metals and alloys is always less than 1% or 2% even at high temperatures, and the experimental data are less accurate than that. Finally, the order-disorder contribution to the heat capacity was obtained by subtracting the classical value of $3k$ per atom from the experimental data. Again, the error introduced by this procedure is less than that in the experimental data.

There are several possible origins for the discrepancy between the theories and experiment. The most obvious one is that the theoretical calculation of the number of complexions is not accurate. In the Bragg–Williams approximation, the site occupation probabilities are taken to be independent of each other. In the quasi-chemical approximation this is improved somewhat by counting pairs, but these pairs are then taken to be independent. The actual number of complexions is difficult to compute since the occupation probabilities of all sites are related.

Another source of the discrepancy is the nearest neighbor approximation for the interaction energies. Even if we grant the possibility of expressing the energy in pairwise form through pseudopotential theory, interactions among atoms that are farther apart than nearest neighbors should be taken into account. These can have an appreciable effect on the order-disorder transition.

A third source of error that has been considered is the neglect of the vibrational contribution to the free energy, and this has been reported to give an improvement by raising the maximum heat capacity. But because of the insensitivity of the heat capacity to the vibrational distribution function in the region of the transition temperature, it is hard to see how including the lattice vibrations could have a major effect on the theoretical results.

Exercises

8.1 Take the critical temperature of β-brass to be 740K and find

A. the ordering energy, and
B. the temperature at which the long-range order parameter is equal to 0.1.

8.2 If the long-range order parameter of a 50-50 order-disorder alloy is 0.05, what is the average fraction of unlike (AB) pairs?

8.3 Show that, for a 50-50 order-disorder alloy, as the temperature approaches the critical temperature the rate of decrease of the long-range order with temperature approaches infinity.

8.4 If the ordering energy of a 50-50 AB alloy with the CsCl structure is 1/2 the energy of an AB bond, use the Bragg–Williams approximation to

A. show that the energy of the completely disordered crystal is 1.5 times the average energy of pure A and pure B, and

B. show that an ordering energy of 1/2 that of an AB bond is completely unrealistic by computing the critical temperature, assuming that the average cohesive energy of A and B is 5 eV/atom.

8.5 The fully ordered BCC structure of 50-50 β-brass consists of a set of parallel planes alternately occupied by Cu atoms (cube centers) and Zn atoms (cube corners). Consider a plane P midway between a Cu plane (on the left) and a Zn plane (on the right). On the right-hand side of the plane P interchange all Cu atoms and Zn atoms. This forms a Cu-Cu antiphase domain boundary. A Zn-Zn antiphase domain boundary is formed by interchanging all Cu and Zn atoms on the left-hand side of the P plane. Using the simple pair-bond energy assumptions of Bragg–Williams theory and the critical temperature of 740K, compute the energy change (per atom in a plane) in going from a perfectly ordered crystal to a crystal having one Cu-Cu and one Zn-Zn antiphase domain boundary (given an atomic surface density of 1.15×10^{15}; atoms/cm^2 = 1.15×10^{19} atoms/m^2).

8.6 Consider a crystal of 50-50 β-brass containing edge dislocations such that the extra half planes are (111) planes. The dislocations are formed by transporting either an α half plane or a β half plane from a (111) surface to the interior of the crystal. Derive a formula for the average energy per dislocation as a function of long-range order, assuming the number of dislocations formed by moving an α half plane equals the number formed by moving a β half plane. Neglect atomic relaxation and elastic effects. (Hint: The key to this problem is in carefully setting it up with an initial drawing so that the α and β planes are properly and consistently identified.) Note that this order-disorder contribution to the dislocation energy is proportional to the area of the half plane. This is different than the elastic or atomic relaxation contributions which are proportional to the length of the dislocation line.

Notes

1. Norvell, J.C., and J. Als-Nielsen; 1970; *Physics Review;* vol. 2, p. 277.
2. The data were taken from Hultgren, Ralph, P.D. Desai, D.T. Hawkins, M. Gleiser, and J. Chipman; 1973; *Selected Values of the Thermodynamic Properties of Binary Alloys;* American Society for Metals, Metals Park, Ohio.

9

Magnetic Order

9.1 Magnetic response

When a material is placed in a magnetic field, it can respond in a number of ways, all of which have their origin in the interaction of the field with the angular momenta of the components of the atoms. This response is measured by the isothermal magnetic susceptibility χ_T, which is defined by

$$\chi_T = \left(\frac{\partial M}{\partial H}\right)_T \qquad (9.1.1)$$

where H is the applied external field and M is the magnetization, defined as the magnetic moment per unit volume. (For anisotropic crystals, the susceptibility is a tensor that is related to the crystal structure. This effect will be ignored.)

There is always a diamagnetic response. In this case, the applied magnetic field induces currents in the electron systems of the atoms with an associated magnetic field that opposes the external field. The induced magnetic moments in a diamagnetic material therefore are in the opposite direction of the applied field, and the diamagnetic susceptibility is negative. The diamagnetic effect is small and essentially independent of temperature and will not be treated here.

The atoms or electrons in a material may have permanent magnetic moments of their own, as is the case for conduction electrons in a metal, or for atoms or ions with an odd number of electrons or partially filled inner shells. A magnetic dipole tends to line up in the direction of an applied field, so the magnetic susceptibility is positive. That is, the moments in a paramagnetic material tend to line up in the same direction as the applied field.

In some materials, the magnetic moments on different atoms interact with each other to such an extent that the direction of a magnetic dipole is strongly influenced by the direction of the magnetic dipoles on its neighbors. If the interaction is such as to align the moments in the same direction, the material is ferromagnetic. Iron is the most important example of a ferromagnetic material. It is body-centered cubic, and its magnetization can be understood by assuming that each site can exist in one of two spin states, corresponding to two magnetic moments pointing in opposite directions. At low temperatures the interaction between these states causes the magnetic moments to line up in the same direction, thereby giving rise to strong internal fields. As the temperature is raised, increasing thermal agitation and entropy oppose the interaction energy, thereby decreasing the number of moments that are aligned in the same direction. Ferromagnetism in iron is therefore an order-disorder phenomenon analogous to that in β-brass, in which each site can have one of two kinds of atoms. When a ferromagnet is completely ordered, all moments on atomic sites are in the same direction. When it is completely disordered, the two possible spin states are distributed at random on the lattice.

In a paramagnetic material the interaction among the magnetic moments is so small that they can be treated as independent, and any long-range ordering is the result of the response of the material to an external magnetic field. The field applies a torque that induces the moments to line up in the same direction. But in ferromagnetic materials, the interaction among the atomic moments is so strong that magnetic ordering occurs in the absence of any external field and produces very large magnetizations. Paramagnetic effects still exist in ferromagnets, but they are much smaller than the ferromagnetic magnetization.

The response of a ferromagnet to external magnetic fields is complex. It is highly nonlinear and even depends on the magnetic and thermal history of the material. A major reason for this is that the ferromagnet consists of regions, called domains, whose net magnetic moments are in different directions. When the material is fully magnetized, all the atomic moments in a domain are parallel, but the moments are not parallel for all domains. At the domain boundaries, there is a mismatch of the orientation of the moments and a consequent domain boundary energy. Domain boundaries in ferromagnets are analogous to antiphase boundaries in order-disorder alloys. If the domains are mobile, the domain structure is the result of the balance between the energy of the interior of the domain and the domain boundary energy, but it also depends on the history and microstructure of the material. For example, a strong external magnetic field can line up the domains so that their net moments are parallel. The relaxation of domains to lower energy configurations can be inhibited by grain boundaries or inclusions, while high temperatures can increase the mobility and therefore accelerate the relaxation of domains.

At a macroscopic level therefore, the net magnetization is a complex phenomenon dependent on details of history and microstructure. But within a domain these effects can be ignored because the direction of the moments, for a given crystal structure, is determined only by their magnetic interactions (and the external field, if present) as modified by the randomizing tendencies of the temperature. The statistical mechanical theory of ferromagnetism developed by ignoring the existence of domains therefore applies to the individual domains.

Ferromagnetic ordering and order-disorder in alloys are both examples of cooperative second-order phase changes, and we will exploit the similarities between the two phenomena.

The analogy between order-disorder alloys and ferromagnets is the result of adopting the Ising model of a ferromagnet, in which it is assumed that the magnetic energy is the result of the interaction of nearest neighbor spins. These can take on only two values and adjacent spins can only be parallel or antiparallel. This is equivalent to a binary order-disorder alloy whose lattice points can only have A or B atoms and adjacent sites can consist of either like or unlike pairs. But the Ising Hamiltonian is a truncated approximation of the energy of the spin interactions. The full Hamiltonian has a more complex form than that of the scalar nearest-neighbor interaction energy of order-disorder theory. Also, interacting spins can have relative directions that are not necessarily parallel or antiparallel. Despite these caveats, the analogy is useful.

The body-centered cubic structure of iron can be divided into two sublattices just as in the case of β-brass, and the completely magnetized state of iron is that state in which each sublattice is occupied by moments arising from spin states that have the same direction. An antiferromagnetic material is one in which the fully ordered state corresponds to two equivalent sublattices having moments in opposite directions. This means that, in the fully ordered state, each lattice site is surrounded by moments of opposite sign. If the two sublattices are not equivalent, the substance is ferrimagnetic.

9.2 Paramagnetism of independent moments

The total angular momentum of an atom is determined by the sum of the orbital and electron spin, where the sum is performed according to the quantum rules for angular momenta. The total angular momentum gives rise to a magnetic moment whose z component is given by

$$m_z = -g\mu_B M_S \tag{9.2.1}$$

where $\mu_B = e\hbar/2m_e$ is the Bohr magneton, m_e is the electron mass, and g is the gyromagnetic ratio, which depends on the total spin: it is 1 for a moment that arises only from orbital electronic motion and 2 for a pure spin moment. M_S is the azimuthal quantum number of the total angular momentum. It can take on the values from $-S$ to $+S$ in integral steps, S being the total spin quantum number, which is always either an integer or a half integer.

The z-axis can always be chosen to be in the direction of the field, and the energy of the atomic moment in the applied field is

$$u_m = gM_S\mu_B H \tag{9.2.2}$$

where H is the magnitude of the applied field.

These energy states determine the statistical mechanical average for the total magnetization per unit volume of an assembly of identical independent moments. These moments may be attached to atoms, ions, or molecules, but we will refer to them as atoms. The total magnetization is just the atomic moment given by (9.2.1), times the number of atoms per unit volume n. Since the energy and the magnetization are proportional to the azimuthal quantum number, it is only necessary to find the statistical mechanical average of M_S. This average is given by

$$\overline{M_S} = \frac{\sum_S M_S e^{-gM_S\mu_B H/kT}}{\sum_S e^{-gM_S\mu_B H/kT}} \tag{9.2.3}$$

The sums are taken over the allowed values of M_S ranging from $-S$ to $+S$.

Note that only the magnetic energy has been included in equation (9.2.3). The rest of the crystal energy, which includes vibrational energy contributions, has been left out because we take them to be independent of the applied magnetic field, and they therefore appear in the same way in the numerator and denominator and cancel out. Using (9.2.3) assumes that the magnetic field does not affect the lattice vibrations and that it affects the potential energy only through the additive term given by (9.2.2). An analogous assumption was used in order-disorder theory, in which it was assumed that the lattice vibrations could be treated independently of the degree of order.

The denominator in (9.2.3) is the magnetic contribution to the partition function. To facilitate its evaluation, the partition function is written as follows:

$$Z = \sum_{M_S=-S}^{M_S=S} e^{-M_S y} = \sum_{M_S=-S}^{M_S=S} x^{M_S} \tag{9.2.4}$$

with x and y defined as

$$y = g\mu_B H/kT$$

$$x = e^{-y} \quad (9.2.5)$$

The partition function is written this way so that it can be summed by using the formula for the sum of a geometric series, as can be seen by writing it out explicitly:

$$Z = x^{-S} + x^{-S+1} + x^{-S+2}$$
$$= x^{-S}(1 + x + x^2 + \ldots + x^{2S}) \quad (9.2.6)$$

The bracketed expression is just a geometric series whose sum is known, so (9.2.6) is $Z = (x^{-S} - x^{S+1}/1 - x) = (e^{Sy} - e^{-(S+1)y}/1 - e^{-y}) = (e^{(S+1/2)y} - e^{-(S+1/2)y}/e^{y/2} - e^{-y/2})$, and therefore the partition function is

$$Z = \frac{\sinh[(S+1/2)y]}{\sinh(y/2)} \quad (9.2.7)$$

The average $\overline{M_S}$ follows directly from the partition function by using the same technique as that leading to equation (4.4.4). Differentiation of equation (9.2.4) with respect to y gives

$$\frac{dZ}{dy} = \sum_{M_S=-S}^{M_S=S} M_S e^{-M_S y} \quad (9.2.8)$$

Dividing this by Z gives $-\overline{M_S}$, as is evident by comparing (9.2.8) to (9.2.3):

$$\overline{M_S} = -\frac{dZ}{Zdy} = -\frac{d\ln Z}{dy} = \frac{1}{Z}\sum_{M_S=-S}^{M_S=S} M_S e^{-M_S y} \quad (9.2.9)$$

To get our final answer, it is now only necessary to differentiate the logarithm of (9.2.7) and equate it to $\overline{M_S}$. But rather than the reduced variable y, it is customary to use a variable z defined by

$$z = Sy = gS\mu_B H/kT \quad (9.2.10)$$

so that (9.2.7) becomes

$$Z = \frac{\sinh[z(2S+1)/2S]}{\sinh(z/2S)} \quad (9.2.11)$$

The variable z is introduced because it is more convenient for comparing the quantum result to the classical limit.

Since $S(d\ln Z/dz) = d\ln Z/dy$, differentiation of the logarithm of (9.2.7) gives

$$\frac{d\ln Z}{dy} = S\left\{\left(\frac{2S+1}{2S}\right)\frac{\cosh[z(2S+1)/2S]}{\sinh[z(2S+1)/2S]} - \frac{1}{2S}\frac{\cosh(z/2S)}{\sinh(z/2S)}\right\} = \overline{M_S} \quad (9.2.12)$$

The function in the braces is called the *Brillouin function* and is defined by

$$B_S = \frac{2S+1}{2S}\coth[z(2S+1)/2S] - \frac{1}{2S}\coth(z/2S) \quad (9.2.13)$$

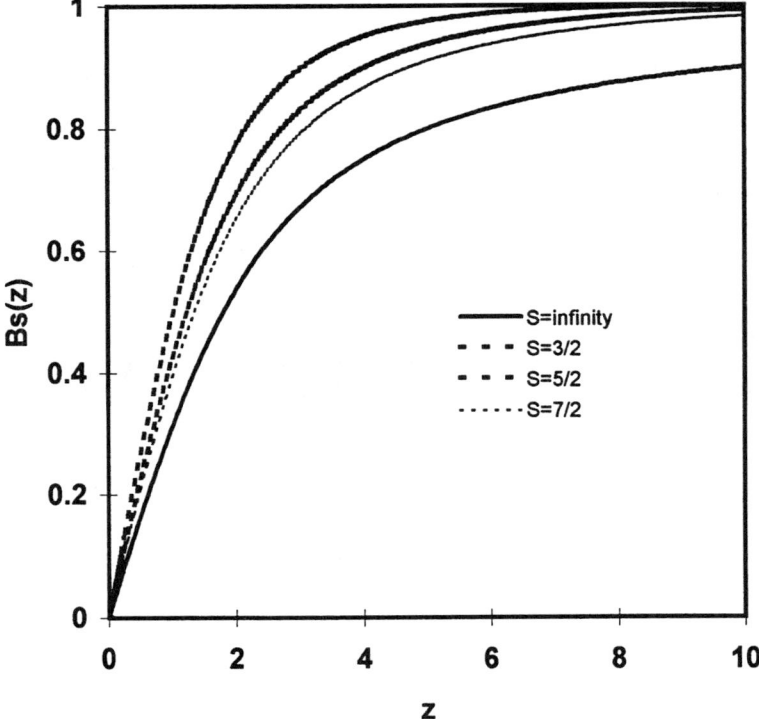

Figure 9.1. Brillouin function.

$\overline{M_S}$ is therefore written as

$$\overline{M_S} = -SB_S(z) \tag{9.2.14}$$

The statistical mechanical average for the magnetization per atom is obtained by taking the average of (9.2.1) and using (9.2.14). Multiplying the result by the number of atoms per unit volume gives the paramagnetic magnetization as

$$M_p = n\overline{m_z} = -g\mu_B \overline{M_S} = nSg\mu_B B_S(z) \tag{9.2.15}$$

The form of the Brillouin function is given in figure 9.1, which shows that the magnetization approaches a maximum value for high z; that is, at low temperatures there is a saturation value for the magnetization.

The classical limit of the quantum result is obtained by neglecting quantization so that the angular momenta can take on any value rather than just a discrete set. This is equivalent to letting the quantum number S become very large. For large S, $(2S+1)/2S \to 1$, so the first term in the Brillouin function (9.2.13) goes to $\coth(z)$. The limit of the second term is readily obtained from the definition of the hyperbolic cotangent.

$$\coth X = \frac{e^X + e^{-X}}{e^X - e^{-X}} \tag{9.2.16}$$

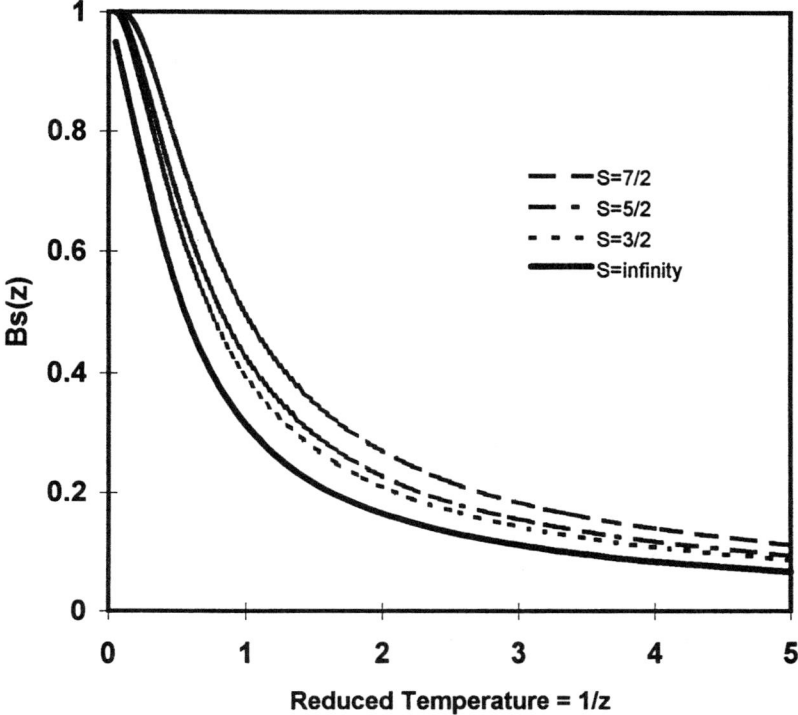

Figure 9.2. Direct temperature dependence of the Brillouin function.

For small X, $e^X \to 1 + X$ and $e^{-X} \to 1 - X$, so $\coth X \to 1/X$. In the Brillouin function, the argument becomes small as S becomes large, so applying (9.2.16), the second term in (9.2.13) goes to $1/z$ and we get

$$B_S(z) \to B_\infty(z) = L(z) = \coth z - \frac{1}{z} \qquad (9.2.17)$$

This limiting function, $L(z)$, is called the *Langevin function* and is shown as the curve labeled "$S = $ infinity" in figure 9.1. Note that the role of S in the definition of the variable z was ignored in getting the limit of the second term. The reason for this is that the classical result was obtained by applying statistical mechanics without quantization to a classical system with an arbitrary unit moment. In making the connection to the quantum result, we see that $S\mu_B$ plays the role of the unit moment. To make the classical limit consistent with the quantum result, we want the unit moment to have the same value in both cases, and this is done by formally taking $S = 1$ in the definition of z in equation (9.2.17).

The temperature dependence is displayed by plotting the Brillouin function against $1/z$ as in figure 9.2, which shows that the magnetization decreases rapidly with increasing temperature.

The asymptotic magnetization is reached only at quite low temperatures because the Bohr magneton is small, and even at fields as high as 10^4 gauss z

is still only of the order of 10^{-2} at ordinary temperatures and does not approach unity until the temperature decreases to about a few degrees Kelvin. For ordinary temperatures and magnetic fields that are not beyond the usual laboratory range, it is therefore sufficiently accurate to take z as small and expand the hyperbolic cotangent function accordingly. The power series expansion is

$$\coth X = \frac{1}{X} + \frac{X}{3} - \frac{X^3}{45} - \ldots \tag{9.2.18}$$

Applying this to the Brillouin function, the first term in the expansion cancels out, so ignoring terms in z^3 and higher gives

$$B_S(z) = \frac{(S+1)}{3S} z \quad \text{(small } z\text{)} \tag{9.2.19}$$

The high-temperature result for the magnetization is now obtained by putting (9.2.19) into (9.2.15) and using the definition of z in (9.2.10), with the result that

$$M_p = nS(S+1)(g\mu_B)^2 \frac{H}{3kT} \quad \text{(high } T\text{)} \tag{9.2.20}$$

The coefficient of H in (9.2.20) is the paramagnetic susceptibility of a system of independent atoms at temperatures that are not too low and fields that are not too high. The paramagnetic susceptibility is independent of the field only in the high-temperature, low-field approximation. In general, it is a function of the field and the temperature as determined by the Brillouin function. Experimentally, it is found that the paramagnetic susceptibility is indeed inversely proportional to the absolute temperature, a relation that is known as Curie's law.

Note that although we have labeled this the high-temperature approximation, it is valid down to quite low temperatures for ordinary magnetic fields. At very low temperatures and/or very high fields, z becomes very high and the Brillouin function approaches unity, as shown by its definition (9.2.16), so (9.2.15) gives the magnetization as

$$M_p = nSg\mu_B \quad \text{(low } T, \text{ high } H\text{)} \tag{9.2.21}$$

This is the paramagnetic saturation magnetization, which is the result of all moments being aligned in the same direction such that further increases in the field cannot increase the magnetization.

9.3 Paramagnetism of free electrons

In a metal, half the electrons have a spin equal to 1/2 and the other half have spin $-1/2$. Despite this equality, there is a paramagnetic response of the electron gas because the energy of electrons with spin parallel to an applied field H is increased by $\mu_B H$ while those with spin opposite to the field have their energy decreased by $-\mu_B H$. The effect of the external field on the electron distribution is shown in figure 9.3, in which the energy of the electrons is plotted against the density of states. The density of states in energy is still given by equation (9.3.1) [same as equation (6.3.3)]:

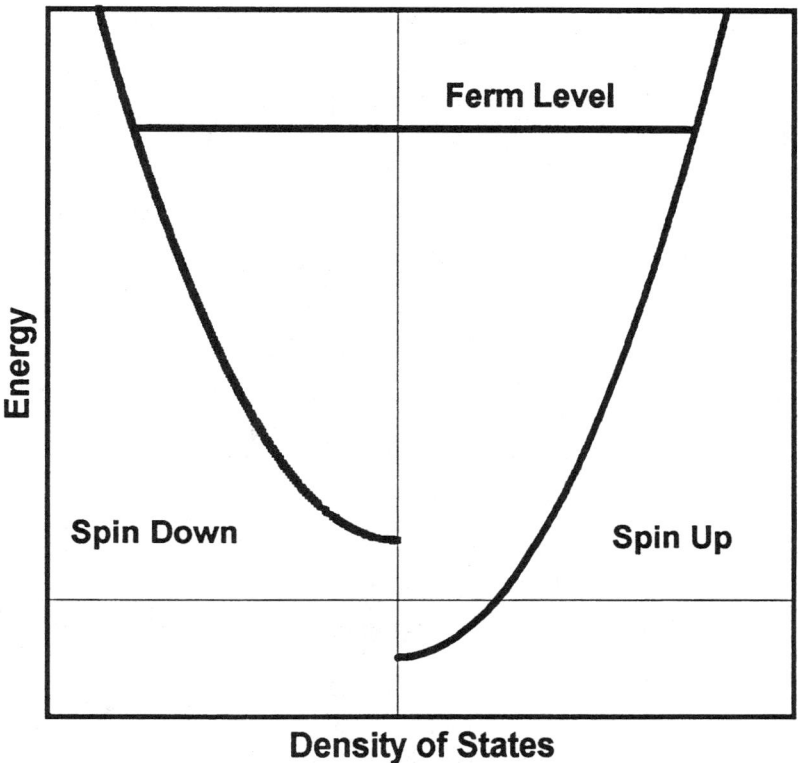

Figure 9.3. Energy shift due to spin orientation of free electrons in a magnetic field.

$$\omega(\varepsilon) = CV\sqrt{\varepsilon} \qquad (9.3.1)$$

with

$$C = \frac{8\pi\sqrt{2m^3}}{h^3} \qquad (9.3.2)$$

The Fermi level μ_0 is related to the number density $n = N/V$ of electrons by

$$\mu_0 = \left(\frac{3N}{2CV}\right)^{2/3} \qquad (9.3.3)$$

The effect of temperature on the Fermi level is very small and can be ignored, so the number of electrons can be computed by taking the Fermi function to be unity up to the Fermi level, after which it is zero. The number of electrons with spins parallel to the field is then

$$N_+ = \frac{1}{2}\int_{-\mu_B H}^{\mu_0} \omega(\varepsilon + \mu_B H)d\varepsilon = \frac{1}{2}\int_0^{\mu_0 + \mu_B H} \omega(\varepsilon)d\varepsilon \qquad (9.3.4)$$

while the number of electrons with spins opposite to the applied field is

$$N_- = \frac{1}{2}\int_{\mu_B H}^{\mu_0}\omega(\varepsilon+\mu_B H)d\varepsilon = \frac{1}{2}\int_0^{\mu_0-\mu_B H}\omega(\varepsilon)d\varepsilon \qquad (9.3.5)$$

The excess number of electrons with spins parallel to the field is $(N_+ - N_-)$. Taking the difference of (9.3.4) and (9.3.5) gives

$$N_+ - N_- = \frac{CV}{2}\int_{\mu_0+\mu_B H}^{\mu_0-\mu_B H}\omega(\varepsilon)d\varepsilon \qquad (9.3.6)$$

The term $\mu_B H$ is always small relative to the Fermi energy, and the integral is centered on the Fermi energy, so its value is

$$N_+ - N_- = \frac{1}{2}\omega(\mu_0)[(\mu_0+\mu_B H)-(\mu_0-\mu_B H)] = \omega(\mu_0)\mu_B H \qquad (9.3.7)$$

The magnetization is just this number of excess spins in the parallel direction times the moment of an electron, which is just the Bohr magneton. Therefore, if M_e is the magnetization of the conduction electrons,

$$M_e = \omega(\mu_0)\mu_B^2 H \qquad (9.3.8)$$

From equation (9.3.1) for the density of states and using (9.3.3) to express the result in terms of the number of electrons per unit volume, (9.3.8) becomes.

$$M_e = \frac{3n\mu_B^2}{2\mu_0}H \qquad (9.3.9)$$

The coefficient of H is the paramagnetic susceptibility of a system of n free electrons per unit volume, which is independent of temperature. It is easy to verify from (9.3.9) that the susceptibility of conduction electron is much smaller than the paramagnetic susceptibility of ions at laboratory temperatures.

9.4 Ferromagnetism: mean field theory

Consider a ferromagnet in which only nearest neighbor interactions of the spins exist. The magnetic energy at a site is determined by the interaction of its spin with the magnetic field arising from the particular configuration of spins on its z nearest neighbors, as well as by any external fields that may be present. In the effective mean field approach of Weiss, each such particular configuration of neighbors is replaced by its statistical mechanical average. That is, it is assumed that each spin is subject to an effective internal magnetic field that is proportional to the overall magnetization of the crystal. If M_f is the magnetization arising from ferromagnetic interactions, then the internal field at each site is assumed to be

$$H_i = \gamma M_f \qquad (9.4.1)$$

where γ is a constant that measures the strength of the internal field.

Using (9.4.1) reduces our problem to one that is identical to that for independent paramagnetic atoms. Formally, the mean field assumption is equivalent to treating each atom as if it were an independent magnetic moment. The only difference is that, instead of using only the external field H, the internal field H_i is added to it and the total field $(H + H_i)$ appears in the Brillouin function. Therefore, equation (9.2.15) for the ferromagnetic mean field case becomes

$$M_f = nSg\mu_B B_S(x) \tag{9.4.2}$$

but, with z of equation (9.2.10) now being replaced by x, which is defined by

$$x = Sg\mu_B(H + \gamma M_f)/kT \tag{9.4.3}$$

The first point to note is that, since the maximum value the Brillouin function can have is unity, there is a maximum value for the magnetization M_f given by $nSg\mu_B$. This saturation magnetization is independent of the external field and is the same as that for a paramagnet in extremely high external fields. Just as in the paramagnetic case, once all moments are in the same direction, no further increase in field can increase the magnetization.

The second important point is that there is a temperature above which the spontaneous magnetization vanishes. To see this, apply the high-temperature approximation to equation (9.4.2). The result is just like equation (9.2.20) except that H is replaced by $(H + \gamma M_f)$, so

$$M_P = nS(S+1)(g\mu_B)^2 \frac{H + \gamma M_f}{3kT} \quad \text{(high } T\text{)} \tag{9.4.4}$$

It is convenient to define a constant with the dimensions of temperature, T_c, by

$$T_c = \frac{n\gamma}{3k} S(S+1)(g\mu_B)^2 \tag{9.4.5}$$

so that, using this definition, (9.4.4) gives

$$M_f = \frac{T_c}{T}(M_f + H/\gamma) \quad \text{(high } T\text{)} \tag{9.4.6}$$

or, solving for the magnetization,

$$M_f = \frac{C}{T - T_c} H \quad \text{(high } T\text{)} \tag{9.4.7}$$

where C is a constant defined by $C = T_c/\gamma$. Equation (9.4.7) shows that the high-temperature approximation can hold only for temperatures above T_c.

If there is no external field, the total ferromagnetic magnetization reduces to the spontaneous magnetization, which we will call M_f^s, and (9.4.7) gives $M_f^s(T - T_c) = 0$ (high T), which states that the spontaneous magnetization is zero at temperatures above T_c.

Since there is no spontaneous magnetization above T_c, (9.4.7) is a paramagnetic response to the external field. This equation is the Curie–Weiss law. T_c is called the *Curie temperature* and marks the transition of a ferromagnet from the ferromagnetic to the paramagnetic state as the temperature is raised above T_c.

To examine this transition a little more closely again use the high-temperature (low x) approximation, but now retain the first three terms in the expansion given by (9.2.18). The Brillouin function is then approximated by

244 STATISTICAL MECHANICS OF SOLIDS

$$B_S(x) = \frac{(S+1)}{3S}x - \left(\frac{1}{2S}\right)^4[(2S+1)^4 - 1]\frac{x^3}{45} \tag{9.4.8}$$

Put this into (9.4.2) and, using the definition of the Curie temperature, write the result as

$$M_f = \frac{T_c}{T}\left(M_f + \frac{H}{\gamma}\right) - K\left(\frac{T_c}{T}\right)^3\left(M_f + \frac{H}{\gamma}\right)^3 \tag{9.4.9}$$

where K is a constant that is readily evaluated from the definition of T_c. When the external field is zero, M_f becomes the spontaneous magnetization M_f^s, and upon solving for M_f^s, (9.4.9) gives

$$M_f^s = \frac{1}{K}\left(\frac{T_c - T}{T}\right)^{1/2} \tag{9.4.10}$$

This is the asymptotic form for the temperature dependence of the spontaneous magnetization as the temperature approaches the Curie temperature. There is clearly no solution for the spontaneous magnetization for temperatures higher than T_c, thereby verifying that there is a transition from the ferromagnetic to the paramagnetic state at the Curie temperature. If the external field is retained, a solution above T_c can be shown to exist which would reduce to the Curie–Weiss law.

It is easy to verify from equation (9.4.10) that as the temperature approaches T_c, the temperature derivative goes to

$$\frac{dM_f^s}{dT} \rightarrow \frac{1}{2K(T_c - T)^{1/2}} \tag{9.4.11}$$

The derivative becomes infinite as T goes to T_c. This behavior is characteristic of a second-order phase transition. Although the mean field theory correctly displays the Curie–Weiss law and the existence of a second-order ferromagnetic to paramagnetic transition, the agreement of experiment with the functional form (9.4.10) and the exponent in (9.4.11) is only approximate.

Let us now examine the low-temperature limit. At low temperatures x is large, and we can write the hyperbolic cotangent from (9.2.16) as follows:

$$\coth(X) = \frac{1 + e^{-2X}}{1 - e^{-2X}} = 1 + 2e^{-2X} \tag{9.4.12}$$

where only the first two terms of the result of the division have been retained. Applying (9.4.12) to the Brillouin function gives $B_S(x) = 1 - 1/Se^{-x/S}[1 - (2S + 1)e^{-2x}]$, so neglecting the third term, (9.4.2) becomes

$$M_f = nSg\mu_B\left(1 - \frac{1}{S}e^{-x/S}\right) \tag{9.4.13}$$

Since x is inversely proportional to T, this result states that at low temperatures the magnetization differs from unity by a term of the form $e^{-\text{const}/T}$. Experimentally, it is found that at low temperatures the magnetization varies with temperature as $(1 - T^{-3/2})$. Thus, although the mean field theory provides an approximate overall picture of ferromagnetism, it does not give the correct low-temperature limit for the magnetization.

9.5 The Ising model for ferromagnetism

In the Ising model of a ferromagnetic crystal, it is assumed that the spin at an atomic site can take on one of only two values. The two different spin states are labeled +1 (up) and −1 (down), and in the absence of any external field, the ferromagnetic energy is written as

$$E\{s\} = -J\sum s_i s_j \qquad (9.5.1)$$

where s_i and s_j can only have values +1 or −1, $E\{s\}$ is the energy for a given distribution of spins on the lattice sites, J is a constant, and the summation is taken over all nearest neighbor pairs in the crystal because it is assumed that the spins are subject to nearest neighbor interactions only.

Equation (9.5.1) states that nearest neighbors with parallel spins (s_i and s_j have the same sign) have an energy of interaction $-|J|$ while antiparallel nearest neighbors (s_i and s_j have opposite signs) have an interaction energy equal to $+|J|$. For ferromagnetic materials J is positive, so parallel spins are energetically favored. (For negative J, antiparallel spins would be favored and the crystal would be an antiferromagnet.)

The energy given by (9.5.1) is a simplification of the Hamiltonion given by an analysis of the exchange energy including spin. The correct Hamiltonian takes the vector character of the spins into account, whereas (9.5.1) is equivalent to retaining only the z components of the spin angular momentum and neglecting interactions among the cross components of the spins. Without this simplification, the theory would be extremely difficult to deal with. The Ising model itself is troublesome enough.

The statistical mechanical problem is to use the energy given by (9.5.1) along with a count of the number of complexions to get the degree of magnetization and the thermodynamic properties of a ferromagnet. The Ising model can actually be solved exactly in one and two dimensions but not in three dimensions. The one-dimensional result shows that there is no spontaneous magnetization. The two-dimensional case can be solved for certain lattices in the absence of an external field and does exhibit spontaneous magnetization.

The crystal is taken to be monatomic with all lattice sites being geometrically equivalent. That is, the atoms on the sites differ only by their spin states, and every site has the same number and arrangement of neighbors. This condition is fulfilled by iron, which is the prototypical ferromagnetic material. Equation (9.5.1) is then formally equivalent to equation (8.4.18) except for a constant term that can be defined away to zero by a proper choice of reference states. Since (8.4.18) is the energy of an order-disorder alloy, the three-dimensional ferromagnetic Ising model is equivalent to nearest-neighbor order-disorder theory. All the statistical mechanics based on (9.5.1) will have the same structure as that for the 50–50 order-disorder alloy. Iron and β-brass both have a body-centered cubic lattice, so the analogy is complete. It is only necessary to identify the role of the spontaneous magnetization, and this is done as follows:

Let

N = total number of atoms (lattice sites)
N_+ = number of atoms with spin up ($s_i = 1$)
N_- = number of atoms with spin down ($s_i = -1$)

Clearly,

$$N = N_+ + N_- \qquad (9.5.2)$$

and the fraction of excess spin-up atoms is defined by

$$R_m = \frac{N_+ - N_-}{N} \qquad (9.5.3)$$

The spontaneous magnetization is proportional to the excess number of up spins per unit volume. For electron spin, the magnitude of the magnetic moment is equal to the Bohr magneton, and it can be positive or negative depending on the sign of the spin. The spontaneous magnetization of the crystal (net moment per unit volume) is therefore

$$M_s = \mu_B(N_+ - N_-)/V \qquad (9.5.4)$$

Now take N_+ to correspond to the number of "rightly" occupied sites ($N_{A\alpha} + N_{B\beta}$) in the 50–50 AB alloy and let N_- correspond to the number of "wrongly" occupied sites ($N_{A\beta} + N_{B\alpha}$). From equations (8.5.6) and (8.5.7), the long-range order parameter for the AB alloy can be written as

$$R = \frac{(N_{A\alpha} + N_{B\beta})}{N} - \frac{(N_{A\beta} + N_{B\alpha})}{N} \qquad (9.5.5)$$

Therefore, R corresponds to the fraction of excess spin-up (positive spin) atoms, and the Ising theory of a ferromagnet is equivalent to the AB order-disorder theory if we make the following replacements:

$$Qv_{AB}/2 \to 0$$

$$v/2 \to J$$

$$(N_{A\alpha} + N_{B\beta}) \to N_+$$

$$(N_{A\beta} + N_{B\alpha}) \to N_-$$

$$R \to \frac{N_+ - N_-}{N} = R_m \qquad (9.5.6)$$

R_m is defined to be the magnetic ordering, so the magnetization (9.5.4) now reads

$$M_s = n\mu_B R_m \qquad (9.5.7)$$

with n being the number of atoms per unit volume. R_m is just the magnetization per atom in units of Bohr magnetons.

The entire theory of the order-disorder transition can be translated to the ferromagnetic case. For example, the Bragg–Williams approximation as expressed by equations (8.6.10)–(8.6.14), (8.6.22), and (8.6.28) are completely applicable if R is taken to be R_m and ϕ is defined as $\phi = kT/zJ$. The critical temperature for the ferromagnetic transition in the Bragg–Williams approximation is then

$$T_c = zJ/k \qquad (9.5.8)$$

The units of spontaneous magnetization are taken to be in Bohr magnetons per unit volume, so equation (9.5.7) divided by $n\mu_B$ is the same as the Bragg–Williams equation for R versus ϕ in the order-disorder theory of β-brass.

The calculation of the spontaneous magnetization in the mean field approximation starts with equation (9.4.2), for which it is assumed that $S = 1/2$ and $g = 2$, so dividing (9.4.2) by $n\mu_B$ to get the magnetization in reduced units gives

$$m_s = B_{1/2}(x) \tag{9.5.9}$$

where x is given by putting $H = 0$ in equation (9.4.3) since we are computing spontaneous magnetization. That is,

$$x = \mu_B \gamma M_f / kT \tag{9.5.10}$$

The critical temperature (9.4.5) is now given by

$$T_c = n\mu_B^2 \gamma / kT \tag{9.5.11}$$

Since $Sg = 1$ and $H = 0$, x becomes [from (9.4.3)]

$$x = \frac{m_s}{\phi} \tag{9.5.12}$$

where $m_S = M_S/n\mu_B$ is the magnetization in units of Bohr magnetons per atom and ϕ is the reduced temperature T/T_c. The Brillouin function for $S = 1/2$ is

$$B_{1/2}(x) = 2\coth 2x - \coth x = \tanh x \tag{9.5.13}$$

The last equality is the result of using the identity

$$\coth(2x) = \frac{1 + \tanh^2 x}{2\tanh x}$$

Equation (9.5.9) therefore reduces to

$$m_s = \tanh x \tag{9.5.14}$$

Equation (9.5.14) is identical in form to equation (8.6.21). The Bragg–Williams approximation to the Ising model is therefore fully equivalent to the mean field theory.

A comparison of experimental data[1] for the magnetization of iron, nickel, and cobalt with the mean field theory as computed from (9.5.14), taking $S = 1/2$, is shown in figure 9.4. The agreement is quite good (except, of course, in the low-temperature limit, which does not show up on the scale of figure 9.3). In fact, the theory clearly describes the magnetization of ferromagnets better than it describes the degree of long-range order in CuZn. The reason for this is that long-range interactions among atoms in brass are more important than long-range interactions among spins in ferromagnets. The nearest neighbor approximation is therefore better in the ferromagnetic case than in the order-disorder case.

Since ferromagnetism is analogous to the order-disorder transition, a similar heat capacity effect should be expected and in fact does exist, as shown by the experimental data for iron in figure 9.5. Just as for the order-disorder case, the difference between the heat capacities at constant volume and constant pressure is neglected and the classical value of $3k$ is subtracted out.[2]

The shape of the heat capacity curve is very similar to that for the order-disorder transition. It shows a very large rise and a tail above the Curie

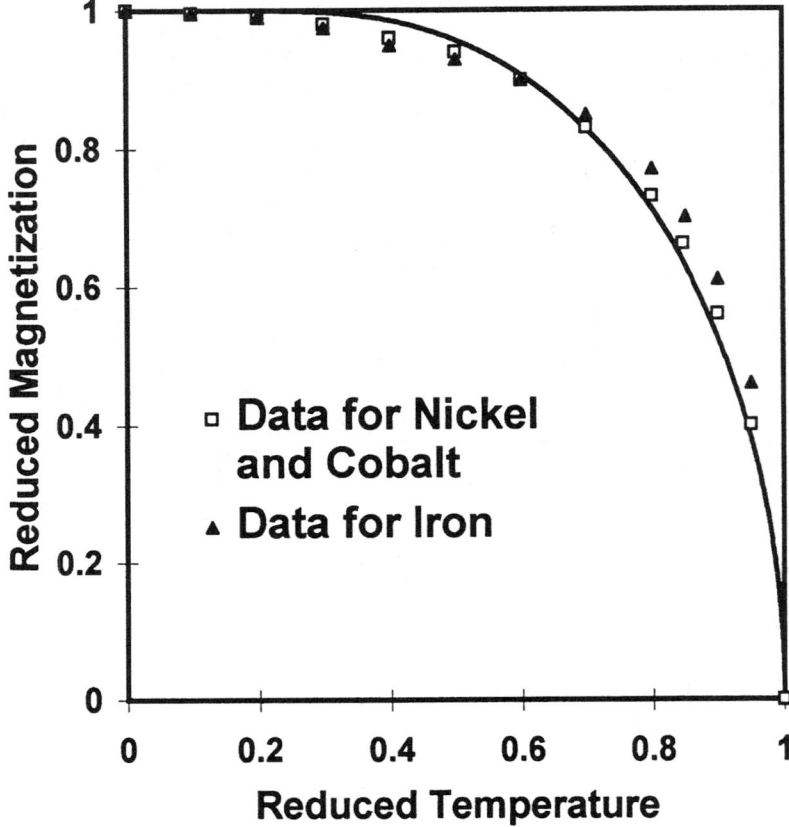

Figure 9.4. Comparison of mean field theory to experiment.

temperature that is the result of short-range magnetic order that is not captured by the mean field theory. The points on the plot are restricted to the temperature range for which the α (body-centered cubic) phase is stable. At higher temperatures, iron transforms to the γ (face-centered cubic) phase and this complicates the interpretation of the heat capacity data.

9.6 Antiferromagnetism: mean field theory

Manganese oxide is a prototypical antiferromagnetic material. It has the NaCl structure consisting of two interpenetrating face-centered cubic structures, one occupied by the Mn^{2+} ions and the other by the O^{2-} ions. The Mn^{2+} ions carry magnetic moments whose directions lie in (111) planes, and in the fully ordered state all moments in a given (111) plane are lined up in the same direction while the moments in adjacent (111) planes are in the opposite direction. The interaction between a moment of one Mn^{2+} ion and the nearest Mn^{2+} neighbors on adjacent Mn^{2+} planes is such as to give a lower energy for antiparallel alignment. Note that this interaction is equivalent to a negative exchange

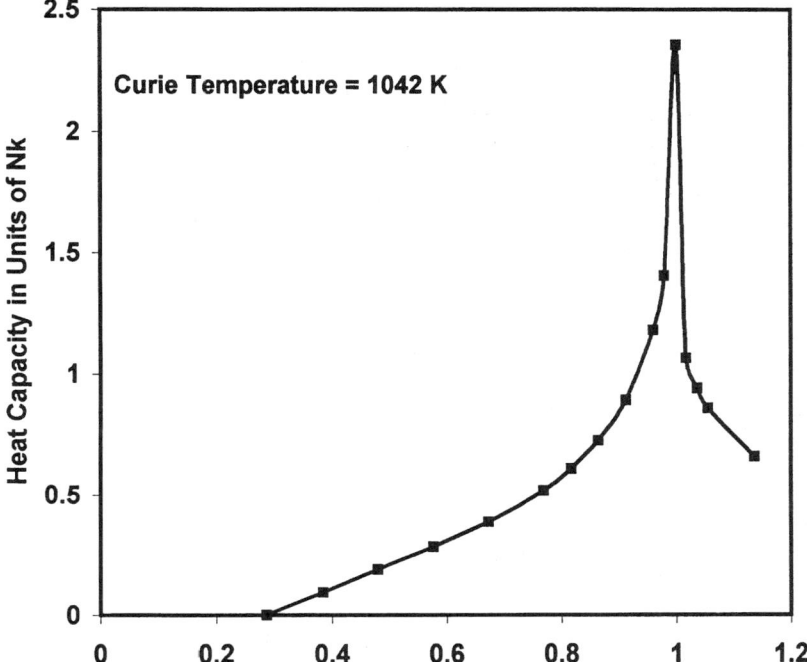

Figure 9.5. Heat capacity of iron due to magnetic ordering.

energy that acts through the Mn^{2+} nearest neighbor O^{2-} ions. The existence of the O^{2-} ions can be ignored when visualizing the geometry of the magnetic interactions. An Mn^{2+} ion on a given (111) plane in the Mn^{2+} face-centered cubic structure has six Mn^{2+} neighbors in its own plane and six more on neighboring Mn^{2+} planes.

In MnF_2, the Mn^{2+} ions are at the sites of a body-centered cubic structure embedded in the rutile structure of the salt. The body-centered cubic structure consists of two interpenetrating simple cubic sublattices, each of which is ferromagnetic when fully ordered. The ordering on these sublattices is opposite to each other, so the crystal is antiferromagnetic. In this case, the interactions of the ions that are nearest neighbors are antiferromagnetic, and the ferromagnetic ordering is the result of next nearest neighbor interactions.

Within each of the two equivalent sublattices, labeled A and B, there are ferromagnetic interactions between each ion and z_a neighbors, while across sublattices there are antiferromagnetic interactions between an ion and z_b neighbors. In the mean field approach, the magnetic moment on each ion is subject to an internal field that has two components, one arising from interactions with moments on its own sublattice and the other from interactions on the other sublattice. Thus, the internal magnetic fields H_A and H_B at ions on the A and B sublattices are given by

$$H_A = \gamma_f M_A - \gamma_a M_B \qquad (9.6.1)$$

$$H_B = \gamma_f M_B - \gamma_a M_A \qquad (9.6.2)$$

In these equations, M_A and M_B are the magnetizations of the A sublattice and the B sublattice, respectively, and γ_f and γ_a are positive constants. The first term is positive because A-A interactions and B-B interactions are ferromagnetic and tend to line up adjacent moments in the same direction, while the second term is negative because A-B interactions are antiferromagnetic and tend to line up the moments in opposite directions. Since the two sublattices are equivalent, the same mean field constants γ_f and γ_a appear in both equations. The magnetization of each sublattice can be treated in exactly the same way as the ferromagnetic case with internal fields given by (9.6.1) and (9.6.2). The result for the A sublattice is that instead of (9.4.2) we get

$$M_A = nSg\mu_B B_S(x_A) \tag{9.6.3}$$

but with x of equation (9.4.3) now being replaced by

$$x_A = gS\mu_B(H + \gamma_f M_A - \gamma_a M_B)/kT \tag{9.6.4}$$

Similarly, the magnetization for the B sublattice is

$$M_B = nSg\mu_B B_S(x_B) \tag{9.6.5}$$

with x_B defined by

$$x_B = gS\mu_B(H + \gamma_f M_B - \gamma_a M_A)/kT \tag{9.6.6}$$

instead of equation (9.4.3). (As before, H is the magnitude of the external field.)

The total magnetization of the material is the sum of that for the two sublattices:

$$M_T = M_A + M_B \tag{9.6.7}$$

In the absence of an external magnetic field, the total magnetization is zero because there are just as many ions with a given spin as with an opposing spin. This is obviously true for the fully ordered state but is also true for partially ordered states because the probability of finding a given spin on one sublattice is equal to the probability of finding the opposite spin on a site of the other sublattice. But the magnetization on each sublattice is described by molecular field theory in the same way as ferromagnetism because equations (9.6.3) and (9.6.5) taken separately are completely analogous to the ferromagnetic case. This means that there is a critical temperature above which the spontaneous magnetization vanishes on each sublattice and the crystal is paramagnetic instead of antiferromagnetic. In the paramagnetic region, the first term in the high-temperature expansion of the Brillouin function, equation (9.4.8), is sufficient, so (9.6.3) and (9.6.5) are well approximated by

$$M_A = nSg\mu_B\left[\frac{(S+1)x_A}{3S}\right] \tag{9.6.8}$$

$$M_B = nSg\mu_B\left[\frac{(S+1)x_B}{3S}\right] \tag{9.6.9}$$

Putting in the definitions of x_A and x_B, these equations become

$$M_A = \frac{C}{T}(H + \gamma_f M_A - \gamma_a M_B) \tag{9.6.10}$$

$$M_B = \frac{C}{T}(H + \gamma_f M_B - \gamma_a M_A) \quad (9.6.11)$$

the constant C being defined by

$$C = \frac{nS(S+1)}{3k} g^2 \mu_B^2 \quad (9.6.12)$$

Equations (9.6.10) and (9.6.11) are easily put in the form

$$M_A\left(1 - \frac{C}{T}\gamma_f\right) + \frac{C}{T}\gamma_a M_B = \frac{C}{T}H \quad (9.6.13)$$

$$M_B\left(1 - \frac{C}{T}\gamma_f\right) + \frac{C}{T}\gamma_a M_A = \frac{C}{T}H \quad (9.6.14)$$

Adding the last two equations and solving for the total magnetization $M_T = M_A + M_B$ gives

$$M_T = \chi H \quad (9.6.15)$$

where χ is the paramagnetic susceptibility defined by

$$\chi = \frac{2C}{T + T_N} \quad (9.6.16)$$

T_N is a constant with the dimension of temperature and is given by

$$T_N = C(\gamma_a - \gamma_f) \quad (9.6.17)$$

T_N is called the *Neel temperature*. Since, in an antiferromagnetic crystal, the antiferromagnetic coupling constant is greater than the ferromagnetic coupling constant, the Neel temperature is positive. Note the similarity to the Curie–Weiss law for the paramagnetic susceptibility of ferromagnetic materials. The forms of (9.4.7) and (9.6.16) are the same except for the sign in the numerators.

In the absence of an external field, the total magnetization of an antiferromagnet is zero, so it does not show the critical temperature behavior typical of a ferromagnet. However, the temperature dependence of the heat capacity is similar to that of a ferromagnet and displays the typical rapid rise and nearly immediate drop at a critical temperature that is characteristic of a second-order phase transition. The reason for this is that on each sublattice of an antiferromagnet the atomic moments are in the same direction as the molecular internal field, so the energy has the same sign on both sublattices. In fact, with respect to the energy and heat capacity, each sublattice can be treated just as if it were a ferromagnet. It would suffice to compute the energy of one sublattice and then multiply by 2 to get the total energy.

9.7 Spin waves

The theory of ferromagnetism worked out so far gives a good physical picture of the consequences of strong interactions among magnetic dipoles and describes the magnetic ordering transition and the high-temperature magnetization reasonably well. However, both molecular field theory and the Ising model give incorrect results for the temperature dependence of the spontaneous magnetization at low temperatures. A different approach, based on analysis of the spatial variation of spin orientations, provides a powerful tool for the study of magnetic ordering and reproduces the correct low-temperature result.

Let us recall that the energy for the Ising model, given by (9.5.1), is a simplification of the exchange Hamiltonian for a system of interacting spins. Quantum theoretic analysis for a system of N spins gives this Hamiltonian, in zero external field, as

$$\hat{H} = E\{\mathbf{S}\} = -2J\sum_{i,j} \mathbf{S}_i \cdot \mathbf{S}_j \qquad (9.7.1)$$

where the $\mathbf{S}_i$ is now the quantum mechanical spin operator associated with the spins on the ith lattice point. As usual, all but nearest neighbor interactions are ignored and the sum is over all nearest neighbor pairs.

At absolute zero all the spins are parallel in a ferromagnet, and the vector product of the spin operators on two adjacent sites becomes the scalar operator S_j^2. This is the ground state energy of the ferromagnet. Since only nearest neighbor interactions are nonzero, and all sites are equivalent, all $\mathbf{S}_i$ have the same magnitude S and the ground state Hamiltonian is proportional to S^2. As the temperature is raised above absolute zero, thermal fluctuations will kick some spins out of their parallel positions, giving rise to excited states. A low-lying excited state would be one in which just a single spin would reverse its orientation and be antiparallel to all other spins. At low temperatures most spins are in a parallel configuration, with only a few being out of line. But it turns out to be energetically more favorable if the transition from one spin direction to its opposite is gradual rather than abrupt. That is, the difference in orientation between neighbors is relatively small, and the orientation of spins changes gradually as we go through the crystal, at least at low temperatures. This can be described in terms of a wave of spin orientation existing in the crystal. The spatial variation of the spin orientation is then a superposition of regular sinusoidal waves just as the thermal motion of an atom in a crystal can be described as the superposition of displacement waves. Magnetization can be described in terms of these spin waves just as the thermal properties of crystals can be described in terms of normal mode lattice waves.

A semiclassical approach to the origin of spin waves starts with equation (9.7.1) but treats the $\mathbf{S}_i$ as if they were classical vectors rather than quantum mechanical operators such that multiplying them by $\hbar$ gives their angular momentum. The magnetic dipole moment of the ith spin is

$$\mu_i = -g\mu_B \mathbf{S}_i \qquad (9.7.2)$$

The energy of a magnetic moment in a magnetic field is the scalar product of the dipole moment and the applied field such that

$$U_i = \mu_i \cdot \mathbf{H}_i \qquad (9.7.3)$$

and summing over the lattice gives

$$U = \frac{1}{2}\sum_i \mu_i \cdot \mathbf{H}_i \qquad (9.7.4)$$

where $\mathbf{H}_i$ is the field at i arising from the moments of its neighbors. The factor of 1/2 is introduced to prevent double counting of the interactions. For the sum of (9.7.3) over the lattice to have the form (9.7.1), the internal field at i must be

$$\mathbf{H}_i = -\frac{2J}{g\mu_B}\sum_j \mathbf{S}_j \qquad (9.7.5)$$

From classical mechanics, $\mathbf{H}_i$ exerts a torque on the moment at i, which is given by $\boldsymbol{\mu}_i \times \mathbf{H}_i$, and this torque is equal to the rate of change of the angular momentum. That is, the equation of motion for the angular momentum is

$$\hbar \frac{d\mathbf{S}_i}{dt} = \mathbf{m}_i \times \mathbf{H}_i \qquad (9.7.6)$$

Using (9.7.2) and (9.7.5), this becomes

$$\hbar \frac{d\mathbf{S}_i}{dt} = 2J\mathbf{S}_i \times \sum_j \mathbf{S}_j \qquad (9.7.7)$$

This is the equation of motion we want to solve to get semiclassical spin wave solutions. To do this, let us write (9.7.7) in component form:

$$\hbar \frac{dS_i^x}{dt} = 2J\left(S_i^y \sum_j S_j^z - S_i^z \sum_j S_j^y\right) \qquad (9.7.8)$$

$$\hbar \frac{dS_i^y}{dt} = 2J\left(S_i^x \sum_j S_j^z - S_i^z \sum_j S_j^x\right) \qquad (9.7.9)$$

$$\hbar \frac{dS_i^z}{dt} = 2J\left(S_i^x \sum_j S_j^y - S_i^y \sum_j S_j^x\right) \qquad (9.7.10)$$

In the ground state, all spins are in the same direction, and if we label this the z-direction, then for all spins $S_i = S_i^z = S$. For these low-lying states the deviation from the ground state is small (low temperatures). Then the x- and y-components of the spins are small compared to the z-components. This means that in equations (9.7.8)–(9.7.10) all products of x- and y-components can be neglected and z-components can be replaced by the ground state value S. These equations then become

$$\hbar \frac{dS_i^x}{dt} = 2JS\left(zS_i^y - \sum_j S_j^y\right) \qquad (9.7.11)$$

$$\hbar \frac{dS_i^y}{dt} = 2JS\left(zS_i^x - \sum_j S_j^x\right) \qquad (9.7.12)$$

$$\hbar \frac{dS_i^z}{dt} = 0 \qquad (9.7.13)$$

In these equations, z is the usual notation for the number of nearest neighbors to a central site and is not to be confused with the z-component of a Cartesian coordinate.

Wavelike solutions exist for these equations, as can be seen by assuming special solutions for S_i^x and S_i^y that have the form of lattice waves:

$$S_i^x = u e^{i(\mathbf{k}\cdot\mathbf{R}_i - \omega t)} \qquad (9.7.14)$$

$$S_i^y = v e^{i(\mathbf{k}\cdot\mathbf{R}_i - \omega t)} \qquad (9.7.15)$$

Just as for the expansion of vibrational displacements in terms of lattice waves, the **k**s are vectors in reciprocal lattice space with $3N$ values given by $\mathbf{k} = (2\pi/N^{1/3})(n_1\mathbf{b}_1 + n_2\mathbf{b}_2 + n_3\mathbf{b}_3)$, where n_1, n_2, n_3 are integers with values ranging from 0 to $N^{1/3}$ and $\mathbf{b}_1$, $\mathbf{b}_2$, $\mathbf{b}_3$ are the reciprocal lattice unit vectors defined by $\mathbf{a}_i \cdot \mathbf{b}_j = \delta_{ij}$.

Now follow the usual procedure of substituting (9.7.14) and (9.7.15) into (9.7.11) and (9.7.12) to determine the conditions under which such solutions can exist. The result of the substitution is

$$i\omega u + \frac{2JS}{\hbar} v \left[z - \sum_j e^{i\mathbf{k}\cdot(\mathbf{R}_j - \mathbf{R}_i)} \right] = 0 \qquad (9.7.16)$$

$$i\omega v + \frac{2JS}{\hbar} u \left[z - \sum_j e^{i\mathbf{k}\cdot(\mathbf{R}_j - \mathbf{R}_i)} \right] = 0 \qquad (9.7.17)$$

These are two equations in the two unknowns u and v. They have a solution if the determinant of the coefficients is zero. Therefore, we must have

$$\begin{vmatrix} i\omega & \frac{2JS}{\hbar}\left(z - \sum_\mathbf{l} e^{i\mathbf{k}\cdot\mathbf{l}}\right) \\ \frac{2JS}{\hbar}\left(z - \sum_\mathbf{l} e^{i\mathbf{k}\cdot\mathbf{l}}\right) & i\omega \end{vmatrix} = 0 \qquad (9.7.18)$$

$\mathbf{l} \equiv \mathbf{R}_j - \mathbf{R}_i$ is the vector connecting a lattice site to its jth neighbor. Evaluating the determinant gives

$$\hbar\omega(\mathbf{k}) = 2JS\left(z - \sum_\mathbf{l} e^{i\mathbf{k}\cdot\mathbf{l}} \right) \qquad (9.7.19)$$

The frequency ω has been given the argument **k** because (9.7.19) shows that there are $3N$ solutions, one for each **k** vector in the first Brillouin zone.

Since the frequency is real, only the real part of the exponential is retained, so

$$\hbar\omega(\mathbf{k}) = 2JS\left[z - \sum_\mathbf{l} \cos(\mathbf{k}\cdot\mathbf{l}) \right] \qquad (9.7.20)$$

This is the dispersion relation for spin waves. The spin waves are a complete set of special solutions of the equations of motion, so the general solution for the spin at any site is a linear combination of the spin waves.

For small wave number vectors, the series expansion of the cosine gives an approximation to the second order as

$$\cos(\mathbf{k}\cdot\mathbf{l}) = 1 - \frac{1}{2}(\mathbf{k}\cdot\mathbf{l})^2$$

$$= 1 - \frac{1}{2}(kl\cos\theta_l)^2 \qquad (9.7.21)$$

θ_l being the angle between the vectors **k** and **l**. In this approximation, (9.7.20) is

$$\hbar\omega(\mathbf{k}) = JS\sum_l (kl\cos\theta_l)^2 \qquad (9.7.22)$$

For cubic crystals the vectors **l** all have the same magnitude l, and if a is the edge of the elementary unit cell, $l = a/2$. Also, the average of $(\cos\theta_l)^2$ is $1/3$, and we finally get the long-wavelength approximation to the dispersion relation as

$$\hbar\omega(\mathbf{k}) = 2JS(ka)^2 \qquad (9.7.23)$$

Just as phonons are the result of quantized lattice vibrations, the quantization of spin waves leads to excitations with energy $\hbar\omega(\mathbf{k})$, and the energy of magnetization is a sum of the energies of these quantized excitations. A rigorous quantum mechanical analysis shows this to be the case. The quantized spin wave excitations obey Bose–Einstein statistics and are called *magnons*. The Bose–Einstein statistics tell us that the number of magnons with energy centered on $\hbar\omega(\mathbf{k})$ and **k** vectors in the range **k** to $\mathbf{k} + d\mathbf{k}$ is

$$n(\mathbf{k})d\mathbf{k} = \frac{V}{8\pi^3} \frac{1}{e^{\hbar\omega/kT} - 1} d\mathbf{k} \qquad (9.7.24)$$

Remember that the magnons are excitations above the ground state, and that in the ground state all spins are parallel so the magnetization has its maximum value. The magnons represent spins that are in the opposite direction and give rise to a demagnetization. The spontaneous magnetization is therefore obtained by computing the moment per unit volume for the magnons and subtracting the result from the magnetization for the completely ferromagnetic state. The moment per unit volume for magnons is readily obtained by multiplying (9.7.24) by the atomic moment $g\mu_B$, dividing by V, and integrating over all wave number vectors **k**:

$$M_g = \frac{g\mu_B}{8\pi^3} \int_{BZ} \frac{1}{e^{\hbar\omega/kT} - 1} d\mathbf{k} \qquad (9.7.25)$$

The integration should be performed over the Brillouin zone, but an easy way to get a good approximation is to recognize that the integrand is large for low frequencies and becomes smaller for increasing frequencies. In fact, at low temperatures the integrand goes to zero as ω increases. We therefore approximate (9.7.25) by integrating over spherical coordinates and replacing the upper Brillouin zone limits by infinity. Therefore, (9.7.25) can be written as

$$M_g = \frac{g\mu_B}{2\pi^2} \int_0^\infty \frac{k^2}{e^{\hbar\omega/kT} - 1} dk \qquad (9.7.26)$$

This is still a complicated integral to evaluate, but for the low-temperature approximation (9.7.23), it can be written as

$$M_g = \frac{g\mu_B}{2\pi^2} \int_0^\infty \frac{k^2}{e^{Ak^2} - 1} dk \qquad (9.7.27)$$

where A is defined by

$$A = \frac{2JSa^2}{kT} \tag{9.7.28}$$

Now the integral can be put into a standard form by a transformation of variables given by

$$x = Ak^2 \tag{9.7.29}$$

so that (9.7.27) becomes

$$M_g = \frac{g\mu_B}{4\pi^2 A^{2/3}} \int_0^\infty \frac{x^{1/2} dx}{e^x - 1} \tag{9.7.30}$$

The integral has the value $(0.0587) 4\pi^2$, so using the definition of A in (9.7.28), equation (9.7.30) becomes

$$M_g = 0.0587 \frac{g\mu_B}{a^3} \left(\frac{kT}{2JS}\right)^{3/2} \tag{9.7.31}$$

In the completely ferromagnetic state, the magnetization is $ng\mu_B S$, and subtracting (9.7.31) from this gives the total magnetization as

$$M_g = ng\mu_B S - 0.0587 \left(\frac{kT}{2JS}\right)^{3/2}$$

$$= ng\mu_B S \left[1 - \frac{0.0587}{rS} \left(\frac{kT}{2JS}\right)^{3/2}\right] \tag{9.7.32}$$

$r = na^3$ is just the ratio of the volume of the elementary cube to the atomic volume. It is 2 and 4 for the body-centered cubic and face-centered cubic structures, respectively.

Equation (9.7.32) states that at low temperatures the spontaneous magnetization of a ferromagnet approaches its maximum value as $T^{3/2}$ as the temperature is decreased, a result that agrees with experiment.

Exercises

9.1 If the Debye temperature of a paramagnetic salt is 400K, and there are 10 times as many atoms as magnetic moments, at what temperature is the lattice heat capacity equal to the paramagnetic heat capacity of the system of independent moments if the external field is 10,000 gauss? Use low-temperature approximations. (Note: This requires an approximate numerical calculation.)

9.2 Find the formulas for the entropy of the system of independent spins in a paramagnetic salt in a magnetic field H. Compare this to the entropy of the system in the absence of the magnetic field. Use the low-temperature approximation throughout. (Note: This is the basis of cooling materials to very low temperatures by adiabatic demagnetization.)

9.3 As stated in section 9.5, the ferromagnetic Ising model is isomorphic with the order-disorder ising model. Write equation (8.7.6) for the high temperature approximation to the temperature dependence of the order parameter in the second moment approximation in the notation appropriate to a ferromagnet. Show that the magnetization becomes zero with a temperature dependence of $\propto T^{-1/2}$ as the temperature approaches the critical temperature. Do the same for the Bragg–Williams approximation.

9.4 Iron has a ferromagnetic critical temperature of 1040K and a Wigner–Seitz radius of 1.41 angstroms. Compute the mean field parameter. Note that the ratio of the internal to the applied magnetic field is high.

Notes

1. Data were taken from *American Institute of Physics Handbook* (1957).
2. The data were taken from Hultgren, Ralph, P.D. Desai, D.T. Hawkins, M. Gleiser, and J. Chipman; 1973; *Selected Values of the Thermodynamic Properties of the Elements*; American Society for Metals, Metals Park, Ohio.

10

Phase Equilibria

10.1 Phase equilibria in one-component systems

Consider a one-component system in equilibrium in field-free space so that the only external force acting on it is pressure and its only thermodynamic variables are temperature, pressure, and volume. Over the temperature and pressure ranges of interest, the system can exist in gas, liquid, and solid phases, but assume that there is only one solid phase. For each phase, the pressure, temperature, and volume are connected by an equation of state. The first step in understanding the possible equilibria among these phases is to determine the conditions of temperature and pressure under which the phases can coexist. Quite a lot can be learned by simple applications of thermodynamics.

From the phase rule, if the solid, liquid, and gas phases exist simultaneously, then there are no degrees of freedom. That is, there is only one pair of values of temperature and pressure at which the three phases can coexist. In a pressure–temperature diagram, this pair of values is represented by a point called the *triple point*.

If two phases coexist, then there is one degree of freedom, which means that either the pressure or the temperature, but not both, can be freely chosen. This corresponds to a curve in the pressure–temperature diagram for each pair of phases solid-liquid, solid-vapor, and liquid-vapor. For brevity, let us call these the S-L, S-V, and L-V curves. The three curves obviously meet at the triple point since each pair of curves (S-L and S-V, S-V and L-V, S-L and L-V) must intersect.

If only one phase is present, then there are two degrees of freedom and there must be three areas in the phase–temperature diagram, one for each possible phase, within which the pressure and temperature can be independently varied. These considerations lead to pressure–temperature phase diagrams for one-component systems of the type shown in figure 10.1.

We can go further by applying the condition that the chemical potentials in two phases at equilibrium must be equal. Now assume that two phases A and B are in equilibrium and that an infinitesimal amount of material is transferred from phase A to phase B because of an infinitesimal change in pressure and temperature. From the Gibbs–Duhem equation (1.15.10), the chemical potential in each phase is

$$d\mu_A = \overline{S}_A dT - \overline{V}_A dP \qquad (10.1.1)$$

$$d\mu_B = \overline{S}_B dT - \overline{V}_B dP \qquad (10.1.2)$$

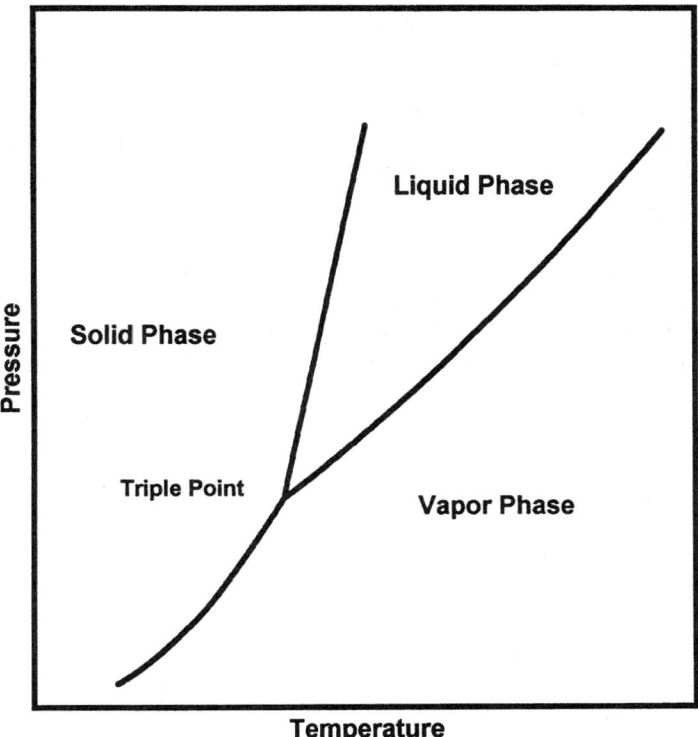

Figure 10.1. Pressure–temperature phase diagram for a one-component system.

where the bars indicate that the quantities are per molecule. That is,

$$\overline{S}_A = \frac{S_A}{N_A}, \quad \overline{V}_A = \frac{V_A}{N_A}$$
$$\overline{S}_B = \frac{S_B}{N_B}, \quad \overline{V}_B = \frac{V_B}{N_B} \tag{10.1.3}$$

The chemical potentials in the two phases must be equal after as well as before the transfer, so their changes (10.1.1) and (10.1.2) are equal and therefore

$$\overline{S}_B dT - \overline{V}_B dP = \overline{S}_A dT - \overline{V}_A dP \tag{10.1.4}$$

or

$$\frac{dP}{dT} = \frac{\Delta \overline{S}}{\Delta \overline{V}} \tag{10.1.5}$$

$\Delta \overline{S}$ and $\Delta \overline{V}$ are the entropy and volume changes per molecule for the phase transformation of A to B. From equation (1.15.6),

$$\mu_A = \overline{H}_A - T\overline{S}_A$$
$$\mu_B = \overline{H}_B - T\overline{S}_B \qquad (10.1.6)$$

$\overline{H}_A$ and $\overline{H}_B$ being the enthalpies per molecule in the phases A and B, respectively. From the equality of the chemical potentials in the two phases, (10.1.6) gives $\Delta \overline{S} = \Delta \overline{V}/T$ and (10.1.5) becomes

$$\frac{dP}{dT} = \frac{\Delta \overline{H}}{T\Delta \overline{V}} \qquad (10.1.7)$$

$\Delta \overline{H} = \overline{H}_B - \overline{H}_A$ is the enthalpy, or latent heat, per molecule of the transformation. Equation (10.1.7) is the Clapeyron equation. When applied to the solid-liquid transition in our one-component system, (10.1.7) gives

$$\frac{dP}{dT} = \frac{\Delta \overline{H}_f}{T\Delta \overline{V}_f} \qquad (10.1.8)$$

while for the liquid-vapor and the solid-vapor transitions

$$\frac{dP}{dT} = \frac{\Delta \overline{H}_v}{T\Delta \overline{V}_v} \qquad (10.1.9)$$

$$\frac{dP}{dT} = \frac{\Delta \overline{H}_s}{T\Delta \overline{V}_s} \qquad (10.1.10)$$

$\Delta \overline{H}_f$, $\Delta \overline{H}_v$, $\Delta \overline{H}_s$, and $\Delta \overline{V}_f$, $\Delta \overline{V}_v$, $\Delta \overline{V}_s$ are the enthalpies and volumes of fusion, vaporization, and sublimation, respectively. Equations (10.1.8)–(10.1.10) define the melting point, boiling point, and sublimation curves, respectively.

Since the enthalpy of transformation is independent of the path, we have

$$\Delta \overline{H}_s = \Delta \overline{H}_v + \Delta \overline{H}_f \qquad (10.1.11)$$

In general, $\Delta \overline{H}_s > \Delta \overline{H}_v > \Delta \overline{H}_f > 0$ and $\Delta \overline{V}_s > \Delta \overline{V}_v > \Delta \overline{V}_f > 0$, so the slopes of the coexistence lines in the pressure–volume diagram are positive. In a few special cases, such as antimony and water, the molecular volume is greater in the solid phase than in the liquid phase, in which case the melting point curve has a negative slope.

For temperature ranges that are not too large, the enthalpy of fusion and volume of fusion can both be taken as independent of temperature and (10.1.8) can be integrated between two temperatures to give an approximate expression for the melting point as

$$P = P_0 + \frac{\Delta H_f}{\Delta V_f} \ln \frac{T}{T_0} \qquad (10.1.12)$$

An alternate approximation can be made by inverting (10.1.8) to read

$$\frac{dT}{dP} = \frac{T\Delta \overline{V}_f}{\Delta \overline{H}_f} \qquad (10.1.13)$$

Numerical values of the right-hand side of (10.1.13) are usually quite small. That is, it takes a lot of pressure to change the melting point by one degree. It is therefore often sufficiently accurate to compute (10.1.13) for a given temperature and use that value to get the effect of pressure on melting temperatures near that value. For temperatures near the melting point T_m, the change in melting point as a function of change in pressure can therefore be written as

$$\Delta T_m = \frac{T_m \Delta \overline{V}_f}{\Delta \overline{H}_f} \Delta P \qquad (10.1.14)$$

The solid-vapor and liquid-vapor Clapeyron equations can also be integrated approximately by recognizing that, for temperatures not too near the critical point, the volume per molecule in the vapor is much larger than in either the liquid or the solid, so both the volumes of sublimation and vaporization can be replaced by the vapor volume. Then, if the vapor is assumed to be dilute enough to follow the ideal gas law, we have

$$\Delta \overline{V}_s = \Delta \overline{V}_v = \overline{V}_g = \frac{kT}{P} \qquad (10.1.15)$$

$\overline{V}_g$ being the volume per molecule in the gas phase. Putting (10.1.15) into equations (10.1.9) and (10.1.10) gives

$$\frac{d \ln P}{dT} = \frac{\Delta \overline{H}_v}{kT^2} \qquad (10.1.16)$$

$$\frac{d \ln P}{dT} = \frac{\Delta \overline{H}_s}{kT^2} \qquad (10.1.17)$$

These are the Clausius–Clapeyron equations for vaporization and sublimation, respectively. If the enthalpies of vaporization and sublimation do not vary much with temperature, these equations can be readily integrated.

Note that the basic constituent of the one-component system has been referred to as a molecule. Of course, the above development applies to atomic as well as molecular systems. In what follows, the general term "molecule" will still be used, but the theories will be written as if the system were composed of atoms. This means that internal degrees of freedom of molecules will be ignored. A diatomic molecule, for example, will be assumed to move as a unit, with respect to either translation or vibration. If the internal degrees of freedom are the same in the two phases, then the theory applies unchanged. This would be the case for vibrations in a diatomic molecule for which the chemical bond in the molecule is much stronger than the intermolecular interactions. Since the effect of the intermolecular forces on the intramolecular vibration frequencies is small, the molecule has a vibrational energy, and therefore a vibrational partition function, that is very nearly the same in the solid, liquid, and vapor phases. The situation is a bit more complex for rotational degrees of freedom. In the solid, the rotation is often restricted by nearest neighbor interactions and does not come into play. However, the molecule can rock back and forth in its restricted position, giving rise to librational degrees of freedom. In the gas phase, the molecule can rotate freely and the rotational term must be included in the partition function while librations are excluded. No general rule can be given for the contributions of rotations to the partition function of a liquid. The rotations of most molecules will be restricted while those of some molecules may not be. Even for the restricted rotations, there is a difference from the solid because we expect the amplitudes of librations are greater in the liquid.

The following developments can to be applied to molecular solids by adding in the effects of the internal degrees of freedom. Assuming a complete separation of internal and external motions, this means that when the partition function that neglects the internal motions is obtained, it is simply multiplied by partition functions representing the internal degrees of freedom.

10.2 The van der Waals model

Real gases follow the ideal gas law only at low densities or high temperatures. To go beyond the ideal gas representation, it is necessary to include nonzero intermolecular interactions in the theory of the equation of state. A simple approach is provided by the van der Waals theory of molecular gases, which is based on the following assumptions:

1. The potential energy of any two molecules in the gas arises form a central-force, pairwise potential that depends only on the distance between the molecules and is the same for all pairs.
2. The potential energy of the gas is a sum of the interactions for all pairs of molecules.
3. The pair potential consists of a hard sphere repulsion and a relatively long-range attraction such as the attractive part of the Lennard–Jones potential.
4. For their motion in the gas, only the volume outside of the hard-sphere radii is available to the molecules.
5. The configurational partition function can be evaluated by expanding the Boltzmann factor about the mean energy and retaining just the first term. This is the same procedure as followed in the Bragg–Williams order-disorder theory.

These assumptions amount to applying two simple corrections to the ideal gas law to take into account the repulsive and the attractive forces acting between molecules. The repulsive force is included by making the volume occupied by the molecules, regarded as hard spheres, unavailable to the molecular centers. Thus, the volume in the ideal gas law is replaced by the "free" volume, which is less than the actual volume. The attractive forces are included by recognizing that, when they are present, less pressure needs to be exerted on the gas to maintain a given volume. The pressure in the ideal gas law must therefore be increased by an amount that corrects for the decrease resulting from the intermolecular attractions. Since two molecules are involved in an interaction, this increase in gas pressure is proportional to the square of the density. From these simple considerations, the ideal gas law is modified to read

$$\left(P + a\frac{N^2}{V^2}\right)(V - Nb) = NkT \tag{10.2.1}$$

where a and b are constants representing the attractive and repulsive effects of the intermolecular forces, respectively.

The van der Waals model is unexpectedly rich in results. Not only does it provide a reasonable equation of state for gases that are not too dense, but also it provides a description of the gas-liquid phase transition that displays all the qualitative features of real gas-liquid equilibria. This is done by simply ignoring the fact that the model was designed for the gas phase only and using it to also describe liquids.

An elementary statistical mechanical derivation based on the above assumptions gives not only the equation of state, but also all thermodynamic functions.

The molecular masses and temperature conditions for the systems we are interested in are such that the semiclassical approximation (2.14.20) for the partition function is quite accurate:

$$Z_{cl} = \frac{Z_q}{\Lambda^{3N}} \qquad (10.2.2)$$

where $\Lambda \equiv h/(2\pi mkT)^{1/2}$ is the thermal wavelength and Z_q is the configurational partition function defined by

$$Z_q = \frac{1}{N!}\int e^{-\phi(q)/kT} dq \qquad (10.2.3)$$

The first task is to evaluate the integral in (10.2.3). In a van der Waals gas, the potential energy is assumed to be a sum of pairwise interactions that consist of a hard sphere repulsion and a relatively long-range attraction such as the attractive part of the Lennard–Jones potential. That is,

$$\phi(q) = \sum_{<ij>} u(r_{ij}) \qquad (10.2.4)$$

where the sum is over all pairs and the central, pairwise potential between two molecules i and j a distance r_{ij} apart is

$$\begin{aligned} u(r_{ij}) &= \infty & \text{for } r < R \\ u(r_{ij}) &= -\frac{A}{r_{ij}^6} & \text{for } r > R \end{aligned} \qquad (10.2.5)$$

R is the hard sphere radius of the molecule. A schematic plot of the van der Waals potential is shown in figure 10.2.

The configurational partition function (10.2.3) is then

$$Z_q = \frac{1}{N!}\int e^{-\sum_{<ij>} u(r_{ij})/kT} dq \qquad (10.2.6)$$

Now use assumption 5 above and replace the sum of potential functions over all pairs by its average value. This can be done as follows.

First, consider the interaction of one molecule, labeled α, with all other molecules in the gas. Draw a spherical shell around this molecule at a distance r_α from its center and define a function $\rho_\alpha(r)$ such that the number of molecules in the shell is

$$\frac{(N-1)}{V} 4\pi r_\alpha^2 \rho_\alpha(r) dr \qquad (10.2.7)$$

$\rho_\alpha(r)$ is the radial distribution function and measures the deviation from a random distribution of molecules since, if the molecules were distributed totally at random, $\rho_\alpha(r)$ would be unity. The potential energy of the central molecule in the field of all the others is therefore

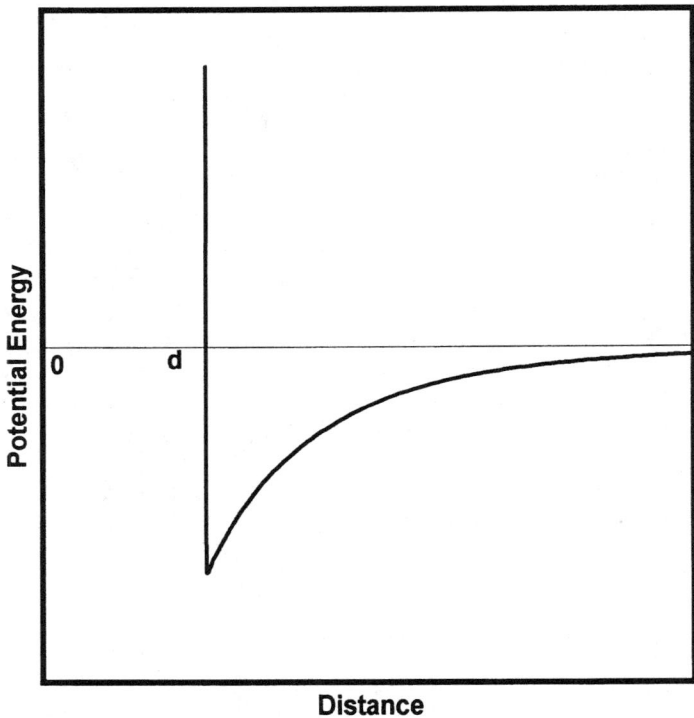

Figure 10.2. The van der Waals potential.

$$\frac{N}{V}\int_0^\infty 4\pi r^2 \rho_\alpha(r_\alpha)u(r_\alpha)dr_\alpha \quad (10.2.8)$$

(Unity has been neglected relative to N.) Equation (10.2.8) is the average interaction of the α molecule with the rest of the gas because the radial distribution function describes the average density of molecules as a function of distance from the central molecule. Since all molecules are alike, the total average potential energy of interaction is obtained by multiplying (10.2.8) by the number of molecules and dividing by 2 to avoid double counting. That is,

$$\phi(q) = \frac{N^2}{V}\overline{W} \quad (10.2.9)$$

where

$$\overline{W} = \frac{1}{2}\int_0^\infty 4\pi r^2 \rho_\alpha(r_\alpha)u(r_\alpha)dr_\alpha \quad (10.2.10)$$

and the configurational integral (10.2.6) becomes

$$Z_q = \frac{1}{N!}e^{-N^2\overline{W}/VkT}\iiint \ldots d\mathbf{r}_1 d\mathbf{r}_2 d\mathbf{r}_3 \ldots \quad (10.2.11)$$

Because of the assumption of hard sphere repulsions, each integral in (10.2.11) ranges only over the free volume of the gas. This excludes a volume Nb, where b is the volume excluded per molecule. Thus, each integral in (10.2.11) has the value $(V - Nb)$, and

$$Z_q = \frac{1}{N!} e^{-N^2\overline{W}/VkT} (V - Nb)^N \tag{10.2.12}$$

and the partition function for the van der Waals model is

$$Z_{vdw} = \frac{(V - Nb)^N}{N! \Lambda^{3N}} e^{-N^2\overline{W}/VkT} \tag{10.2.13}$$

The Helmholtz free energy is

$$A = -kT \ln Z_{vdw} = NkT \ln\left(\frac{N\Lambda^3}{V - Nb}\right) - NkT + \frac{N^2\overline{W}}{V} \tag{10.2.14}$$

Stirling's approximation has been used on the factorial. The pressure is

$$P = -\left(\frac{\partial A}{\partial V}\right)_T = \frac{N^2\overline{W}}{V^2} + \frac{NkT}{(V - Nb)} \tag{10.2.15}$$

Remember that $\overline{W}$ is an average attractive potential, so it is a negative quantity. Replacing it with a positive quantity according to $a \equiv -\overline{W}$, (10.2.15) is seen to be the van der Waals equation given by (10.2.1), in which, in terms of the volume per molecule, $\overline{V}$ is

$$P = -\frac{a}{\overline{V}^2} + \frac{kT}{(\overline{V} - b)} \tag{10.2.16}$$

The van der Waals equation can be put into a reduced, universal form by finding the pressure, temperature, and volume for which both the first and second derivatives of the pressure with respect to volume vanish. That is, we look for $P = P_c$, $T = T_c$, $\overline{V} = \overline{V}_c$ such that

$$\left(\frac{\partial P}{\partial \overline{V}}\right)_c = \left(\frac{\partial^2 P}{\partial \overline{V}^2}\right)_c = 0 \tag{10.2.17}$$

Differentiating (10.2.16) and setting the results equal to zero gives

$$\left(\frac{\partial P}{\partial \overline{V}}\right)_c = \frac{2a}{\overline{V}_c^3} - \frac{kT_c}{(\overline{V}_c - b)^2} = 0 \tag{10.2.18}$$

$$\left(\frac{\partial^2 P}{\partial \overline{V}^2}\right)_c = \frac{2kT_c}{(\overline{V}_c - b)^3} - \frac{6a}{\overline{V}_c^4} = 0 \tag{10.2.19}$$

P_c, T_c, $\overline{V}_c$ are called the *critical pressure*, *temperature*, and *volume*, respectively, and they define a critical point for the van der Waals equation of state for reasons that will soon be apparent. At the critical point, the van der Waals equation (10.2.16) is

$$P_c = -\frac{a}{V_c^2} + \frac{kT_c}{(V_c - b)} \tag{10.2.20}$$

Equations (10.2.18)–(10.2.20) are readily solved to give the critical parameters in terms of the van der Waals constants as

$$P_c = \frac{a}{27b^2}, \quad T_c = \frac{8a}{27kb}, \quad V_c = 3b \tag{10.2.21}$$

If a reduced pressure, temperature, and volume are defined by

$$\tilde{P} \equiv \frac{P}{P_c}, \quad \tilde{V} = \frac{V}{V_c}, \quad \tilde{T} = \frac{T}{T_c} \tag{10.2.22}$$

then, from (10.2.16),

$$\tilde{P} = \frac{8\tilde{T}}{3\tilde{V} - 1} - \frac{3}{\tilde{V}^2} \tag{10.2.23}$$

or

$$\left(\tilde{P} + \frac{3}{\tilde{V}^2}\right)(3\tilde{V} - 1) = 8\tilde{T} \tag{10.2.24}$$

Note that these results do not depend on the precise form of the attractive potential, but only on the fact that it exists.

The van der Waals equation is cubic in the volume, so for a given pressure, there can be one, two, or three distinct unequal roots, as shown in figure 10.3 for different reduced temperatures.

Above the critical temperature, $\tilde{T} = 1$, the isotherm is single valued, corresponding to a single (gas) phase for all values of pressure and volume. When the critical temperature is unity, there is an inflection point as defined by equation (10.2.17). Below the critical temperature, there is a range of volumes for which the pressure has three values for each volume.

Let us examine this case in more detail, using figure 10.4. Above the maximum, there is only one pressure for each volume, the volume is low, and it takes a large increase in pressure to produce a small change in volume. In this region, the curve is the P–V isotherm for a liquid. Below the minimum, the curve is again single valued, the volume is high, the compressibility is high, and the isotherm is that for a gas.

Above the critical point, the liquid and gas phases can change into each other by continuous changes in temperature and pressure. This is not possible for solid-vapor or solid-liquid transitions. The liquid and the gas are both fluid phases with full translational and rotational symmetry, so there is no discontinuous symmetry change in going from one to the other. A solid, however, has a lower symmetry since it is a discrete periodic structure whose translational and rotational symmetries are restricted. A fluid with full symmetry cannot continuously be transformed to a solid whose symmetry is much lower.

Between the minimum and the maximum, only two of the three pressures for a given volume can correspond to physical reality. The reason for this is that between the minimum and maximum pressures, the van der Waals curve states that the volume increases with increasing pressure. This is contrary to the stability criterion of equation (1.14.21).

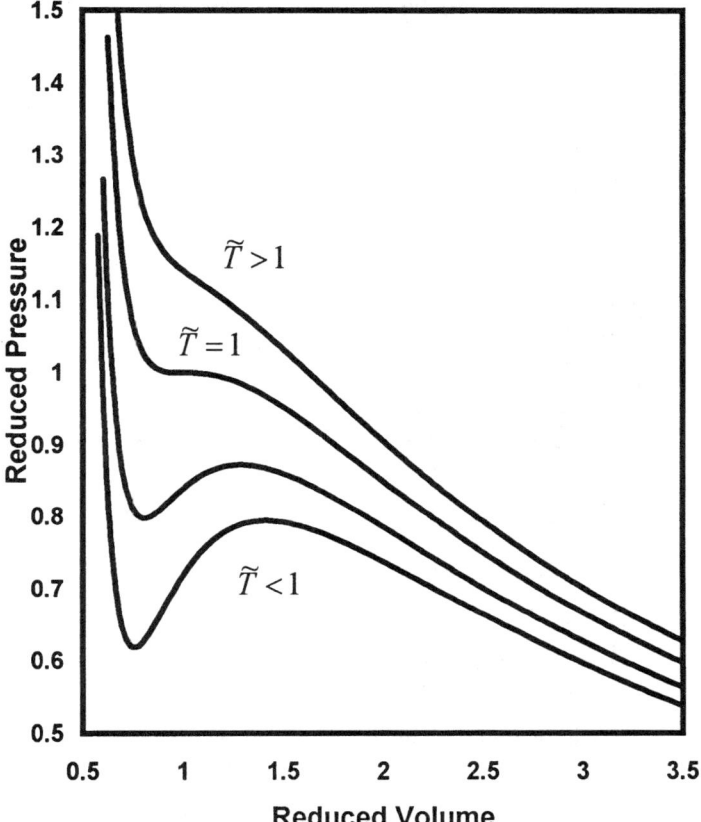

Figure 10.3. Isotherms for a van der Waals fluid.

Although there are two volumes for every pressure that do not violate the stability condition, only one pair of volumes between the minimum and maximum corresponds to an equilibrium between the liquid and the gas phase since only one pair satisfies the requirements of equality of chemical potentials and equality of pressures. These equilibrium volumes can be found by a construction proposed by Maxwell in 1875, just two years after van der Waals's thesis in which he presented his equation from considerations based on the classical kinetic theory of gases. This construction starts with the volume as a function of pressure as shown in figure 10.5, in which a line parallel to the volume axis is drawn between the gas and liquid phases starting at a point labeled 1, going through the point M, and terminating at the point 2.

From the Gibbs–Duhem equation (1.15.10) for constant temperature, the change in chemical potential of a one-component system is related to the change in pressure by

$$d\mu = \overline{V} dP \tag{10.2.25}$$

Therefore, the line integral of (10.2.25) along the van der Waals isotherm between the two points 1 and M gives

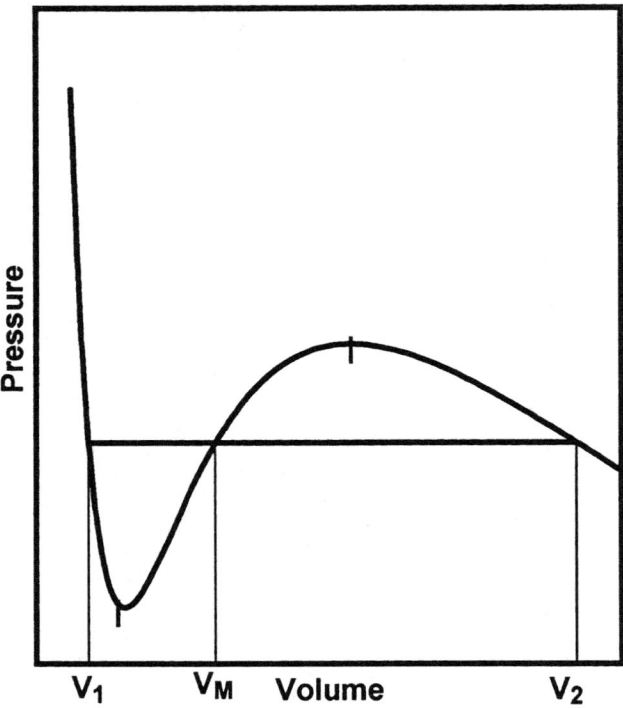

Figure 10.4. *P–V* isotherm for a van der Waals fluid below Critical temperature T_c.

$$\mu_M - \mu_1 = \int_{V_1}^{V_M} \overline{V} dP \qquad (10.2.26)$$

and the line integral from M to 2 is

$$\mu_2 - \mu_M = \int_{V_M}^{V_2} \overline{V} dP \qquad (10.2.27)$$

But the line integral around a closed curve is just the area within the curve, and since the pressure is constant along the line parallel to the pressure axis and contributes nothing to the area integral, (10.2.26) is the area enclosed by the curve from 1 to M back to 1, while (10.2.27) is the area within the curve going from M to 2 back to M. This means that the chemical potentials at 1 and 2 can be equal only if the two areas are equal. This is the Maxwell equal area construction. For a given temperature, it fixes the only pressure at which the two phases are in equilibrium. In figure 10.4 the tie line drawn to give equal areas gives the equilibrium pressure for the coexistence of the liquid and vapor. (Note that there is some lack of rigor in this derivation because it is assumed that equilibrium thermodynamic relations are valid even in the volume range where the system is unstable.)

Although the points between 1 and the minimum, and between the maximum and 2, are not in thermodynamic equilibrium, they do not violate the stability condition of equation (1.14.21) since the volume decreases with pressure in these regions. These are regions of metastability: the first region

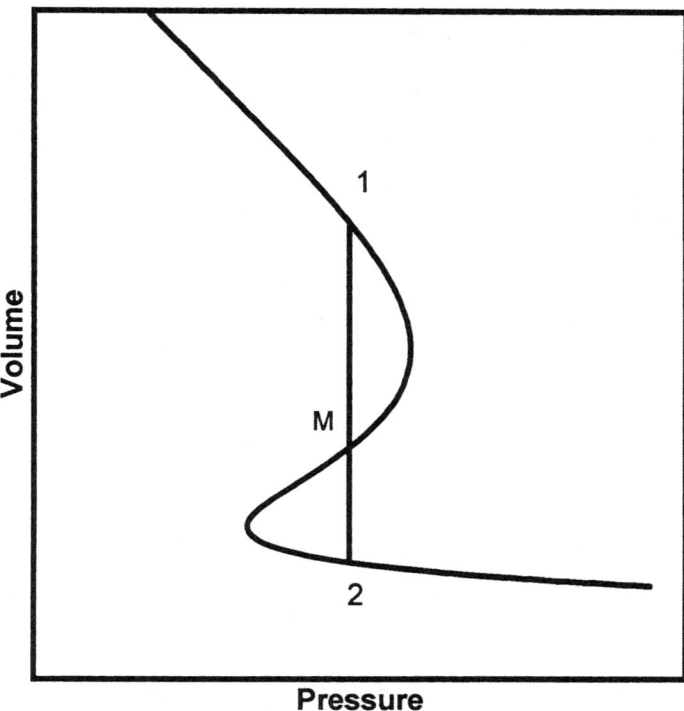

Figure 10.5. *V–P* isotherm for a van der Waals fluid below the critical temperature.

corresponds to possible superheating in which the system remains liquid above the boiling point, and the second region represents possible supercooling in which the system remains gaseous below the boiling point.

Of course, there is a construction similar to figure 10.4 for every temperature, and the locus of all points like V_1 and V_2 can be found. Also, the locus of all points at which the maxima and minima occur can be obtained by setting the derivative of the pressure with respect to volume equal to zero. The phase diagram thus obtained is shown in figure 10.6. The maximum is the critical point above which there is only one fluid phase. The region below the critical point and to the left of the solid curve is the region in which the system is a liquid, while below the critical point and to the right of the solid curve is the vapor region. This solid curve is called the *binodal*, and it separates two phase from one phase regions. The dotted curve is called the *spinodal*. Inside the spinodal the system is unstable, while the area between the two curves are ranges of pressure and volume for which the system is metastable.

If a system is in the liquid region below the critical point, increasing the temperature at constant pressure will move the system to the equilibrium curve, some of the liquid will vaporize, and two phases will coexist. As more heat is put into the system, all the liquid will evaporate and the system moves into the gas region. (This presumes that no superheating or supercooling takes place.) An interesting result is that a liquid can be converted to a gas without ever going through a two-phase condition by first raising the pressure at constant volume to bring the system above the critical point, then raising the volume at constant pressure to a volume well to the right, and then lowering

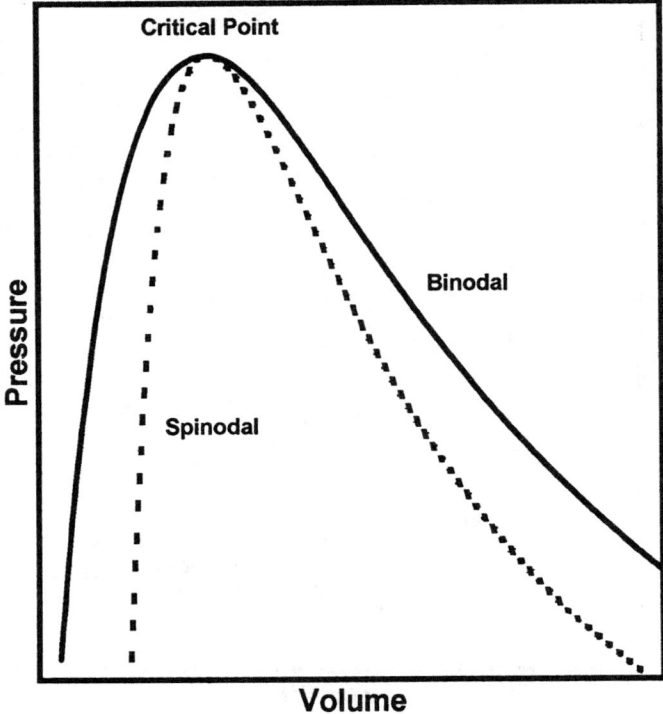

Figure 10.6. The van der Waals phase diagram.

the pressure at constant volume to a point below the critical point and to the right of the binodal.

10.3 Sublimation

An attractive feature of the van der Waals theory is that a single model describes both the liquid and the gas phases. From a specific assumption about molecular interactions, a single partition function is derived that leads directly to a phase transition. It displays a specific case giving rise to a phase change and shows that the possibility of phase equilibria is built right into statistical mechanics.

In many cases phase equilibria are more easily studied by adopting different models for each phase and finding the conditions under which the separate models yield equilibrium among the phases. A relatively simple example is the equilibrium between a gas and a solid below the critical point. The $P-T$ sublimation curve can be derived by adopting the harmonic model (or one of its approximations) for a solid, and a gas model, such as the ideal gas or the van der Waals gas, for the vapor.

Consider a monomolecular solid in equilibrium with its vapor. It is assumed that there is no association in the vapor, so that it consists of free molecules. The solid will be represented by the Debye theory while the gas will be assumed to be ideal and all internal degrees of freedom of the molecules are ignored. Using this model, the conditions under which the chemical potentials of the two phases are equal can be obtained.

The Helmholtz free energy for the model is obtained by using the Debye distribution function in equation (4.3.15). Adding PV_c to this and dividing by the number of molecules gives the chemical potential of the crystal as

$$\mu_c = \overline{E}_0 + 9kT\left(\frac{T}{\Theta_D}\right)^3 \int_0^{\Theta_D/T} x^2 \ln(1-e^{-x})dx + P\overline{V}_c \qquad (10.3.1)$$

where μ_c is the chemical potential of the crystal $\overline{V}_c$ is its volume per molecule and $\overline{E}_0$ is the zero point energy per molecule of the crystal.

The chemical potential of the gas is given by equations (3.4.2) and (3.4.18) as

$$\mu_g = kT\ln\left(\frac{N_g}{V_g}\right) - \frac{3}{2}kT\ln\left(\frac{2\pi mkT}{h^2}\right) = kT\ln\left(\frac{P\Lambda^3}{kT}\right) \qquad (10.3.2)$$

N_g and V_g are the number of molecules and the volume of the gas phase.

Now all that needs to be done is to equate the two chemical potentials. But first let us transform the integral in the expression for the chemical potential of the solid using an integration by parts to get

$$\int_0^{\Theta_D/T} x^2 \ln(1-e^{-x})dx = \frac{1}{3}\left(\frac{\Theta_D}{T}\right)^3 \ln(1-e^{-\Theta_D/T}) - \frac{1}{3}\int_0^{\Theta_D/T} \frac{x^3}{e^x - 1}dx$$

so that (10.3.1) becomes

$$\mu_c = \overline{E}_0 + 3kT\ln(1-e^{-\Theta_D/T}) - k\Theta_D D_E\left(\frac{\Theta_D}{T}\right) + P\overline{V}_c \qquad (10.3.3)$$

with $D_E(\Theta_D/T)$ being the Debye energy function defined by equation (4.8.5). Equating (10.3.2) and (10.3.3) gives

$$kT\ln\left(\frac{P\Lambda^3}{kT}\right) = \overline{E}_0 + 3kT\ln(1-e^{-\Theta_D/T}) - k\Theta_D D_E\left(\frac{\Theta_D}{T}\right) + P\overline{V}_c \qquad (10.3.4)$$

which is the equilibrium pressure–temperature relation between a solid and its vapor. This is an integrated form of the Clapeyron equation for sublimation (10.1.10), as can be shown by differentiating it to get

$$kT^2 \frac{d\ln P}{dT} = \frac{5kT}{2} - \overline{E}_0 - P\overline{V}_c + T\overline{V}_c \frac{dP}{dT} - 3k\Theta_D D_E\left(\frac{\Theta_D}{T}\right) \qquad (10.3.5)$$

In this differentiation, it is assumed that the volume of the crystal is constant. For real crystals, the thermal expansion is small, and for the harmonic crystal model it is zero. This assumption is therefore consistent with Debye theory and is also not too far from reality.

From equation (3.4.35), the enthalpy of the ideal gas per molecule is $5kT/2$, and from (4.8.7) the second and last terms add up to the negative of the crystal energy, which when added to the third term gives the negative of the crystal enthalpy. Therefore, (10.3.5) is

$$kT^2 \frac{d\ln P}{dT} = T\overline{V}_c \frac{dP}{dT} + \Delta \overline{H}_s \qquad (10.3.6)$$

where the heat of sublimation is the enthalpy of the gas minus the enthalpy of the crystal, per molecule:

$$\Delta \overline{H}_s \equiv \overline{H}_g - \overline{H}_c \qquad (10.3.7)$$

Making use of the ideal gas law for the vapor in the left-hand side of (10.3.6) and solving for the derivative of the pressure gives

$$\frac{dP}{dT} = \frac{\Delta \overline{H}_s}{T(\overline{V}_g - \overline{V}_c)} \qquad (10.3.8)$$

This is the Clapeyron equation for sublimation. It has been derived from the Debye theory with the heat of sublimation given explicitly by the model. The Debye theory is therefore consistent with thermodynamics for the vapor-solid phase transition, and (10.3.4) is indeed an integration of the Clapeyron equation.

The volume per molecule in the gas phase is much larger than in the crystal, so using the ideal gas law, (10.3.8) can be written as

$$\frac{d\ln P}{dT} = \frac{\Delta \overline{H}_s}{kT^2} \qquad (10.3.9)$$

which is just the Clausius–Clayperon equation (10.1.17).

Note that the zero point energy is negative, so the enthalpy of sublimation is positive. Of course, the zero of energy must always be taken to be the same for both phases. This is automatic in the above development since the zero of energy for the harmonic crystal was taken to be that of molecules at infinite separation in the gas phase.

10.4 The liquid state

The theory of liquids is complicated by the fact that liquids do not have a rigid periodic structure like that of a crystal, nor the dilute nature of a gas, which can be described by pair interactions of randomly moving molecules. The intermediate nature of a liquid, in which the density is similar to that in solids, but molecular mobilities are much greater, makes it more difficult to arrive at rigorous theories for which numerical computations can be made. Formally correct theories have been constructed, but these are restricted to liquids in which the molecules interact according to pairwise central forces. Calculations from such theories require considerable simplification, and they are difficult to extend to noncentral force systems.

There are two classes of simplified models that are useful. One class is based on theories of gases, and the other on theories of solids. The van der Waals model treats the liquid as if it were a dense gas and is successful in qualitatively describing the liquid-vapor transition. Here, our interest is in the solid-liquid transition, and we therefore adopt a model that is closer to the solid state.

A relatively simple approach, called the *method of significant structures*, was developed in 1964 by Eyring et al.,[1] which assumes from the start that a liquid has a mixture of solidlike and gaslike characteristics. While lacking the rigor of more advanced methods, this approach has the advantage of providing a good physical description of liquids from which calculations can readily be made. The results are in adequate agreement with experiment. The development below is a variation of Eyring's model.

The method starts with the observation that X-ray analysis of liquids shows that they exhibit strong short-range order and that the nearest neighbor distances between molecules are very close to those of the solid. For increasing separation distances, the correlation between molecular positions decreases rapidly. That is, each molecule has an immediate environment that is similar to that in the solid state, but there is no long-range order. Also, the density increase on melting varies from about 10% to 20%. This is not too far from the difference in the density of closely packed and randomly packed spheres, which is about 14%. Furthermore, the heat of fusion is a small fraction of the heat of vaporization, being about 2–5% for metals, and about 5–20% for molecular solids. This accords with the idea that vaporization is the result of complete separation of the molecules and that the liquid intermolecular distances are not too different from those of the solid.

All this is true for systems that are not too near the critical point. As the critical point is approached, the distinction between liquid and vapor vanishes and the heat of vaporization approaches zero. Our interest is in melting at temperatures and pressures far below the critical point, and the models we construct are restricted to such conditions.

But, while the short-range order in a liquid is close to that of a solid, molecules in the liquid are much more mobile, as shown by its fluidity and much greater diffusion coefficients. Also, a number of the lattice sites in solids near the melting point are known to be vacant, the fraction of vacant sites in metals being of the order of 10^{-3}, so if a liquid is solidlike, it must contain more vacancies than the crystal phase. If the decrease in density on melting is an indication of vacancy concentration, the liquid state must contain about an order of magnitude more vacancies than the solid. The simplest physical model of a liquid in accord with this description describes the liquid as containing a number of holes, or vacancies, and a mixture of molecules that are solidlike with molecules that are gaslike. Adopting the viewpoint that molecules adjacent to vacancies are gaslike, their number is approximately the number of vacancies times the coordination number, which is usually about 10–12. The fraction of gaslike molecules in the liquid can then be as much as 10%. This leads to the following model.

During any small increment of time, most molecules are surrounded by a number z_l of nearest neighbors. While this number may be less than that in the solid, It is sufficient to enclose the central molecule in a cage within which it has vibrational degrees of freedom as in a solid. But the liquid also contains a number of holes of molecular size (vacancies). The molecules next to these holes can easily escape into it. They therefore have translational degrees of freedom like molecules in a gas. Of course, the particular molecules that are solidlike or gaslike change with time, but in keeping with the ergodic hypothesis, the time average is replaced by an ensemble average, so the continually changing structure can be replaced by the most probable structure. The partition function for the liquid is then taken to be the product of two partition functions: one for a solid containing N_s molecules and N_v vacancies, and the other for a gas containing N_g molecules, $N_g + N_s = N_l$ being the total number of molecules in the liquid. The partition function representing the "solid" is taken to be just that for the harmonic crystal.

The partition function of the liquid is then

$$Z = Z_g Z_s \qquad (10.4.1)$$

so the Helmholtz free energy is

$$A_l = -kT \ln Z_g - kT \ln Z_s \tag{10.4.2}$$

The "solid" partition function, Z_s, is given by equation (4.3.12), the product being taken over $3N_s$ vibrational degrees of freedom. The high-temperature approximation is adequate, so expanding the exponential and retaining only the first two terms, (4.3.12) gives

$$Z_s = e^{-\bar{E}_0^s/kT} \prod_{j=1}^{3N_s} \frac{kT}{h\nu_j} \tag{10.4.3}$$

The corresponding Helmholtz free energy, in the continuum notation, is

$$A_s = N_s \bar{E}_0^s + N_s kT \int_0^\infty g(\nu) \ln\left(\frac{h\nu}{kT}\right) d\nu \tag{10.4.4}$$

$\bar{E}_0^s$ being the zero point energy per solidlike molecule.

Using the Debye distribution function gives

$$A_s = N_s \bar{E}_0^s + 3N_s kT \ln\left(\frac{\Theta_D^l}{T}\right) \tag{10.4.5}$$

where Θ_D^l is a Debye characteristic temperature for the solidlike portion of the liquid phase.

From the semiclassical partition function for an ideal gas, equation (3.4.18), the partition function for the "gas," is

$$Z_g = \frac{(V_g)^{N_g}}{\Lambda^{3N_g}} \frac{e^{-N_g \bar{E}_0^g/kT}}{N_g!} \tag{10.4.6}$$

This differs from the ideal gas partition function through the exponential factor because the gaslike molecules move in an average potential $\bar{E}_0^g$, which is assumed to be constant. In the ideal gas, the background potential was assumed to be zero.

We assume that the volume available to the "gas" is the total volume of the "liquid" minus the volume of the "solid," so

$$V_g = V_l - V_s \tag{10.4.7}$$

This is often called the "free volume" since it is the volume over which molecules in the liquid are treated as a gas. From the physical picture of the liquid as a solid-vacancy-gas mixture, the number of gaslike molecules is the number of vacancies times the coordination number, and the "gas" volume is the volume of a vacancy plus its nearest neighbors, all multiplied by the number of vacancies. That is,

$$N_g = N_v z_l \tag{10.4.8}$$

and

$$V_g = N_v(z_l \bar{V}_l + \bar{V}_v) \tag{10.4.9}$$

$\bar{V}_l = V_l/N_l$ being the volume per molecule of the liquid. In (10.4.9) it is assumed that the volume per atom of the neighbors to the vacancy is just that of the liquid as a whole.

In the Eyring theory the vacancy formation volume, $\overline{V}_v$, is taken to be the molecular volume of the liquid. If there is not too much relaxation of molecules around the vacancy from the positions they have when there is no vacancy, and if this relaxation is confined to first nearest neighbors, this is not a bad approximation, but it cannot be adopted as a general rule. For large, spherical molecules that interact via pairwise central forces with strong repulsions and whose attractive range is short relative to intermolecular distances, very little relaxation occurs and the vacancy formation volume is very close to the molecular volume. The C_{60} fullerene molecule is an excellent example of this. However, it is known from theory and experiments on point defects that the relaxations around vacancies in metals with small ion cores, such as sodium, are quite large even in the solid state. In such cases, the "gaslike" volume per molecule is somewhat larger then the overall molecular volume of the liquid. But no great error is introduced if the vacancy formation volume is replaced by the liquid molecular volume in (10.4.9) because it contributes less than 10% to the total volume of gaslike molecules. Also, while the vacancy formation volume is less than the liquid molecular volume, the molecular volume of gaslike molecules (around the vacancy) is somewhat greater because of the presence of the vacancy, so there is some resulting compensation of errors. Thus, we take (10.4.9) to be

$$V_g = N_v(z_l + 1)\overline{V}_l \quad (10.4.10)$$

From the theory of point defects, the vacancy concentration is determined by the Gibbs free energy of vacancy formation, G_v^f, by

$$\frac{N_v}{N_l + N_v} = e^{-G_v^f/kT} \quad (10.4.11)$$

Using (10.4.8) and (10.4.9), equation (10.4.6) gives the Helmholtz free energy for the gaslike molecules as

$$A_g = -kT \ln Z_g = N_g kT \ln\left[\frac{z_l \Lambda^3}{e(z_l + 1)\overline{V}_l}\right] + N_g \overline{E}_0^g \quad (10.4.12)$$

The sum of (10.4.5) and (10.4.12) gives the Helmholtz free energy. To get the Gibbs free energy, two terms must be added. The first is the usual PV term. But if the vacancy model is taken seriously, this is not enough; the theory of point defects tells us that each vacancy contributes an amount $-kT$ to the Gibbs free energy [see equation (15.3.22)]. Adding these two terms to the sum of (10.4.5) and (10.4.12) gives the Gibbs free energy as

$$G_l = N_g kT \ln\left[\frac{z_l \Lambda^3}{e(z_l + 1)\overline{V}_l}\right] + N_g \overline{E}_0^g + N_s \overline{E}_0^s$$
$$+ 3N_s kT \ln\left(\frac{\Theta_D^l}{T}\right) + PN_l \overline{V}_l - \frac{N_g}{z_l} kT \quad (10.4.13)$$

Dividing through by the total number of molecules gives the chemical potential of the liquid as

$$\mu_l = f_g kT \ln\left[\frac{z_l \Lambda^3}{e(z_l + 1)\overline{V}_l}\right] + f_g \overline{E}_0^g + f_s \overline{E}_0^s$$
$$+ 3f_s kT \ln\left(\frac{\Theta_D^l}{T}\right) + P\overline{V}_l - \frac{f_g}{z_l} kT \quad (10.4.14)$$

276   STATISTICAL MECHANICS OF SOLIDS

f_g and f_s being the fraction of gaslike and solidlike molecules, respectively.

For liquids of low density, the vacancy formation free energy is low so that the number of vacancies, and therefore the number of gaslike molecules, is high. For high densities, the formation free energy is high, the number of vacancies is small, and most of the molecules are solidlike. Equation (10.4.14) therefore exhibits a continuous change from solidlike to gaslike behavior as a function of decreasing density.

10.5 Communal entropy

The statistical number of complexions for molecules in a gas is fundamentally different than for molecules occupying lattice sites in a crystal. In the gas, the molecules can be anywhere throughout the gas volume and therefore are indistinguishable. This is the origin of the divisor $N!$ in the semiclassical configurational integral. In a lattice, however, molecules can be labeled, so the $N!$ does not appear because there is only one way of placing molecules on a lattice in a perfect crystal.

The gas therefore has a configurational entropy term arising solely from the indistinguishability of molecules. In a monomolecular gas, using Stirling's approximation for the factorial, the partition function is

$$Z_g = \frac{(\overline{V})^N}{\Lambda^{3N}} e \qquad (10.5.1)$$

where $\overline{V}$ is the volume per molecule. If the molecules were fixed on a lattice, the volume would not be shared by all the molecules. Each molecule would move in its own volume formed by the cage of its nearest neighborhood and the $N!$ term would be absent. Then the partition function would be

$$Z_g = \frac{(\overline{V})^N}{\Lambda^{3N}} \qquad (10.5.2)$$

Calculating the entropies corresponding to the two partition functions (10.5.1) and (10.5.2), for example, from the negative derivative of the Helmholtz free energy with respect to temperature, shows that they differ by a term

$$S_{\text{com}} = kN \qquad (10.5.3)$$

which arises directly from the factor e in (10.5.1). This is called the *communal entropy*.

There is no obvious way of deciding whether or not liquids have communal entropy. The most reasonable description is that at low temperatures and high densities molecules are restricted to their surrounding cages and there is little or no communal entropy. This would certainly be expected in glasses, where molecular mobility is very low. As the temperature goes up or density goes down, the molecules become more mobile and at least some of them can escape from their immediate location. The communal entropy would gradually increase to its full gaslike value.

In the model presented above, it is assumed that there is a communal entropy for the gaslike molecules moving in the free volume of the liquid given by $N_g k$. In this model there is an additional configurational entropy resulting from the presence of vacancies. The number of ways of distributing N_l vacancies and N_v molecules on $N_v + N_l$ sites is

$$W_v = \frac{(N_l + N_v)!}{N_l! N_v!} \qquad (10.5.4)$$

and, within the approximation that the number of vacancies is much less than the number of molecules, the corresponding entropy is $N_v k$. The sum of the communal entropy and that arising from the distribution of vacancies can be regarded as an effective communal entropy given by

$$S_{com} = N_g\left(1 + \frac{1}{z_l}\right)k \qquad (10.5.5)$$

The vacancies contribute only about 10% to this, which is within the error range of the model.

10.6 Vibrations and melting

It is obvious that the crystal vibrations play an important role in the process of fusion. As the temperature of a solid increases, the increasing vibrational amplitudes have two related effects: the vacancy concentration increases and the molecules are less tightly bound to their neighbors. In the vicinity of a vacancy, the mobility becomes high and molecules find it easier to exchange places. The vacancy and its neighbors are like an incipient puddle of liquid. At the melting point, the vibrations have such a large amplitude that the long-range order of the solid is destroyed and the system becomes liquid. It is therefore of interest to estimate the vibrational amplitudes at the melting point. To do this, harmonic vibration theory will be used even though there is an appreciable anharmonic effect at the melting point.

Let A_k be the normal modes of a monatomic crystal. Then the displacement of an atom at **R** is given by

$$\mathbf{u} = \frac{1}{\sqrt{N}} \sum_k A_k e^{i\mathbf{k}\cdot\mathbf{R}} \qquad (10.6.1)$$

The sum is taken over all **k**-vectors, which is equivalent to summing over all normal mode frequencies.

The magnitude of the square of the displacement is

$$u^2 = \frac{1}{N} \sum_k A_k^2 \qquad (10.6.2)$$

The statistical mechanical average of (10.6.2) gives the average for the square of the total atomic displacement:

$$\overline{u^2} = \frac{1}{N} \sum_k \overline{A_k^2} \qquad (10.6.3)$$

From the theory of the simple harmonic oscillator of angular frequency ω, the amplitude and the total energy are related by

$$E = \frac{1}{2} M\omega^2 A^2 \qquad (10.6.4)$$

so the mean square amplitude of the **k**th oscillator is related to its energy by

$$\overline{A_k^2} = \frac{2}{M\omega_k^2} \overline{E_k} \qquad (10.6.5)$$

At high temperatures, each oscillator has an energy kT, so using this value for the average energy $\overline{E_k}$, substitution of (10.6.5) into (10.6.3) gives

$$\overline{u^2} = \frac{2kT}{MN} \sum_k \frac{1}{\omega_k^2} \qquad (10.6.6)$$

Now convert the sum to an integral in the usual way through the use of the frequency distribution function. Using the Debye model, this is given by equations (4.7.19) and (4.7.23), so in the Debye theory we can write

$$\overline{u^2} = \frac{18T\hbar^2}{Mk\Theta_D^2} \qquad (10.6.7)$$

For purposes of calculation, let us divide (10.6.7) by the square of the nearest neighbor distance λ. Then, if the nearest neighbor distance is expressed in Angstroms and the atomic mass M is expressed in units of the proton mass, evaluating the constants in (10.6.7) gives

$$\frac{\overline{u^2}}{\lambda^2} = 866 \frac{T}{M\lambda^2\Theta_D^2} \qquad (10.6.8)$$

At the melting point T_m, define f_m to be the square root of the ratio on the left hand side of (10.6.8) and write

$$f_m = 29.6 \frac{1}{\lambda\Theta_D} \sqrt{\frac{T_m}{M}} \qquad (10.6.9)$$

f_m can be computed from the known Debye temperatures, melting points, nearest neighbor distances, and atomic masses.

An alternate approach to estimating the amplitude of the vibrations at the melting point starts from a consideration of the forces holding the crystal together. The potential energy of a crystal as a function of lattice parameter is shown in figure 5.1. It is reasonable to assume that the same general form holds for the interaction between individual atoms. This is certainly true for central pairwise forces, and for any crystal, the displacements are harmonic for small amplitudes, so the parabolic form for the atomic interactions near equilibrium is a good approximation. Let us represent the force between two atoms by a curve that is the negative derivative of the usual type of potential function. Such a curve, computed from the universal energy equation of (5.1.16), is shown in figure 10.7.

The force is zero at the equilibrium separation, negative (repulsive) for smaller values, and positive (attractive) for larger values. The important point to note is that, if the displacement becomes large enough to reach the point of maximum force, then any further increase in the distance between the two atoms results in a *decreasing* force of attraction and the interatomic distance tends to increase. Clearly, vibrational amplitudes that approach the maximum in the force curve should disrupt the crystal structure. It follows that, at the melting point, the fractional atomic displacements should be near the value at maximum interatomic force. This value is easy to calculate from the universal

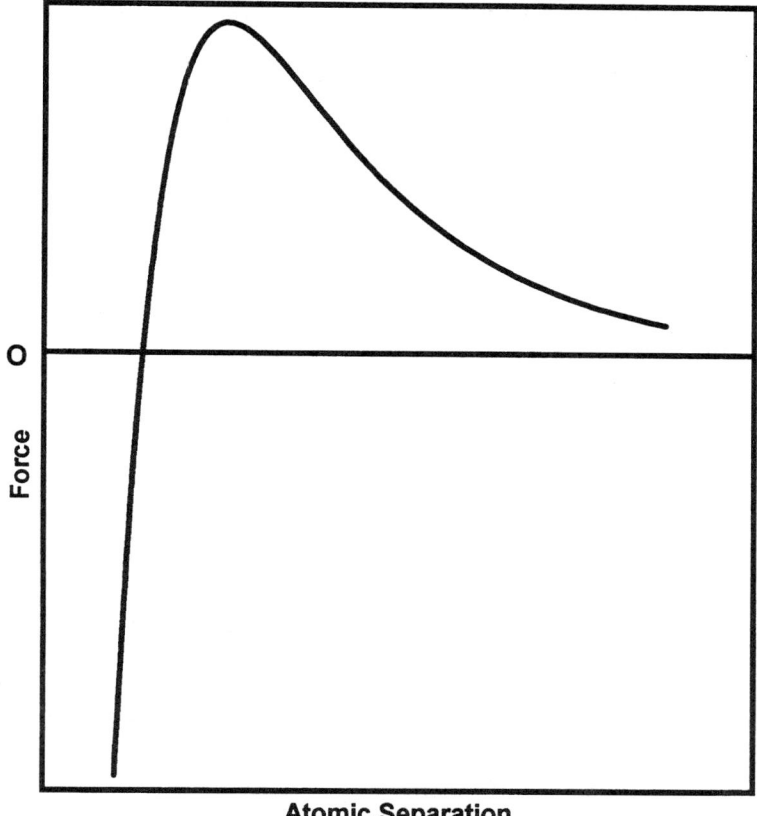

Figure 10.7. Force between two atoms simulated from the universal energy curve.

energy equation. The maximum in the force curve occurs at $a = 0.9$, so from the definition of a given by equation (5.1.15), we have

$$f_m = \frac{r_s - r_s^0}{r_s^0} = \frac{0.9}{\eta^0} \tag{10.6.10}$$

with

$$\eta^0 = \frac{r_s^0}{L} \tag{10.6.11}$$

and L is defined by equation (5.1.14).

Since Debye temperatures and the parameters in the universal energy equation are known, the fractional vibrational atomic displacement can be estimated in two ways: from equations (10.6.9) and (10.6.10). The results of such computations for a number of metals are shown in table 10.1, the second column being computed from equation (10.6.9) and the third from equation (10.6.10). The agreement between the two methods of calculation is remarkable considering that harmonic theory is being used for displacements that are

Table 10.1: The Ratio of Root Mean Square Vibrational Amplitudes to the Nearest Neighbor Distance for Some Simple Metals

Metal	f_m (force)	f_m (RMS)
Li	0.18	0.29
Na	0.21	0.24
Al	0.15	0.19
K	0.19	0.23
Ca	0.17	0.20
V	0.19	0.19
Cr	0.17	0.16
Fe	0.16	0.17
Ni	0.17	0.18
Cu	0.17	0.17
Rb	0.22	0.22
Sr	0.16	—
Mo	0.16	0.15
Pd	0.16	0.14
Ag	0.16	0.15
Cs	0.19	0.22
Ba	0.16	0.20
Ta	0.19	0.18
W	0.16	0.16
Pt	0.15	0.14
Au	0.13	0.13
Pb	0.16	0.14

RMS = root mean square.

anharmonic and lends strong support to the idea that melting occurs when the atomic vibrational displacements reach a critical fraction of the interatomic distance. Table 10.1 shows that the amplitude of vibration at the melting point is about a quarter to one half the half-neighbor distance. This represents a large excursion from the mean position and indicates that the atomic vibrations are rather violent at the melting point.

Equation (10.6.9) is a form of the Lindemann equation, which was proposed as an empirical relation between melting point and Debye temperature in 1910. Before 1975, there was no theoretical foundation for the Lindemann formula, but David Goodstein gave a problem in his 1975 text[2] based on the same arguments that led to equation (10.6.9), and this was explicitly presented in Andrew Zangwill's text in 1988.[3]

Now let us consider the surface of a solid. Because an atom in the surface has fewer nearest neighbors than in the bulk, the normal modes of vibrations associated with the surface will have lower frequencies and the vibrational amplitudes of the surface atoms will be greater than those in the bulk. A rough estimate is that, since there are half as many nearest neighbors to an atom in the surface, the force constants will have about half the value of those in the bulk, so the frequencies will be less than those in the bulk by a factor of about $\sqrt{2}$ and the root mean amplitude will be about 40% more than in the solid. This is consistent with experimental estimates that compare X-ray diffraction and LEED data, which show that the vibrational amplitudes in the surface of metals are 50–100% greater than in the bulk solid.[4] The atoms on a solid surface therefore have vibrational amplitudes at the melting point that are comparable to the amplitudes in the liquid phase.

As the temperature increases, the amplitudes of the vibrations and the mean displacements of the atoms increase. This effect is greater for the surface atoms than for the bulk atoms because they are more weakly bound. When the surface displacements become large enough, the crystalline structure breaks down and the surface layer melts. Now the atoms in the second layer are more weakly bound and will have larger displacement amplitudes than those in the bulk, although the difference will not be as great as for the first layer. The second layer will then melt, but at a higher temperature than the melting temperature of the first layer. This effect propagates into the surface, but after a few layers the thermodynamic melting temperature is needed to get amplitudes that cause further melting.

For the model of a liquid as a mixture of gaslike and solidlike molecules, there are two different kinds of molecular motions to be considered. For molecules around a vacancy, it is assumed that the vibrational amplitudes have become so large that the molecules move rather freely and have gaslike properties. The large amplitudes associated with surfaces support this idea. For the solidlike molecules, there are two different factors that make the vibrations different from those in the crystal. The first is that the number of nearest neighbors in the liquid may be different from that in the solid; the second is that the interatomic spacings are different (usually higher), so the atoms are vibrating in a potential well different from that in the solid.

10.7 Melting

The fusion curve relating pressure to temperature is obtained by equating the chemical potential of the liquid and solid states at the melting point. The chemical potential of the liquid is given by equation (10.4.14). Again adopting the high-temperature Debye approximation, but this time for the crystal, the Helmholtz free energy is given by a form just like equation (10.4.5). Adding the PV term to get the Gibbs free energy and dividing through by the number of molecules, the chemical potential of the solid is

$$\mu_c = \overline{E}_0^c + 3kT\ln\left(\frac{\Theta_D^c}{T}\right) + P\overline{V}_c \qquad (10.7.1)$$

$\overline{E}_0^c$, $\overline{V}_c$, and Θ_D^c are the zero point energy per molecule, the volume per molecule, and the Debye temperature, respectively, in the crystal.

Equating (10.7.1) to (10.4.14) at the melting point T_m and zero pressure gives the zero pressure melting point in terms of the material parameters of the system through the relation

$$f_g\left\{\ln\left[\frac{(z_l+1)}{ez_l\Lambda^3}\right] + \frac{1}{z_l} - 3\ln\left(\frac{\Theta_D^l}{T_m}\right)\right\} + 3\ln\left(\frac{\Theta_D^l}{\Theta_D^c}\right) + \frac{H_m}{kT_m} = 0 \qquad (10.7.2)$$

where H_m is the zero point contribution to the heat of melting per molecule defined by

$$H_m = -(\overline{E}_0^c - f_g\overline{E}_0^g - f_s\overline{E}_0^s) \qquad (10.7.3)$$

Equation (10.7.2) can be used to estimate the fraction of gaslike atoms in the liquid. To do this, a value of the Debye temperature for solidlike molecules in

Table 10.2: Fraction of Atoms in Liquid Metals That Have Gaslike Properties: Computed from Equation (10.7.2)

Metal	Fraction of "Gas" Atoms
Na	0.11
Mg	0.09
Al	0.14
K	0.11
Cr	0.12
Mn	0.13
Fe	0.12
Co	0.14
Ni	0.13
Cu	0.13
Zn	0.13

the liquid must be known. From the discussion in section 10.6, a reasonable assumption is that this is lower than that in the solid by at least $\sqrt{2}$. For purposes of calculation, we will indeed assume that $\Theta_D^l = \Theta_D^c/\sqrt{2}$. Since the Debye temperature appears as a logarithmic function, errors in its estimation do not lead to serious errors in the final results. With this assumption, f_g can be estimated from known values of the molecular volume, Debye temperature, melting point, and heat of fusion. The results are shown in table 10.2 for a number of metals. The results are much the same for all metals, and imply that the vacany concentration in liquids is about 10^{-2}, which is about an order of magnitude greater than in the solids, as expected. Using these values for the fraction of gaslike molecules, the melting curve of P–T_m could be computed by equating the chemical potentials at nonzero pressure provided the volume of melting is known.

10.8 Regular solution theory of binary alloys

When two metals are mixed and brought to equilibrium, the resulting system can have a variety of forms. The two components could retain their identities so that the system is just a mixture of the pure components, or they might dissolve in each other to form a solid solution with no trace of the original metals. These are two extreme cases corresponding to zero mutual solubility to complete miscibility. In general, the system will consist partly of pure metals and partly of solid solutions and/or intermetallic compounds. Often, there are several kinds of structures and compounds that can form, so a binary phase diagram can be rather complex. But the principles of binary phase equilibria can be illustrated by analyzing a simple system in which the metals either stay in their original state or form solid solutions with a single structure. It is assumed that the solutions have no long-range order and that the crystal energy is a sum of pairwise nearest neighbor interactions.

An analysis of simple 50-50 order-disorder alloys has been given in chapter 8, and the methods presented there can readily be adapted to arbitrary compositions. The Bragg–Williams approximation is sufficient to bring out the important factors controlling alloy formation. Equation (8.4.8) is valid for the energy for any configuration of A and B atoms distributed on a lattice, with pairwise, nearest neighbor interactions. However, since the crystal is now assumed to be completely random, there is no need to identify sublattices and the superscript on the total number of AB pairs is no longer needed. Thus, we write for the energy of the alloy

$$U = -v_{AA}Q_{AA} - v_{BB}Q_{BB} - v_{AB}Q_{AB} \tag{10.8.1}$$

where $-v_{AA}$ = energy of an AA pair, $-v_{BB}$ = energy of a BB pair, and $-v_{AB}$ = energy of an AB pair. The subscripts no longer refer to sublattices and the Qs are the total number of the three kinds of pairs in a binary solid solution of arbitrary composition. Just as in order-disorder theory, the minus signs are introduced to make the vs positive constants.

The equilibrium composition is obtained by minimizing the free energy of a random solution of arbitrary composition. Since the volume is being maintained constant and the pressure is assumed to be zero, the Helmholtz and Gibbs free energies are equal.

In a random solution, the probability that a site is occupied by an atom of a particular kind is just the atom fraction. That is, if x is the probability that a site contains an A atom, and $1 - x$ is the probability that it contains a B atom, then

$$x = c_A = \frac{N_A}{N_A + N_B} = \text{atomic fraction of A} \tag{10.8.2}$$

$$1 - x = c_B = \frac{N_B}{N_A + N_B} = \text{atomic fraction of B} \tag{10.8.3}$$

N_A and N_B being the number of A and B atoms in the solution, respectively.

The number of AA pairs is obtained by writing the probability that a site contains an A atom, multiplying by the number of nearest neighbor pairs connected to it and by the probability that a neighbor also contains an A atom. The result must then be divided by 2 to correct for double counting. The same procedure holds for BB and AB pairs except that in the case of unlike pairs the result is not divided by 2. That is,

$$\text{No. of AA pairs} = Q_{AA} = \frac{Nzx^2}{2} \tag{10.8.4}$$

$$\text{No. of BB pairs} = Q_{BB} = \frac{Nz(1-x)^2}{2} \tag{10.8.5}$$

$$\text{No. of AB pairs} = Q_{AB} = Nzx(1-x) \tag{10.8.6}$$

$N = N_A + N_B$ being the total number of atoms. Putting (10.8.4)–(10.8.6) in (10.8.1) gives the energy of the homogeneous alloy phase as

$$\begin{aligned} U(x) &= -v_{AA}Q_{AA} - v_{BB}Q_{BB} - v_{AB}Q_{AB} \\ &= -v_{AA}\frac{Nzx^2}{2} - v_{BB}\frac{Nz(1-x)^2}{2} - v_{AB}Nzx(1-x) \end{aligned} \tag{10.8.7}$$

A slight rearrangement of this gives

$$U(x) = -\frac{Nz}{2}\left[xv_{AA} + (1-x)v_{BB} + 2x(1-x)\left(v_{AB} - \frac{v_{AA}+v_{BB}}{2}\right)\right] \tag{10.8.8}$$

The Helmholtz free energy is the energy minus TS, the entropy term, which is obtained from the statistical count. The number of ways of putting N_A atoms of type A and N_B of type B on N sites is

$$W(x) = \frac{N!}{N_A! N_B!} = \frac{N!}{(Nx)![N(1-x)]!} \qquad (10.8.9)$$

so, using Stirling's approximation, the entropy is

$$S(X) = k \ln W(x) = -Nk[x \ln x + (1-x) \ln(1-x)] \qquad (10.8.10)$$

which is just the entropy of mixing of an ideal, random solution.

The free energy of the alloy is

$$A(x) = U(x) - TS(x) = -\frac{Nz}{2}[x v_{AA} + (1-x) v_{BB} + 2x(1-x) v]$$
$$+ NkT[x \ln x + (1-x) \ln(1-x)] \qquad (10.8.11)$$

where v, which may be called the "alloying energy," is defined by

$$v = v_{AB} - \frac{v_{AA} + v_{BB}}{2} \qquad (10.8.12)$$

A regular solution is one whose free energy is defined by equation (10.8.11).

The first point to note is that the first two terms in (10.8.11) refer to the energies of pure A and B. That is, for a mixture of pure A metal containing N_A atoms and pure B containing N_B atoms, the energy of the mixture is

$$U_A(x) + U_B(1-x) = -\frac{Nz}{2}[x v_{AA} + (1-x) v_{BB}] \qquad (10.8.13)$$

and since there is no entropy of mixing for the pure metals, the difference in free energy between the homogenous alloy and the mixture is

$$\Delta A(x) = A(x) - A_A(x) - A_B(1-x)$$
$$= -Nzx(1-x)v + NkT[x \ln x + (1-x) \ln(1-x)] \qquad (10.8.14)$$

The entropy of mixing is always positive, and its contribution to the free energy is negative for all x. If v is positive, then the attraction between unlike atoms is greater than the average attraction between like atoms, the free energy is negative for all x, and the system forms solid solutions for all compositions. If v is zero, then there is no preference for any particular type of pairs, the entropy term dominates, the free energy is always negative, and again, there is solid solution formation for all concentrations. For negative v, like atoms are more strongly attracted than unlike atoms. The range of solid solubility then depends on temperature since the relative values of thermal energy and alloying energy determines the sign of the free energy. If the alloying energy is not too negative, then at a sufficiently high temperature the entropy term dominates, the free energy is negative, and solid solutions will form. At low temperatures, or for alloying energies that are strongly negative, solid solutions will not form and the system consists of a mixture of the pure metals.

The free energy as a function of composition is shown for a series of temperatures in figure 10.8. The units are arbitrary: the energy of pure A is defined to be zero. The straight line labeled M_x represents the free energy of mixtures

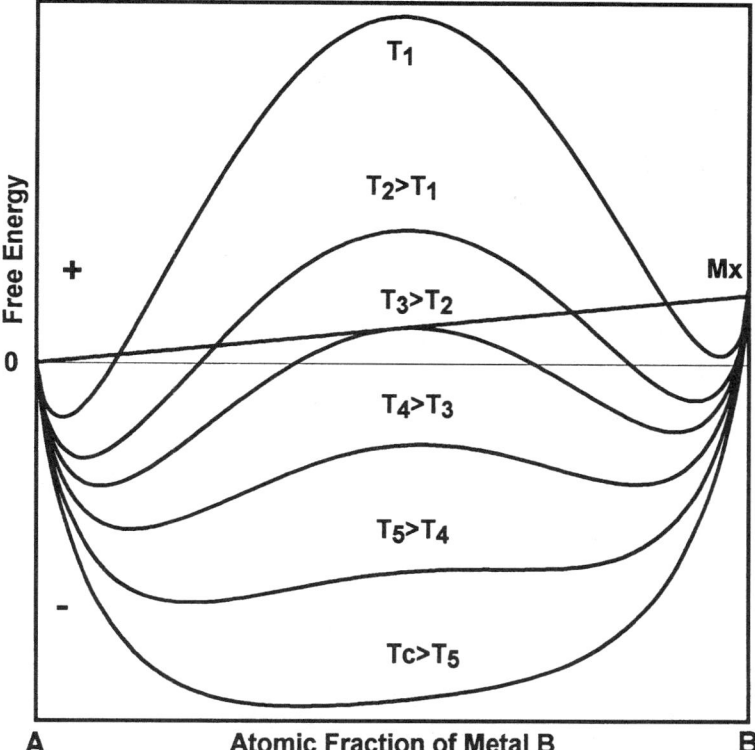

Figure 10.8. Free energy versus composition for a binary alloy: regular solution theory.

of pure A and pure B and is a simple linear function of composition. The free energy of the solid solution displays two minima for all temperatures until a maximum temperature is reached (represented by T_c in the figure) where the two minima coalesce.

The two minima in each free energy curve for temperatures below T_c represent solid solutions that are stable with respect to alloys of other compositions. If, for a given temperature, these minima are lower than the free energy of a mixture of the pure metals, then they represent equilibrium and the stable system is a mixture of the two alloys with the compositions at the free energy minima.

For temperatures equal to or higher than T_c, there is only one minimum, which becomes wider and flatter with increasing temperature. This indicates that at high temperatures the two metals form solid solutions over the entire range of composition except very near to pure B, where the entropy of mixing is insufficient to overcome the positive free energy of pure B. T_c is a critical temperature above which a single solid solution phase exists over a wide range of composition and below which the system separates into two solid solution phases. This is summarized in the phase diagram shown in figure 10.9, which is a plot of the temperature versus composition obtained from the minima in figure 10.8. The critical temperature corresponds to the maximum in the figure, above which there is only one phase. Below the critical temperature, any point inside the curve represents a temperature and composition at which the system

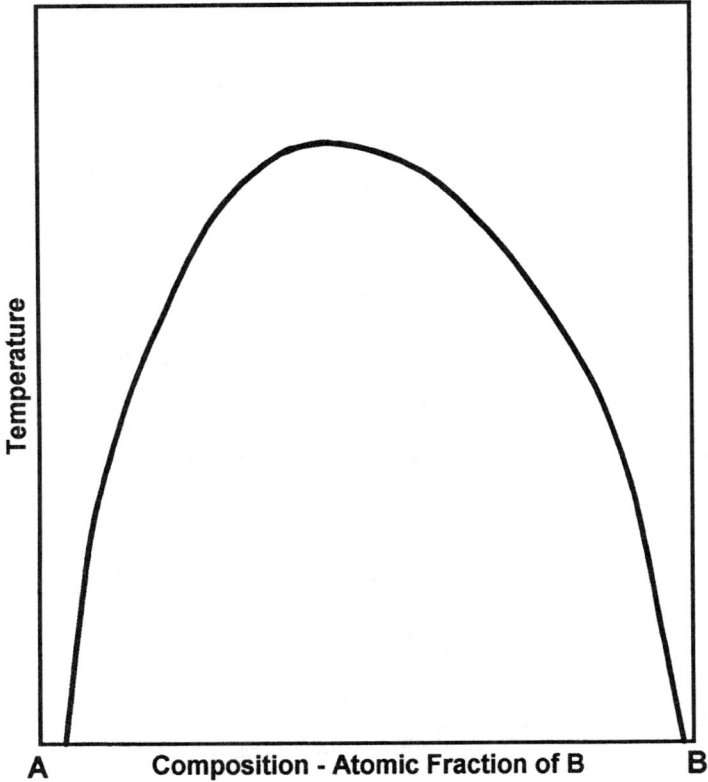

Figure 10.9. Phase diagram for a binary solid solution.

is a mixture of two solid solution phases. For any point to the left of the curve, the system is a mixture of a solid solution and pure A, while points to the right of the curve denote mixtures of a solid solution and pure B.

Exercises

10.1 Assume that in a column of ice, such as in a glacier or ice cap, the pressure varies linearly with height. Derive the differential equation for the variation of the melting point of the ice with distance from the surface. Assume that the heat of fusion and the compressibility of ice and water are constant.

10.2 Find the relation between temperature and pressure for a Debye crystal in equilibrium with its vapor if the vapor is assumed to be a Fermi–Dirac gas instead of a semiclassical gas. Derive the equation for the heat of sublimation and compare it to that in which the vapor is an ideal gas. Use the zero temperature approximation for the Fermi gas.

10.3 Derive equations for the entropy and energy for a van der Waals fluid. Compare the energy result with that of an ideal gas and interpret the result.

10.4 The equation of state of a real gas is often written as a series called the *virial expansion*. This series is $P/kT = n + B_2(T)n^2 + B_3(T)n^3 + \ldots$, where $n = N/V$ is the density and B_j is called the jth virial coefficient. For a van der Waals fluid, find the second virial coefficient in terms of the van der Waals parameters. Use reduced pressures, volumes, and temperatures throughout and derive the second virial coefficient at the critical temperature.

10.5 An early model took liquids to be like solids but with a lower Debye temperature. That is, the fraction of atoms with "gaslike" properties was taken to be zero. With this model, estimate the ratio of the liquid Debye temperature to that in the solid for magnesium from the fact that the heat of melting is 0.05 eV and the melting point is 923 K.

10.6 For a two-component regular solution, find the equation for the most stable compositions as a function of temperature.

Notes

1. Eyring, H., D. Henderson, B.J. Stover, and E.M. Eyring; 1964; *Statistical Mechanics and Dynamics*; chapter 12; John Wiley, New York.
2. Goodstein (1975, p. 223).
3. Zangwill, Andrew; 1988; *Physics at Surfaces*; Cambridge University Press, Cambridge; p. 117.
4. See ibid.

11

Critical Exponents and the Renormalization Group

11.1 Equivalent models

The order-disorder transition and ferromagnetism have their origins in quite different manifestations of atomic interactions and are exhibited in quite different kinds of material. But, as shown in chapter 9, both can be described by the Ising model with considerable success. In fact, there are other phenomena that are different from either magnetism or order-disorder that can also be described by similar models. These include adsorption of atoms or molecules on surfaces, absorption of gases in solids, and the attachment of impurities to dislocation cores. A simple description of such phenomena is provided by the lattice gas model, and this model is easily shown to be equivalent to the Ising model. The fact that the same model can describe such varied systems implies that there are some important similarities among them that do not depend on the specific nature of the systems or the interaction energies involved. This, in fact, turns out to be the case and is most clearly apparent in the neighborhood of critical points. The statistical mechanics of the lattice gas is described in appendix 7, and specific applications are given in chapter 12. Here, the model and its connection to the Ising model are described.

Divide the system into cells such that, at most, only one molecule can occupy a given cell and a cell can either be occupied or empty. This is described by defining a parameter that is unity if a cell is occupied by a molecule and zero otherwise. Also, the energy of the system is assumed to be a sum of nearest-neighbor pairwise interactions. Thus, if $-K$ is the energy of interaction of a nearest neighbor pair and if $e_j = 0$ if the jth cell is empty and $e_j = 1$ if the jth cell is occupied, then the energy of the system is

$$W\{r\} = -K \sum_{<i,j>} e_i e_j \qquad (11.1.1)$$

The sum is over all pairs, and $W\{r\}$ is the energy of the system in a particular configuration labeled by the index r. Sometimes, as in adsorption of a gas on a surface, there is an energy of interaction of a molecule with a cell, and a term must be added to (11.1.1) that is the number of occupied cells times the energy of interaction. If this interaction energy is called $-\varepsilon$, then instead of (11.1.1) the energy is

$$W\{r\} = -K\sum_{<i,j>} e_i e_j - \varepsilon \sum_j e_j \qquad (11.1.2)$$

The magnetic Ising model follows from equation (9.5.1), which is the magnetic energy in the absence of any external magnetic field. If an external field exists, then a term must be added to (9.5.1) that is the interaction of spins with the external field:

$$E\{s\} = -J\sum_{<i,j>} s_i s_j - H\sum_j s_j \qquad (11.1.3)$$

with $s_j = +1, -1$ for spins up or down, respectively.

The identification of the lattice gas with the Ising ferromagnet is made by identifying e_j with

$$e_j = \frac{1}{2}(1 + s_j) \qquad (11.1.4)$$

so that when $s_j = +1$, $e_j = 1$ and when $s_j = -1$, $e_j = 0$. With this identification, (11.1.2) immediately takes on a form identical to (11.1.3).

The systems best described by the lattice gas model are those in which the description of cells as being occupied or unoccupied is a natural one, such as ferromagnets or order-disorder alloys, in which the cells are of atomic size. Chemisorption, in which a surface has a fixed number of sites to which at most one atom can be attached and the adsorbed atoms cannot wander about the surface, is also well described as a lattice gas.

11.2 Critical points

For systems that exhibit critical points, experimental data show that physical properties near the critical point vary with temperature according to power laws such as the following.

For the heat capacity of a magnet or an order-disorder alloy:

$$C \propto (T - T_c)^{-\alpha}, \qquad T > T_c$$
$$C \propto (T_c - T)^{-\alpha'}, \qquad T < T_c$$

For spontaneous magnetization of a ferromagnet or degree of order of an order-disorder alloy:

$$M_f \propto (T_c - T)^{\beta}, \qquad T < T_c$$
$$M_f \propto (T - T_c)^{\beta}, \qquad T > T_c$$

For the susceptibility of a ferromagnet:

$$\chi_T \propto (T - T_c)^{-\gamma}, \qquad T > T_c$$
$$\chi_T \propto (T_c - T)^{-\gamma'}, \qquad T < T_c$$

The α, α', γ, γ', β, β' are called *critical exponents*. It is conventional to use the prime when the critical temperature is approached from above. The values are given for three different systems in table 11.1.[1]

Table 11.1: Values of Critical Exponents

System	$\alpha = \alpha'$	$\beta = \beta'$	$\gamma = \gamma'$
Fe	0.12 ± 0.01	0.34 ± 0.02	1.333 ± 0.015
Ni	0.1 ± 0.03	0.33 ± 0.03	1.32 ± 0.02
β-Brass	0.1 ± 0.1	0.305 ± 0.005	—
Theory	0.11	0.33	1.24

It is an experimental fact that the critical exponents for a given property are the same for an approach to T_c from above or below. Also, it is found that within the limits of experimental error there is a relation among the critical exponents that is

$$\alpha + 2\beta + \gamma = 2 \qquad (11.2.1)$$

The long-range order parameter in order-disorder alloys, magnetization and magnetic susceptibility in ferromagnets, and compressibility in gas-liquid systems all show divergent power law behavior near the critical point.

One reason that the study of critical points is important is that it lays bare the commonalities among different systems. These commonalities are most evident in the critical point exponents. For a given property, the critical exponent does not depend on the particular system. Thus, all ferromagnets and all 50–50 binary order-disorder alloys have the same critical exponents. Furthermore, systems exhibiting quite different phenomena often have the same critical point exponents. Order-disorder alloys and ferromagnets, for example, have the same critical point exponents for corresponding properties. This means that the critical exponents represent very general physics that are independent of many of the important characteristics of the systems, such as atomic constitution and interaction energies, that control their detailed behavior. The similarities of temperature dependence near critical points are related to the fact that equivalent models can be used to describe widely different systems, as described above.

Also, the results of calculations of the values of critical point exponents are sensitive to details of the theoretical models used to describe them. Thus, while mean field theory gives an overall magnetization–temperature curve that agrees well with experiment when plotted over the entire range of temperature, this is not true when the neighborhood of the critical temperature is examined on a fine scale. Critical points are therefore sensitive tests of the applicability of theoretical calculations. It is a remarkable fact that this sensitivity is relative to the methods and approximations used to get numerical calculations, not to the fundamental model itself. The evidence is quite strong that the Ising model yields a good description of the important features of criticality if accurate methods of solution can be found. Disagreements of the values of critical exponents with experiment rise from the shortcomings of approximate solutions. Thus, mean field theory gives wrong answers for these values, but accurate numerical calculations based on the Ising model can do quite well.

Let us start by examining the neighborhood of critical points in mean field theories. Expanding the hyperbolic tangent for small values of the argument in equation (8.6.21) [or equation (9.5.14)] gives the temperature dependence of the order parameter (or the magnetization) for high temperatures just below the critical point as

$$R = \tanh\left(\frac{R}{\phi}\right) = \frac{R}{\phi} - \frac{1}{3}\left(\frac{R}{\phi}\right)^3 \tag{11.2.2}$$

from which

$$R = \sqrt{3}\,\frac{T}{T_c}\left(\frac{T_c - T}{T_c}\right)^{1/2} \tag{11.2.3}$$

where R is either the long-range order parameter for a 50–50 binary alloy, or the magnetization of a ferromagnet. Since T/T_c is close to unity and does not vary much, the temperature dependence of the order parameter is controlled by the square root. The mean field critical exponent for long-range order (or magnetization) is therefore 1/2, but the experimental value is close to 1/3.

The magnetic susceptibility for temperatures above the critical point is readily obtained from equation (9.4.6) as follows:

$$\frac{\partial M_f}{\partial H} = \chi_T = \chi_T \frac{T_c}{T}\left(1 + \frac{1}{\gamma}\frac{\partial M_f}{\partial H}\right) = \chi_T \frac{T_c}{T}\left(1 + \frac{1}{\gamma\chi_T}\right) \tag{11.2.4}$$

which gives the susceptibility as

$$\chi_T = \frac{T_c}{\gamma(T - T_c)} \tag{11.2.5}$$

Mean field theory therefore gives unity for the critical exponent for the susceptibility and is in obvious disagreement with experiment, which gives 4/3.

Critical exponents are also important in fluid phase transformations and, in fact, are analogous to those in magnetic and order-disorder systems. Just as the susceptibility is the response of the system to an external magnetic field, so the compressibility is the response of a system to external pressure. That is, the susceptibility and compressibility are thermodynamically analogous quantities.

The van der Waals model is a mean field theory in the same sense as the Bragg–Williams model. The compressibility of the van der Waals fluid near the critical point can be obtained by starting with the reduced van der Waals equation and expanding it as a power series in the volume about the critical volume. That is, we compute, to the second order,

$$\tilde{P} = \left(\frac{\partial \tilde{P}}{\partial \tilde{V}}\right)_{\tilde{V}=1}(1 - \tilde{V}) + \frac{1}{2}\left(\frac{\partial^2 \tilde{P}}{\partial \tilde{V}^2}\right)_{\tilde{V}=1}(1 - \tilde{V})^2 \tag{11.2.6}$$

Getting the derivatives from (10.2.23), (11.2.6) becomes

$$\tilde{P} = 6(1 - \tilde{T})(1 - \tilde{V}) + 9(\tilde{T} - 1)(1 - \tilde{V})^2 \tag{11.2.7}$$

so the bulk modulus is

$$B = -V\left(\frac{\partial P}{\partial V}\right)_T = -\frac{P_c}{V_c}\left(\frac{\partial \tilde{P}}{\partial \tilde{V}}\right)_T = 6(\tilde{T} - 1)[1 + 3(1 - \tilde{V})]\frac{P_c}{V_c} \tag{11.2.9}$$

The reciprocal of this is the compressibility, which to first order is

$$\kappa = \frac{V_c}{6P_c} \frac{T_c}{T_c - T} \qquad (11.2.10)$$

so the critical exponent for the compressibility of a van der Waals fluid is unity, just as for the susceptibility of a mean field magnet.

Note that as the fluid approaches the critical point, the compressibility diverges and it becomes easy for liquid to vaporize into gas and for gas to condense into liquid. This is the origin of the critical opalescence observed near the critical point, since the system can form droplets or vaporize with very small fluctuations in temperature or pressure. It is analogous to the rapid change from order to disorder in a binary alloy as the temperature increases to its critical value.

Although mean field theory yields divergencies in physical properties near the critical point, the calculated critical exponents do not agree with experimental values. The critical exponents obtained from mean field theory of $\alpha = 0$, $\beta = 0.5$, and $\gamma = 1$ are not in accord with the data shown in table 11.1.

A theory for critical exponents should show that the physical quantities indeed diverge according to a power law near the critical temperature, that they are the same for temperature changes above and below the critical point, that they are connected by equation (11.2.1), and that they do not depend on the atomic constitution of the system. This task has been accomplished by scaling and renormalization theory in a general and straightforward way, as shown below. A complete theory would also permit the calculation of specific values of the critical exponents. While this can also be done with scaling and renormalization theory, the calculations require heavy mathematical and numerical work.

11.3 Landau theory and the Kirkwood expansion

A description of second-order phase transformations was developed by Landau as a series expansion of the free energy in the order parameter. The theory is quite general in that it only requires that an order parameter can be defined. This includes not only order-disorder and magnetic systems but also any system that can be modeled as a lattice gas and even gas-liquid systems. In the latter case, the order parameter is taken to be the difference between the liquid and gas density, which is high at low temperatures, low at high temperatures, and vanishes at the critical point. Here, the discussion is first restricted to Ising lattice systems.

Landau assumed that the Gibb's free energy can be written as a power series in the order parameter. For symmetry, only even powers are allowed. To see this for the case of ferromagnets or order-disorder (O-D) alloys, remember that the order parameter is just the magnetization per atom given by

$$\frac{\overline{M}}{N} = \frac{N^+ - N^-}{N} \qquad (11.3.1)$$

and the corresponding order parameter for O-D alloys is

$$R = \frac{(N_{A\alpha} + N_{B\beta})}{N} - \frac{(N_{A\beta} + N_{B\alpha})}{N} \qquad (11.3.2)$$

In both cases, an exchange in the occupation of lattice sites changes the sign of the order parameter but nothing else. All physical equations must therefore be even functions of the order parameter.

To the fourth order, the Landau expansion for the Ising model is

$$G(R, T) = G_0(T) + a(T)R^2 + c(T)R^4 \tag{11.3.3}$$

The first term in the expansion is just the free energy at zero order. In general, the coefficients are functions of pressure as well as temperature.

A considerable amount of information can be extracted from this simple idea, as shown in Landau and Lifshitz (1958). Here, the critical exponent for the magnetization (order-disorder parameter) will be derived as an example of the use of Landau theory, and its connection to the Kirkwood second moment expansion given in chapter 8 will be noted.

At the transition point, $a(T)$ must be zero because, below the transition temperature, there must be a minimum in free energy as a function of order, and this can only occur if $a(T) < 0$. But above the transition temperature a zero value of the order parameter must be a stable state, and this can only happen if $a(T) > 0$. Otherwise, there would be a minimum at some positive value of the order parameter that is lower than that at zero order. Thus, at the transition temperature, $a(T)$ must vanish, and this determines the critical temperature as

$$a(T_c) = 0 \tag{11.3.4}$$

Near the critical temperature, it is assumed that the coefficients in (11.3.3) can be expanded in the temperature difference $(T_c - T)$. Actually, since the higher order terms drop off rapidly as the critical temperature is approached, we expand only the coefficient of the quadratic term, retain only the first two terms, and write (11.3.3) as

$$G(R, T) = G_0(T) + [a_0 + a_1(T - T_c)]R^2 + c(T_c)R^4 \tag{11.3.5}$$

a_0, a_1, and $c(Tc)$ being constants. The order parameter at thermodynamic equilibrium is obtained by minimizing the free energy with respect to R to get

$$a_0 + 2a_1(T - T_c)R + 4c(T_c)R^3 = 0 \tag{11.3.6}$$

Note that at the critical temperature and above the long-range order parameter must be zero, so $a_0 = 0$. Then

$$R^2 = \frac{a_1(T_c - T)}{2c} \tag{11.3.7}$$

so the critical exponent is 1/2 and Landau theory gives the same result as mean field (Bragg–Williams) theory. In general, all critical exponents obtained from the Landau expansion to the fourth order are the same as those derived from mean field theories.

Note that the Kirkwood expansion is just like this. Up to the second moment approximation, the Kirkwood expansion for the Helmholtz free energy is given by (8.5.24) and (8.7.1) as

$$A_c = -\frac{NkT}{2}[2\ln 2 - (1+R)\ln(1+R) - (1-R)\ln(1-R)]$$

$$+ W(0) - \frac{N_z}{4}R^2 - \frac{Nzv^2(1-R^2)^2}{16kT} \tag{11.3.8}$$

For magnets, R is the magnetization in units of the Bohr magneton and the ordering energy v is twice the exchange energy.

Now expand (11.3.8) to the fourth order in R. The log terms in the entropy reduce to

$$(1+R)\ln(1+R) + (1-R)\ln(1-R) = R^2 + \frac{1}{6}R^4 \qquad (11.3.9)$$

so (11.3.8) gives, for the free energy per atom (or spin),

$$\overline{A}_c = \overline{W}(0) - kT \ln 2 - \frac{zv^2}{16kT} + \left(\frac{kT}{2} + \frac{zv^2}{8kT} - \frac{zv}{4}\right)R^2$$
$$+ \left(\frac{kT}{12} + \frac{zv^2}{16kT}\right)R^4 \qquad (11.3.10)$$

Comparing this to the Landau expansion (11.3.3) shows that

$$a(T) = \frac{kT}{2} + \frac{zv^2}{8kT} - \frac{zv}{4}$$
$$c(T) = \frac{kT}{12} - \frac{zv^2}{16kT} \qquad (11.3.11)$$

All the results of Landau theory have their counterpart in the Kirkwood expansion. From (8.7.6), for example, it is easy to show that the critical exponent for the order-disorder parameter is 1/2. The Kirkwood expansion has somewhat less generality since it was derived for the Ising model, but it has the advantage of providing explicit expressions for the thermodynamic properties.

A similar approach to that of Landau can be applied to the liquid-vapor transition for a monocomponent fluid by expanding the pressure as a function of reduced temperature and reduced volume near the critical point. Remembering that the first and second derivatives of pressure with respect to volume vanish at the critical point, and neglecting all terms beyond the third-order term in the volume, this gives

$$P(\tilde{T}, \tilde{V}) = P_c - A_1(\tilde{T}-1) + A_2(\tilde{V}-1)(\tilde{T}-1) - A_3(\tilde{V}-1)^3 \qquad (11.3.12)$$

Very close to the critical point, the first two terms give a sufficient approximation, and since the higher order terms approach zero faster than the first-order term, they are approximately zero relative to the first-order term. That is, it is sufficiently accurate to take

$$A_2(\tilde{V}-1)(\tilde{T}-1) - A_3(\tilde{V}-1)^3 = 0 \qquad (11.3.13)$$

There are three solutions of (11.3.13) for the volume. One is the critical point at which $\tilde{V} = 1$, and the other two are the solutions of the quadratic equation (11.3.13). These are

$$\tilde{V} = 1 \pm \sqrt{\frac{A_2(\tilde{T}-1)}{A_3}} \qquad (11.3.14)$$

The positive and negative signs correspond to volumes above and below the critical point, which we denote by $\tilde{V}_G$ and $\tilde{V}_L$, respectively. Since the reciprocal of the reduced volume is the reduced density, the difference in the densities of the liquid and vapor phases near the critical point is

$$\tilde{n}_L - \tilde{n}_G = \frac{1}{\tilde{V}_L} - \frac{1}{\tilde{V}_G} = \frac{1}{1+\sqrt{\frac{A_2(\tilde{T}-1)}{A_3}}} - \frac{1}{1-\sqrt{\frac{A_2(\tilde{T}-1)}{A_3}}}$$

$$= \frac{\sqrt{\frac{A_2(\tilde{T}-1)}{A_3}}}{\left[1+\sqrt{\frac{A_2(\tilde{T}-1)}{A_3}}\right]^2} \tag{11.3.15}$$

Therefore, since the radical can be neglected relative to unity in the denominator near the critical point,

$$\tilde{n}_L - \tilde{n}_G \propto (\tilde{T}-1)^{1/2} \tag{11.3.16}$$

so the critical exponent for the liquid-vapor transition is 1/2. Note that this is a mean field result and is equal to the mean field critical exponent for magnetization. Both experiment and more sophisticated theories show that the critical exponent for fluid densities, magnetization, and order-disorder in alloys are the same and close to 1/3.

11.4 Fluctuations and correlation length

Let us take a closer look at the relation between fluctuations and the divergence of physical properties near critical points. This is most easily done in the language of the Ising model for ferromagnets, although the results are more generally applicable.

For a magnetic system with magnetization $\overline{M}$ in the presence of an external field H, a change in the magnetization $d\overline{M}$ results in a magnetic work term $Hd\overline{M}$. The field is analogous to pressure and the moment is analogous to volume.

The isothermal susceptibility is defined by

$$\chi_T = \left(\frac{\partial \overline{M}}{\partial H}\right)_T \tag{11.4.1}$$

For a system of N spins, each of which can be ± 1 (up or down), and each having the same moment μ, the magnetization $\overline{M}$ is the statistical mechanical average of all possible arrangements of up and down spins. For a particular spin configuration, the magnetization is the total magnetic moment per unit volume. That is,

$$M\{n_j\} = \frac{\mu}{V}\sum_{j=1}^{N} s_j, \quad s_j = \pm 1 \tag{11.4.2}$$

the sum being over all dipoles. The energy for the configuration $\{s_j\}$ in an external field H is

$$-M\{n_j\}H = -\frac{\mu H}{V}\sum_{j=1}^{N} s_j, \quad s_j = \pm 1$$

where $+\mu s_j$ is defined as the magnetic moment in the direction of the field. The partition function is

$$Z = \sum_{\{n_j\}} e^{M\{n_j\}H/kT} \tag{11.4.3}$$

and the magnetization of the system is

$$\overline{M} = \frac{1}{Z} \sum_{\{n_j\}} M\{n_j\} e^{M\{n_j\}H/kT} \tag{11.4.4}$$

so the susceptibility is

$$\chi_T = \left(\frac{d\overline{M}}{dH}\right)_T$$

$$= \frac{1}{ZkT} \sum_{\{n_j\}} M\{n_j\}^2 e^{-M\{n_j\}H/kT} - \sum_{\{n_j\}} M\{n_j\} e^{-M\{n_j\}H/kT} \frac{\partial}{\partial H}\left(\frac{1}{Z}\right) \tag{11.4.5}$$

or

$$\chi_T = \frac{1}{kT}\left(\overline{M^2} - \overline{M}^2\right) = \frac{1}{kT}\overline{\Delta M^2} \tag{11.4.6}$$

where

$$\Delta M = M - \overline{M} \tag{11.4.7}$$

This is completely analogous to the relation between heat capacity and fluctuations in energy given by equation (2.17.4). The heat capacity and the magnetic susceptibility are called *response functions* since they measure the response of the system to a change in external fields (temperature and magnetic field).

The magnetic susceptibility is directly related to fluctuations in magnetic moments. From (11.4.2), equation (11.4.6) is

$$\chi_T = \frac{\mu^2}{VkT}\left(\overline{\sum_{i,j=1}^{N} s_i s_j} - \overline{\sum_{i=1}^{N} s_i \sum_{j=1}^{N} s_j}\right)$$

or

$$\chi_T = \frac{\mu^2}{VkT} \sum_{i,j=1}^{N}\left(\overline{s_i s_j} - \overline{s_i}\,\overline{s_j}\right) \tag{11.4.8}$$

The divergence of the susceptibility means that fluctuations must be large near the critical point. The reason for this is that near the critical point the difference in energy among the possible distributions is of the order of the thermal energy, so large excursions from the mean distribution can take place without much cost in energy. The susceptibility is expected to increase rapidly as the temperature is increased from $T < T_c$ toward T_c, and this is observed.

Completely analogous reasons account for the rapid increase in the heat capacity of order-disorder alloys as the temperature approaches the critical temperature from above. The ordering energy is about equal to the thermal energy, and large relative fluctuations in energy can occur, giving rise to increasing heat capacity.

The summand in (11.4.8) is a measure of the correlation of spins on different sites and leads to the definition of the pair correlation function as

$$C_{ij} = \overline{s_i s_j} - \overline{s_i}\,\overline{s_j} \tag{11.4.9}$$

If the spins are completely uncorrelated, then the average of the product equals the product of the averages and $C_{ij} = 0$. The greater the degree of correlation, the larger the pair correlation function. If (11.4.9) is written in the form

$$C_{ij} = \overline{(s_i - \overline{s_i})(s_j - \overline{s_j})} \tag{11.4.10}$$

It is obvious that the correlation function measures the deviation of site occupation from its average value.

If the system has translational symmetry and is isotropic, then the correlation function depends only on the magnitude of the distance between the two sites. That is,

$$C_{ij} = C(i - j) = C(q) \tag{11.4.11}$$

where $q = |i - j|$.

For a fully ordered crystal, (11.4.11) is obviously true. For partially ordered crystals, it is also true since the macroscopic crystal properties are independent of the choice of the lattice sites. Equation (11.4.8) can then be expressed in terms of the correlation function as

$$\chi_T = \frac{\mu^2}{VkT} \sum_{i,j=1}^{N} C_{ij} = \frac{N\mu^2}{2VkT} \sum_{q=1}^{N} C(q) \tag{11.4.12}$$

where the $N/2$ accounts for the fact that q can be measured from each of the N spins with the index i, but each i must not be counted twice.

At very high temperatures, when there is no order at all, the correlation function is zero. As the temperature approaches the critical temperature from above, the correlation function increases, diverging at the critical point.

The correlation function measures the fluctuations. If there is complete disorder, the thermal energy dominates and there is little fluctuation from the average. If there is complete order, again there is little fluctuation because the ordering energy dominates. Near the critical point, the system is rapidly changing from complete disorder to complete order; the fluctuations and the correlation function are large.

The concept of a correlation length is associated with that of the pair correlation function. At very high temperatures the distance over which there is a correlation between spins is extremely small. This becomes large as the critical temperature is approached from above, going to infinity for a fully ordered crystal.

Some insight into the dependence of correlation on the distance between sites can be obtained by assuming that the correlation function falls off exponentially with distance in the form

$$C(q) = \frac{1}{q} e^{-q/\xi} \tag{11.4.13}$$

This defines ξ as the correlation length in units of the lattice spacing. Now let us make the additional assumption that q is a continuous rather than a discrete variable. This is not a bad assumption when $q \gg 1$, and since the temperature range of interest is near the critical point where the correlation length is large, the results are valid except for small q.

Now substitute (11.4.13) into (11.4.12) and replace the summand by an integral accounting for the three dimensionality by taking the differential of volume to be $4\pi q^2 dq$:

$$\chi_T = \frac{2N\mu^2\pi}{VkT}\int_0^\infty q e^{-q/\xi}dq = \frac{2N\mu^2\pi}{VkT}\xi^2 \qquad (11.4.14)$$

Combining this with (11.2.5) and solving for the correlation length gives

$$\xi = \left(\frac{kTT_c}{2n\pi\gamma\mu^2}\right)^{1/2}\frac{1}{(T-T_c)^{1/2}} \qquad (11.4.15)$$

For a system with spin 1/2 and $g = 2$, equation (9.4.5) is

$$T_c = 3n\frac{\mu_B^2\gamma}{k} \qquad (11.4.16)$$

($n = N/V$, the number density of spins). Equation (11.4.15) therefore becomes

$$\xi = \left(\frac{3}{2\pi}\right)^{1/2}\left(\frac{T}{T-T_c}\right)^{1/2} \qquad (11.4.17)$$

so the critical exponent for the correlation length is 1/2 in mean field theory. More accurate treatments of mean field theory give the same critical exponent. Also, the critical exponent is the same for $T < T_c$ as well as for $T > T_c$, although the proportionality constants are not the same. Experimental values are close to 2/3, which is in agreement with theories that go beyond the mean field approximation.

Equations (11.4.11) and (11.4.13) show that the correlation function is the same in an expanded or contracted lattice as in the original lattice if both the distance q and the correlation length ξ are measured in terms of the new lattice spacing. Also, equation (11.4.14) shows that changing the scale of the correlation length does not change the temperature dependence of the divergence of the susceptibility near the critical point. That is, if the lattice is expanded or contracted, the correlation function, the correlation length, and the susceptibility each scale with the changes of length. While these results were obtained for mean field theory, they turn out to be generally true.

The correlation length as a function of T/T_c for $T > T_c$ is shown in figure 11.1. At high temperatures the correlation length is small, but as the temperature approaches the critical temperature it diverges rapidly. Clearly, because parts of the system that are farther apart than the correlation length are essentially independent, blocks of crystal that have dimensions of the correlation length look alike. Also, such blocks are representative of the whole system.

11.5 The monatomic Ising chain

The three-dimensional Ising model has not been solved exactly and to date can only be studied by various approximate methods. The Kirkwood and

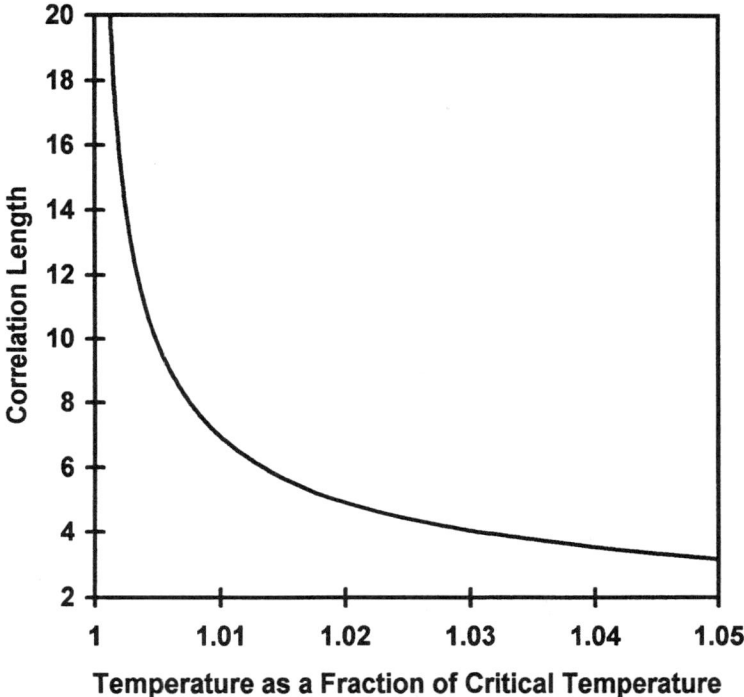

Figure 11.1. Mean field correlation length as a function of temperature.

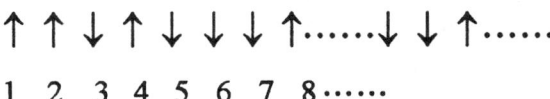

Figure 11.2. One-dimensional array of interacting spins.

quasi-chemical approximations are examples of such methods. However, exact solutions exist for both the one-dimensional Ising chain and the two-dimensional Ising lattice. The solution of the two-dimensional problem is complex and requires lengthy advanced mathematical manipulation, but the one-dimensional problem is straightforward.[2] It is reproduced here primarily for the insight it gives into the correlation length. The Ising chain of N spins is shown in figure 11.2.

In the nearest-neighbor Ising approximation, adjacent spins have an interaction energy $-J$ if the spins are alike and $+J$ if they are unlike. The energy of the chain for a particular configuration of spins is therefore

$$E\{s\} = -J\sum_{i=1}^{N-1} s_j s_{j+1} \qquad (11.5.1)$$

where the spin variables can take on one of the two values $s_j = \pm 1$ and the chain is assumed to have free ends. The partition function is

300 STATISTICAL MECHANICS OF SOLIDS

$$Z = \sum_{\{s\}} e^{J \sum_{j=1}^{N-1} s_j s_{j+1}/kT} \tag{11.5.2}$$

the sum being over all possible sets of spins $\{s\} = \{s_1, s_2, s_3, \ldots\}$. This partition function can be evaluated exactly as follows. Write out the exponential and sum over all configurations explicitly as

$$Z = \sum_{s_1=\pm 1} \sum_{s_2=\pm 1} \cdots \sum_{s_{N-1}=\pm 1} e^{J(s_1 s_2 + s_2 s_3 + \ldots + s_{N-1} s_N)/kT} \tag{11.5.3}$$

The variable s_1 occurs only in the first term of the exponential, and (11.5.3) can be written as

$$Z = \sum_{s_1=\pm 1} e^{J s_1 s_2/kT} \sum_{s_{N-1}=\pm 1} \cdots e^{J(s_2 s_3 + \ldots + s_{N-1} s_N)/kT} \tag{11.5.4}$$

But the first sum is independent of s_2. That is,

$$\sum_{s_1=\pm 1} e^{J s_1 s_2/kT} = e^{J s_2/kT} + e^{-J s_2/kT} = 2\cosh\left(\frac{J}{kT}\right) \tag{11.5.5}$$

whether $s_2 = +1$ or -1. Continuing this process with the other sums, up to the $(N-1)$st spin, gives

$$Z = \left[2\cosh\left(\frac{J}{kT}\right)\right]^{N-2} \sum_{s_{N-1}=\pm 1} \sum_{s_N=\pm} e^{J s_{N-1} s_N/kT}$$

$$= 2\left[2\cosh\left(\frac{J}{kT}\right)\right]^{N-1} \tag{11.5.6}$$

The Helmholtz free energy is therefore

$$A = -N/kT \ln\left[2\cosh\left(\frac{J}{kT}\right)\right] \tag{11.5.7}$$

where only terms proportional to N have been retained because for large N they are the only ones that matter.

The pair correlation function can be evaluated by the following device. Define a function $\hat{Z}(J_1, J_2, \ldots)$ by

$$\hat{Z}(J_1, J_2, \ldots J_{N-1}) = \sum_{\{s\}} e^{\sum_{j=1}^{N-1} J_j s_j s_{j+1}/kT} \tag{11.5.8}$$

This is just like the partition function (11.5.2) except that each spin pair is assigned its own energy. When the algebra is done, we will set all $J_l = J$.

Now choose the spin $j = 1$ and differentiate (11.5.8) with respect to J_1 to get

$$\frac{\partial \hat{Z}}{\partial J_1} = \frac{1}{kT} \sum_{\{s\}} s_1 s_2 e^{\sum_{j=1}^{N-1} J_j s_j s_{j+1}/kT} \tag{11.5.9}$$

Differentiate with respect to J_2, and then with respect to J_3 up to J_q. The result is

CRITICAL EXPONENTS AND THE RENORMALIZATION GROUP 301

$$\frac{\partial^q \hat{Z}}{\partial J_1 \partial J_2 \ldots \partial J_q} = (kT)^{-q} \sum_{\{s\}} (s_1 s_2)(s_2 s_3) \ldots (s_{q-1} s_q) e^{\sum_{j=1}^{N-1} J_j s_j s_{j+1}/kT}$$

$$= (kT)^{-q} \sum_{\{s\}} (s_1 s_q) e^{\sum_{j=1}^{N-1} J_j s_j s_{j+1}/kT} \tag{11.5.10}$$

where the last equality is the result of the fact that $s_j s_j = 1$. Obviously,

$$\hat{Z} = 2 \prod_{p=1}^{N-1} \left[2 \cosh\left(\frac{J_p}{kT}\right) \right] \tag{11.5.11}$$

which is obtained from (11.5.8) just as (11.5.7) is obtained form (11.5.2). Because $(\partial/\partial u) \cosh u = \sinh u$, differentiating (11.5.11) gives

$$\frac{\partial^q \hat{Z}}{\partial J_1 \partial J_2 \ldots \partial J_q} = 2(kT)^{-q} \prod_{j=1}^{q} \left[\sinh\left(\frac{J_j}{kT}\right) \right] \prod_{p=q+1}^{N-1} \left[2 \cosh\left(\frac{J_p}{kT}\right) \right] \tag{11.5.12}$$

Combining (11.5.10) and (11.5.12) gives

$$\sum_{\{s\}} (s_1 s_q) e^{\sum_{j=1}^{N-1} J_j s_j s_{j+1}/kT} = 2 \prod_{j=1}^{q} \left[\sinh\left(\frac{J_j}{kT}\right) \right] \prod_{p=q+1}^{N-1} \left[2 \cosh\left(\frac{J_p}{kT}\right) \right] \tag{11.5.13}$$

Now set the $J_p = J_j = J$ and divide through by the partition function (11.5.6) to get

$$\frac{1}{Z} \sum_{\{s\}} (s_1 s_q) e^{J \sum_{j=1}^{N-1} s_j s_{j+1}/kT} = \left[\tanh\left(\frac{J_j}{kT}\right) \right]^q \tag{11.5.14}$$

The left-hand side of (11.5.14) is just the pair correlation function (11.4.9) because the mean value of the spin for the one-dimensional case is always zero. That is,

$$C(1, q) = \left[\tanh\left(\frac{J}{kT}\right) \right]^q \tag{11.5.15}$$

As expected, this depends only on the distance between the two sites.

The correlation function decreases with increasing separation of the two sites, and the rate of decrease is greater for higher values of J/kT. That is, the correlation is weaker for higher temperatures. If a correlation length ξ is defined by the relation (but now for a one-dimensional system)

$$C(1, q) = e^{-q/\xi} \tag{11.5.16}$$

then using (11.5.15) allows us to write

$$\xi = -\left\{ \ln\left[\tanh\left(\frac{J}{kT}\right) \right] \right\}^{-1} \tag{11.5.17}$$

Plots of the correlation function for different values of J/kT are shown in figure 11.3. Each curve is labeled with its correlation length, longer correlation lengths corresponding to lower temperatures.

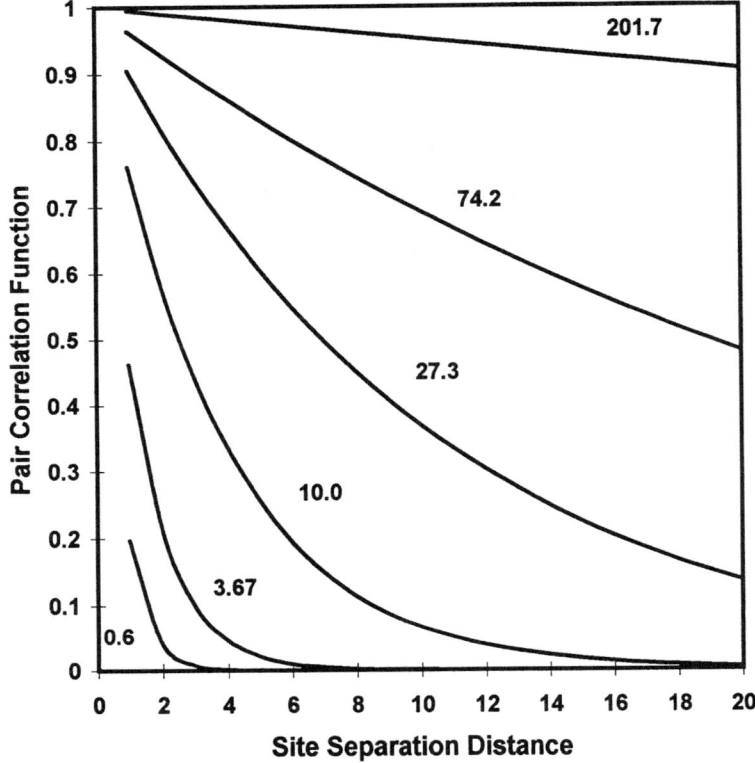

Figure 11.3. Correlation function for the one-dimensional Ising model.

The correlation function decreases with distance between sites in a quasi-exponential manner, the rate of decrease being more rapid for higher temperatures. Just as in the three-dimensional case, the correlation is small at high temperatures, but as the temperature is lowered the correlation among spins extends over many lattice spacings. While these results were obtained for a one-dimensional case, their general features are applicable in two and three dimensions as well.

From the properties of the tanh, the correlation length goes to zero as J/kT goes to zero and becomes infinite as J/kT becomes infinite. That is, for zero interaction constant or infinite temperature there is no correlation, while for zero temperature or infinitely strong interactions there is complete correlation.

11.6 Renormalization of the one-dimensional Ising model

The most important physical result of the development so far is that near the critical point the details of the site occupancy at the level of individual lattice sites cannot matter very much for the physical properties near the critical point. What is important is the interaction of large blocks or patches of the crystal interacting over a large distance. This is because the correlation factor and the correlation length diverge as the critical temperature is approached. To see this clearly, recall that near the critical temperature the system is highly disordered such that in any given cluster of sites there is a close balance between the number of spin-up and spin-down sites. As the critical point is approached, the size of the clusters for which this is true increases and what matters is the overall degree of this small imbalance rather than which particular sites have spins up or down. In some sense, it should be possible to average out a lot of information about site occupancy and still retain essential physical information near the critical point. This in fact turns out to be the case. While an exact solution of the three-dimensional Ising model is still not attainable, and even the two-dimensional case requires lengthy and complicated mathematics, a great deal of progress can be made by approximate methods that suppress the detailed site occupancy information. The methodology for doing this is called *renormalization theory*. It derives its name from the fact that it attempts to find a partition function that refers to the interaction of blocks or clusters of spins that look just like the partition function that refers to individual site spins except that the interaction constants and temperature are replaced by other quantities. The use of these new quantities "renormalizes" the partition function.

The greatest value of renormalization theory is, of course, in three dimensions where exact solutions are not available. However, the basic ideas are most easily introduced by using a renormalization procedure to solve the one-dimensional Ising lattice.

The energy of the chain for a given configuration of spins $\{s_i\}$ is given by equation (11.5.1), and the partition function is given by (11.5.3), but now it is convenient to use periodic boundary conditions for which site N is next to site 1.

The spins on odd-numbered sites interact only with spins on even-numbered sites. Thus, the only terms in the partition function that include site number 1 are those that couple site 1 to site 2 and site N. The contribution of site 1 to the partition function is therefore

$$\sum_{s_1=\pm 1} e^{\beta J(s_1 s_2 + s_1 s_N)} = 2\cosh[\beta J(s_2 + s_N)] \tag{11.6.1}$$

with $\beta = 1/kT$. This can be put in a form that looks like a term in the original partition function. That is, constants $a(1)$ and $b(2)$ can be found such that

$$2\cosh[\beta J(s_2 + s_N)] = e^{a(1)+b(1)s_2 s_N} \tag{11.6.2}$$

To find the constants a and b, consider the two possible cases: the two spins are the same such that $s_2 s_N = 1$, $s_2 + s_N = \pm 2$, or the two spins are different such that $s_2 s_N = -1$, $s_2 + s_N = 0$. For the first case,

$$2\cosh(2\beta J) = e^{a(1)+b(1)} \tag{11.6.3}$$

while for the second case,

$$e^{a(1)-b(2)} = 2 \tag{11.6.4}$$

Solving (11.6.3) and (11.6.4) for $a(1)$ and $b(2)$ gives $b(2) = \tfrac{1}{2}\ln[\cosh(2\beta J)]$, $a(1) = \ln 2 + b(1)$. An index 1 has been included in these equations because we will apply this procedure successively. Define a new interaction constant by

$$J_1 = \frac{b(1)}{\beta} = \frac{1}{2\beta}\ln[\cosh(2\beta J)] \tag{11.6.5}$$

so that (11.6.2) becomes

$$2\cosh[\beta J(s_2 + s_N)] = e^{a(1)+\beta J_1 s_2 s_N} \qquad (11.6.6)$$

A similar procedure can be carried out for every odd-numbered site in the chain so that each odd-numbered site sums to a similar form and the contribution of the odd-numbered sites to the partition function is

$$\sum_{s_1,s_3,s_5\ldots} \ldots = e^{Na(1)/2+\beta J_1 \sum_i s_{2i}s_{2i+2}} \qquad (11.6.7)$$

the sum in the exponential being only over all even-numbered sites.

The partition function for the entire chain is obtained by putting (11.6.7) into (11.5.3). Only the even-numbered sites in the partition sum survive, so

$$Z = e^{Na(1)/2} \sum_{s_2=\pm 1} \sum_{s_4=\pm 1} \ldots \sum_{s_{N-2}=\pm 1} e^{\beta J_1 \sum_i s_{2i}s_{2i+2}} \qquad (11.6.8)$$

which looks just like an Ising partition function with a new interaction constant. This means that the same treatment can be applied to (11.6.8). That is, we pick out every other site and sum over them just as in getting (11.6.7). A new interaction constant, J_2, is defined that has the same relation to J_1 as J_1 has to J in equation (11.6.6).

This process is called *decimation* and can be carried out indefinitely until the entire chain is summed. At each stage of decimation the interaction constant is related to that of the preceding stage by

$$J_n = \frac{b(n-1)}{\beta} = \frac{1}{2\beta}\ln[\cosh(2\beta J_{n-1})] \qquad (11.6.9)$$

An important point is that, because

$$\beta J_n = \frac{1}{2}\ln[\cosh(2\beta J_{n-1})] \leq \beta J_{n-1} \qquad (11.6.10)$$

the renormalized interaction constant is never greater than the old one and for small β (high temperature) it is less than the old one. Thus,

$$J_n \leq J_{n-1} \qquad (11.6.11)$$

so for all n and at high temperatures, the interaction constant gets progressively smaller with increasing decimation steps.

Note that when the argument of the cosh is infinity, the cosh is infinity, so the equality holds in (11.6.11). Likewise, when the argument of the cosh is 0, 2cosh = 2, so again the equality holds in (11.6.11). Thus, at zero temperatures and at infinite temperature, all successive βJ_n are equal. These are called *fixed points*. For any βJ_n between zero and infinity, (11.6.11) shows that as $n \to \infty$, the interaction constant goes to zero so the βJ_n goes to the fixed point zero. This is a stable fixed point. The fixed point at infinity is unstable since large values of βJ_n get smaller with increasing n. They are closely related to critical points.

Let us go through this explicitly. Rewrite (11.6.8) as

$$Z = e^{Na(1)/2}Z(1) \qquad (11.6.12)$$

where

$$Z(1) \equiv \sum_{s_2=\pm 1} \sum_{s_4=\pm 1} \sum_{s_{N-2}=\pm 1} \ldots e^{\beta \sum_i J_1(s_{2i}s_{2i+2})} \quad (11.6.13)$$

is the partition function of the decimated chain. Renumber the sites in (11.6.13) so they have consecutive indices from 1 to $N/2$ and apply the decimation procedure again to give

$$Z(1) = e^{Na(2)/4} \sum_{s_2=\pm 1} \sum_{s_4=\pm 1} \sum_{s_{N/2-2}=\pm 1} \ldots e^{\beta \sum_i J_2(s_{2i}s_{2i+2})} \quad (11.6.14)$$

The factor 1/4 in the exponential outside the sums is the result of the fact that only half as many sites are involved in the second decimation. The multiple sum in (11.6.14) is defined to be a partition function for the second decimation by

$$Z(1) \equiv e^{Na(2)/4} Z(2) \quad (11.6.15)$$

Continuing this process, the partition function becomes

$$Z = e^{Na(1)/2} e^{Na(2)/4} e^{Na(3)/8} \ldots \quad (11.6.16)$$

so the free energy per spin is

$$\frac{\beta A}{N} = -\frac{1}{2}\left[a(1) + \frac{a(2)}{2} + \frac{a(3)}{4} \ldots\right] = -\frac{1}{2}\sum_{j=1}^{\infty}\left(\frac{1}{2}\right)^{j-1} a(j) \quad (11.6.17)$$

and the recursion relation for the $a(j)$ is

$$a(j) = \ln[\cosh(2\beta J_{j-1})] \quad (11.6.18)$$

with

$$2\beta J_n = \ln[\cosh(2\beta J_{n-1})] \quad (11.6.19)$$

The hyperbolic cosine approaches unity as its argument approaches zero, and the terms in (11.6.17) therefore converge to zero as $n \to \infty$. This is a critical point at which correlation is complete. Decimation moves the system closer to criticality.

The sum is taken to infinity because we want the free energy in the thermodynamic limit of $N \to \infty$. Also, the terms converge, so a very large N is equivalent to infinity in the summation.

The decimation procedure has resulted in a series of convergent terms that can be truncated at any point to give a solution in terms of renormalized interaction constants, each higher order of truncation giving a more accurate solution.

11.7 The Kadanoff construction

In 1966 Kadanoff proposed a method to take advantage of the fact that near the critical point physical properties are not sensitive to details of site occupancy. This method is a generalization of the decimation procedure used above for the one-dimensional Ising chain. It consists of choosing a set of sites around each lattice site and treating these clusters as if they were single sites on a lattice of expanded size. From this "renormalized" lattice it is possible to get the relations among the critical exponents without explicit calculation. As an example, consider a square lattice whose sites are occupied by spins (or atoms) which are labeled 0 and can be either up or down. Now group the sites into sets of four sites as shown in figure 11.4A. Consider each such set as a new "spin" (or "atom") that we now label x, and place these on a lattice with twice the lattice spacing of the original lattice. This is called the *Kadanoff construction*, and the new lattice is said to be a renormalization of the original lattice.

The renormalized lattice of figure 11.4B can be renormalized again, and indeed the process can be continued indefinitely, giving a series of Kadanoff constructions that consecutively wipe out more and more of the site-specific details.

Now consider a d-dimensional Ising lattice, with an external magnetic field H, and Hamiltonian

$$E\{s\} = -J\sum_{<ij>} s_i s_j - \mu H \sum_i s_j, \qquad s_j = \pm 1 \qquad (11.7.1)$$

$E\{s\}$ being the energy for a specific configuration of spins denoted by $\{s\}$. Let us renormalize this in the same sense as the square lattice. If the initial lattice spacing is a, the first renormalization results in a lattice with a larger spacing ba where b is the lattice renormalization ratio (which had the value 2 for the square lattice treated above) and the number of original lattice sites in each of the blocks that define the renormalized lattice is b^d. Thus, if the number of sites in the original lattice is N, the number of sites in the renormalized lattice is $N_1 = N/b^d$.

A block spin variable in the renormalized lattice is defined as some function of the site variables in the original lattice by

$$s_k^{(1)} = f\{s_j^k\} \qquad (11.7.2)$$

where s_j^k is the spin on the jth site of the kth block in the original lattice. The function f can be defined in a variety of ways. Three common methods are the average, in which the block spin is taken to be the arithmetic mean of the spins in the block, the majority rule, in which the block is assigned a spin equal to that of the majority of spins, and the decimation rule in which the block spin is the same as that on a particular site in the block. This is the method that was used above for the one-dimensional Ising chain.

Let us define the block spins by

$$s_k^{(1)} = \frac{1}{b^d |\bar{s}_i|} \sum_{j \in k} s_j \qquad (11.7.3)$$

where the sum is over all sites in the block k and $\bar{s}_i$ is the average spin of the block, defined by

$$\bar{s}_i = \frac{1}{b^d} \sum_{j \in k} s_j \qquad (11.7.4)$$

This is just the majority rule since

$$s_k^{(1)} = \frac{1}{b^d |\bar{s}_i|} \sum_{j \in k} s_j = \frac{\bar{s}_i}{|\bar{s}_i|} = \pm 1 \qquad (11.7.5)$$

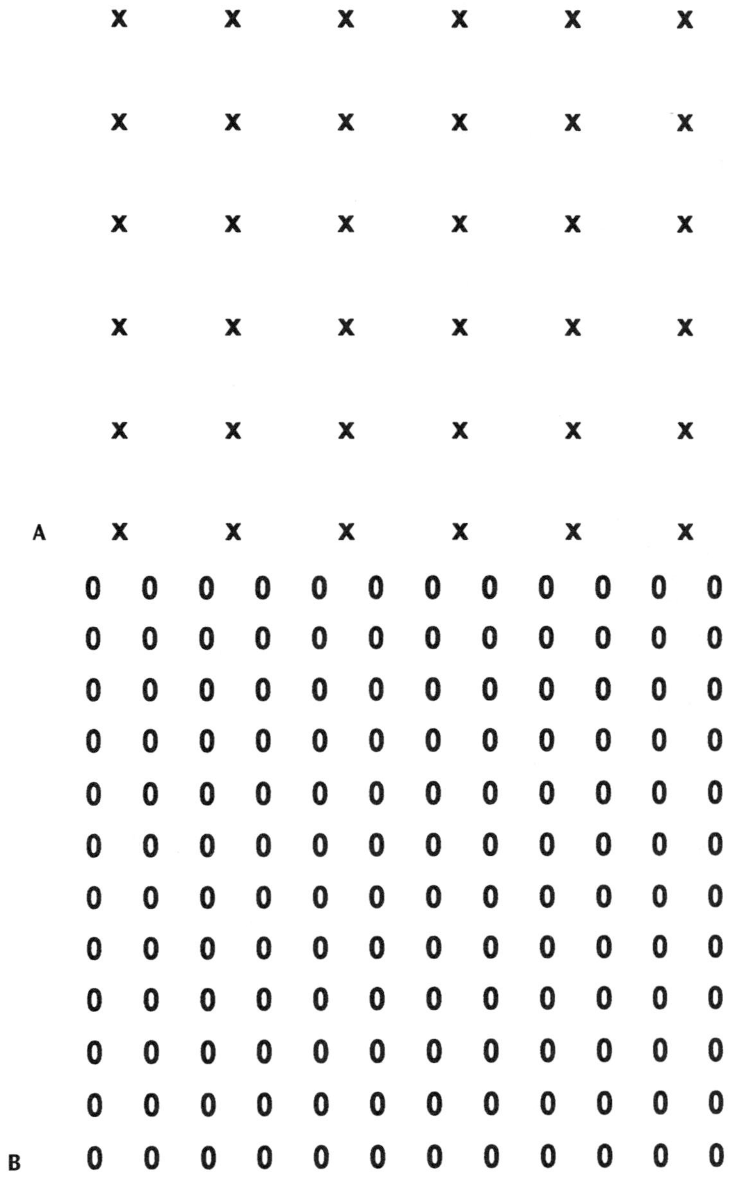

A x   x   x   x   x   x

   x   x   x   x   x   x

   x   x   x   x   x   x

   x   x   x   x   x   x

   x   x   x   x   x   x

   x   x   x   x   x   x

0 0 0 0 0 0 0 0 0 0 0 0
0 0 0 0 0 0 0 0 0 0 0 0
0 0 0 0 0 0 0 0 0 0 0 0
0 0 0 0 0 0 0 0 0 0 0 0
0 0 0 0 0 0 0 0 0 0 0 0
0 0 0 0 0 0 0 0 0 0 0 0
0 0 0 0 0 0 0 0 0 0 0 0
0 0 0 0 0 0 0 0 0 0 0 0
0 0 0 0 0 0 0 0 0 0 0 0
0 0 0 0 0 0 0 0 0 0 0 0
0 0 0 0 0 0 0 0 0 0 0 0
B 0 0 0 0 0 0 0 0 0 0 0 0

Figure 11.4. Block renormalization of a square lattice. B, Original lattice. A, Lattice after renormalization.

The Kadanoff construction is useful only if the energy of the renormalized lattice can be expressed in terms of block spins. To do this, we make an assumption:

Assumption 1: The block spins have the same form for the Hamiltonian as the site spins. That is,

$$H\{s^{(1)}\} = -J_1 \sum_{<kl>} s_k^{(1)} s_l^{(1)} - \mu_1 H_1 \sum_k s_k^{(1)} \quad (11.7.6)$$

where J_1 and H_1 are new constants, to be determined, which refer to the interaction among blocks and of blocks with the external field. This assumption could be regarded as a condition imposed on the block spins that defines the new interaction constants.

It turns out that it is not possible in general to construct a block Hamiltonian that has the same functional form as the original because the lattice renormalization can result in interactions that are farther apart than nearest neighbor blocks and that couple more than two sites at a time. Equation (11.7.6) is the result of ignoring all but nearest-neighbor block interactions. Thus, accepting this assumption is equivalent to ignoring all but nearest neighbor interactions among blocks in the renormalized lattice.[3]

Consider the correlation length: in the original lattice it is expressed in units of the lattice spacing a as

$$\xi = \xi(a) \quad (11.7.7)$$

In the renormalized lattice it is expressed in terms of the renormalized lattice spacing ba as

$$\xi_1 = \xi_1(ba) \quad (11.7.8)$$

Because the system after renormalization is the same as the original system, the two correlation lengths must be the same at the same physical point. But a length described in units of the old lattice spacing is equivalent to a point decreased by a factor $1/b$ in units of the new lattice. This means that

$$\xi_1 = \frac{1}{b}\xi \quad (11.7.9)$$

when each length is measured in units of its own lattice spacing.

The magnetic field in the renormalized lattice is also scaled, as can be seen by summing (11.7.3) over all blocks to get

$$\sum_{\{s_k\}} s_k^{(1)} = \frac{1}{b^d |\bar{s}_j|} \sum_{\{s_j\}} s_j \quad (11.7.10)$$

and recalling that the energy of interaction of the spins with the field is

$$-\mu H \sum_{\{s_j\}} s_j = -\mu H b^a |\bar{s}_i| \sum_{\{s_k\}} s_k^{(1)} \quad (11.7.11)$$

so in the new lattice, the block spins interact with the external field as if it had the value

$$H_1 = H b^d |\bar{s}_j| \quad (11.7.12)$$

In the renormalized lattice the correlation length is smaller. This is equivalent to the renormalized system being at a higher temperature and farther away from the critical point. The renormalized system therefore has a different temperature, as well as different interaction constants from the original. The temperature always appears in combination with the interaction constants in the Hamiltonian, so it is convenient to define a reduced Hamiltonian from (11.7.6) by multiplying through by β and defining reduced interaction constants by

$$\beta \tilde{E}\{s\} = -\tilde{J} \sum_{<ij>} s_i s_j - \tilde{H} \sum_i s_j, \qquad s_j = \pm 1 \qquad (11.7.13)$$

where

$$\tilde{J} = \beta J$$

$$\tilde{H} = \mu \beta H \qquad (11.7.14)$$

are a reduced reciprocal temperature and a reduced magnetic field, respectively. Similarly, for the renormalized lattice

$$H\tilde{E}\{s^{(1)}\} = -J_1 \beta \sum_{<kl>} s_k^{(1)} s_l^{(1)} - \mu_1 H_1 \beta \sum_k s_k^{(1)}$$

$$= -\tilde{J}_1 \sum_{<kl>} s_k^{(1)} s_l^{(1)} - \tilde{H}_1 \sum_k s_k^{(1)} \qquad (11.7.15)$$

Note that the partition function of the renormalized lattice is

$$Z_1 = \sum_{\{s^{(1)}\}} e^{-\tilde{E}\{s^{(1)}\}} \qquad (11.7.16)$$

the sum being over all possible sets of values of *block* spins. The free energy corresponding to (11.7.16) is

$$A_1 = N_1 f_1(\tilde{J}, \tilde{H}_1) \qquad (11.7.17)$$

with $f_1(\tilde{J}_1, \tilde{H}_1)$ being the free energy per block of the new lattice. In the original lattice, the free energy is

$$A = Nf(\tilde{J}, \tilde{H}) \qquad (11.7.18)$$

where $f(\tilde{J}, \tilde{H})$ is the free energy per spin site. If the free energy per site of the physical system is to be the same in the original and the renormalized representation, then

$$\frac{A_1}{N} = \frac{A_1}{N_1 b^d} = \frac{f_1}{b^d} = \frac{A}{N} = f \qquad (11.7.19)$$

so the free energy per site scales as follows:

$$f_1(\tilde{J}_1, \tilde{H}_1) = b^d f(\tilde{J}, \tilde{H}) \qquad (11.7.20)$$

In the analysis of critical phenomena, physical properties near the critical point vary with temperature according to powers of the difference between temperature and the critical temperature. It makes sense to recognize this by defining a parameter t by

$$t = \frac{T - T_c}{T_c} \tag{11.7.21}$$

The free energy can always be written as a function of this parameter rather than the temperature itself, so we can write

$$f(\tilde{J}, \tilde{H}) = f(t, \tilde{H}) \tag{11.7.22}$$

Similarly, for the renormalized lattice

$$f_1(\tilde{J}_1, \tilde{H}_1) = f_1(t_1, \tilde{H}_1) \tag{11.7.23}$$

where t_1 refers to the new lattice:

$$f_1(t_1, \tilde{H}_1) = b^d f(t, \tilde{H}) \tag{11.7.24}$$

Now we introduce another assumption.

Assumption 2: The reduced temperature and reduced magnetic field scale according to

$$t_1 = b^m t \tag{11.7.25}$$

$$\tilde{H}_1 = b^n \tilde{H} \tag{11.7.26}$$

At this stage, this can be regarded either as an assumption or as a definition of the quantities m and n. But m and n must be greater than unity since the renormalized temperature and magnetic field are greater than that in the original lattice, so this much, at least, is an assumption that requires future theoretical justification. With these scaling assumptions, (11.7.24) becomes

$$f(t, \tilde{H}) = b^{-d} f_1(t b^m, \tilde{H} b^n) \tag{11.7.27}$$

Thus far, there has been no restriction on the value of b, and it can be given any value greater than unity. It is convenient to choose b as

$$b^m |t| = 1 \tag{11.7.28}$$

since $|t|$ is always less than unity, $b^m > 1$, and since it is assumed that $m > 0$, then $b > 1$.

For comparison of theory with the experimental scaling laws, we define two new constants in terms of m and n by

$$\Delta = \frac{n}{m} \tag{11.7.29}$$

and

$$\alpha = 2 - \frac{d}{m} \tag{11.7.30}$$

Using these definitions, (11.7.27) becomes

$$f(t, \tilde{H}) = |t|^{2-\alpha} f_1\left(1, \frac{\tilde{H}}{|t|^\Delta}\right), \quad t > 1 \qquad (11.7.31)$$

$$f(t, \tilde{H}) = |t|^{2-\alpha} f_1\left(-1, \frac{\tilde{H}}{|t|^\Delta}\right), \quad t < 1 \qquad (11.7.32)$$

These scaling equations are sufficient to get the relations among the critical exponents for the heat capacity, the magnetization, and the susceptibility. The heat capacity per spin is just the second derivative of the free energy per spin:

$$C_v = -T\left(\frac{\partial^2 f}{\partial T^2}\right)_V = -\frac{T}{T_c^2}\left(\frac{\partial^2 f}{\partial t^2}\right)_V \qquad (11.7.33)$$

From (11.7.31), as the magnetic field goes to zero, $f_1(1, \tilde{H}/|t|^\Delta)$ goes to a constant, so the second derivative becomes proportional to $-\alpha$, which is the critical exponent for the heat capacity:

$$C_V \sim |t|^{-\alpha} \qquad (11.7.34)$$

Note that this result holds both for temperatures above and below the critical temperature because the same answer is obtained from (11.7.32).

The magnetization is just the derivative of the free energy with respect to the magnetic field. Differentiation of (11.7.30) with respect to $\tilde{H}$ then shows that the magnetization varies as

$$M \sim |t|^{2-\alpha-\Delta} \qquad (11.7.35)$$

so the critical exponent for the magnetization is

$$\beta = 2 - \alpha - \Delta \qquad (11.7.36)$$

The susceptibility is the derivative of the magnetization with respect to the magnetic field, which is the second derivative of the free energy with respect to the magnetic field, so as the field goes to zero,

$$\chi_T \sim |t|^{2-\alpha-2\Delta} \qquad (11.7.37)$$

Again, (11.7.36) and (11.7.37) hold on both sides of the critical point. The critical exponent for the susceptibility is therefore

$$\gamma = 2 - \alpha - 2\Delta \qquad (11.7.38)$$

From (11.7.38) and (11.7.36) we get

$$\alpha + 2\beta + \gamma = 2 \qquad (11.7.39)$$

This is the experimental result we were trying to account for.

11.8 The renormalization group

The above development is based on Kadanoff's argument that a renormalized Hamiltonian, and therefore a renormalized free energy, must exist for the renormalized lattice that reproduces the essential physics near the critical point. However, there is a problem with both of the assumptions arising from this plausibility argument. For the first assumption, it has already been pointed out that renormalization of a pairwise, nearest-neighbor Ising Hamiltonian is not recovered when the lattice is renormalized according to the Kadanoff construction. It has also been pointed out that the assumption of the scaling of the reduced temperature and magnetic field was yet to be justified. Both of these defects are removed by the generalization introduced by K. G. Wilson in 1971 known as the *renormalization group*, for which he was awarded the Nobel prize. The theory is applicable to more complicated Hamiltonians than the simple pairwise Ising model we have been using. (Actually, the use of more general Hamiltonians simplifies the theory.) Thus, let us consider systems that have N spin variables and Hamiltonians of the form

$$\mathcal{H} = \beta E\{s\} = N K_0 - \tilde{H}\sum_i s_i - \tilde{J}_1 \sum_{<ij>} s_i s_j - \tilde{J}_2 \sum_{<ijk>} s_i s_j s_k + \ldots \quad (11.8.1)$$

The first term is independent of spins and is included for purposes of symmetry because, after a renormalization, constant terms can appear, as was the case for the one-dimensional Ising chain. The second term is the interaction with the external field, and the succeeding terms are the interactions of the spins with each other, $<ij>$ denoting a sum over all pairs of spins, $<ijk>$ denoting a sum over all triplets, and so on. At this stage, it is not assumed that the pairs, triplets, and so forth, are nearest neighbors. Equation (11.8.1) can be written compactly as

$$\mathcal{H}[K_n] = \sum_n K_n S_n\{s\} \quad (11.8.2)$$

where the K_n are constants and the $S_n\{s\}$ are functions such that S_1 depends on the spins occupying each point on the lattice, S_2 depends on the pairwise occupations, S_3 on the triplet occupations, and so on. Now let us renormalize the lattice by a Kadanoff construction. The set of constants $[K_n]$ in the original lattice is then changed to a new set $[K_n^{(1)}]$ in the renormalized lattice. Let this change be represented by the symbol R_1. That is, R_2 is an operator such that

$$[K_n^{(1)}] = R_1[K_n] \quad (11.8.3)$$

R_1 is called the *renormalization operator* for the first Kadanoff construction. A sequence of renormalization operators can be defined, one for each of a sequence of Kadanoff constructions, each renormalizing the lattice from the previous construction. Thus, if $[K_n^{(2)}]$ is the set of coupling constants for a second renormalization, then

$$[K_n^{(2)}] = R_2[K_n^{(1)}] \quad (11.8.4)$$

and in general, for the rth renormalization,

$$[K_n^{(r)}] = R_r[K_n^{(r-1)}] \quad (11.8.5)$$

All quantities in (11.8.5) have the same interpretation as the corresponding quantities in (11.8.2), except that they refer to the renormalized lattice.

In the preceding section, the Kadanoff lattice renormalization changed the scale of the lattice by a factor b. It is not necessary to have the same change of scale for successive renormalizations, and we will sometimes recognize this by calling the change of scale for the rth renormalization b_r. Clearly, the same lattice is obtained from two renormalizations with two changes of scales as from one renormalization with a combined change of scale. That is, if $[K_n^{(r+1)}] = R_{r+1}[K_n^{(r)}]$, then $[K_n^{(r+1)}] = R_{r+1}R_r[K_n^{(r-1)}]$, so the renormalization operator for a change of scale of $b_r b_{r'}$ equals the product of the individual operators, or

$$R_{r,r'} = R_r R_{r'} \tag{11.8.6}$$

Equation (11.8.5) is the recursion relation for successive renormalization and is the basis for calculating successive renormalized coupling constants from the constants of the original lattice.

The set of all renormalization operators is called the renormalization group since (11.8.6) is a group property. However, the R_r do not form a true group because there is no inverse renormalization operation, and they would more appropriately be called a semigroup.

Repeated scaling gives ever smaller correlation lengths, when measured in the renormalized lattices, so the correlation lengths accompanying a sequence of M renormalizations, each of which has the same scale factor b, are

$$\xi[K_n] = b\xi[K_n^{(1)}] = b^2\xi[K_n^{(2)}] = b^3\xi[K_n^{(3)}] = \ldots = b^M\xi[K_n^{(M)}] \tag{11.8.7}$$

There are at least two values of the correlation length that remain constant upon renormalization: zero and infinity. If the set of coupling constants for which either of these two cases is true is labeled $[K_n^*]$, then these constants maintain their values upon renormalization such that

$$\left[K_n^*\right] = R_r\left[K_n^*\right] \tag{11.8.8}$$

The sets of constants for which (11.8.8) is satisfied define the fixed points. The fixed point for which the correlation length is zero is called a *trivial fixed point*, while that for which the correlation is infinity is called a *critical fixed point*. The nomenclature arises from the fact that the correlation length approaches infinity as the system approaches criticality.

In a sequence of renormalizations, the set of coupling constants may or may not approach a critical point. All sets that approach a critical point upon continued renormalization are said to be in the basin of that critical point, and the changing values of the set of coupling constants are said to flow toward the critical point. This nomenclature arises from thinking of the coupling constants as parameters defining a geometric space and the set of constants as a vector in this parameter space.

11.9 Scaling and the renormalization group

The scaling relations obtained by the semi-intuitive treatment of the Kadanoff construction are easily obtained more rigorously from the theory of the renormalization group. Let us assume that, just as in the preceding section, the only parameters of interest are the reduced temperature and reduced magnetic field, and denote them by t_r and $\tilde{H}_r$ for the rth renormalization.

Just as in the preceding section, we are looking for a relation between the free energy of the original and the renormalized lattice. To this end, define the partition function corresponding to the Hamiltonian (11.8.2). This is

$$Z[K_n] = \sum_{\{s\}} e^{-\mathcal{H}[K_n]} \qquad (11.9.1)$$

the sum being over all possible configurations of the spins s_j. The corresponding free energy per spin site in units of kT is

$$f[K_n] = \frac{\beta A}{N} = -\frac{1}{N} \ln Z[K_n] \qquad (11.9.2)$$

In the renormalized lattice, the partition function is

$$Z_1[K_n^{(1)}] = \sum_{\{s^{(1)}\}} e^{-\mathcal{H}[K_n^{(1)}]} \qquad (11.9.3)$$

with a corresponding free energy per site given by

$$f_1[K_n^{(1)}] = \frac{\beta^{(1)} A_1}{N_1} = -\frac{1}{N_1} \ln Z[K_n^{(1)}] \qquad (11.9.4)$$

If renormalization is to represent the physical system, then the free energy per spin must be the same before and after renormalization, so

$$f[K_n] = b^{-d} f_1[K_n^{(1)}] \qquad (11.9.5)$$

This is equivalent to (11.7.20) except that, because of the use of a series of pair, triplet, and so on, interactions, the difficulty of generating new coupling constants that had no counterpart before renormalization is removed.

Let us consider the same case as before, namely, when the free energy is a function only of reduced temperature and external magnetic field. Then (11.9.5) is

$$f(\tilde{T}, \tilde{H}) = b^{-d} f_1(\tilde{T}^{(1)}, \tilde{H}^{(1)}) \qquad (11.9.6)$$

The renormalized temperature and magnetic field are functions of the original T and $\tilde{H}$. Denote this functional relationship by defining two functions $R_T^{(1)}$ and $R_H^{(1)}$ such that

$$\tilde{T}^{(1)} = R_T^{(1)}(\tilde{T}, \tilde{H})$$
$$\tilde{H}^{(1)} = R_H^{(1)}(\tilde{T}, \tilde{H}) \qquad (11.9.7)$$

At a fixed point, the temperature and external field do not change upon renormalizing, so (11.9.7) gives

$$\tilde{T}^* = R_T^{(1)}(\tilde{T}^*, \tilde{H}^*)$$
$$\tilde{H}^{(1*)} = R_H^{(1)}(\tilde{T}^*, \tilde{H}^*) \qquad (11.9.8)$$

Close to a fixed point, (11.9.7) can be expanded in a Taylor series about T^* and $\tilde{H}^*$. Retaining only the linear term, this gives

$$\tilde{T}^{(1)} = \left(\frac{\partial R_{\tilde{T}}^{(1)}}{\partial T}\right)_{\tilde{T},\tilde{H}=T^*\tilde{H}^*}(\tilde{T}-\tilde{T}^*) + \left(\frac{\partial R_{\tilde{T}}^{(1)}}{\partial \tilde{H}}\right)_{T,\tilde{H}=T^*\tilde{H}^*}(\tilde{H}-\tilde{H}^*) \qquad (11.9.9)$$

$$\tilde{H}^{(1)} = \left(\frac{\partial R_{\tilde{H}}^{(1)}}{\partial \tilde{H}}\right)_{\tilde{T},\tilde{H}=T^*\tilde{H}^*}(\tilde{H}-\tilde{H}^*) + \left(\frac{\partial R_{\tilde{H}}^{(1)}}{\partial \tilde{H}}\right)_{\tilde{T},\tilde{H}=T^*\tilde{H}^*}(\tilde{T}-\tilde{T}^*) \qquad (11.9.10)$$

The coefficients in (11.9.9) and (11.9.10) define the transformation matrix for renormalization near a fixed point. Although the derivatives are constants with respect to temperature and magnetic field, they obviously depend on the lattice scaling parameter b.

Let us assume that this matrix is symmetric.[4] That is, the temperature and external magnetic field are not linked. Then, (11.9.9) and (11.9.10) reduce to

$$\tilde{T}^{(1)} = \Gamma_{\tilde{T}}^{(1)}(\tilde{T} - T^*) \qquad (11.9.11)$$

$$\tilde{H}^{(1)} = \Gamma_{\tilde{H}}^{(1)}(\tilde{H} - \tilde{H}^*) \qquad (11.9.12)$$

where

$$\Gamma_{\tilde{T}}^{(1)} = \left(\frac{\partial R_{\tilde{T}}^{(1)}}{\partial \tilde{T}}\right)_{\tilde{T},\tilde{H}=T^*\tilde{H}^*} \qquad (11.9.13)$$

and

$$\Gamma_{\tilde{H}}^{(1)} = \left(\frac{\partial R_{\tilde{H}}^{(1)}}{\partial H}\right)_{\tilde{T},\tilde{H}=T^*\tilde{H}^*} \qquad (11.9.14)$$

Now apply the renormalization group property by taking a second renormalization, denoted by (2), and compare the two renormalizations (1) and (2) to the renormalization (1, 2) that takes the system directly from the original to the second renormalized lattice. Then it must be true that

$$T^{(2)} = \Gamma_{\tilde{T}}^{(1,2)}(\tilde{T} - T^*) = \Gamma_{\tilde{T}}^{(2)}\Gamma_{\tilde{T}}^{(1)}(\tilde{T} - T^*) \qquad (11.9.15)$$

$$\tilde{H}^{(2)} = \Gamma_{\tilde{H}}^{(1,2)}(\tilde{H} - \tilde{H}^*) = \Gamma_{\tilde{H}}^{(2)}\Gamma_{\tilde{H}}^{(1)}(\tilde{H} - \tilde{H}^*) \qquad (11.9.16)$$

such that

$$\Gamma_{\tilde{T}}^{(1,2)} = \Gamma_{\tilde{T}}^{(2)}\Gamma_{\tilde{T}}^{(1)} \qquad (11.9.17)$$

$$\Gamma_{\tilde{H}}^{(1,2)} = \Gamma_{\tilde{H}}^{(2)}\Gamma_{\tilde{H}}^{(1)} \qquad (11.9.18)$$

as would be expected from the group property of a series of renormalizations. For simplicity, let us modify the notation slightly to make the Γ transformation coefficients explicit functions of the scale factor b, and let us give the scale factor its index to identify the number of the renormalization. Also, it is obvious that (11.9.17) and (11.9.18) hold for any two renormalizations r and r', and that they all have the same functional form of the scaling length. That is, (11.9.17) and (11.9.18) become

$$\Gamma_{\tilde{T}}(b_r b_{r'}) = \Gamma_{\tilde{T}}(b_r)\Gamma_{\tilde{T}}(b_{r'}) \qquad (11.9.19)$$

$$\Gamma_{\tilde{H}}(b_r b_{r'}) = \Gamma_{\tilde{H}}(b_r)\Gamma_{\tilde{H}}(b_{r'}) \qquad (11.9.20)$$

The simplest way these equations can hold is for the transformation coefficients to be powers of the scaling factor. That is, taking the scaling factor to be the same for every renormalization, we must have

$$\Gamma_{\tilde{T}} = b^m \qquad (11.9.21)$$

$$\Gamma_{\tilde{H}} = b^n \qquad (11.9.22)$$

Putting these in (11.9.11) and (11.9.12), we recover the scaling relations (11.7.25) and (11.7.26), and therefore the relation among the critical exponents given by equation (11.7.39).

Let us pause to look at what has been accomplished here. First, it has been shown that the physical parameters indeed diverge as the temperature approaches the critical point according to power laws. Second, the critical exponents for a given property are the same for approaches to criticality both from below and from above the critical temperature. Third, the observed algebraic relation among the critical exponents has been derived. Finally, an enormous amount of universality has been found for critical phenomena. There was nothing of a specific nature that depends on the particular atomic constitution, so the results are valid for a wide variety of systems. They are not even restricted to magnetic systems since the arguments hold equally well for lattice gases and order-disorder systems. Clearly, the general features of the approach to criticality depend on the general properties of correlations over large distances. In fact, the critical exponents and the relations among them depend only on the dimensionality of the system and the symmetry of the Hamiltonian. All systems that have the same dimensionality and for which the Hamiltonian have the same symmetry are said to belong to the same universality class. Specific numerical values for the physical parameters certainly depend on the constitution of the system, but the functional forms of the approach to criticality depend only on the universality class.

11.10 Numbers

The foregoing analysis has yielded an enormous amount of physical insight without any numerical calculation. However, a comparison of theoretical values of the critical exponents with experiment is important to demonstrate that the theory is correct in detail as well as conceptually. Both series expansion methods and renormalization calculations have been developed for this purpose.

Except for the Ising chain, numerical computations based on renormalization are difficult because the recursion relations for the partition function and the Hamiltonian become quite complicated. Nevertheless successful calculations have been performed. A relatively straightforward and efficient approach combines renormalization group theory with Monte Carlo methods. Also, advanced series expansion methods that go beyond the Kirkwood expansion in chapter 8 have been worked out. The agreement of Monte Carlo and series expansion computations with each other and experiment is additional evidence that critical exponent theory is very successful The values listed as "Theory" in table 11.1 were computed by these methods.

Exercises

11.1 From the Landau theory, get the relation for the temperature dependence of the order parameter to the second order in $(T_c - T)$.

11.2 From the relation between the Landau expansion and the Kirkwood second moment approximation, find the variation of the long-range order parameter as a function of temperature near the critical point by minimizing the Landau free energy.

11.3 Show that the relation among critical exponents, equation (11.7.39), is correct for mean field theory of a three-dimensional Ising lattice.

Notes

1. Tables of values of critical exponents for a variety of experimental systems and theoretical calculations are given in Stanley (1971), Yeomans (1992), Binney et al. (1993), and Chaikin and Lubensky (1995).
2. See Plischke and Bergersen (1971) for the solution of the one-dimensional Ising ferromagnet including the external magnetic field.
3. An explicit demonstration of this for the two-dimensional square lattice can be found in Yeoman (1992, pp. 132–136).
4. This is often true. In those cases for which it is not, the scaling laws we are about to derive are the same and can be obtained from an analysis of the non-symmetric transformation matrix. See, for example, Goldenfeld (1992, p. 255).

12

Surfaces and Interfaces

12.1 Basic concepts

The boundary between any two phases is not sharp, but contains a transition region in which there is a continuous change in composition from one phase to the other. Of course, all the laws of thermodynamics apply to the system as a whole, and at equilibrium, the temperature, pressure, and chemical potentials are constant throughout the system. But because of the spatial variation of the densities of the components, the energy per unit volume (and other thermodynamic functions per unit volume) varies with position over a finite distance in the transition region. The total energy is therefore different than that of the sum of the energies of the two phases taken separately.

Two ways of assigning thermodynamic properties to a planar interface separating two phases were introduced by J. Willard Gibbs. In the first method, the system is divided into three parts by constructing two planar surfaces parallel to the interface, as shown in figure 12.1, which represents two phases, A and B, in contact across a planar interface in a large container. It is assumed that the system is in thermodynamic equilibrium and all external fields, including gravitation, are zero. The planes are chosen in such a way that the concentration of components varies sensibly with position only between them.

Thus, to the left of the plane AA' the system has the properties of the bulk phase A, while to the right of plane BB' it has the properties of bulk phase B. In between the planes AA' and BB', the composition varies from that in one phase to that in the other. Since the range of interatomic forces is short, the thickness of the region that defines the interface is small. But thermodynamic quantities can be defined for that region and identified with the interface by first choosing some mathematical surface that is parallel to and near the region of the physical interface. It turns out that all interfacial thermodynamic quantities are independent of the precise location of the mathematical surface if they are defined as follows.

Consider a fictitious system in which each phase is homogeneous right up to the mathematical surface. That is, each phase has the same energy, free energy, entropy, and composition density that it has in the bulk phase right up to the surface. For any component i, the number of molecules in the interface is defined as the number of molecules in excess of that which would exist if the two phases kept their bulk composition right up to the mathematical surface. That is, the number of molecules in the interface, N_i^σ, is defined by

$$N_i^\sigma = (N_i - n_i^A V_A - n_i^B V_B) \qquad (12.1.1)$$

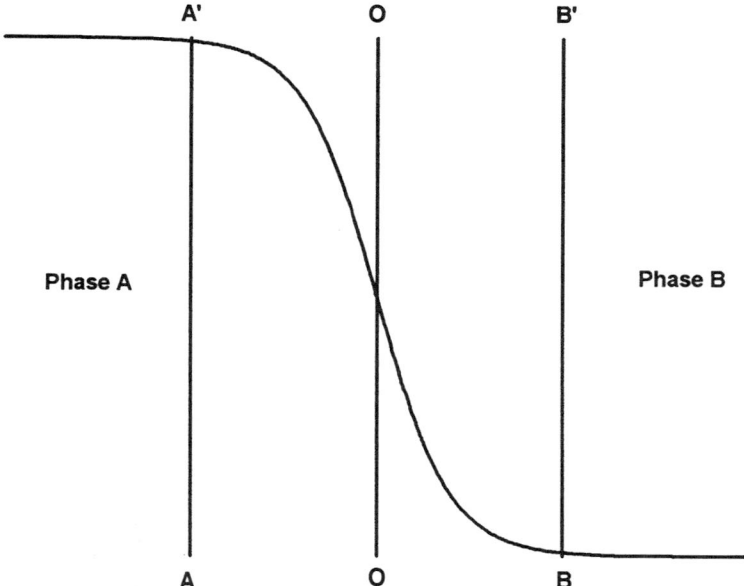

Figure 12.1. The interface between two phases.

N_i is the total number of molecules of component i, n_i^A and n_i^B are the bulk concentrations of component i in phases A and B, respectively, measured as if each phase had its bulk properties right up to the Gibbs reference surface, and V_A, V_B are the volumes of the two phases, measured up to the mathematical surface separating the two phases. Note that since these are the actual volumes, there is no excess volume associated with the interface. The area of the interface will be denoted by σ, and interfacial quantities will be identified by a superscript σ.

Similarly, let the Gibbs free energies for the two phases in this fictitious system be G_A and G_B. The Gibbs free energy of the interface G^σ is then defined by

$$G^\sigma = G - G_A - G_B \qquad (12.1.2)$$

where G is the total Gibbs free energy of the real system. Similar definitions hold for all other extensive thermodynamic functions. Note that such definitions are independent of the precise choice of the mathematical surface because, if the chosen surface is moved parallel to the interface, what is gained on one side by the translation is lost on the other. Also, because of the way the surface is defined, there is no excess volume associated with the interface.

An important consequence of this definition is that it leads to a fully consistent thermodynamics of interfaces. Because the laws of thermodynamics are linear and they hold for each quantity on the right-hand side of (12.1.2), then by subtraction they must hold for the difference on the left-hand side.

For systems in which there is very little mutual solubility between the phases, it is more convenient to use the alternative approach that defines a geometric surface of separation between the two phases in terms of one component of the system. A definition of the surface is constructed by choosing one of the components of the system to use as a reference and labeling it component 1. Now construct a surface such that the total amount of component 1 on each side of the interface is the same as if the bulk phases extended right up to the interface with no change in concentration of this reference component. This is called the *Gibbs reference surface*. A crystal with a perfect planar surface in contact with a gas is a good approximation to this, the solid surface being the reference plane. This is a special case of the previous definition, and an excess free energy is defined just as in equation (12.1.2). (Note that, by definition, the excess amount of the reference component at the interface $N_1^\sigma = 0$.) All other interfacial thermodynamic quantities are defined relative to the Gibbs reference surface.

The reference phase boundary, and therefore the excess interface concentrations, depend on which component is defined to have a zero excess interface concentration. Although the choice of the reference component is arbitrary, it is often dictated by convenience and clarity of interpretation. For a pure solid or liquid in contact with a gas phase, for example, the best choice of reference component is that of the pure condensed phase because the variation of concentration with position near the interface is quite small for the condensed phase. If one of the phases is a dilute solid or liquid solution while the other phase is a gas or a concentrated solution, the best choice is the major component of the dilute solution.

When an interface is present, the total free energy of the system contains a work term that depends on the area of the interface. The reason for this arises from the fact that the interatomic forces in the bulk are different than those at the interface. Thus, if the interface is extended, work must be done to bring atoms from the bulk to the interface. A simple analysis of this can be made by referring back to figure 12.1. The Gibbs surface is somewhere between the two planes AA' and BB', and an interfacial phase is sometimes defined as the product of the area of this surface and the distance between AA' and BB'. There is a degree of ambiguity in this definition, and it is not needed to construct a consistent thermodynamics of interfaces.

Let a uniform pressure be the only external force acting on the system. Then in any plane parallel to the Gibbs surface, all properties are uniform and the force per unit area is just the pressure P. In a plane perpendicular to the Gibbs surface, however, the force is not the same at all points because the composition varies in that plane. Now increase the area of the interface by $d\sigma$. This can be done in three ways. The first is to add both A and B material to the A and B sides of the container in figure 12.1 while keeping the length of the system constant. This increases the interface area while maintaining the composition of the system constant and increasing the amount of the constituents. Alternatively, the area can be increased while maintaining the amount of material constant by transferring material in such a way as to decrease the systems length while increasing its cross-sectional area. Finally, if an external stress is applied to the system, it will deform, thereby changing the interfacial area. In general, the area can be changed by any of these processes, or any combination of them, and the thermodynamic formalism must reflect this.

An interfacial area increase will be accompanied by a corresponding increase in volume with a corresponding PV work done on the system given and a corresponding change in the amount of constituents dN_i. Since atoms must be brought from the bulk phases to the interfacial phase when the area is increased, or since work must be done if the area changes in response to a stress, there is additional work that is proportional to the increase in interface area. The work associated with an increase of the area of the interface phase is proportional to the increase in area, so we write

$$dW = \gamma d\sigma \qquad (12.1.3)$$

where the proportionality constant γ has the dimensions of energy per unit area, or force per unit length, and is called the *interfacial tension*.

The interfacial tension γ is assumed to be independent of the surface area and to be the same for any of the three processes that can change the interfacial area. This assumption is obviously true if the area is increased by transferring material in the absence of internal stresses and the interface separates two fluid phases. The molecules in fluids are mobile, and they can readjust to give equilibrium states that require that the energy per unit area of the interface be a constant. Liquids cannot support shear, so the only stress that can be applied is hydrostatic pressure. In principle, the surface tension is not constant with respect to changes in pressure because the resulting changes in volume change the intermolecular distances and therefore the intermolecular interaction. However, if the pressure is low enough to be within the limits of linear elasticity, this effect can be neglected. For fluid systems, it will be assumed that the interfacial tension is the same and independent of interfacial area no matter how the area is altered. From the third method of changing the interfacial area, it is clear that γ is the force acting on a unit line parallel to, and in the interface.

Interfaces that include solid phases must be treated with more care. Not only can they react to external forces anisotropically, but also internal stresses can arise from the structural mismatch at interfaces. This is true both for the interface between two different solids and for the interface between two different crystallographic planes of the same solid. It is often the case that the extension of a solid-solid interface by any of the three methods described above maintains the local structure at the interface constant. One way of doing this is to have dislocations take up the crystallographic mismatch at the interface so that the energy per unit area is independent of area. For some phenomena, the effect of internal stresses and strains can be ignored. This is often the case for solid-fluid interfaces since the solid surface is then affected by the other phase only in a minor way. In the physical adsorption of inert gases on solid surfaces, for example, it is sufficient to treat the solid surface as being homogeneous and isotropic in two dimensions.[1]

For a solid-liquid system, there are three interfaces: between the solid and the liquid, between the liquid and its vapor, and between the solid and the vapor. Each interface has its own surface tension, which is identified with appropriate subscripts. Consider such a system in which the liquid is a drop on a solid surface that does not chemically react with it. In general, the drop forms an angle θ with the solid, as illustrated in figure 12.2.

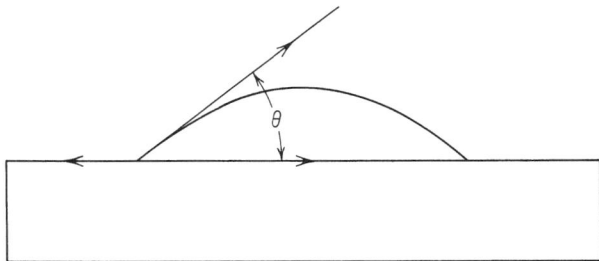

Figure 12.2. Surface forces at a liquid drop on a solid surface.

The three interfacial tensions are forces per unit length acting parallel to each interface, and at equilibrium, the forces must sum to zero. This means that the surface tension of the solid-vapor interface must balance the forces from the solid-liquid and liquid-vapor interfaces. That is,

$$\gamma_{sv} - \gamma_{sl} = \gamma_{lv} \cos\theta \qquad (12.1.4)$$

This is Young's equation.

For a contact angle of π the liquid becomes a spherical drop (neglecting gravity), while a contact angle of 0 corresponds to the spreading of the liquid into a flat film. For this latter case, wetting, there is no solid-vapor or liquid-vapor interface and the total interfacial free energy of the system is γ_{sl}. If this is lower than the solid-vapor ($\gamma_{sv} + \gamma_{lv}$) then the liquid will spread and wet the solid. This leads to the definition of the spreading coefficient w as

$$w \equiv (\gamma_{sv} + \gamma_{lv}) - \gamma_{ls} \qquad (12.1.5)$$

If the spreading coefficient is negative, the liquid will remain as a spherical drop on the surface. (Actually, because of the influence of gravity, it will form a lens, rather than a sphere.) But if the spreading coefficient is positive, the liquid will spread spontaneously; $w = 0$ defines the wetting transition. Note that using (12.1.4) gives the spreading coefficient as

$$w = \gamma_{lv}(1 + \cos\theta). \qquad (12.1.6)$$

so for complete wetting, $w = \gamma_{lv}$.

12.2 Thermodynamics of interfaces

Let us consider a system for which internal stresses and anisotropy, as well as external fields, either do not exist or can be ignored. Then, when an interface is present, the interfacial work term (12.1.3) must be added to the differential of energy, so equation (1.13.6) must be modified to read

$$dU = TdS - PdV + \sum_i \mu_i dN_i + \gamma d\sigma \qquad (12.2.1)$$

Using (12.2.1) in the definitions of the Helmholtz and Gibbs free energies gives

$$dA = -SdT - PdV + \sum_i \mu_i dN_i + \gamma d\sigma \qquad (12.2.2)$$

$$dG = -SdT + VdP + \sum_i \mu_i dN_i + \gamma d\sigma \qquad (12.2.3)$$

and therefore

$$\gamma = \left(\frac{\partial U}{\partial \sigma}\right)_{S,V,N_i} = \left(\frac{\partial A}{\partial \sigma}\right)_{T,V,N_i} = \left(\frac{\partial G}{\partial \sigma}\right)_{T,P,N_i} \qquad (12.2.4)$$

Note that now chemical potentials must be defined as derivatives in which the interfacial area is held constant. That is,

$$\mu_i = \left(\frac{\partial U}{\partial N_i}\right)_{S,V,N_{j\ne i},\sigma} = \left(\frac{\partial A}{\partial N_i}\right)_{T,V,N_{j\ne i},\sigma} = \left(\frac{\partial G}{\partial N_i}\right)_{T,P,N_{j\ne i},\sigma} \qquad (12.2.5)$$

Let us recall equation (1.15.8) and derive the analogous equation when an interface is present. Integration of (12.2.1) by Euler's theorem gives

$$U = TS - PV + \sum_i \mu_i N_i + \gamma\sigma \qquad (12.2.6)$$

and using the definition of the Gibbs free energy $G = U - TS + PV$ gives

$$\gamma = \frac{1}{\sigma}\left(G - \sum_i \mu_i N_i\right) \qquad (12.2.7)$$

But from (1.15.8), the sum in (12.2.7) is just the Gibbs free energy of the system in the absence of the interface. The surface tension is therefore seen to be the free energy of the surface per unit area. There is nothing on the right-hand side of (12.2.7) that refers to the mathematical interface, so this verifies that the surface tension does not depend on how its location is chosen.

For the Gibbs free energy, substitute its sum according to (12.1.2) into (12.2.7) and for the number of molecules use (12.1.1) in the form $N_i = N_i^\sigma + N_i^A + N_i^B$ to get $G^\sigma + G_A + G_B = \gamma\sigma + \Sigma_i\mu_i N_i^\sigma + \Sigma_i\mu_i N_i^B + \Sigma_i\mu_i N_i^A$, which reduces to

$$G^\sigma = \gamma\sigma + \sum_i \mu_i N_i^\sigma \qquad (12.2.8)$$

The difference between the interfacial and bulk concentrations of the systems components is closely related to the surface tension through the Gibbs adsorption isotherm, which can be obtained from (12.2.8). Start with the differential of (12.2.8):

$$dG^\sigma = \gamma d\sigma + \sigma d\gamma + \sum_i \mu_i dN_i^\sigma + \sum_i N_i^\sigma d\mu_i \qquad (12.2.9)$$

Also, from

$$dG = dG^\sigma + dG_A + dG_B \qquad (12.2.10)$$

it follows that

$$dG^\sigma + dG_A + dG_B = -(S^\sigma + S_A + S_B)dT + (V_A + V_B)dP + \sum_i \mu_i d(N_i^\sigma + N_i^A + N_i^B) + \gamma d\sigma \qquad (12.2.11)$$

or

$$dG^\sigma = -S^\sigma dT + \sum_i \mu_i dN_i^\sigma + \gamma d\sigma \qquad (12.2.12)$$

and comparing (12.2.12) with (12.2.9) shows that, at constant temperature,

$$\sigma d\gamma = -\sum_i N_i^\sigma d\mu_i \qquad (12.2.13)$$

Dividing through by the surface area gives

$$d\gamma = -\sum_i \Gamma_i d\mu_i \qquad (12.2.14)$$

where $\Gamma_i = N_i^\sigma/\sigma$ is the surface concentration of species i. This is the relation between the surface tension and the excess surface concentrations in terms of the chemical potentials of the components of the system. Equation (12.2.14) is the Gibbs adsorption isotherm.

For a two-component, two-phase system in which one of the components is present in small amounts, the Gibbs surface is chosen such that the major component has zero excess surface concentration. In this case, (12.2.14) reduces to

$$d\gamma = -\Gamma_2^\sigma d\mu_2 \qquad (12.2.15)$$

the subscript 2 labeling the minority component. From the theory of dilute solutions, the chemical potential of the component present in small amounts is related to its concentration c_2 by $\mu_2 = \text{const} + kT\ln c_2$, and (12.2.15) becomes

$$\Gamma_2 = -\frac{c_2}{kT}\left(\frac{\partial \gamma}{\partial c_2}\right)_T \qquad (12.2.16)$$

This is the Gibbs adsorption isotherm for a dilute two-component system. Note that if increasing the concentration decreases the surface tension, the solute will be preferentially adsorbed at the surface, and vice versa.

For small changes in the concentration of a dilute solution, the derivative can be replaced by a finite ratio so that (always remembering that the temperature is kept constant)

$$\Gamma_2 = -\frac{c_2 \Delta\gamma}{kT \Delta c_2}$$

and for very small concentrations, $\Delta\gamma = \gamma - \gamma_0$ and $\Delta c_2 = c_2$, so this becomes

$$\Gamma_2 = -\frac{\Delta\gamma}{kT} \qquad (12.2.17)$$

That is, for very dilute solutions the change in surface tension is directly proportional to the excess surface concentration.

The Gibbs adsorption isotherm is applied to the case of the adsorption of a pure gas on a solid by taking the solid to be the reference component. We assume the gas has a very low solubility in the solid. In this case, the Gibbs surface is practically identical to the solid surface. Equation (12.2.14) then reduces to a form similar to that of (12.2.15) but with the excess surface concentration being that of the gas. The chemical potential can be taken as that of the gas phase, and if the pressure is not too high, this is proportional to $kT\ln P$, and equation (12.2.16) gives

$$d\gamma = -kT\Gamma_g \frac{dP}{P} \qquad (12.2.18)$$

Γ_g is the excess surface concentration of gas, that is, the amount of gas adsorbed on the surface per unit area.

Consider an adsorbed film on a substrate in which the film material is insoluble. The surface tension of the pure substrate is γ_0 and the surface tension when the film is present is γ. The film pressure π is defined by

$$\pi = \gamma_0 - \gamma \qquad (12.2.19)$$

This definition ensures that the sign convention is the same as that for three dimensions. That is, a positive pressure means a force acting *on* the system. The surface pressure is the change in the force per unit length on a line parallel to the surface when adsorbate is present. It is also the negative of the excess free energy of the surface with adsorbate over the free energy of the pure surface.

For liquid substrates, the film pressure can be measured by floating a bar on the surface and measuring the force on it, the adsorbate being confined by the bar and the edges of a pan containing the substrate. Two-dimensional phase diagrams can be obtained by measuring the film pressure as a function of the surface area. Typical systems that have been experimentally studied in this way consist of fatty acids floating on water. In these systems, the carboxyl groups are attracted to the water, leaving hydrocarbon chains sticking out of the surface, which interact via van der Waals forces.

For a gas adsorbed on an inert solid substrate, experimental values of the spreading pressure can be obtained as a function of the amount of gas on the surface from adsorption isotherms by using the integrated form of the Gibbs adsorption isotherm (12.2.18):

$$\gamma - \gamma_0 = -kT \int_0^P \Gamma_g d\ln P \qquad (12.2.20)$$

If the surface area is increased by an amount $d\sigma$ for both the clean surface and that containing adsorbate, while holding temperature and volume constant, then the free energy of adsorption is increased by

$$dA_{ads} = \gamma d\sigma - \gamma_0 d\sigma \qquad (12.2.21)$$

where A_{ads} is the Helmholtz free energy of adsorption. This gives another definition of the film pressure:

$$\pi = -\left(\frac{\partial A_{ads}}{\partial \sigma}\right)_{T,V} \qquad (12.2.22)$$

The area can also be increased while holding the temperature and pressure constant, in which case the spreading pressure is the derivative of the Gibbs free energy with respect to area at constant temperature and pressure:

$$\pi = -\left(\frac{\partial G_{ads}}{\partial \sigma}\right)_{T,P} \qquad (12.2.23)$$

These equations are often applied to two dimensional phases, since they are correct in principle even if there is no substrate. The subscript ads is then dropped.

12.3 Thermodynamics of adsorption on solid surfaces

Consider a solid surface in equilibrium with a monocomponent gas that does not dissolve to any appreciable extent in the solid. Because of the existence of surface forces, gas will be adsorbed on the surface and the nature of this adsorption will depend on the nature of the interaction forces between the gas and the solid. There are two primary modes of adsorption: physical and chemical. Physical adsorption is the result of van der Waals type forces. It is characteristic of the adsorption of the rare gases on solid surfaces, but also occurs in many other systems in which there is no chemical reaction between the solid and the gas, for example, the adsorption of nitrogen on alkali halide or oxide surfaces. The binding energy of the gas molecule to the surface is weak, of the order of 10 kcal/mole, and the adsorption takes place rapidly. Also, the surface film of gas that results from the adsorption can be a monolayer or can consist of a number of molecular layers, and the adsorbed molecules generally have a high transverse mobility on the surface. The adsorbed molecules have the same chemical formula as in the gas. That is, if the gas is diatomic, then the adsorbed molecules are also diatomic and there is very little distortion of the geometry of the molecule. Chemisorption, on the other hand, is the result of the formation of chemical bonds between the surface atoms and the gas molecules. This binding energy is high, of the order of 100 kcal/mole, and the adsorption is often slow because an activation energy of chemical reaction must be overcome. The surface film is a monolayer, and its molecules have no surface mobility. Also, the adsorbed molecule sometimes loses the character it had in the gas phase. Diatomic hydrogen on metals, for example, dissociates into atoms that react with the surface atoms of the solid.

For a given solid-gas system, the lower the temperature and higher the pressure, the more gas will be adsorbed on the surface. Both the adsorption isotherm (amount adsorbed as a function of pressure at constant temperature) and the adsorption isobar (amount adsorbed as a function of temperature at constant pressure) depend on the heats of adsorption. Some care must be taken in defining these heats since they depend on the experimental conditions of measurement.

In an experiment in which a small amount of gas is transferred from the gas phase onto the solid surface while both the pressure and the temperature are held constant, the heat evolved is given by the first law of thermodynamics as

$$dQ = TdS = -dU - PdV \quad (12.3.1)$$

The PdV term is the work done on the system as a result of the decrease in volume accompanying the adsorption of N^σ molecules of gas on the surface.

From the ideal gas law, $PdV = -kT\, dN^\sigma$ so from (12.3.1) a differential heat of adsorption can be defined as

$$q_{st} = \frac{dQ}{dN^\sigma} = \frac{TdS}{dN^\sigma} = -\frac{dU}{dN^\sigma} + kT \quad (12.3.2)$$

If the experiment is done at constant temperature and volume so that no PdV work is done, then the kT term does not appear and instead of (12.3.2), equation (12.3.1) leads to a differential heat q_d given by

$$q^d = -\frac{dU}{dN^\sigma} = q_{st} - kT \quad (12.3.3)$$

Both q_d and q_{st} are differential heats of adsorption since they are the heats evolved per atom for an infinitesimal increment of gas adsorbed. From (12.3.2) and (12.3.3) it is clear that q_{st} is the enthalpy change per molecule accompanying adsorption while q_d is the energy change. q_{st} is called the *differential isosteric heat of adsorption*.

The connection between the heat of adsorption and the adsorption isotherm is readily obtained by considering two equilibrium states of the system that are at different pressures and temperatures but contain the same distribution of gas molecules in the surface and gas phase. That is, we consider a small reversible change in pressure and temperature, dP and dT, while holding the amount of gas adsorbed on the surface constant. For equilibrium in the initial state, the Gibbs free energy of the gas phase must equal that of the adsorbed surface phase. But this must also be true for the final state, and this means that the change in free energies per molecule in the gas and adsorbed phases must be equal. That is, $dG_g/N_g = dG^\sigma/N^\sigma$, or

$$-s_g dT + v_g dP = -s^\sigma dT \tag{12.3.4}$$

s_g and v_g being the entropy and volume per molecule in the gas phase, and s^σ the entropy per molecule in the adsorbed phase. The left-hand side of (12.3.4) is just a bulk thermodynamics result, while the right-hand side follows from (12.2.12) for the case of constant surface concentration and constant surface area. Therefore,

$$\left(\frac{\partial P}{\partial T}\right)_{N^\sigma} = \frac{s_g - s^\sigma}{v_g} \tag{12.3.5}$$

Equation (12.3.5) refers to a phase change from the gas to the adsorbed phase, and the entropy change is therefore equal to the enthalpy change of the system divided by the temperature, so (12.3.5) becomes

$$\left(\frac{\partial P}{\partial T}\right)_{N^\sigma} = \frac{h_g - h^\sigma}{v_g T} \tag{12.3.6}$$

h_g and h^σ being the enthalpy per molecule in the gas phase and adsorbed phase, respectively, so $(h_g - h^\sigma)$ is just the isosteric differential heat of adsorption of equation (12.3.2). Equation (12.3.6) is just the Clapeyron equation for the equilibrium between the gas and the adsorbed phase. Using the ideal gas law to replace v_g, (12.3.6) becomes

$$\left(\frac{\partial \ln P}{\partial T}\right)_{N^\sigma} = \frac{q_{st}}{kT^2} \tag{12.3.7}$$

where $q_{st} = (h_g - h^\sigma)$ is the isosteric differential heat of adsorption. This is the Clausius–Clapeyron equation for the gas-adsorbed layer equilibrium.

The differential heat of adsorption can be determined from sets of isotherms by using (12.3.7) directly, but it is more convenient to have an integral form. If the differential heat of adsorption is assumed to be independent of temperature, this form easily follows by integrating (12.3.7) between two states (P_1, T_1) and (P_2, T_2) to get

$$\left(\ln \frac{P_2}{P_1}\right)_{N^\sigma} = \frac{q_{st}}{k}\left(\frac{1}{T_1} - \frac{1}{T_2}\right) \tag{12.3.8}$$

so the differential heat of adsorption is readily obtained from points on two different isotherms corresponding to the same amount of gas adsorbed.

q_{st} is the enthalpy change on adsorbing an incremental amount of gas and varies with the amount adsorbed. This variation can arise from two sources: first, the adsorbed molecules interact with each other and the energy of this interaction varies with surface concentration; second, the adsorption sites on the surface may not, and usually do not, have the same energy, so the heat of adsorption is different for different sites. The integral heat of adsorption is defined as the total heat released for adsorbing the total amount of gas up to a given amount adsorbed and is just the integral of the differential heat:

$$Q_{st} = \int_0^{N^\sigma} q_{st} dN^\sigma \tag{12.3.9}$$

Equation (12.2.19), which defines the surface pressure, is still valid, with γ_0 being the surface tension of the pure solid and γ being the surface tension of the solid surface containing adsorbed gas. Since the surface tension of the pure solid is independent of pressure, $d\pi = -d(\gamma - \gamma_0)$, so for this case (12.2.18) can be written in integral form as

$$\pi = kT \int_0^P \Gamma_g d\ln P \tag{12.3.10}$$

The surface pressure of a gas on a solid can therefore be obtained experimentally from measurements of the amount of gas adsorbed on the surface as a function of pressure.

Some care must be taking in performing the numerical (or graphical) integration necessary because at low pressures γ/P becomes very large. But it is easy to show that for physical adsorption at low pressures the adsorbed gas follows the two-dimensional ideal gas law:

$$\pi\sigma = N^\sigma kT \tag{12.3.11}$$

so an analytic integration can be performed in the low-pressure region.

12.4 Adhesion and cohesion

The process of surface formation can be schematically represented as in figure 12.3, which represents the separation of two phases A and B in contact across an interface to give two phases separated an infinite distance from each other. Each phase has its own surface, with none of the phase of the other material adsorbed on it.

This separation process can be represented as $A|B \rightarrow A| + B|$, which simply indicates that an interface has been broken to yield two surfaces. The energy needed to effect this separation is called the *energy of adhesion* and is obviously the difference between the interfacial energy and the surface energies. That is, if E_A^σ, E_B^σ, E_{AB}^σ are the surface energy of phase A, surface energy of phase B, and interfacial energy of the interface, respectively, then the energy of adhesion $E_{\text{adh}}(AB)$ is

$$E_{\text{adh}}(AB) = E_A^\sigma + E_B^\sigma + E_{AB}^\sigma \tag{12.4.1}$$

where the energies have been identified by subscripts in an obvious fashion.

If the two phases are the same so that initially there is no interface, then sep-

Figure 12.3. Separation of two semi-infinite phases across an interface to form two surfaces.

aration across a plane to form two surfaces is represented as $AA \to A| + A|$, and the corresponding energy is just that required to form two surfaces of A. The energy of separation is then called the *energy of cohesion* and is given by

$$E_{\text{coh}}(AA) = 2E_A^\sigma \qquad (12.4.2)$$

Clearly, this is just a special case of (12.4.1) because when $A = B$ the interfacial energy is zero and adhesion reduces to cohesion.

It is unfortunate that common usage refers to both the energy defined by (12.4.2) and the energy required to separate a condensed phase into its constituent molecules to infinity as the energy of cohesion. To avoid confusion, we will use the term *energy of surface cohesion*.

In a similar fashion, the free energies and entropies of adhesion and surface cohesion are given in terms of the interfacial or surface quantities as

$$G_{\text{adh}}(AB) = (\gamma_A + \gamma_B - \gamma_{AB})\sigma \qquad (12.4.3)$$

$$S_{\text{adh}}(AB) = S_A^\sigma + S_B^\sigma + S_{AB}^\sigma \qquad (12.4.4)$$

Again, the equations for surface cohesion are special cases of these equations given by setting the interfacial quantities equal to zero.

One reason for relating the adhesion (or surface cohesion) quantities to the interfacial (or surface) quantities is that it leads naturally to a method of computing the surface quantities because the adhesion energy is just the sum of all intermolecular interactions across the interfacial plane. (For surface cohesion, the surface cohesive energy is just the sum of all intermolecular interactions across an arbitrary plane in the homogeneous phase.) A method of doing this was first given in 1939 by Fowler and Guggenheim, who actually treated only the liquid-vapor interface, though the method is easily generalized to include any two phases whose molecules interact according to pairwise central forces. This was done in 1954[2] and subsequently further developed and applied to a number of systems.[3]

Consider two semi-infinite phases with plane surfaces of equal area that are parallel to each other and a distance z apart as in figure 12.4. Take a molecule at point X in phase B at a distance f from a slab of thickness df in phase A. The slab df is parallel to the surfaces of the two phases. On this slab, trace an annular ring whose inner and outer circumferences, on the side closest to the molecule at X, are at distances r and $r + dr$ from X. This ring has a radius $r\sin\theta$ and width $dr/\sin\theta$, θ being the angle between f and r, so the volume of the ring is $2\pi rdrdf$.

Let $\varepsilon_{AB}(r)$ be the energy of interaction of a molecule of A with a molecule of B when the two molecules are a distance r apart and let n_A be the average number density of A molecules in the annular ring. Then the energy of interaction of the molecule x with the annular ring is $2\pi n_A \varepsilon_{AB}(r)rdrdf$ and the energy of interaction of molecule x with the entire semi-infinite phase A is obtained by first integrating over r from f to infinity and then over f from j to infinity to get

$$2\pi \int_j^\infty df \int_f^\infty A^\varepsilon{}_{AB}(r)rdr$$

The total interaction of phase B with phase A is then obtained by multiplying the above result by n_B, the average number density of B molecules in the slab dj and integrating from $j = z$ to infinity. This gives the energy of adhesion of the two phases when z is the equilibrium distance between the two parallel surfaces as

330 STATISTICAL MECHANICS OF SOLIDS

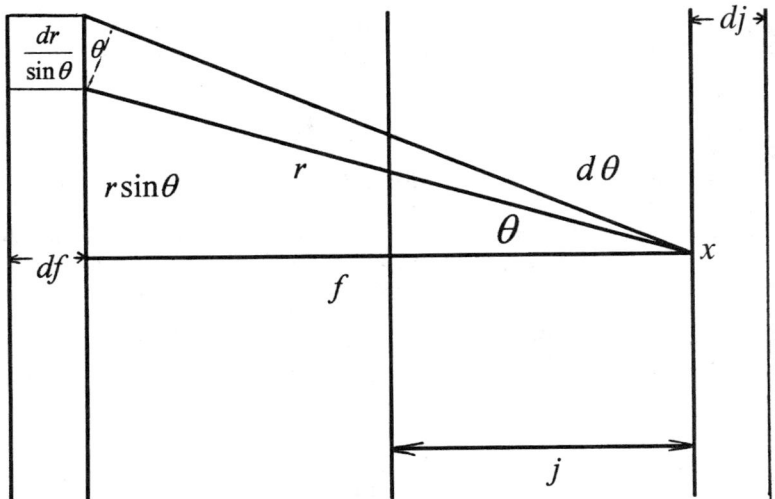

Figure 12.4. Interaction of two phases across a plane surface.

$$E_{\text{adh}}(AB) = 2\pi \int_z^\infty n_B(j)dj \int_j^\infty df \int_f^\infty n_A(r)\mathcal{E}_{AB}(r)rdr \quad (12.4.5)$$

The difficulty in computing the energy of adhesion from this equation, even when the intermolecular potential energy is known, is that the local average densities are functions of distance from the surfaces. In fact, a complete theory for fluid-fluid interfaces would relate the average densities to pair correlation functions, which would be different in the bulk and near the surfaces. For interfaces with a solid phase, the integrals over the solid phase should be replaced by a lattice sum. Also, the calculation takes into account only the potential energy of interaction and ignores any entropy changes arising from the change in the vibrational frequencies near the surface. Nevertheless, approximations to (12.4.5) are very useful in exposing the origins of surface properties and the relations among them.

The simplest approximation that can be made is to assume that the average densities are constant and equal to their bulk values right up to the surfaces. Then they come out of the integral and (12.4.5) becomes

$$E_{\text{adh}}(AB) = n_A n_B K_{AB}(z) \quad (12.4.6)$$

where

$$K_{AB}(z_{AB}) \equiv 2\pi \int_{z_{AB}}^\infty dj \int_j^\infty df \int_f^\infty \mathcal{E}_{AB}(r)rdr \quad (12.4.7)$$

and $z = z_{AB}$ is the perpendicular distance between the plane going through the average centers of the molecules in surface A to the corresponding plane in surface B. The energy of separation is then just the energy of adhesion, which is related to the surface energies by (12.4.1) and therefore

$$E_A^\sigma + E_B^\sigma - E_{AB}^\sigma = n_A n_B K_{AB}(z_{AB}) \quad (12.4.8)$$

If A and B are the same phase, the energy of cohesion becomes the energy of surface cohesion and instead of (12.4.8) we have

$$2E_A^\sigma = n_A^2 K_{AA}(z_{AA}) \tag{12.4.9}$$

for phase A and

$$2E_B^\sigma = n_B^2 K_{BB}(z_{BB}) \tag{12.4.10}$$

for phase B. The Ks are integrals just like (12.4.7) except that they contain the intermolecular potentials and distances for a single phase.

Equations (12.4.8)–(12.4.10) provide a relationship between the interfacial and surface energies in terms of intermolecular distances and potentials. To get this, just take the ratio of the energy of adhesion to the geometric mean of the energies of cohesion and solve for the interfacial energy to get

$$E_{AB}^\sigma = E_A^\sigma + E_B^\sigma - 2\Psi\sqrt{E_A^\sigma E_B^\sigma} \tag{12.4.11}$$

with

$$\Psi \equiv \frac{K_{AB}(z_{AB})}{\sqrt{K_A(z_A) K_B(z_B)}} \tag{12.4.12}$$

The interfacial and surface free energies of adhesion can be treated similarly by remembering that the free energy of adhesion is the work required to separate the two surfaces to infinity. Then the force between the two phases, across a plane parallel to the two surfaces, is obtained in the same way as the energy of adhesion except that the intermolecular potential is replaced by the force of attraction between two molecules in a direction perpendicular to the interface. This is $-\cos\theta(\partial \mathcal{E}_{AB}/\partial r) = -(f/r)(\partial \mathcal{E}_{AB}/\partial r)$.

To obtain the free energy of adhesion, the work required to separate the two phases from z to infinity must be calculated. This is just the integral of the force from z to infinity, so the free energy of adhesion is given by

$$G_{\text{adh}}(AB) = 2\pi \int_{z_{AB}}^\infty dx \int_x^\infty n_B(j) dj \int_j^\infty f df \int_f^\infty n_A(r) \frac{\partial \mathcal{E}_{AB}}{\partial r} dr \tag{12.4.13}$$

Equations for the free energies that are similar to those above hold for the energies, except that wherever surface energy appears it is replaced by the interfacial or surface tension, and the energy integrals are replaced by work integrals. The result corresponding to (12.4.11) is

$$\gamma_{AB} = \gamma_A + \gamma_B - 2\Phi\sqrt{\gamma_A \gamma_B} \tag{12.4.14}$$

with

$$\Phi \equiv \frac{L_{AB}(z_{AB})}{\sqrt{L_A(z_A) L_B(z_B)}} \tag{12.4.15}$$

and

$$L_{AB}(z_{AB}) = 2\pi \int_z^\infty dx \int_x^\infty dj \int_j^\infty f df \int_f^\infty \frac{\partial \mathcal{E}_{AB}}{\partial r} dr \tag{12.4.16}$$

$L_A(z_{AA})$ and $L_B(z_{BB})$ have the same form as (12.4.16) except that the intermolecular force and distance between planes are for AA and BB interactions, respectively.

The utility of (12.4.11) and (12.4.14) lies in the fact that, for phases that are similar, both Ψ and Φ can be expected to be close to unity, and if the intermolecular potential is known, it can be calculated explicitly. The results are particularly simple for inverse power potentials. For a $6 - m$ potential of the form

$$\mathcal{E}_{AB}(r) = -\frac{A_{AB}}{r^6} + \frac{B_{AB}}{r^m}, \quad m > 6 \qquad (12.4.17)$$

the requirement that the force at equilibrium be zero yields a relation between attractive and repulsive constants given by

$$B_{AB} = \frac{6 A_{AB} d_{AB}^{m-6}}{m} \qquad (12.4.18)$$

d_{AB} being the equilibrium distance between an A and a B molecule, so the intermolecular potential and force can be written as

$$\mathcal{E}(r) = -A_{AB}\left(\frac{1}{r^6} - \frac{6 d_{AB}^{m-6}}{m r^m}\right) \qquad (12.4.19)$$

$$\frac{\partial \mathcal{E}}{\partial r} = 6 A_{AB}\left(\frac{1}{r^7} - \frac{d_{AB}^{m-6}}{r^{m+1}}\right) \qquad (12.4.20)$$

Put these in (12.4.7) and (12.4.16) and perform the integrations. Also, make the assumption that $z_{AB} = d_{AB}$, $z_{AA} = d_{AA}$, and $z_{BB} = d_{BB}$. That is, we assume that the equilibrium distance of approach between two planes of molecules is the same as the equilibrium distance of approach of two isolated molecules. Then

$$K_{AB}(d_{AB}) = \frac{2\pi A_{AB}}{d_{AB}^3} K_E \qquad (12.4.21)$$

$$L_{AB}(d_{AB}) = \frac{12\pi A_{AB}}{d_{AB}^3} K_F \qquad (12.4.22)$$

with expressions for $K_A(d_{AA})$, $K_B(d_{BB})$, $L_A(d_{AA})$, and $L_B(d_{BB})$ that are similar. K_E and K_F are constants that are the same for the three interfaces $A|A$, $B|B$, and $A|B$, so if ratios are taken according to (12.4.12) and (12.4.15), the constants drop out and two parameters can be defined:

$$\Psi \equiv \frac{A_{AB}}{\sqrt{A_{AA} A_{BB}}} \frac{d_{AA}^2 d_{BB}^2}{d_{AB}^4} \qquad (12.4.23)$$

$$\Phi \equiv \frac{A_{AB}}{\sqrt{A_{AA} A_{BB}}} \frac{d_{AA}^3 d_{BB}^3}{d_{AB}^6} \qquad (12.4.24)$$

This development displays the origin of the interfacial tension in the intermolecular forces. It is not sufficiently accurate to yield good numerical results for the interfacial energies or tensions separately (except in special cases such as the surface tension of nonpolar spherical molecules whose intermolecular energy is accurately described by a simple inverse power law and whose pair distribution functions are known.) However, it has been quite successful in describing the relation between interfacial and surface tensions by equation (12.4.14) because taking ratios often cancels out deficiencies in the theory. For nonpolar liquids of spherical molecules whose sizes are not too different, $\Phi \approx 1$. Values of Φ for such systems can be computed from (12.4.24). If the intermolecular attractive constants are not known, their ratio in (12.4.24) is taken to be unity and the ratio containing the intermolecular distances is calculated by assuming that the distance of closest approach for unlike molecules is the arithmetic mean of those for the like molecules. This is a hard sphere model for the diameters and is rather good for spherical molecules.

12.5 Critical point and critical exponent for surface tension

At the critical point, the interface between the liquid and vapor phases of a liquid vanishes, so the interfacial energy and interfacial tension will both go to zero. The above theory can be used to relate this to the liquid and vapor densities and thereby get the critical exponent for interfacial tension from that for the fluid density. Start with equation (12.4.6) as applied to the liquid-vapor interface of a one-component system and solve for the interfacial energy. That is,

$$E_N^\sigma = E_l^\sigma + E_v^\sigma - n_l n_v K_N(z_N) \tag{12.5.1}$$

Note that the subscripts now refer to different phases of the same material, each phase being a monomolecular system with the same intermolecular potential within and across the phases. Remember that the surface energy is one half the energy of surface cohesion, so (12.5.1) can be written as

$$E_N^\sigma = \frac{1}{2}[n_l^2 K_{ll}(z_{ll}) + n_v^2 K_w(z_w)] - n_l n_v K_N(z_N) \tag{12.5.2}$$

If the K's are all assumed to be the same, thereby ignoring the differences in their dependece on molecular diameters, (12.5.2) can be written as

$$E_N^\sigma = \frac{1}{2}K(n_l - n_v)^2 \tag{12.5.3}$$

The interfacial energy is therefore proportional to the square of the difference in densities between the liquid and vapor phase.

From equation (11.3.16), the critical exponent for the fluid density in mean field theory is 1/2. More generally, it is written as β (which experiment and more accurate theory gives as close to 1/3), so the interfacial energy approaches the critical point with a critical exponent equal to twice that of the fluid density transition and of magnetization. That is,

$$E_N^\sigma = 2K(n_l - n_v)^2 \propto (\tilde{T} - 1)^{2\beta} \tag{12.5.4}$$

The critical exponent for the interfacial tension is easily obtained by starting from the thermodynamic relation

$$\left(\frac{\partial (A/T)}{\partial T}\right)_V = -\frac{U}{T^2} \tag{12.5.5}$$

All thermodynamic relations apply to interfacial quantities, so let us use (12.5.5) for the surface tension near the critical point and write

$$\left(\frac{\partial(\gamma/T)}{\partial T}\right)_V = -\frac{E_N^\sigma}{T^2} \propto \frac{(\tilde{T}-1)^{2\beta}}{T^2} \qquad (12.5.6)$$

For a temperature near the critical point (12.5.6) can be integrated (keeping the volume of the system constant) to give

$$\gamma \propto T \int_T^{T_c} \frac{(\tilde{T}-1)^{2\beta}}{T^2} d\tilde{T} \qquad (12.5.7)$$

Since the temperature T is very close to the critical temperature, it can be taken equal to T_c, so

$$\gamma \propto \frac{1}{T_c} \int_T^{T_c} (\tilde{T}-1)^{2\beta} d\tilde{T} \propto (\tilde{T}-1)^{2\beta+1} \qquad (12.5.8)$$

For the mean field value of $\beta = 1/2$ the critical exponent for the surface tension is 2, but for the more accurate value of $\beta = 1/3$ it is 1.67.

At first glance, it would seem that the critical exponent for the surface tension could be obtained in exactly the same way as for the surface energy. That is, instead of starting with (12.4.6), simply start with the analogous equation for interfacial tensions. These include integrals that are again functions of the product of densities, so the critical exponent would be the same as for the energy. However, these integrals contain the force between two molecules rather than the energy, and they represent a process in which two phases are separated to infinity. Applying this process to a vapor means that a fictitious wall must be constructed to contain the vapor as it is removed from the liquid surface. The calculation of the integrals for vapor phases is therefore far from straightforward, and the calculation of the surface tension from thermodynamic relations is preferred.[4] This difficulty does not arise in calculating the energy integrals because all that is required in this case is to add up interactions in phases that are in contact.

An examination of equations (12.5.4) and (12.5.8) shows that the surface tension is related to the liquid and vapor densities by

$$\gamma \propto (n_l - n_v)^{(2\beta+1)/\beta} \qquad (12.5.9)$$

or,

$$\frac{\gamma^{\beta/2\beta+1}}{(n_l - n_v)} = \text{constant} \qquad (12.5.10)$$

The constant is independent of temperature, but is different for different substances. If the mean field value of 1/2 is used for β, the exponent in (12.5.10) becomes 1/4 and the result is Mcleod's equation, whereas if we take $\beta = 1/3$, the exponent is 1/5.

Although (12.5.10) grew out of an analysis near the critical point, it is found to hold even at temperatures far from the critical point for many liquids. Because of this, an interesting quantity can be defined. Let us neglect the vapor density since this is always at least several orders of magnitude lower than the liquid density at ordinary temperatures and pressures, and replace the number density of the liquid by the molecular volume v. Call the result P_{or}. That is, if we take $\beta = 1/2$,

$$P_{or} \equiv \nu\gamma^{1/4} \qquad (12.5.11)$$

P_{or} is called the *parachor*, and it is of interest because it is an additive molecular property in the sense that the parachor of a molecular liquid is the sum of parachors of its constituents. This means that a set of numbers can be found for each atom, each group, and each type of bond (carbon-carbon double bond, triple bond, six-membered carbon ring) that add up to give the parachor of any molecule. "Group parachors" can be constructed from the atomic parachors that correctly reproduce the molecular parachors for a very large number of substances. This is useful in estimating the surface tension. As an example, the surface tension of polymers, both in the solid and in the melt, is important for polymer uses as well as for processing. The additivity of the parachor has been found to be valid for a large number of polymer systems.[5]

Altough the parachor was originally defined as in equation (12.5.11), the theory of critical exponents shows that it would be better to use an exponent of 1/5 and define a parachor by

$$P'_{or} \equiv \nu\gamma^{1/5} \qquad (12.5.12)$$

The fact that experimental data have been successively analyzed with an exponent of 1/4 does not invalidate the theory leading to an exponent of 1/5 because the ratio of the two parachors is proportional to $\gamma^{1/20}$, and this varies slowly among liquid systems because most of them have surface tensions that are not too different. That is, the experimental data cannot distinguish between the two exponents.

The derivation of (12.5.9) is based on that of Fowler and Guggenheim (1956) and leads to a critical temperature exponent of 1.67 for the surface tension. But this derivation did not treat the dependence of the surface energies on density accurately. A more accurate treatment (Rowlinson and Widom, 1982) yields a critical exponent of 1.26.

12.6 Monolayer adsorption: Langmuir isotherm

Let us adopt the simplest model of a solid surface, which is that it consists of atomic sites all of which are equivalent, and consider a system in which this surface is in equilibrium with a pure ideal gas.

Because of the existence of interatomic forces, some of the gas molecules will be adsorbed on the surface. The simplest result of this interaction would be that each surface site could adsorb one molecule, that when a molecule is adsorbed it could not move laterally over the surface, and that the maximum number of molecules that could be adsorbed equals the number of surface sites. These conditions are satisfied by the phenomenon of chemisorption at low pressures, for which the energy of adsorption is very high because it is the result of the formation of a chemical bond between the surface atoms and the adsorbate molecules. The low pressure and the high energy of adsorption make the interaction between adsorbed molecules and molecules in the gas phase sufficiently small that multilayer adsorption is negligible. Also, the lateral interactions among molecules can be neglected since these are assumed to be of the van der Waals type and are small relative to the energy of interaction with the surface.

Our task is to get the fraction of monolayer coverage as a function of pressure. To do this, assume that the energy of an adsorbed molecule relative to a molecule at rest in the gas phase is the same for all adsorbed molecules. Further assume that the molecules are bound to the surface as three-dimensional harmonic oscillators, but with a vertical frequency that differs from the two equal lateral frequencies. The partition function for a molecule on the surface is therefore

$$z_s = z_p^2 z_v e^{W_0/kT} \tag{12.6.1}$$

where z_p and z_v are the vibrational partition functions for the lateral and vertical vibrations, respectively, and $-W_0$ is the energy of adsorption including the zero point energy of the vibrations. The adsorption energy is negative, so W_0 is a positive quantity.

The vibrational partition functions are

$$z_p = \frac{1}{1 - e^{-h\nu_p/kT}} \tag{12.6.2}$$

$$z_v = \frac{1}{1 - e^{-h\nu_v/kT}} \tag{12.6.3}$$

ν_p and ν_v being the vibration frequencies in the directions parallel and perpendicular to the surface. These expressions are obvious special cases of equation (4.3.12). Let the number of adsorbed molecules be N^σ and let M equal the number of adsorption sites on the crystal surface. The partition function Z_a for the adsorbed phase is then the number of ways of distributing N^σ molecules among M sites (with a site either being empty or containing one molecule) times the product of partition functions given by (12.6.1) for all adsorbed molecules:

$$Z_a = \frac{M!}{N^\sigma!(M - N^\sigma)!} \left(z_p^2 z_v e^{W_0/kT}\right)^{N^\sigma} \tag{12.6.4}$$

If the molecule has internal degrees of freedom, such as intramolecular vibrations, these can be included simply by multiplying (12.6.4) by the appropriate partition functions.

As usual, the Helmholtz free energy is $A = -kT\ln Z$, so if we take the logarithm of (12.6.4), use Stirling's approximation on the factorials, and multiply by $-kT$, we get the Helmholtz free energy of the adsorbed phase as

$$A_a = -N^\sigma W_o - N^\sigma kT \ln(z_p^2 z_v) + M\ln M$$
$$- kT[N^\sigma \ln N^\sigma - (M - N^\sigma)\ln(M - N^\sigma)] \tag{12.6.5}$$

The chemical potential is obtained from (12.6.5) by differentiation with respect to N^σ:

$$\mu = \left(\frac{\partial A_a}{\partial N^\sigma}\right)_{T,V,\sigma} = -W_o - kT\ln(z_p^2 z_v) - kT\ln\left(\frac{M - N^\sigma}{N^\sigma}\right) \tag{12.6.6}$$

At equilibrium, the chemical potential of the adsorbate must equal that of the gas phase. This is just the Gibbs free energy per molecule, which from equation (3.4.19) is

$$\mu g = kT\left[\ln\left(\frac{N_g}{V_g}\right) - \frac{3}{2}\ln\left(\frac{2\pi mkT}{h^2}\right)\right] \tag{12.6.7}$$

Equating (12.6.6) and (12.6.7), using the ideal gas law $PV_g = N_g kT$, and solving for the fractional coverage N^σ/M gives the Langmuir adsorption isotherm:

$$\frac{N^\sigma}{M} = \frac{P}{P_0 + P} \qquad (12.6.8)$$

The same result can be obtained from the lattice gas theory of appendix 7 by taking the cells in equation (A.7.13) to represent sites on the surface and assume they all have the same binding energy, ignoring all cell-cell interaction. The vibrational contribution to (A.7.13) is represented by Einstein modes perpendicular and parallel to the surface.

P_0 is a constant that depends only on temperature and is given by

$$P_0 = e^{-W_0/kT} \left(\frac{h^2}{2\pi m kT} \right)^{3/2} \frac{z_p^2 z_v}{kT} \qquad (12.6.9)$$

From (12.6.8), P_0 is the pressure at which half the sites on the surface are occupied. At low pressures, $P \ll P_0$ and the fractional coverage is proportional to the pressure. At high pressure, $P \gg P_0$ and the fractional coverage approaches unity. The form of the Langmuir isotherm is shown in figure 12.5. Experimental data for chemisorption of simple gases on solids are well represented by the Langmuir isotherm.

The extension of this model to the adsorption of several species is trivial because we neglect the interactions among molecules on the surface as well as in the gas phase. The total partition function is then the product of partition

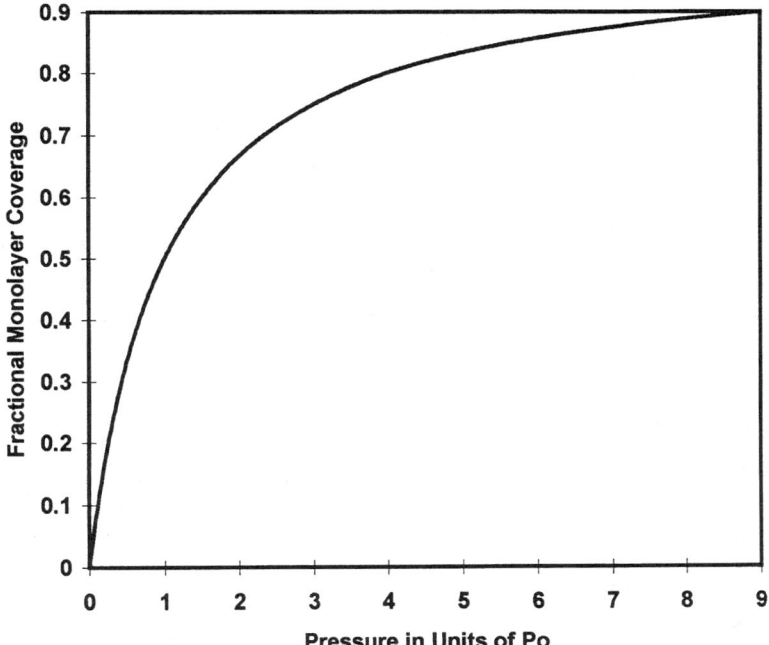

Figure 12.5. The Langmuir adsorption isotherm.

functions for all the species, the total Helmholtz free energy is the sum of the individual free energies, and the total pressure is the sum of the partial pressures. The adsorption isotherm for each species is just like the Langmuir isotherm except that the pressure is replaced by the partial pressure of that species. If N_i^σ is the number of molecules of type i adsorbed on the surface, then the total number of adsorbed molecules is

$$N = \sum_i N_i^\sigma \tag{12.6.10}$$

and the relation between the fraction of the sites covered by molecules of type i and the partial pressure of species i is

$$\frac{N_i^\sigma}{M} = \frac{P_i}{P_{i0} + P_i} \tag{12.6.11}$$

where P_{io} is just like (12.6.9) except that it refers to the molecule of species i.

Because of the assumed independence of the molecules, if the surface has several different kinds of sites they can be treated independently. A surface, for example, may contain steps so that a molecule that adsorbs at the corner of a step has a different energy of adsorption and vibrational energy than a molecule adsorbed at a site on a flat region. Each set of sites will then yield an isotherm just like that of (12.6.8) except that N^σ becomes N_j^σ, the number of molecules adsorbed on the sites of type j, and M becomes M_j, the number of such sites.

It often happens that the adsorbed species is not the same molecule as in the gas phase because the strong interaction with the surface induces chemical reaction. This is the case with many catalysts. An important example is the adsorption of hydrogen on metals. In the gas phase, hydrogen has the molecular form H_2 but on the surface it usually has the atomic form H. The derivation leading to (12.6.8) must then be modified because the chemical potentials given by (12.6.6) and (12.6.7) must refer to the same amount of material. Since each molecule contains two atoms, the chemical potential of the gas phase given by (12.6.7) must be equated to twice that for the adsorbed phase given by (12.6.6). Doing this and solving for $N_H^\sigma/(M - N_H^\sigma)$ gives

$$\frac{N_H^\sigma}{M - N_H^\sigma} = \left(\frac{P}{P_h}\right)^2 \tag{12.6.12}$$

where N_H^σ is the number of hydrogen atoms on the surface and the constant P_h is given by a form identical to (12.6.9) with all quantities referring to the hydrogen molecule:

$$P_h = e^{-W_0/2kT} \left(\frac{h^2}{2\pi mkT}\right)^{3/4} \left(\frac{z_p^2 z_v}{kT}\right)^{1/2} \tag{12.6.13}$$

The mass, of course, is that of the hydrogen molecule. Note that (12.6.13) is just the square root of (12.6.9) because of the factor of 1/2 in the exponent of (12.6.13), which is a direct result of the hydrogen molecule's dissociation into atoms.

Solving (12.6.12) for N_H/M gives the fractional coverage of the M sites with hydrogen atoms as

$$\frac{N_H}{M} = \frac{P^2}{P_h^2 - P^2} \tag{12.6.14}$$

The dissociation of the molecule into atoms on the surface has a strong effect on the form of the adsorption isotherm.

12.7 Monolayer adsorption: mobile layer

If the molecules adsorbed on a surface are not tightly bound, they can move laterally over the surface with two degrees of translational freedom. This is the case when the binding arises from the physical van der Waals interactions rather than from chemical reaction. The binding is then not strong enough to localize the molecules at specific sites. But at low pressures a monolayer model can still be used, so it is of interest to determine the adsorption isotherm for a monolayer in equilibrium with its vapor.

The simplest model for a mobile layer is that of the two-dimensional ideal gas given in chapter 2. To get the adsorption isotherm, the partition function for the adsorbed phase is needed, and this can be obtained from the ideal gas assumption. This means that, in the gas and surface phases, the total partition functions are just products of molecular partition functions corrected, of course, for the indistinguishability of the molecules in each phase as given in section 3.9.

For a one-component ideal gas, the partition function is just the product of molecular partition functions for three degrees of translation, corrected for indistinguishability:

$$Z_g = \frac{z_g^{N_g}}{N_g!} \tag{12.7.1}$$

N_g is the number of molecules in the gas phase, and z_g is the partition function for the molecule, which is given by

$$Z_g = V\left(\frac{2\pi mkT}{h^2}\right)^{3/2} \tag{12.7.2}$$

For the surface phase, the partition function is a product of two-dimensional partition functions, but it also contains the energy of adsorption $-W$. Thus, we have

$$Z_\sigma = \frac{z_\sigma^{N^\sigma}}{N_\sigma!} \tag{12.7.3}$$

N^σ is the number of molecules in the surface phase, and z_σ is the partition function for a molecule on the surface, which is given by

$$z_\sigma = e^{W/kT}\sigma\left(\frac{2\pi mkT}{h^2}\right) \tag{12.7.4}$$

Now go through a familiar drill. Write down the Helmholtz free energy for the surface phase from $A_\sigma = -kT \ln Z_\sigma$, differentiate with respect to the number of surface molecules to get the chemical potential, and set the resulting chemical potential of the surface phase equal to that for the gas phase as required by equilibrium. Then solve for the number of molecules on the surface (using Stirling's approximation and the ideal gas law). The result is

$$\frac{N^\sigma}{\sigma} = \frac{P}{kT} e^{W/kT} \sqrt{\frac{h^2}{2\pi mkT}} \qquad (12.7.5)$$

This isotherm is very different from the Langmuir isotherm. The number of molecules adsorbed is proportional to pressure at all pressures, not only for $P \to 0$. In fact, (12.7.5) imposes no limit on the amount that can be adsorbed. The difference between (12.7.5) and the Langmuir isotherm is a stark example of the strong effect of indistinguishability on physical results.

In practice, the mobile monolayer has limited applicability to real systems and can be used only at very low pressures. A major reason for this is that, for the monolayer to be mobile, the interaction of the molecules with the surface must be weak. In fact, this interaction is of the same magnitude as that between gas molecules when they come into contact. A molecule approaching the surface can therefore bind to a molecule already on the surface with a probability comparable to that for binding at the surface, thereby giving rise to multilayer adsorption. Only at low coverage, when the probability is controlled by the amount of available surface rather than by the binding energies, can the mobile monolayer assumption be expected to hold accurately.

12.8 Multilayer adsorption: BET isotherm

In multilayer adsorption, some of the sites on a surface are empty and others have only one molecule on them. But there are also sites for which there are two or more molecules stacked on top of each other, and there are patches of the surface that have multilayer islands on them. This poses a difficult statistical mechanical problem because the interaction energies of the molecules with each other in both the vertical and lateral directions vary with their positions in the layers. The problem is enormously simplified if the lateral interactions are ignored, and we will do this. Taking the vertical variations into account still leaves too many unknown parameters, but at least a formal solution can be obtained. However, the usual assumptions will be adopted that lead to the Brunauer, Emmett, and Teller (BET) isotherm right away. These assumptions state that the first layer interacts with the surface site with an energy that is characteristic of the interaction between the solid and a gas molecule. But we take all molecules on top of those in the first layer to have the same interaction with molecules beneath them, and this interaction is characteristic of the molecular interactions in the liquid phase. Our system is then a surface containing M equivalent sites in equilibrium with a gas of a single species. The sites on the surface can have any number of molecules stacked on them, and since the energies of the molecules are independent of each other, each molecule can be assigned a molecular partition function. The molecules adsorbed directly on the surface have a partition function of the form

$$z_1 = e^{E_1/kT} z_1(\nu_1) \qquad (12.8.1)$$

while all other molecules have partition functions given by

$$z_l = e^{E_L/kT} z_L(\nu_L) \qquad (12.8.2)$$

In these equations, E_1 and E_L are the interaction energies of molecules on the surface layer and in all other layers, respectively, and the vibrational contributions to the partition functions are designated by $z_1(\nu_1)$ and $z_L(\nu_L)$.

If lateral interactions are ignored, then the existence of adjacent patches of adsorbate is irrelevant and our system is one in which only the stacking of molecules on a site need to be considered. In fact, the BET assumptions result in two classes of molecules (those that are adsorbed in the first layer and those that are adsorbed in all other layers), and this greatly simplifies the statistical count for the number of complexions. If N_1 is the number of molecules in the first layer, then the number of ways of distributing N_1 indistinguishable molecules among M sites, such that there is only one molecule per site, is just the Fermi–Dirac statistical count and is given by

$$W_1 = \frac{M!}{N_1!(M-N_1)!} \tag{12.8.3}$$

Let the number of molecules in all layers but the first be called N_L, so that the total number of adsorbed molecules is

$$N^\sigma = N_1 + N_L \tag{12.8.4}$$

Now let us count the number of ways of putting down the N_L molecules on the surface. These molecules all have the same partition function, and any number of them can go on top of those in the first layer, so the number of complexions is the number of ways of putting N_L indistinguishable molecules in N_1 boxes with no restrictions on the number of molecules per box. This is just the Bose–Einstein statistical count. Therefore,

$$W_L = \frac{(N_L + N_1 - 1)!}{N_L!(N_1 - 1)!} \tag{12.8.5}$$

The canonical partition function is obtained from the above equations first by multiplying (12.8.3) and (12.8.5) together and multiplying the result by the molecular partition functions (12.8.1) and (12.8.2), each raised to the power of the number of molecules to which they refer. The result is then summed over all sets of integers N_1 and N_L consistent with a given N^σ and M. There are two conditions that must be satisfied when performing this sum. The first is given by (12.8.4), and the second is that the number of molecules in the first layer must be equal to or less than the number of sites. These restrictions on N_1 and N_L make it hard to compute this sum, but they can be removed by using the grand canonical partition function, thereby greatly simplifying the calculation. The grand partition function is just the canonical partition function multiplied by the Boltzmann factor of the chemical potential and then summed over all values of N^σ. That is,

$$Q = \sum_{N^\sigma=0}^{\infty} e^{N\mu/kT} W_1 W_L z_1^{N_1} z_L^{N_L} \tag{12.8.6}$$

Since summing over all N^σ means that N_1 can take all values from 0 to M, and N_L can take all values from 0 to infinity, (12.8.6) is

$$Q = \sum_{N_1=0}^{M} \frac{M!(z_1\lambda)^{N_1}}{N_1!(M-N_1)!(N_1-1)!} \sum_{N_L=0}^{\infty} \frac{(N_L+N_1-1)!(z_1\lambda)^{N_L}}{N_L!} \tag{12.8.7}$$

where (12.8.3) and (12.8.5) have been used for the number of complexions and absolute activity is defined as usual by

342 STATISTICAL MECHANICS OF SOLIDS

$$\lambda = e^{\mu/KT} \tag{12.8.8}$$

From equation (A.4.6),

$$\sum_{j=0}^{\infty} \frac{(j+C)!}{j!} x^j = \frac{C!}{(1-x)^{C+1}} \tag{12.8.9}$$

so the second sum in (12.8.7) is

$$\frac{(N_1-1)!}{(1-z_L\lambda)^{N_1}} \tag{12.8.10}$$

and putting this in (12.8.7) gives

$$Q = \sum_{N_1=0}^{M} \frac{M!(z_1\lambda)^{N_1}}{N_1!(M-N_1)!(1-z_L\lambda)^{N_1}} \tag{12.8.11}$$

But this sum is just the binomial expansion of $(1+B)^M$ where B is

$$B = \frac{z_1\lambda}{1-z_L\lambda} \tag{12.8.12}$$

so the grand partition function reduces to

$$Q = (1+B)^M \tag{12.8.13}$$

The adsorption isotherm is now easily obtained since the statistical mechanical average for the total number of molecules in the adsorbed phase is (see section 2.15)

$$\overline{N^\sigma} = kT \left(\frac{\partial \ln Q}{\partial \mu} \right)_{T,M} \tag{12.8.14}$$

The differentiation is readily carried out with the aid of equations (12.8.8), (12.8.12), and (12.8.13), with the result that

$$n^\sigma \equiv \frac{\overline{N^\sigma}}{M} = \frac{cx}{(1-x)(1-x+cx)} \tag{12.8.15}$$

where n^σ is the number of adsorbed molecules per surface site and x and c are given by

$$x = z_L\lambda, \quad cx = z_1\lambda \tag{12.8.16}$$

Equation (12.8.15) can be expressed in terms of the gas pressure because the adsorbed phase is in equilibrium with the gas phase, so the absolute activity in (12.8.16) is the same as that of the gas. Taking the gas to be ideal, this is

$$\lambda = \frac{P}{kT} \left(\frac{h^2}{2\pi mkT} \right)^{3/2} = AP \tag{12.8.17}$$

A being defined as the coefficient of the pressure in (12.8.17). Putting this in (12.8.15) gives the BET adsorption isotherm as

$$n^\sigma = \frac{A_1 P}{(1 - A_L P)(1 + A_1 P - A_L P)} \qquad (12.8.18)$$

the constants A_1 and A_L being given by

$$A_1 = z_1 A, \quad A_L = z_L A \qquad (12.8.19)$$

The form of equation (12.8.18) is shown in figure 12.6 along with experimental data for the adsorption of nitrogen on $BaSO_4$.[6] At low pressures, the curve is similar to that of the Langmuir isotherm, while at high pressures the amount adsorbed increases very rapidly with pressure. In fact, equation (12.8.18) shows that the amount adsorbed goes to infinity as $A_L P$ goes to unity. But this is just the phenomenon of condensation, so the pressure P_0 for which $A_L P_0 = 1$ is just the saturation vapor pressure. That is, the pressure at which the liquid condenses is given by

$$P_0 = \frac{1}{A_L} \qquad (12.8.20)$$

and therefore (12.8.18) can be written as

$$n^\sigma = \frac{cP}{(P_0 - P)[1 + (c-1)P/P_0]} \qquad (12.8.21)$$

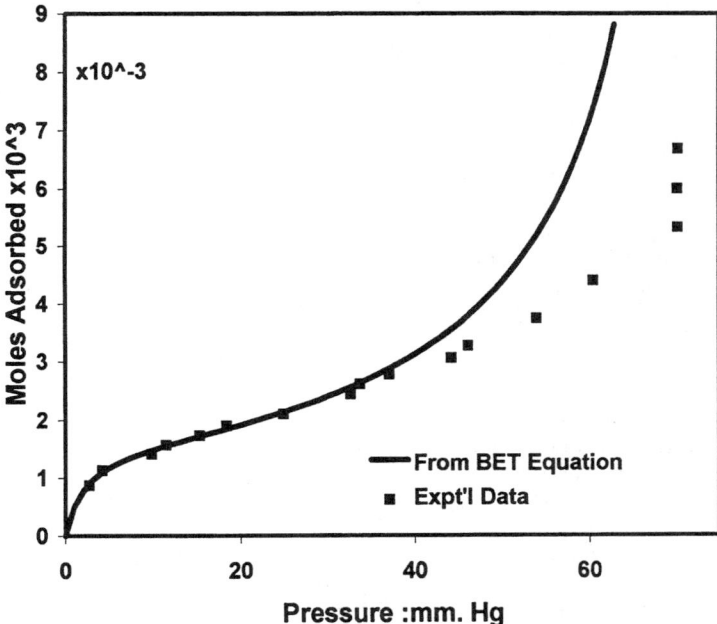

Figure 12.6. Adsorption isotherm for nitrogen on barium sulfate at 77.5K: comparison of experimental data with BET theory.

which explicitly shows that there is condensation of the liquid when the saturation vapor pressure is reached.

While we were after a theory of multilayer adsorption, we have also arrived at a theory of the vapor liquid transition and shown that condensation can be regarded as the final result of multilayer adsorption. The constant c is related to condensation because, within the assumptions of the theory, it is defined by the free energy difference between a molecule in the liquid and a molecule in the first monolayer, as can be seen by writing it out in terms of its definition by equations (12.8.1), (12.8.2), and (12.8.16), which give

$$c = e^{(E_1-E_L)/kT} \frac{z_L(v_L)}{z_l(v_l)}$$
$$= e^{(A_L-A_l)/kT} \quad (12.8.22)$$

$(A_L - A_1)$ being the free energy difference per molecule between the liquid phase and the first adsorbed layer. At low pressures, P can be neglected relative to P_0 in the first factor in the denominator of (12.8.21), and then the equation reduces to the Langmuir monolayer form.

Equation (12.8.21) is often written in the following linear form so that a comparison with experiment can be made by measuring the amount of gas adsorbed as a function of pressure at a constant temperature:

$$\frac{P}{n^\sigma(P_0-P)} = \frac{1}{c} + \frac{(c-1)}{c}\frac{P}{P_0} \quad (12.8.23)$$

Equation (12.8.23) can be used to measure the specific surface area of powders from experimental adsorption isotherms. This is easily done because, if s and t are the slope and intercept, respectively, then it follows that

$$M = \frac{1}{s+t} \quad (12.8.24)$$

But this is just the number of sites, which is the number of molecules adsorbed for a full monolayer, so if the area occupied by a molecule sitting on the surface is known, the area of the powder can be determined. The area of the molecule sitting on the surface is calculated from the mean interatomic distances in the liquid phase, which are obtained from liquid density data and the analysis of adsorption on a variety of surfaces. This was done for the data of Morabito (see note 6), with a resulting surface area of $4.152\,m^2/gm$ for the $BaSO_4$ powder. The solid curve in figure 12.6 was computed from the values of $M = 0.00158$ moles and $c = 36.64$. The agreement of the theory with experiment is quite good up to a pressure given by $P/P_0 \approx 1/3$. This is typical of the results for nonreacting gases on inert powders. The linear plot used to determine the surface area and isotherm constants is taken only for $P/P_0 < 1/3$.

A great many experimental isotherms have been measured for the adsorption of gases such as nitrogen, oxygen, and the rare gases on finely divided powders of solids such as titanium dioxide, barium sulfate, and lead chromate. Plots of the results according to (12.8.23) display excellent linearity in pressure ranges corresponding to $P/P_0 > 0.05$ and $P/P_0 < 0.35$. The failure of the theory at very low pressures arises from surface heterogeneity. There are a small number of sites for which the deviation of the energy of adsorption from the average is so large that the assumption that all sites have the same energy is completely wrong. The theory fails at high pressures because of the assumption that all layers above the first have the same adsorption energy. This is obviously incorrect. The second layer still feels the influence of the surface, and this influence falls off only gradually with distance from the surface. The energy of adsorption decreases with each layer until it becomes equal to the energy of condensation. There is experimental evidence that it takes about four or five layers before the energy of adsorption levels off to that for condensation.

On an atomic scale, solid surfaces are not uniform. The surfaces of even the most carefully prepared single crystals contain steps and ridges and can expose more than one crystal plane orientation. With care, surfaces can be prepared in which the atoms form a two-dimensional periodic array, but these usually contain rows and furrows of atoms that have different local surroundings, so there are differences of energy among the various kinds of sites. The energy of a solid surface therefore varies from point to point on any solid surface, and the degree of this variation is of great intrinsic and technological interest. This inhomogeneity is much more extreme for solid powders that are prepared by grinding or deposition from the vapor or from solution.

12.9 Segregation of impurities at interfaces

Consider an interface, which might be a free surface; the boundary between two fluids, a fluid and a solid, or two different solids; or the grain boundary in a single phase. Let the system contain an impurity that, because of the difference in the atomic environment, will have a concentration at the interface that is different than that in the bulk.

To be specific, let the interface be a planar grain boundary in a metal that contains some impurity at a low concentration. The simplest model for segregation of the impurity to the grain boundary is that of the ideal lattice gas in which the boundary contains M sites and the bulk phase contains L sites, each of which can accommodate one impurity, and the impurity atoms do not interact with each other. The procedure consists of writing the formulas for the two-dimensional ideal lattice gas to represent the grain boundary and the three-dimensional ideal lattice gas to represent the bulk. Equating the chemical potentials in these two expressions yields the grain boundary concentration as a function of the bulk concentration.

Let the binding potential energies of the impurity to a cell in the bulk and in the grain boundary be $-w_B$ and $-w_G$, respectively. Also, assume that the atoms are Einstein oscillators. In the bulk all three degrees of freedom have the same frequency v_B, while in the grain boundary each atom has one vertical mode with frequency v_v and two modes parallel to the surface, each with frequency v_P. The energy for an atom in a cell of the bulk phase is therefore

$$v_B = -w_B + \frac{3}{2} h v_B \tag{12.9.1}$$

and for the grain boundary

$$v_G = -w_G + \frac{1}{2} h v_v + h v_P \tag{12.9.2}$$

Now write equation (A.7.13) for the bulk phase, taking all $v(k)$ to be the same and equal to v_B of equation (12.9.1). The result is

$$C_B \equiv \frac{N_I}{L} = \overline{e_B} = \frac{1}{1 + e^{-\beta(\mu_B + w_B - 3hv_B/2)}} \tag{12.9.3}$$

N_I being the total number of impurity atoms in the bulk phase such that $C_B = N_I/L$ is the probability that a cell in the bulk contains an impurity.

Similarly, the probability that a cell in the grain boundary is occupied is given by lattice gas theory as

$$C_G \equiv \frac{N_G}{L} = \overline{e_G} = \frac{1}{1 + e^{-\beta(\mu_G + w_G - h v_V/2 - h v_P)}} \tag{12.9.4}$$

C_B and C_G are the atomic concentrations in the bulk and boundary phases, respectively.

Solving (12.9.3) and (12.9.4) for the chemical potentials and using $\mu_G = \mu_B$ gives the concentration at the grain boundary relative to that in the bulk as

$$\frac{C_G}{1 - C_G} = \frac{C_B}{1 - C_B} e^{\beta W_S} \tag{12.9.5}$$

where W_S is the segregation energy defined by

$$W_S \equiv w_G - w_B + \frac{h}{2}(3v_B - 2v_P - v_V) \tag{12.9.6}$$

Equation (12.9.5) is the Mclean isotherm, which has been used to analyze data on grain boundary segregation with some success.

If the concentration of impurity in the bulk phase is small, then it can be neglected relative to unity in the denominator of the right-hand side of (12.9.5), and solving for the boundary concentration gives

$$C_G = \frac{C_B}{C_0 + C_B} \tag{12.9.7}$$

with

$$C_0 \equiv e^{-\beta W_S} \tag{12.9.8}$$

Equation (12.9.7) has the same form as the Langmuir isotherm given by equation (12.9.8), with the bulk concentration of impurity playing the role of the pressure. This form is observed for the segregation of oxygen to grain boundaries in molybdenum. However, data for the segregation of a number of impurities to grain boundaries in iron indicate that both lateral and multilayer interactions are important and the Langmuir–Mclean approach is too simple. The data have been analyzed by the same types of isotherms as those used for gas adsorption on solid surfaces. An interesting note is that in some systems, multilayer segregation at grain boundaries leads to second phase formation just as the BET isotherm leads to liquid condensation on a solid surface.[7]

Exercises

12.1 Assume that the surface tension of a dilute solution varies linearly with concentration according to $\gamma = 40 - 400c$, where the surface tension is in ergs/cm² and the concentration is in moles/cm³. Show that the excess surface concentration varies linearly with concentration and compute the proportionality constant at room temperature (300 K).

12.2 Provided that the concentration is not too low and the chain length is not too high, the surface tension of an aqueous solution of a fatty acid is a logarithmic function of the concentration. That is, $\gamma = a + b\ln c$, where b is about the same for all fatty acids and a is a constant specific for each acid. (This is Szyszkowski's rule, which is valid for fatty acids up to a chain length of six carbon atoms.) Prove that the amount of fatty acid adsorbed at the surface of the solution is approximately independent of the concentration of the solution and is about the same for all acids.

12.3 Consider a dilute protein solution in water that forms a protein layer on the surface having 10^{-6} gm/cm^2 of protein. At 20°C the surface tension of water in the absence of the protein is 72.00 ergs/cm^2 whereas with the protein monolayer the surface tension is 71.90 ergs/cm^2. Compute the molecular weight of the protein using the Gibbs adsorption isotherm for dilute solutions.

12.4 Assume that $\Phi = 1$ in the Girifalco–Good equation (12.4.14). If the spreading coefficient is zero, what is the relation between the two surface tensions γ_s and γ_l in a solid-liquid interface? (Approximate the interfacial tension between a condensed phase and the vapor by the surface tension of the condensed phase.)

12.5 Assume that $\Phi = 1$ in equation (12.4.14). What are the two possible values of the ratio γ_A/γ_B if the interfacial tension is the mean of the two surface tensions?

12.6 Using equation (12.4.14), derive Φ in terms of the ratio of the two surface tensions if $\gamma_{AB} = \gamma_B - \gamma_A$ (with $\gamma_B > \gamma_A$). This last equality is Antonoff's rule for two liquids that are partially soluble in each other. Note that Antonoff's rule states that the spreading coefficient is zero.

12.7 Show that the Lennard–Jones potential [equation (12.4.17) with $m = 12$] can be written as $\mathcal{E} = 4\varepsilon \, [(\sigma/r)^{12} - (\sigma/r)^6]$, where $2^{1/6} \sigma$ is the distance between molecules at which the energy of interaction is a minimum and $-\varepsilon$ is the minimum energy of interaction.

12.8 Consider a gas adsorbed on a heterogeneous surface that consists of three patches, each patch having a different energy of adsorption. Assuming immobile adsorbed molecules, derive the analog of the Langmuir adsorption isotherm theory for this system. Get equations for the total amount adsorbed, and for the ratios of the amount adsorbed on each patch, as a function of pressure.

Notes

1. Treatments of the thermodynamics of interfaces that include surface stresses and strains can be found in Zangwill (1988, p. 8) and Sutton and Balluffi (1995, p. 359).
2. Girifalco, L.A.; 1954; Ph.D thesis, University of Cincinnati.
3. Girifalco, L.A., and Robert Good; 1957; *Journal of Physical Chemistry;* vol. *61*, p. 904; Good, Robert, and L.A. Girifalco; 1960; *Journal Physical Chemistry;* vol. *64*, p. 561.
4. Fowler, R.H., and E.A. Guggenheim; 1956; p. 448.

5. Glasstone, Samuel; 1946; *Textbook of Physical Chemistry*; van Nostrand, New York; p. 526.
6. Morabito, Joseph M.; 1967; Ph.D. thesis, University of Pennsylvania.
7. A summary of both theory and data can be found in Hondros, E.D., and M.P. Seah; 1983; *Physical Metallurgy*; R. Cahn and P. Haasen, Eds.; North-Holland, Amsterdam; p. 855.

13

The Theory of Random Flight

13.1 Introduction

The theory of the random motion of a particle through space has a variety of uses. Two important applications are to molecular or atomic diffusion, and to the statistics of long chain molecules. To clarify the concept of random flight, consider a particle that, starting from a fixed point $\mathbf{R}_o = 0$, can move a distance r in any direction. After the first jump, the particle can again move in any direction, but with the same distance r. That is, the motion of the particle consists of a number of N sequential jumps in an arbitrary direction, but all having the same magnitude. The particle is said to have executed a random flight of N equal-sized steps that are defined by the set of vectors $\{\mathbf{r}\}_N \equiv \mathbf{r}_1, \mathbf{r}_2, \mathbf{r}_3, \ldots, \mathbf{r}_N$, each having a magnitude r, after which it has moved a distance $\mathbf{R}$.

Each step is called a jump, and the distance from the starting point after N jumps is called the *total displacement*. An example of a random flight in two dimensions is shown in figure 13.1. For a general random flight, the jump distances do not necessarily all have the same magnitude.

Two important properties of a random flight that arise in physical applications are the scalar distance $R = |\mathbf{R}|$ between the starting point and end point of the flight, and the probability $P_N(\mathbf{R})$ that a flight will result in the particle having traveled a specific distance $\mathbf{R}$, which will be called the *total displacement vector*. Clearly, these can only be computed if the probability $p_j(\mathbf{r}_j)$ that the particle will move through a vector $\mathbf{r}_j$ on the jth jump is specified. The random flight in one or two dimensions is called a *random walk*, and this term is often loosely applied to the three-dimensional random flight. The analogy between solid state diffusion and polymer chains is obvious from figure 13.1. In diffusion, each jump of the random flight represents the motion of an atom from one lattice position to another, while in the polymer each step represents the position of a monomer unit relative to a previous one in the chain. Each possible random flight for a given number of jumps is called a *conformation* for the N-step flight. The number of conformations is particularly important for polymer statistics since it gives the chain contribution to the entropy of polymer systems.

The meanings of the probabilities in the diffusion and the polymer are analogous but different. For the case of a diffusing particle, the probabilities can be defined either by the repeated flights of a single particle, or by a separate flight for a large number of separate particles. For both cases, each flight is governed by the same probability jumps. In the first definition, assume that the flight is repeated M times, each flight starting from the same origin and having

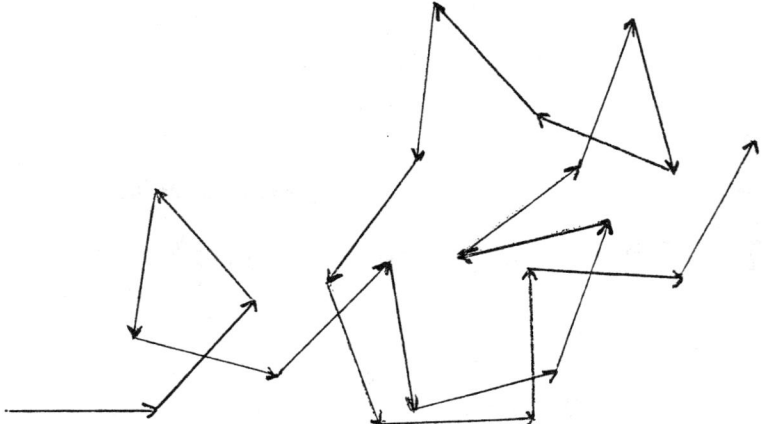

Figure 13.1. Two-dimensional random flight.

the same number of jumps. Then the probability $p_j(\mathbf{r}_j)$ is the limit of the ratio of the number of times the jth jump goes through the vector $\mathbf{r}_j$ to the total number of repetitions M, as $M \to \infty$. Similarly, $P_N(\mathbf{R})$ is the ratio of the number of flights for which the total displacement vector is $\mathbf{R}$ to the total number of flights M, as $M \to \infty$.

For the second definition, it is assumed that a large number of particles all start out under identical conditions and each executes a random flight of the same number of steps. Again, the probabilities are defined as ratios, but this time as ratios of the number of realizations of the vector $\mathbf{r}_j$ in the ensemble of flights to M, as $M \to \infty$, or of the ratio of the number of flights in the ensemble with displacement $\mathbf{R}$ to the total number of flights M, as $M \to \infty$ [for $P_N(\mathbf{R})$]. In analogy with the ergodic theorem, it is assumed that these two definitions yield the same results.

Completely analogous definitions hold for the flight representing polymer chains, but the definition based on an ensemble of chains is clearer and more useful. Note that when the probabilities of interest are used to compute statistical mechanical averages, they are ensemble quantities in the usual sense.

In the general case, it is not necessary that the individual jump probabilities be centrosymmetric or the same for all jumps, although only some simple cases in which the probabilities have convenient mathematical properties lead to results that can be easily calculated.

13.2 The mean square total displacement

Let the probability that, on the jth step, a particle will move through a vector $\mathbf{r}_j$ be denoted by $p_j(\mathbf{r}_j)$, and starting at the position $\mathbf{R} = 0$, let it execute a random flight through vectors $\mathbf{r}_1, \mathbf{r}_2, \mathbf{r}_3, \ldots, \mathbf{r}_N$. The distance traveled by the particle after N steps is given by

$$\mathbf{R} = \sum_{i=1}^{N} \mathbf{r}_i \qquad (13.2.1)$$

and the average vector distance traveled by the particle is

$$\overline{\mathbf{R}} = \sum_{i=1}^{N} p_i(\mathbf{r}_i)\mathbf{r}_i \qquad (13.2.2)$$

For the case in which the probabilities of forward jumps are equal to those for reverse jumps, that is,

$$p_j(\mathbf{r}_j) = p_j(-\mathbf{r}_j) \qquad (13.2.3)$$

the average of the vector defined by (13.2.1) vanishes, but the scalar distance it travels is not zero. It is given by the average of the magnitude of the vector **R**, whose square is

$$\mathbf{R} \cdot \mathbf{R} = \sum_{i=1}^{N} \mathbf{r}_i \cdot \sum_{j=1}^{N} \mathbf{r}_j \qquad (13.2.4)$$

or

$$R^2 = \sum_{i=1}^{N} r_i^2 + \sum_{i,j(i \neq j)}^{N} \mathbf{r}_i \cdot \mathbf{r}_j \qquad (13.2.5)$$

where the products between different jump vectors have been separated from those for the same jump vector.

If the forward and reverse probabilities are not equal, then the mean vector distance is not zero. This case is of interest for the motion of particles in the presence of external fields that bias the jump probabilities in a particular direction as well as for polymers in which steric hindrance restricts the orientation of monomer units.

Taking the average of (13.2.5) and writing out the scalar products in the second term gives

$$\overline{R^2} = \sum_{i=1}^{N} \overline{r_i^2} + \sum_{i \neq j}^{N} \overline{r_i r_j \cos \theta_{ij}} \qquad (13.2.6)$$

with θ_{ij} being the angle between the ith and jth jump vectors as shown in figure 13.2.

In most cases of physical interest, the lengths of the unit jumps in a random flight are independent of each other and uncorrelated with the angle between jump vectors. The average of their product is then equal to the product of their averages and (13.2.6) reduces to

$$\overline{R^2} = \sum_{i=1}^{N} \overline{r_i}^2 + \sum_{i \neq j}^{N} \overline{r_i}\, \overline{r_j}\, \overline{\cos \theta_{ij}} \qquad (13.2.7)$$

If each jump has the same magnitude r, then

$$\overline{R^2} = Nr^2 + r^2 \sum_{i \neq j}^{N} \overline{\cos \theta_{ij}} \qquad (13.2.8)$$

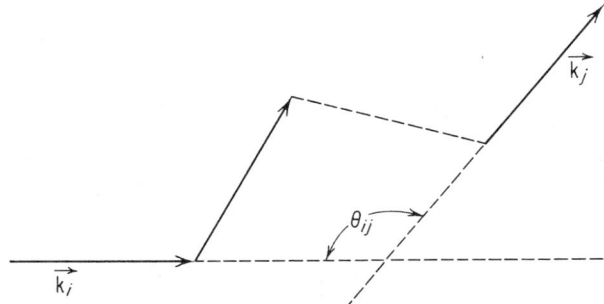

Figure 13.2. Angular relationship between jump vectors.

If, for every possible jump in a given direction, there is also a possible jump in the opposite direction with equal probability, then the averages of the cosine terms vanish and (13.2.8) is

$$\overline{R^2} = Nr^2 \tag{13.2.9}$$

This is the standard result for a simple uncorrelated random flight with unit steps of magnitude r. Its clearest application is to interstitial diffusion in a cubic crystal. The flight is said to be uncorrelated because the probability of a jump does not depend on the probability of the previous jumps.

The root mean square distance defined by (13.2.8) is a measure of the "size" of the random flight since its cube is a measure of the volume over which the flight took place. Note that the root mean square distance increases only as the square root of the number of jumps and is therefore much smaller than the contour length Nr of the jump sequence.

In general, a jump probability can depend on the preceding jump, and the cosine terms then do not necessarily vanish. The mean square displacement is then written from (13.2.7) as

$$\overline{R^2} = f_N Nr^2 \tag{13.2.10}$$

where f_N is called the *correlation coefficient* and is defined by

$$f_N = 1 + \frac{1}{N} \sum_{i \neq j}^{N} \overline{\cos \theta_{ij}} \tag{13.2.11}$$

θ_{ij} is called the *correlation angle* between the jumps i and j. The second term in (13.2.11) can be put in a form that simplifies its calculation for a number of important special cases by writing the sum in terms of partial sums as

$$\sum_{i \neq j}^{N} \overline{\cos \theta_{ij}} = 2 \sum_{i=1}^{N-1} \overline{\cos \theta_{i,i+1}} + 2 \sum_{i=1}^{N-2} \overline{\cos \theta_{i,i+2}} + \ldots$$

$$= 2 \sum_{j=1}^{N-1} \sum_{i=1}^{N-j} \overline{\cos \theta_{i,i+j}} \tag{13.2.12}$$

as is evident from writing out a few terms of the double sum in equation (13.2.11).

Now make use of the following identity from spherical trigonometry:

$$\cos\theta_{i,i+2} = \cos\theta_{i,i+1}\cos\theta_{i+1,i+2} + \sin\theta_{i,i+1}\sin\theta_{i+1,i+2}\cos\alpha_{i,1,2} \qquad (13.2.13)$$

In this formula, $\alpha_{i,1,2}$ is the angle between a plane defined by $\mathbf{r}_i$ and $\mathbf{r}_{i+1}$ and a plane defined by $\mathbf{r}_{i+1}$ and $\mathbf{r}_{i+2}$. For the case of twofold symmetry, for every $\alpha_{i,1,2}$ between two such planes there is an angle $\pi - \alpha_{i,1,2}$. The average of the last term in (13.2.13) then vanishes, so

$$\overline{\cos\theta_{i,i+2}} = \overline{\cos\theta_{i,i+1}\cos\theta_{i+1,i+2}} \qquad (13.2.14)$$

Since the two angles on the right-hand side of (13.2.14) are independent, the average of the product is the product of the averages, and since the average is the same for both angles,

$$\overline{\cos\theta_{i,i+2}} = \overline{\cos\theta}^2 \qquad (13.2.15)$$

Then, from (13.2.14) and (13.2.15),

$$\overline{\cos\theta_{i,i+3}} = \overline{\cos\theta_{i,i+2}\cos\theta_{i+2,i+3}} = \overline{\cos\theta_{i,i+2}} \cdot \overline{\cos\theta_{i+2,i+3}}$$
$$= \overline{\cos\theta_{i,i+2}} \cdot \overline{\cos\theta_{i+2,i+3}} = \overline{\cos\theta}^3$$

so for all j,

$$\overline{\cos\theta_{i,i+j}} = \overline{\cos\theta}^j \qquad (13.2.16)$$

and the correlation coefficient becomes

$$f_N = 1 + \frac{2}{N}\sum_{j=1}^{N-1}(N-j)\overline{\cos\theta}^j = 1 + 2\sum_{j=1}^{N-1}\left(1 - \frac{j}{N}\right)\overline{\cos\theta}^j \qquad (13.2.17)$$

Then, using equations (A.4.1) and (A.4.2) in appendix 4, equation (13.2.17) reduces to

$$f_N = 1 + 2\sum_{j=1}^{N-1}q^j + \frac{2}{N}\sum_{j=1}^{N-1}jq^j = \frac{1+q}{1-q} - \frac{2q}{N}\frac{(1-q^N)}{(1-q)^2} \qquad (13.2.18)$$

where $q \equiv \overline{\cos\theta}$ for simplicity of notation.

For many applications the number of jumps is very large, and it is sufficiently accurate to take the limit as $N \to \infty$. Then (13.2.18) becomes

$$\lim_{N\to\infty} f_N = f = \frac{1+q}{1-q} = \frac{1+\overline{\cos\theta}}{1-\overline{\cos\theta}} \qquad (13.2.19)$$

θ is called the *correlation angle*. Correlation can have a powerful effect on the random flight. This is simply demonstrated by considering a flight in which all correlation angles are the same and using equation (13.12.18) to calculate the correlation factor as a function of the number of jumps for different values of the correlation angle. For $\theta = \pi/2$ the particle executes a random walk on a simple cubic lattice and the correlation factor is unity for all values of N. For

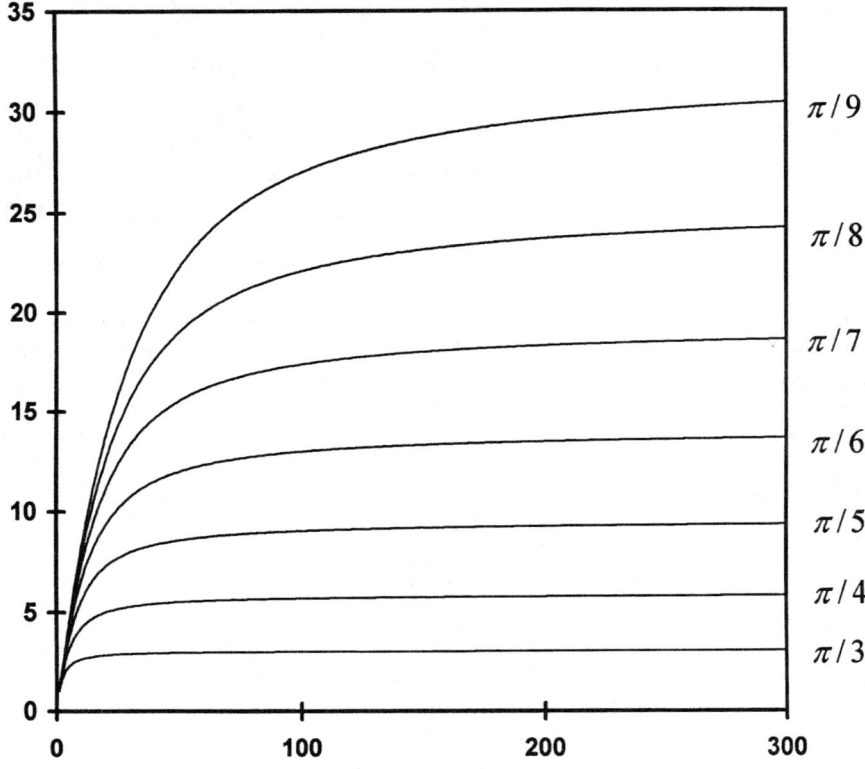

Figure 13.3. Correlation factor for various correlation angles.

smaller angles, however, there is a dependence on N that is stronger as the angle decreases, until at $\theta = 0$ the correlation factor is equal to N.

Figure 13.3 shows the correlation factor as a function of N for angles from $\pi/3$ to $\pi/9$. In every case, the correlation factor approaches a limit with increasing N, becoming essentially constant within a relatively short number of jumps. This limit is close to unity for angles near $\theta = \pi/2$ and increases with decreasing angle. For very small correlation angles, it takes an increasing number of jumps for the correlation factor to approach a constant value, and in the limit of zero angle the correlation factor is just the total number of jumps. The correlation factor as a function of N for small values of θ is shown in figure 13.4.

Clearly, the root mean square distance can be greatly increased by correlation. For polymers, this means that, on the average, molecular chains are less compact the smaller the bond angle between the monomer units.

13.3 Random flight on a lattice

A simple but important example of random flight is the motion of a particle that jumps from one site to another on a lattice. An illustration of this is shown in figure 13.5 for a two-dimensional square lattice. A particle starting from a

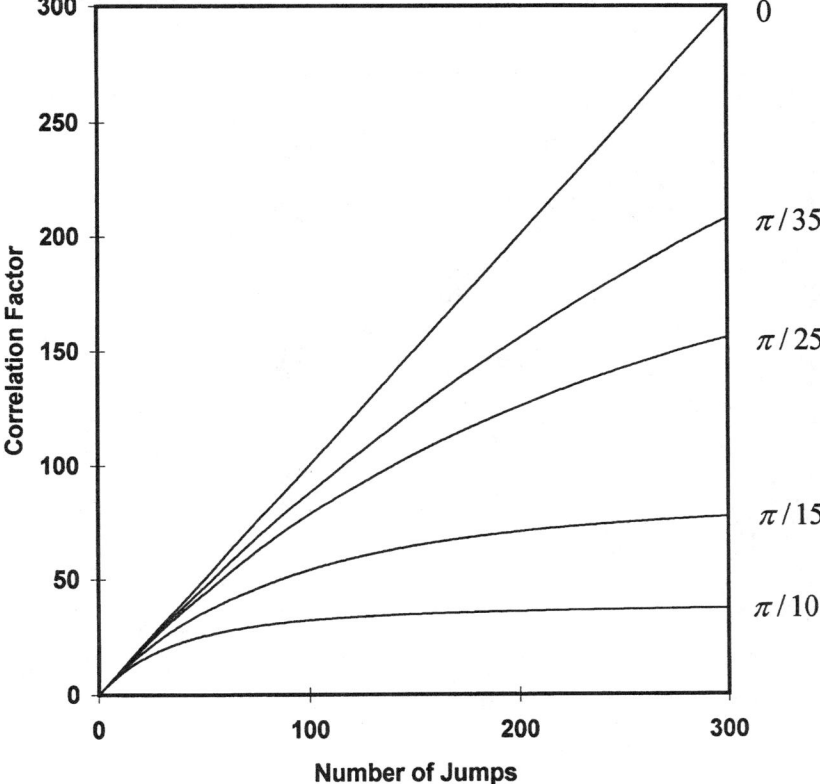

Figure 13.4. Correlation factor for small correlation angles.

lattice point labeled 0 has a given probability that it will jump to an adjacent site on the lattice. In general, the probability can vary with the position of the particle and the direction of the jump. The jump vector $\mathbf{r}_i$ goes from the $(i-1)$th site to an adjacent ith site. Since the projections of the particle motion along the coordinate axes go through a random walk, it is sufficient to analyze the one-dimensional case. This is then applied to each of the three axes of the Cartesian coordinate system and combined to give the two- or three-dimensional case.

A particle going through a random walk along the x-axis has a probability p that it will move to the right and a probability q that it will move to the left each time it makes a jump. It is assumed that these two probabilities can be different, but that they are the same for every jump. Of the total number of N jumps, n_R will be in the positive x-direction and n_L will be in the negative direction. We want the probability that the particle will have gone a distance X in the positive direction after N jumps. If x is the jump distance, then X is given by

$$X = mx \qquad (13.3.1)$$

where m is the net number of steps in the positive x-direction:

356 STATISTICAL MECHANICS OF SOLIDS

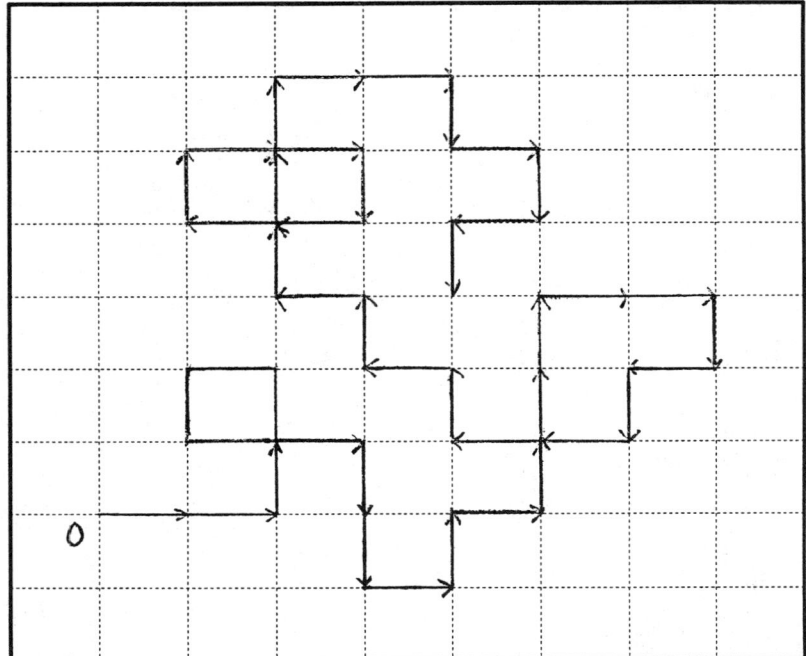

Figure 13.5. Random walk on a two-dimensional lattice.

$$m \equiv (n_R - n_L) \qquad (13.3.2)$$

Note that if N is even (odd) then the difference m is even (odd), as is obvious from the fact that

$$n_R = \frac{1}{2}(N + m)$$
$$n_L = \frac{1}{2}(N - m) \qquad (13.3.3)$$

To get the probability that the sequence of N jumps will contain n_R positive and $n_L = N - n_R$ negative jumps, first get the number of ways of putting N objects into two piles of n_R and n_L each. Multiplying this by the probability that a given sequence will result in n_R positive and n_L negative steps gives the desired probability as

$$P_N(n_R) = \frac{N!}{n_R!(N - n_R)!} p^{n_R} q^{N - n_R} \qquad (13.3.4)$$

This is the binomial, or Bernoulli, distribution, so called because the prefactor is just the coefficient in the binomial expansion

$$(p + q)^N = \sum_{n_R = 0}^{N} \frac{N!}{n_R!(N - n_R)!} p^{n_R} q^{N - n_R} \qquad (13.3.5)$$

Since the two probabilities must sum to unity, this shows that (13.3.4) is normalized.

The mean distance and mean square distance traveled after N jumps are defined by

$$\overline{X} = \sum_{n_R=0}^{N} X P_N(n_R) = x \sum_{n_R=0}^{N} m P_N(n_R) = x\overline{m} \qquad (13.3.6)$$

$$\overline{X^2} = \sum_{n_R=0}^{N} X^2 P_N(n_R) = x^2 \sum_{n_R=0}^{N} m^2 P_N(n_R) = x^2 \overline{m^2} \qquad (13.3.7)$$

These averages can be evaluated by the following device. Consider the binomial expression

$$(p\eta + q\eta^{-1})^N = \sum_{n_R=0}^{N} \frac{N!}{n_R!(N-n_R)!} (p\eta)^{n_R} (q\eta^{-1})^{N-n_R} \qquad (13.3.8)$$

or

$$(p\eta + q\eta^{-1})^N = \sum_{n_R=0}^{N} \eta^m P_N(n_R) \qquad (13.3.9)$$

Now differentiate (13.3.9) with respect to η to get

$$N(p\eta + q\eta^{-1})^{N-1}(p - q\eta^{-2}) = \sum_{n_R=0}^{N} m\eta^{m-1} P_N(n_R) \qquad (13.3.10)$$

and let $\eta = 1$ to get

$$N(p-q) = \sum_{n_R=0}^{N} m P_N(n_R) = \overline{m} \qquad (13.3.11)$$

so from (13.3.6),

$$\overline{X} = \overline{m}x = Nx(p-q) \qquad (13.3.12)$$

The average displacement is proportional to the difference of the forward and reverse probabilities and is zero when the probability of a jump is the same in both directions.

To get the mean square displacement, multiply (13.3.10) by η, again differentiate with respect to η and then set $\eta = 1$. Using the result in (13.3.7) gives

$$\overline{X^2} = \overline{m^2} x^2 = x^2[N + N(N-1)(p-q)]$$
$$= x^2 N[1 + (N-1)(p-q)^2] \qquad (13.3.13)$$

For many applications of physical interest, the number of jumps is very large. Polymer chains can contain thousands of monomer units, and measurable diffusion in crystals involves an enormous number of jumps. It turns out that in the limit of large N, the binomial distribution approaches a Gaussian distribution and can be treated as a continuous function of n_R. Let us show this by taking the logarithm of the binomial distribution and expanding it in a power series about the value of $n_R = \hat{n}_R$ for which the distribution is a maximum, treating n_R as if it were a continuous variable. Then, from (13.3.4),

$$\ln P_N(n_R) = \ln N! - \ln n_R! - \ln(N - n_R)! + n_R \ln p + (N - n_R) \ln q \quad (13.3.14)$$

and the Taylor expansion of this is

$$\ln P_N(n_R) = \ln P_N(\hat{n}_R) + \left(\frac{d \ln P_N}{dn_R}\right)_{\hat{n}_R} (n_R - \hat{n}_R)$$
$$+ \frac{1}{2}\left(\frac{d^2 \ln P_N}{dn_R^2}\right)_{\hat{n}_R} (n_R - \hat{n}_R)^2 + \ldots \quad (13.3.15)$$

To get the derivatives of the factorials, note that

$$\frac{\ln(n_R + 1)! - \ln n_R!}{1} - \ln\frac{(n_R + 1)!}{n_R!} = \ln(n_R + 1) \quad (13.3.16)$$

In the limit of large N, this difference approaches the derivative, so

$$\frac{d \ln n_R!}{dn_R} = \ln n_R \quad (13.3.17)$$

and therefore

$$\frac{d \ln P_N}{dn_R} = -\ln n_R + \ln(N - n_R) + \ln p - \ln q \quad (13.3.18)$$

and the second derivative is

$$\frac{d^2 \ln P_N}{dn_R^2} = -\frac{1}{n_R} - \frac{1}{N - n_R} \quad (13.3.19)$$

The position of maximum probability is readily obtained by setting (13.3.18) to zero. The result is

$$\hat{n}_R = N p \quad (13.3.20)$$

This relation simplifies the expression for the second derivative at the maximum $\hat{n}_R$ to

$$\left(\frac{d^2 \ln P_N}{dn_R^2}\right)_{\hat{n}_R} = -\frac{1}{Npq} \quad (13.3.21)$$

Now put the values of the derivatives in the expansion (13.3.15) to get, up to the second order,

$$\ln P_N(n_R) = \ln P_N(\hat{n}_R) - \frac{1}{2Npq}(n_R - \hat{n}_R)^2 \quad (13.3.22)$$

Equation (13.3.20) is also the probability that the particle has moved n_L steps to the left, and the probability that the net number of steps to the right is m, as well as the probability that it has moved n_R to the right, since these probabilities are all the same. Our objective is to get the probability distribution function for the net distance traveled, so using (13.3.3) we write (13.3.22) as

$$\ln P_N(m) = \ln P_N(\hat{m}) - \frac{1}{8N\,pq}(m-\hat{m})^2 \qquad (13.3.23)$$

The product of the jump probabilities can be related to distances by using (13.3.12) and (13.3.13) which give

$$\begin{aligned}\frac{\overline{x^2} - \overline{x}^2}{\overline{x^2}N} &= 1 + (N-1)(p-q)^2 - N(p-q)^2 \\ &= 1 - (p-q)^2 \\ &= 1 + (p-q)(q-p) = 1 + (2p-1)(2q-1) \\ &= 4pq \end{aligned} \qquad (13.3.24)$$

and substituting this in (13.3.23) gives

$$P_N(X) = P(\hat{X})e^{-(X-\hat{X})^2/2\sigma_x^2} \qquad (13.3.25)$$

where we have adopted a continuum notation with $X = mx$, and the standard deviation σ_x is defined as

$$\sigma_x = \sqrt{\overline{X^2} - \overline{X}^2} \qquad (13.3.26)$$

Note that since the probability distribution is symmetric about its maximum, the maximum is equal to the average. That is, $\hat{X} = \overline{X}$.

The probabilities represented by equation (13.3.25) must sum to unity. In the discrete notation, normalization is achieved by simply summing over all possible jump sequences. In the continuum notation, the segments are as if they were infinitely close together and the sum is replaced by an integral over all possible distances. To be consistent with this, (13.3.25) must be a probability density such that

$$P_N(X)dX = P_N(\hat{X})e^{-(X-\hat{X})^2/2\sigma_x^2}dX \qquad (13.3.27)$$

is the probability that, after N jumps, the particle is between X and $X + dX$.

The pre-exponential constant is determined by the requirement that the probability be normalized over all positive and negative values of the argument. This then converts (13.3.23) to the normalized Gaussian distribution for the probability density:

$$P_N(X) = \frac{1}{\sigma_x \sqrt{2\pi}} e^{-(X-\hat{X})^2/2\sigma_x^2} \qquad (13.3.28)$$

If the probability of a jump is the same for both the forward and reverse directions, then the mean displacement $\bar{X}$ is zero. In diffusion applications, a nonzero mean displacement is the result of forces acting on the system that depend on position, such as gravity, electric forces, or varying internal stresses. In applications to polymers, the difference in forward and reverse probabilities corresponds to an asymmetry in the possible orientations a monomer unit can assume relative to other units in the chain.

The Gaussian is a good approximation to the binomial distribution for surprisingly small values of N, as shown in figure 13.6, which is a comparison of the binomial distribution, represented by the marker points, and the Gaussian distribution, represented by the solid curve, for $N = 100$ and $p = q = 1/2$.

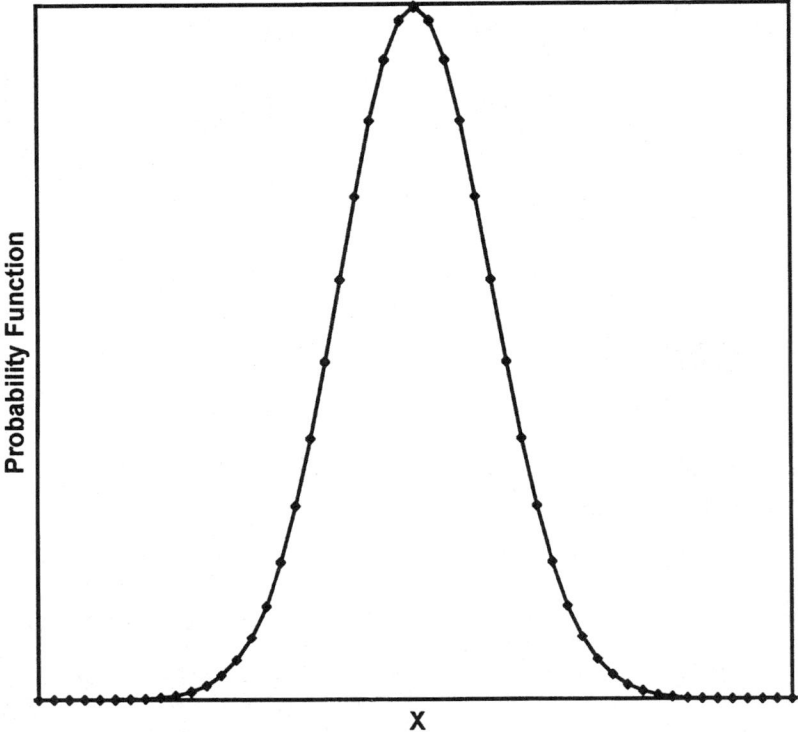

Figure 13.6. Comparison of binomial and Gaussian distributions for N = 100.

The agreement between the curve and the points is excellent. It must be noted that the percentage error for values of the number of steps far from the mean is very large. However, this does not matter since the probability is practically zero for these ranges of X (or m) and contributes very little to any physically important property of the distribution. The continuum approximation is increasingly accurate for increasing numbers of total steps.

The transition to three dimensions is trivial. If the jump vector is expressed in Cartesian coordinates, and the components of the random motions along the coordinate axes are statistically independent, then the probability that the particle will have traveled a distance R after N jumps is

$$P(R) = P(X)P(Y)P(Z) \tag{13.3.29}$$

with

$$R^2 = X^2 + Y^2 + Z^2 \tag{13.3.30}$$

The three-dimensional Gaussian distribution is therefore

$$P(R) = \frac{1}{\sigma_x \sigma_y \sigma_z \sqrt{8\pi}} e^{-(X-\hat{X})^2/2\sigma_x^2} e^{-(Y-\hat{Y})^2/2\sigma_y^2} e^{-(Z-\hat{Z})^2/2\sigma_z^2} \tag{13.3.31}$$

If the random flight is on a cubic lattice with the magnitudes of the components of the unit jumps all being equal, then the standard deviations are the same for all three directions and (13.3.31) is

$$P(R) = \frac{1}{\sigma^3 \sqrt{8\pi^3}} e^{-(R-\hat{R})^2/2\sigma_x^2} \qquad (13.3.32)$$

An interesting point is that, for a body-centered cubic lattice, the unit jump is from a central to a corner atom; the components of the jump therefore all have the same magnitude, and the probabilities for the jump components are *not* independent. In fact, the probability for a forward (or backward) diagonal jump is just the same as the probability that a jump component (in the x-direction, say) will be in the forward (backward) direction. The one-dimensional random walk is therefore fully equivalent to a three-dimensional random flight on a body-centered cubic lattice. The difference between the two is that a particle can move to the right in four ways on a body-centered cubic lattice, but only one way on a one-dimensional lattice. To get the total number of flights in three dimensions, the result for the one-dimensional walk must therefore be multiplied by 4^N. This concept is important in counting the number of conformations in simple random walk models of polymer chains.

Note that one- and two-dimensional random walks are appropriate for modeling diffusion along dislocation cores and on surfaces (or grain boundaries), respectively.

13.4 Reflecting and absorbing barriers

Some important physical phenomena can be modeled by random flights in the presence of reflecting or absorbing barriers. A reflecting barrier is a plane such that whenever the migrating particle hits it, the particle has unit probability of returning to the position it had previous to the jump. An absorbing barrier is a plane such that any particle that meets it vanishes and has zero probability of migrating any farther. If the random flight is on a lattice, the barrier plane must be one of the lattice planes on which the flight takes place.

An example of a physical random flight with a reflecting barrier is the diffusion into a crystal of a layer of impurity or tracer atoms deposited on a crystal surface. Another example is that of a polymer chain adsorbed on a surface. An example of a random flight with an absorbing barrier is the diffusion of atoms through a crystal to a crystal surface, where it escapes into the gas phase or is immobilized by a chemical reaction.

Figure 13.7 illustrates the effect of a barrier by the two-dimensional square lattice. Again, it is sufficient to consider the motion in the x-direction separately, and again, we denote the number of lattice units in the x-direction by m. Let the barrier be at the lattice plane $m = m_B$; without loss of generality, we can take $m_B \geq 0$. If the reflections at the barrier were ignored, then the probability that the particle would have traveled a distance $X = mx$ would be given by (13.3.28), but with the restriction that X is to the left of the barrier. But the presence of the barrier introduces another set of jump sequences since each sequence that reaches the barrier is reflected. The number of these extra sequences can be computed by recognizing the symmetry of the system, which states that a particle at the barrier has the same chance of reaching C as it would have of reaching the mirror image C′ if the barrier were not there. This means that the extra number of sequences can be computed from the probability that a particle reaches the image point C′. This image point is at the position $m_I = 2m_B - m$, or $X_I = 2X_B - X$. Therefore, the probability of reaching the image point

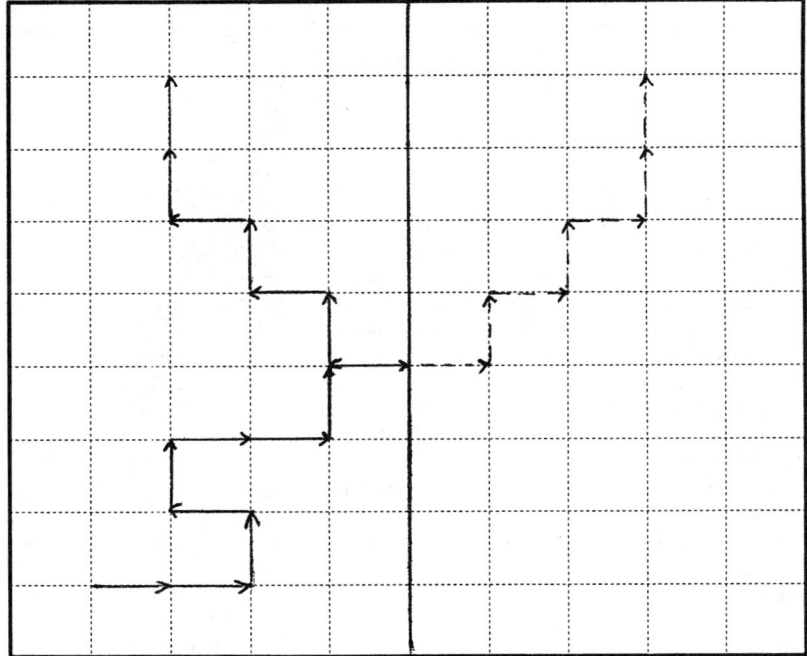

Figure 13.7. Random walk in the presence of a barrier.

$P_N(2m_B - m)$ must be added to the probability $P_N(m)$. That is, the probability of reaching m in the presence of the reflecting barrier is

$$P_N(m \mid m_B) = P_N(m) + P_N(2m_B - m) \tag{13.4.1}$$

In continuum notation for large N this is

$$P_{BR}(X) = \frac{1}{\sigma_x \sqrt{2\pi}} \left[e^{-(X-\hat{X})^2/2\sigma_x^2} + e^{-(2X_B - X - \hat{X})^2/2\sigma_x^2} \right] \tag{13.4.2}$$

The generalization to three dimensions using (13.3.32) is obvious:

$$P_{BR}(X) = \frac{1}{\sigma^3 \sqrt{8\pi^3}} \left[e^{-(R-\hat{R})^2/2\sigma^2} + e^{-(2R_B - R - \hat{R})^2/2\sigma^2} \right] \tag{13.4.3}$$

A subscript BR has been added to identify the fact that the reflecting barrier is present.

If the barrier absorbs particles instead of reflecting them, then each time a particle reaches the barrier it disappears, so a number of sequences that would have brought the particle to C (in the absence of the barrier) are lost. Again, using the fact of symmetry, the number of lost sequences is just the number that would reach the image point C' if the barrier were not there. This means that from the probability computed as if the barrier were not there, we must *subtract* the probability that the particle reaches the image point. Then, instead of (13.4.2) we have

$$P_{BA}(X) = \frac{1}{\sigma_x \sqrt{2\pi}} \left[e^{-(X-\hat{X})^2/2\sigma_x^2} - e^{-(2X_B-X-\hat{X})^2/2\sigma_x^2} \right] \qquad (13.4.4)$$

and in three dimensions,

$$P_{BA}(X) = \frac{1}{\sigma^3 \sqrt{8\pi^3}} \left[e^{-(R-\hat{R})^2/2\sigma^2} - e^{-(2R_B-R-\hat{R})^2/2\sigma^2} \right] \qquad (13.4.5)$$

with the subscript denoting an absorbing barrier.

Note that in all these equations, the distance variable is restricted to the left of the barrier.

13.5 The Markoff method

A general method of analyzing random flights is based on the Markoff theory of random variables as presented by Chandresekar.[1] The method, based on an analysis of the Fourier transform of jump probabilities, is very powerful. It includes all the above presentations as special cases and also leads to a general result of surprisingly wide applicability. Again, we look for the probability that, after N jumps, the particle will have traveled a distance **R**. That is, we want the probability that after N steps, the particle will have arrived in a volume element $d\mathbf{R}$ centered on **R** defined by

$$\mathbf{R} - \frac{d\mathbf{R}}{2} \leq \mathbf{R} \leq \mathbf{R} + \frac{d\mathbf{R}}{2} \qquad (13.5.1)$$

Initially, no restrictions are placed on the individual jump probabilities. The general approach is to assign a probability to each step of the random flight, multiply them together to get the probability for **R**, and integrate over all values of $\mathbf{r}_i$ that satisfy (13.5.1). That is,

$$P_N(\mathbf{R})d\mathbf{R} = \int_\Delta \cdots \int \prod_j^N p_j(\mathbf{r}_j)d\mathbf{r}_j = \prod_{j=1}^N \int_\Delta p_j(\mathbf{r}_j)d\mathbf{r}_j \qquad (13.5.2)$$

where $p_j(\mathbf{r}_j)d\mathbf{r}_j$ is the probability that the jth step will bring the particle into the volume element $d\mathbf{r}_j$ and $P_N(\mathbf{R})$ is the probability that after N steps the particle will be in the volume element $d\mathbf{R}$. The symbol Δ indicates that the integrations are carried out only over the values of $\mathbf{r}_j$ that satisfy the inequality (13.5.1). $P_N(\mathbf{R})$ is the displacement probability for N steps.

Define $A_N(\mathbf{k})$ by

$$A_N(\mathbf{K}) = \int_{-\infty}^{\infty} e^{i\mathbf{k}\cdot\mathbf{R}} p_N(\mathbf{R})d\mathbf{R} = \prod_{j=1}^N \int_{-\infty}^{\infty} e^{i\mathbf{k}\cdot\mathbf{r}_j} p_j(\mathbf{r}_j)d\mathbf{r}_j \qquad (13.5.3)$$

This is just the Fourier transform of the displacement probability and is called the *characteristic function* of the random walk. It has the important property that, because $\mathbf{R} = \Sigma \mathbf{r}_i$, the transform of the total displacement probability is the product of the transforms for the individual step probabilities. From these, the total displacement probability can be computed by taking the inverse Fourier transform to get

$$P_N(\mathbf{R}) = \frac{1}{8\pi^3} \int_{-\infty}^{\infty} e^{-i\mathbf{k}\cdot\mathbf{R}} A_N(\mathbf{k}) d\mathbf{k} \tag{13.5.4}$$

with

$$A_N(\mathbf{k}) = \prod_{i=1}^{N} \int_{-\infty}^{\infty} e^{i\mathbf{k}\cdot\mathbf{r}_i} p_i(\mathbf{r}_i) d\mathbf{r}_i \tag{13.5.5}$$

This reduces the random flight problem to that of finding the Fourier transform of the individual flight probabilities.

A case of particular interest is that in which all the individual jump probabilities are equal, such that $p_i(\mathbf{r}_i) = p(\mathbf{r})$. (Note that these probabilities can be a function of the jump vector, but they are all the *same* function of $\mathbf{r}$.) Then (13.5.5) reduces to

$$A_N(\mathbf{k}) = \left[\int_{-\infty}^{\infty} e^{i\mathbf{k}\cdot\mathbf{r}} p(\mathbf{r}) d\mathbf{r}\right]^N \tag{13.5.6}$$

The analyses of the preceding sections are restricted to cases of this sort.

The Fourier transform, and therefore the displacement probability, can be evaluated for a number of special cases for the jump probability function that are of physical interest. A simple instance is that of the freely jointed flight that is defined by the conditions that each step can take any direction but only one length r_i. For this case, the jump probability of the ith step is the delta function

$$p_i(\mathbf{r}_i) = \frac{1}{4\pi r_i^2} \delta(|\mathbf{r}_i| - r_i) \tag{13.5.7}$$

This just states that on the ith jump the particle can go off at any angle with equal probability, but can only move the prescribed distance r_i. The prefactor normalizes the probability since integrating the delta function in spherical coordinates gives $4\pi r_i^2$. Putting this in (13.5.5) gives

$$A_N(\mathbf{k}) = \prod_{i=1}^{N} \frac{1}{4\pi r_i^2} \int e^{i\mathbf{k}\cdot\mathbf{r}_i} \delta(|\mathbf{r}_i| - r_i) d\mathbf{r}_i \tag{13.5.8}$$

The integral can be evaluated in spherical polar coordinates, letting θ be the angle between the vectors $\mathbf{r}_i$ and $\mathbf{k}$:

$$\frac{1}{4\pi r_i^2} \int_{-\infty}^{\infty} e^{i\mathbf{k}\cdot\mathbf{r}_i} \delta(|\mathbf{r}_i| - r_i) d\mathbf{r}_i$$

$$= \frac{1}{2r_i^2} \int_{0}^{\infty} \int_{0}^{\pi} e^{ikr_i\cos\theta} r_i^2 \delta(|\mathbf{r}_i| - r_i) \sin\theta \, d\theta \, dr_i$$

$$= \frac{1}{2r_i^2} \int_{0}^{\infty} \frac{2r_i \sin kr_i}{k} \delta(|\mathbf{r}_i| - r_i) dr_i = \frac{\sin kr_i}{kr_i} \tag{13.5.9}$$

so (13.5.8) becomes

$$A_N(\mathbf{k}) = \prod_{i=1}^{N} \frac{\sin kr_i}{kr_i} \tag{13.5.10}$$

Put (13.5.10) into (13.5.4) to get

$$P_N(\mathbf{R}) = \frac{1}{8\pi^3} \int_{-\infty}^{\infty} e^{-i\mathbf{k}\cdot\mathbf{R}} \prod_{i=1}^{N} \frac{\sin kr_i}{kr_i} d\mathbf{k}$$

$$= \frac{1}{8\pi^3} \prod_{i=1}^{N} \int_{-\infty}^{\infty} e^{-i\mathbf{k}\cdot\mathbf{r}_i} \frac{\sin kr_i}{kr_i} d\mathbf{k} \qquad (13.5.11)$$

For a flight of equal steps, with all $r_i = r$, (13.5.10) and (13.5.11) become

$$A_N(\mathbf{k}) = \left(\frac{\sin kr}{kr}\right)^N \qquad (13.5.12)$$

$$P_N(\mathbf{R}) = \frac{1}{8\pi^3} \int_{-\infty}^{\infty} e^{-i\mathbf{k}\cdot\mathbf{R}} \left(\frac{\sin kr}{kr}\right)^N d\mathbf{k} \qquad (13.5.13)$$

The function $(\sin x)/x$ is important in diffraction and in other aspects of solid state theory as well as in the theory of random flights. It has a maximum value of unity at $x = 0$ and is small for values of x away from zero, as shown in figure 13.8. This means that for large N the integrand in (13.5.12) is very close to zero except for small values of kr. A series expansion about $kr = 0$ is then useful.

Dividing the series expansion of the sine by its argument gives

$$\frac{\sin kr}{kr} = 1 - \frac{(kr)^2}{3!} + \frac{(kr)^4}{5!} - \dots \qquad (13.5.14)$$

retaining only the first two terms in (13.5.14) gives

$$\left(\frac{\sin kr}{kr}\right)^N = \left[1 - \frac{(kr)^2}{3!}\right]^N = 1 - \frac{N(kr)^2}{3!} \qquad (13.5.15)$$

The right-hand side of (13.5.15) is just the first two terms of the expansion of the exponential function, so

$$\left(\frac{\sin kr}{kr}\right)^N \cong e^{[N(kr)^2]/3!} \qquad (13.5.16)$$

The larger the N, the better this approximation. It is quite good even for rather small values of N.

The difference between the two functions in (13.5.16) decreases rapidly with increasing N and is small even for relatively small values of N. For $N = 20$ the maximum error incurred by using the exponential function is about 5.5%, while for $N = 100$ the maximum error is 0.1%. For a flight of 200 steps, the error reduces to 0.05%.

It is therefore sufficiently accurate to put (13.5.16) into (13.5.13) to get

$$P_N(\mathbf{R}) = \frac{1}{8\pi^3} \int_{-\infty}^{\infty} e^{-i\mathbf{k}\cdot\mathbf{R}} e^{-[N(kr)^2/3!]} d\mathbf{k} \qquad (13.5.17)$$

This integral can be evaluated by using spherical polar coordinates with $\mathbf{k}\cdot\mathbf{R} = kR\cos\theta$ and $d\mathbf{k} = 2\pi k^2 \sin\theta \, d\theta \, dk$ so that

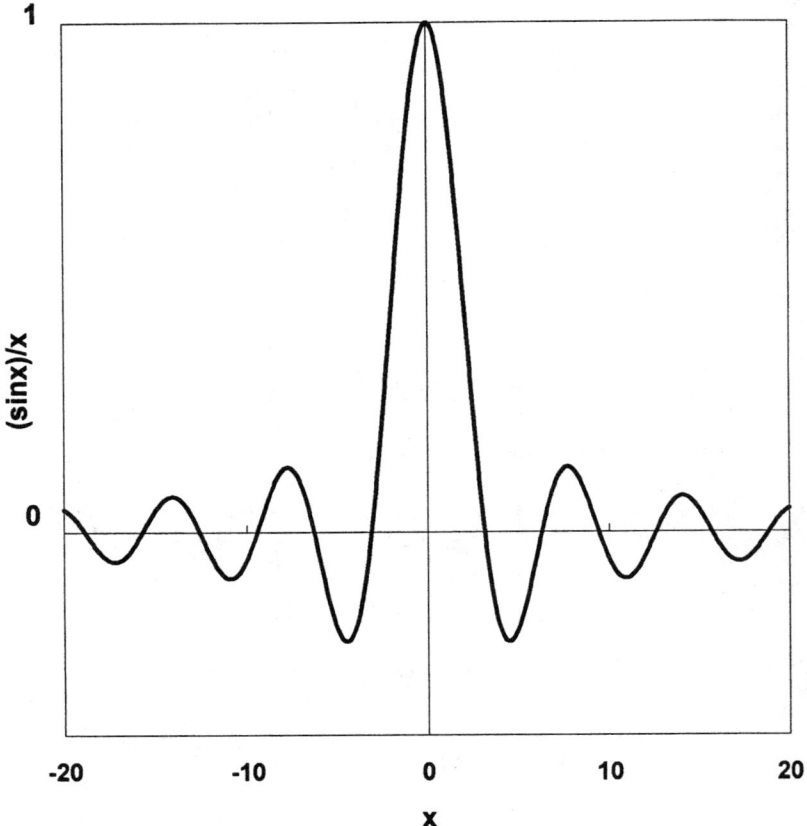

Figure 13.8. The function (sinx)/x.

$$P_N(\mathbf{R}) = \frac{2\pi}{8\pi^3} \int_0^\infty \int_1^{-1} e^{-ikR\cos\theta} d(\cos\theta) e^{-[N(kr)^2/3!]} k^2 dk$$

but

$$\int_1^{-1} e^{-ikRu} du = -\frac{1}{ikr}[\cos(-kR) + i\sin(-kR) - \cos(kR) - i\sin(kR)]$$

$$= \frac{1}{ikr} 2i\sin(kR) = \frac{2\sin(kR)}{kR}$$

and therefore

$$P_N(\mathbf{R}) = \frac{1}{2\pi^2 R} \int_0^\infty \sin(kR) e^{[N(kr)^2]/3!} k\, dk \qquad (13.5.18)$$

This is a known integral.[2] The result is

$$P_N(R) = \left(\frac{3}{2\pi N r^2}\right)^{3/2} e^{3R^2/2Nr^2} \quad \text{(freely jointed chain)} \quad (13.5.19)$$

The vector notation has been dropped because the probability depends only on the magnitude of the displacement vector.

This provides the simplest model of a polymer chain, each jump corresponding to the length of a monomer unit, which is assumed to be able to take on any orientation relative to its adjacent units.

This rather general method can be used for other cases by following the same route as that which led to (13.5.19) but using the appropriate step probabilities.

Another case of interest is that in which the particle can jump in any direction and the probability that the jump will be of a given length is Gaussian. That is, the probability that the ith step will have a length r is

$$p_i(r) = \left(\frac{3}{2\pi \overline{r^2}}\right)^{3/2} e^{-3r^2/2\overline{r^2}} \quad (13.5.20)$$

with $\overline{r^2}$ being the mean square jump distance. It is easy to show that, with this jump probability, the probability that after N steps the particle will be at a distance R from its starting point is

$$P_N(R) = \left(\frac{3}{2\pi N \overline{r^2}}\right)^{3/2} e^{-3R^2/2N\overline{r^2}} \quad (13.5.21)$$

for large N. To arrive at (13.5.21), first substitute (13.5.20) into (13.5.5). The integrals are then readily performed to give

$$A_N(\mathbf{k}) = e^{-k^2 N \overline{r^2}/6} \quad (13.5.22)$$

where the average of the square of the jump distance is defined by

$$\overline{r^2} = \frac{1}{N}\sum_{j=1}^{N} r_j^2 \quad (13.5.23)$$

Putting (13.5.22) into (13.5.4) again gives integrals that are easily performed, with the result given by (13.5.21). This result provides an appropriate model for diffusion in a liquid, for which the jump of a molecule to a new position can take place in any direction and the length of the jump is distributed around a distance for which the probability of the jump is a maximum.

13.6 The general solution

The cases given above, in which a normal distribution of total displacements is obtained as the limit of the product of probabilities for individual displacements, are special cases of the central limit theorem. This theorem states that, given a set of N random events, each with a given probability of occurring, the probability distribution of the compounded event approaches the normal (Gaussian) distribution as N becomes large. This theorem can be understood by considering a generalization of the process leading to equation (13.5.19). Start by examining the characteristic function, given by (13.5.6) and recognizing that the integral in (13.5.6) is always less than unity because, in the absence of the exponential function the integral is unity and the magnitude of the real part of the exponential is always less than unity. The characteristic function is therefore a product of terms that are always less than unity, so as N becomes large, this product gets small.

Expand the exponential in (13.5.6) to get

$$A_N(\mathbf{k}) = \left\{ \int_{-\infty}^{\infty} \left[1 + i\mathbf{k}\cdot\mathbf{r} + \frac{(i\mathbf{k}\cdot\mathbf{r})^2}{2} + \frac{(i\mathbf{k}\cdot\mathbf{r})^3}{3!} + \ldots \right] p(\mathbf{r}) d\mathbf{r} \right\}^N \quad (13.6.1)$$

From the definition of the average and moments of a distribution, (13.6.1) can be written as

$$A_N(\mathbf{k}) = (1 + Q)^N \quad (13.6.2)$$

where Q is defined by

$$Q = \sum_{s=1}^{\infty} \frac{\overline{(i\mathbf{k}\cdot\mathbf{r})^s}}{s!} \quad (13.6.3)$$

and, in keeping with the usual definitions, the sth moment is defined by

$$\overline{(i\mathbf{k}\cdot\mathbf{r})^s} = \int_{-\infty}^{\infty} (i\mathbf{k}\cdot\mathbf{r})^s p(\mathbf{r}) d\mathbf{r} \quad (13.6.4)$$

That (13.6.2) is well approximated by an exponential function for large N can be seen by first using the binomial expansion to get

$$A_N(\mathbf{k}) = 1 + NQ + \frac{N(N-1)Q^2}{2} + \frac{N(N-2)(N-1)Q^3}{3!} + \ldots \quad (13.6.5)$$

Now consider the expansion of the exponential function

$$e^{NQ} = 1 + NQ + \frac{N^2 Q^2}{2} + \frac{N^3 Q^3}{3!} + \ldots \quad (13.6.6)$$

As $N \to \infty$, (13.6.5) approaches (13.6.6), so for large N, the characteristic function becomes

$$A_N(\mathbf{k}) = e^{NQ} \quad (13.6.7)$$

Retaining only the first two terms for Q in (13.6.3), this reduces to

$$A_N(\mathbf{k}) = e^{iN\overline{\mathbf{k}\cdot\mathbf{r}} - N\overline{(\mathbf{k}\cdot\mathbf{r})^2}/2} \quad (13.6.8)$$

To get the probability distribution for the total displacement, substitute (13.6.8) into (13.5.4):

$$P_N(\mathbf{R}) = \frac{1}{8\pi^3} \int_{-\infty}^{\infty} e^{-i\mathbf{k}\cdot(\mathbf{R} - N\overline{\mathbf{r}}) - N\overline{(\mathbf{k}\cdot\mathbf{r})^2}/2} d\mathbf{k} \quad (13.6.9)$$

The only obstacle to evaluating this integral is the quadratic term in the exponential, because the second moment has the form

$$\overline{(\mathbf{k}\cdot\mathbf{r})^2} = k_1^2\overline{x^2} + k_2^2\overline{y^2} + k_3^2\overline{z^2}$$
$$+2k_1k_2\overline{xy} + 2k_1k_3\overline{xz} + 2k_2k_3\overline{yz} \tag{13.6.10}$$

and the mixed (nondiagonal) terms prevent the integral from being separated into a product of integrals, each integrand being a function only of one k-component. However, it is always possible to rotate axes to reduce any quadratic form to a sum of squares. That is, a coordinate system exists for which the last three terms in (13.6.10) vanish. Let us assume that we have chosen such a coordinate system. Then (13.6.9) becomes

$$P_N(\mathbf{R}) = P_N(X)P_N(Y)P_N(Z) \tag{13.6.11}$$

where

$$P_N(X) = \frac{1}{8\pi^3}\int_{-\infty}^{\infty} e^{ik_1(X-N\bar{x})-Nk_1^2\overline{x^2}/2}\,dk_1 \tag{13.6.12}$$

with similar definitions for $P_N(Y)$ and $P_N(Z)$. Equation (13.6.12) is a known integral,[3] so taking the product of (13.6.12) with the similar results for the $P_N(Y)$ and $P_N(Z)$ gives the final result that

$$P_N(R) = \frac{1}{\left(8\pi^3 N^3 \overline{x^2}\,\overline{y^2}\,\overline{z^2}\right)^{1/2}} e^{-\frac{(X-N\bar{x})^2}{2N\overline{x^2}} - \frac{(Y-N\bar{y})^2}{2N\overline{y^2}} - \frac{(Z-N\bar{z})^2}{2N\overline{z^2}}} \tag{13.6.13}$$

If the mean displacement for the individual jumps is zero, then (13.6.13) reduces to a Gaussian form similar to those derived for the special cases given above. In the general case, when there is an asymmetry in the jump probabilities, there is a "drift" displacement, defined by the nonzero averages of the unit jumps, that is superimposed on the Gaussian probability distribution. This occurs, for example, in Brownian motion or gas diffusion in a gravitational field or in solid state diffusion in the presence of nonuniform stresses.

An important special case is that in which there is no drift displacement and the probability has sufficient symmetry that the second moments in the three directions are equal. That is,

$$\bar{x} = \bar{y} = \bar{z} = 0$$
$$\overline{x^2} = \overline{y^2} = \overline{z^2} = \overline{r^2}/3 \tag{13.6.14}$$

This would be the case if the jump probabilities were a function only of the magnitude of the individual jumps, or for a random flight on a lattice where the probability of a jump was the same for the three components of the jump. Using (13.6.14) in (13.6.13) gives

$$R_N(R) = \left(\frac{3}{2\pi N\overline{r^2}}\right)^{3/2} e^{-3R^2/2N\overline{r^2}} \tag{13.6.15}$$

This is just like (13.5.21), which is the result for the case of a Gaussian form for the individual jumps. Equation (13.6.15) shows that the jump probabilities need not be restricted to a Gaussian form to yield a Gaussian distribution for the total displacement, but depends only on the jump probabilities having a sufficient degree of symmetry.

13.7 Self-similarity

A long random flight can be divided into sections, each section consisting of a long sequence of contiguous jumps. That is, if N is large enough, the flight can be subdivided into a large number of sections for each of which the number of jumps is large. For this case the sections "look like" the total flight in the sense that its conformation is similar and that the probability distributions of the displacements are similar. To quantify this, consider a flight of N_1 steps, each with the same symmetric jump probability, and with N_1 large enough that the displacement probability distribution is given by (13.6.15). After the N_1th step, let the particle execute another random flight of N_2 steps whose probability distribution is again given by a form like (13.6.15). This can be continued ϖ times until the flight makes a total of N steps given by

$$N = \sum_{j=1}^{\varpi} N_j \qquad (13.7.1)$$

The probability distribution for the displacement distance for the jth section is

$$P_{N_j}(R_j) = \left(\frac{3}{2\pi N_j \overline{r^2}}\right)^{3/2} e^{-3R_j^2/2N_j \overline{r^2}} \qquad (13.7.2)$$

and the probability distribution for the entire flight is the product

$$P_N(R) = \prod_{j=1}^{\varpi} P_{N_j}(R_j) \qquad (13.7.3)$$

If the N_j are all equal, then (13.7.3) becomes the same as (13.6.15). This means that, for a very long chain, the smaller sections act as if they were unit jumps with a Gaussian jump probability, but each of these sections has a Gaussian displacement probability of its own. As $N \to \infty$, the flight can be subdivided into parts each of which is similar to the whole, and each of these parts can be subdivided into subsections that are similar to the sections. This is the phenomena of self-similarity in which scaling up or down results in systems that are similar to the original system in important respects. For diffusion, this means that the path of a diffusing particle looks the same at any scale of resolution, and for polymers, this means that a subsequence of polymer units looks very much like the total polymer chain. Of course, these statements are meant to apply to the averages of flight paths or conformations, and are valid only if the smallest section considered is not too small.

The above conclusions were based on flight paths whose jump probabilities had a high degree of symmetry, but self-similarity occurs even when this is not the case. Consider, for example, a flight in which the possible jumps are restricted to a small range of angles. This would be the case in a polymer chain in which the relative orientation of two adjacent monomers is restricted by chemical bond angles, as illustrated for in figure 13.9.

The chemical bond requires the angle between the first and second segments to be either θ or $-\theta$. This is also true of succeeding pairs of segments, but it is clear that after a sufficient number of segments, the chain can have any of a wide variety of conformations and can even fold back on itself. In three dimensions, there is the added factor that the orientation of a pair of segments is defined by two angles, θ and ϕ, only one of which is fixed by bond angle

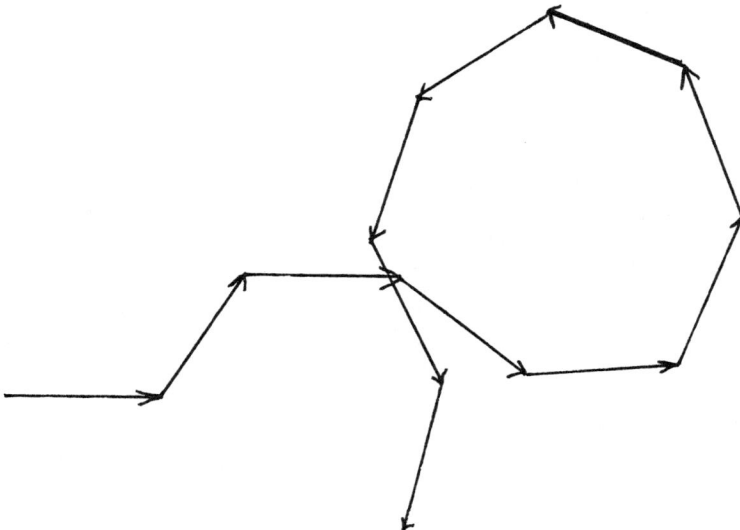

Figure 13.9. Bond angles and coiling in a polymer chain.

requirements. It does not take too many segments in the chain (or steps in a flight) before coiling through almost any region of space can occur. This can be treated by defining a distance (or number of segments) such that the probability of the orientation between two such distances is nearly spherically symmetric. The phenomenon of self-similarity then holds. This is why linear polymers can be modeled as looping strands of cooked spaghetti.

13.8 The diffusion equation from random flights

Let us adopt the multiparticle interpretation of the random flight. That is, it is assumed that a very large number of particles all start from the same point and each executes a random flight, and assume that the number of jumps is very large. Then the probability distribution $P_N(R)$ is taken to be the fraction of the flights for which the total displacement is R. Also, assume that the displacements are sequential jumps in time and let the average number of jumps per unit time be defined by

$$\bar{\Gamma} = \frac{N}{t} \tag{13.8.1}$$

This is called the *average jump frequency*.

It is assumed that the number of jumps and the time are both large enough that continuum approximations are valid. That is, R is always much larger than interatomic distances, and t is always much larger than the time between jumps. It is also assumed that the average jump frequency is independent of time. Equation (13.6.15) now represents diffusion from a point source, and we write it as

$$P(R) = \left(\frac{1}{4\pi Dt}\right)^{3/2} e^{-R^2/4Dt} \tag{13.8.2}$$

where the constant D is defined by

$$D = \frac{1}{6}\overline{\Gamma r^2} \qquad (13.8.3)$$

In keeping with the temporal continuum approximation, the subscript N on the probability distribution has been dropped. $P(R)$ is proportional to the concentration of particles a distance R from a point source at time t after diffusion begins.

If (13.8.2) is differentiated with respect to time, the result is

$$\frac{\partial P}{\partial t} = P(R)\left(\frac{R^2}{4Dt^2} - \frac{3}{2t}\right) \qquad (13.8.4)$$

and taking the Laplacian of (13.8.2) gives

$$\nabla^2 P = P(R)\frac{1}{D}\left(\frac{R^2}{4Dt^2} - \frac{3}{2t}\right) \qquad (13.8.5)$$

Comparing (13.8.4) and (13.8.5) shows that

$$\frac{\partial P}{\partial t} = D\nabla^2 P \qquad (13.8.6)$$

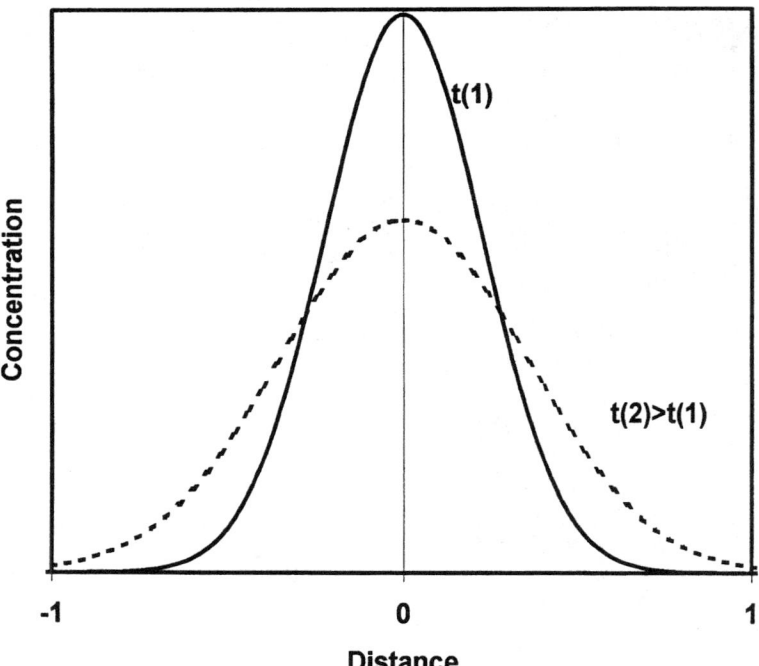

Figure 13.10. Diffusion profile from a point source for two different times.

But this is just the diffusion equation for a system of noninteracting particles in the absence of fields. This shows that (13.8.2) is the solution of the macroscopic diffusion equation for the spreading of a point source. In fact, (13.8.3) can be used to derive the diffusion equation for any starting distribution of particles, thereby connecting the macroscopic diffusion theory to microscopic events. This will be done explicitly in chapter 17, using the concept of transition probabilities.

Equation (13.8.2) shows that the concentration of particles around a point source spreads out with time although it remains Gaussian. This is illustrated in figure 13.10 for one dimension. Two curves are shown, one corresponding to a later time than the other.

Exercises

13.1 Consider a one-dimensional random walk in which the probability p is very much less than the probability q. Show that the probability that the displacement to the right is n_R after N jumps will be, for $n_R \ll N$, $P_N(n_R) = [(N_P)^{n_R}/n_R!]e^{-N_P}$ This is the Poisson distribution for a very large number of events, each having a very small probability. (Hint: start with the binomial distribution, take limits and use Stirling's approximation.)

13.2 Show, for a one-dimensional random walk of equal steps x, that when the probabilities of forward and backward jumps are equal, the mean square fluctuation of the displacement is $\overline{\Delta X^2} \equiv \overline{X^2} - \overline{X}^2 = x^2 N$

13.3 For a one-dimensional random walk, what is the mean square fluctuation if the probability of a forward jump is twice that of a backward jump?

13.4 Derive equation (13.3.4) for the probability distribution of the displacement in a one-dimensional random walk by the Markoff method. *Hint:* To do this, the probability of forward and reverse jumps must be written as a probability density. That is, the probability that the jth jump will result in a displacement between x_j and $x_j + dx_j$ is $P_j(x_j)dx_j = [p\delta(x_j - x) + q\delta(x_j + x)]dx$. Also, you will need the definition of the Dirac delta function as

$$\frac{1}{2\pi}\int_{-\infty}^{\infty} e^{i(x-\eta)}dx = \delta(x - \eta)$$

which is just the orthonormality condition for the complete set of plane waves.

13.5 Consider a very long random flight whose probability distribution function for the displacement is given by a Gaussian with a zero mean displacement. From the concept of self-similarity, we write the probability distribution as another Gaussian with a mean unit displacement of $\tilde{r}$ and a number of jumps $\tilde{N} = N/\omega$. The probability that the total displacement is R must of course be the same for both the standard and the self-similar descriptions. Find the ratio of the mean square jump distance in the self-similar description to that in the standard description by requiring that the total root mean square displacement is the same in both descriptions and also by requiring that distances in the probability density for the self similar description scale according to $1/\omega$.

Notes

1. Chandrasekhar (1954).
2. See, for example, p. 495 of Gradshteyn I.S., and I.M. Ryzhik; 1965; *Table of Integrals, Series and Products*, Academic Press, New York.
3. See p. 307 of Gradshteyn and Ryzhik (ibid.).

14

Linear Polymer Chains

14.1 Polymer chains and random flight

The salient fact about polymer chains is that, on a molecular scale, they are very long. Linear polymers consist of chains whose lengths can vary from hundreds to many thousands of monomer units. The statistical mechanics of systems composed of such large molecules can be quite complex and, if pursued with a degree of rigor and completeness, soon becomes intractable to analytic methods.

But it is the great length of the polymer chain that distinguishes it from other systems, and it is of interest to focus on this fact. Accordingly, it is customary to adopt a simple model that assumes that a polymer chain consists of a large sequence of identical segments joined together at their ends by chemical bonds, each segment being rigid, but having at least some freedom to rotate in space relative to its adjoining segments. Such a chain has many possible conformations, and the analogy to a random flight is clear. Figure 13.1 is then interpreted as a possible conformation of a long-chain molecule, and figure 13.2 then shows the angular relation among adjacent monomer segments. In this model, all the results of random flight theory carry over to polymer statistics.

However, some caveats are in order because polymer segments consist of groups of atoms and are not geometric lines. There are intramolecular interactions among these groups that are not accounted for in the theory of random flights. An important effect of these interactions is that segments cannot go through each other, although this is a possibility that is included in random flight theory. A modification that takes this into account would require the theory of self-avoiding random flights, which is considerably more complex and does not have the neat simplicity of the theory presented in chapter 13. A second important effect is that the interaction of segments modifies the jump probabilities by a Boltzmann factor of the interaction energy, so they are not fully random in the sense defined in chapter 13.

These caveats are not fatal. The conformations of a random flight are surprisingly open, and the number of jumps in which segments cross is not large, so the random flight is a reasonable approximation. Also, the interaction energies can be included in the construction of the partition function. For elementary theories this is done by adopting a mean field approximation equivalent in principle to Bragg–Williams alloy theory. Thus, a simple approach based on random flights for the conformations of polymer chains yields theories that display the physical effects of long chain length in an informative manner. The unique contribution of chain length to thermodynamic properties arises from the relation between the number of possible chain

14.2 Persistence length

Because of the property of self-similarity, discussed in chapter 13, the Gaussian form for the distribution of end-to-end distance has a wide applicability and is sufficient for a coherent description of chain statistics. A formal description of this fact can be expressed in terms of a quantity called the *persistence length*, which is a measure of the minimum distance between two segments at which the correlations between them are negligible.

If the average cosine of the angle between two segments vanishes, there is no correlation between the segments, so a definition of the persistence length starts with equation (13.2.16), which gives the angle between two segments separated by j bond lengths. We rewrite this as

$$\overline{\cos\theta_j} = \overline{\cos\theta}^{\,j} \qquad (14.2.1)$$

where θ_j is the angle between any two segments that are j segments apart in the chain sequence.

Unless $\overline{\cos\theta} = \pm 1$ (corresponding to a rigid linear rod), the right-hand side of (14.2.1) approaches zero as the number of bond lengths increases. For all physically realistic cases, therefore, the correlation between segments decreases as the number of unit segments between them increases and eventually becomes very small. The length at which segments are essentially uncorrelated can be used to define a multisegment, consisting of a number of unit segments. By the principle of self-similarity, the chain can then be thought of as a chain of uncorrelated multisegments. The length of a multisegment is clearly important. A parameter describing this length can be obtained in terms of the average of the cosine of the angle between two unit segments as follows.

For some other separation k, (14.2.1) is

$$\overline{\cos\theta_k} = \overline{\cos\theta}^{\,k} \qquad (14.2.2)$$

Also, $\overline{\cos\theta_{j+k}} = \overline{\cos\theta}^{\,j+k}$, and therefore

$$\overline{\cos\theta_j}\;\overline{\cos\theta_k} = \overline{\cos\theta}^{\,j+k} \qquad (14.2.3)$$

This is just the property of an exponential, so (14.2.2) can be written as

$$\overline{\cos\theta_j} = e^{-j/n_p} = e^{-jl/\tilde{l}} \qquad (14.2.4)$$

with

$$n_p \equiv \frac{1}{\overline{\ln\cos\theta}} \qquad (14.2.5)$$

n_p is a measure of the decrease in the correlation between segments as they get farther apart in the chain. When multiplied by the length of a segment l, it is called the *persistence length*, which is defined by

$$\tilde{l} \equiv n_p l = -\frac{l}{\overline{\ln\cos\theta}} \qquad (14.2.6)$$

The persistence length is characteristic of the linear polymer chain under consideration. The correlation between segments decreases rapidly in units of the persistence length. A quantitative measure of this is obtained by writing, from (14.2.4),

$$\frac{\overline{\cos\theta_j}}{\overline{\cos\theta_1}} = e^{-(j-1)\tilde{l}/l} \qquad (14.2.7)$$

This ratio gives the fractional decrease in average correlation angle with increasing separation between segments, relative to the average correlation angle for adjacent segments. It decreases rapidly when measured in units of the number of segments in a persistence length, $j = \tilde{l}/l$.

For $j = 2$ the correlation between segments, as measured by the average cosine of the correlation angle, has decreased to 36.5% of its value between adjacent segments, has gone down to 13.5% for $j = 3$ and is less than 5% for $j = 4$.

A plot of the persistence length, in units of l, is shown in figure 14.1. For very small angles, the persistence length is very high and the chain takes on the properties of a rigid rod as the correlation angle goes to zero. But the persistence length decreases rapidly with increasing angle until, for angles approaching π, it becomes less than the length of a unit segment, since this represents a freely jointed chain.

In terms of the persistence length, equation (13.2.19) for the correlation factor is given by

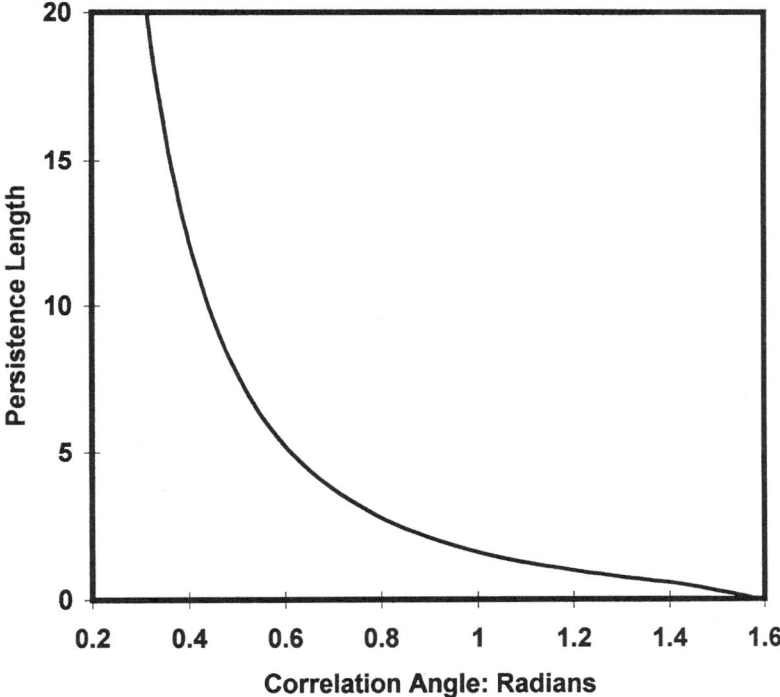

Figure 14.1. Persistence length as a function of correlation angle.

$$f = \frac{1+e^{-1/n_p}}{1-e^{-1/n_p}} \tag{14.2.8}$$

The persistence length is closely related to the effect of correlation on the size of a polymer molecule. The simplest measure of molecular size is the root mean square end-to-end distance given from equation (13.2.10) as

$$\sqrt{\overline{R^2}} = \sqrt{f}\sqrt{Nl^2} \tag{14.2.9}$$

The second factor on the right is the root mean square distance for an ideal freely jointed chain, so the square root of the correlation factor measures the expansion of molecular size arising from correlations in the directions between segments. This expansion is the result of the restrictions on the free random flight imposed by the fact that a segment cannot have every possible orientation relative to its neighbors. Some regions of space are therefore unavailable, so the chain spreads out over a larger volume than would be the case for a freely jointed chain.

The expansion factor, as measured by $\sqrt{f}$, is shown as a function of the average correlation angle in figure 14.2. For high degrees of correlation (small angles) the expansion factor is quite large, but rapidly decreases to unity for large correlation angles.

A physically important point is that, regardless of the correlations among adjacent segments and the bond angle relations between them, a long-enough

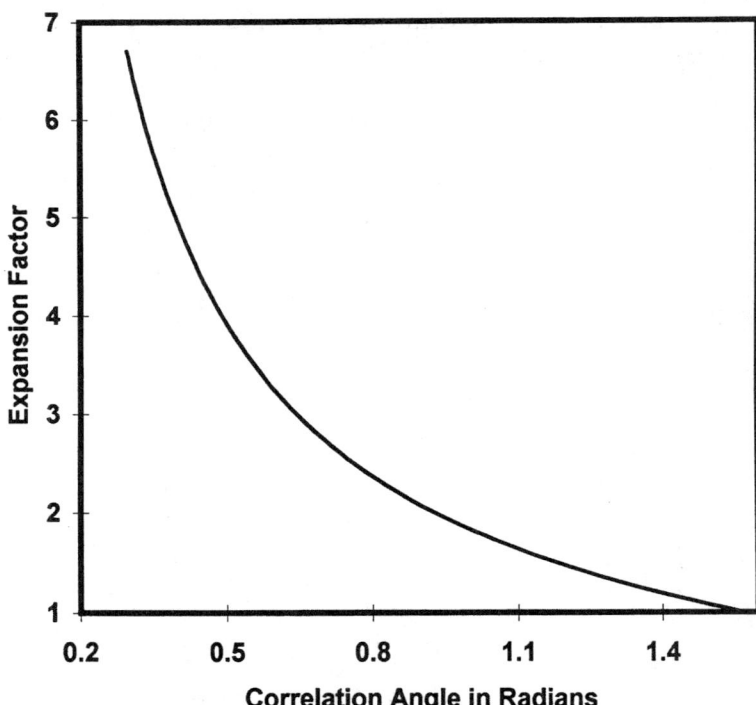

Figure 14.2. Expansion factor as a function of correlation angle.

chain is totally flexible in the sense that it can twist and coil throughout space, provided the scale of resolution examined is larger than the order of the persistence length. A length of the multisegment can be defined to describe this by requiring that, in terms of this new segment, the root mean square distance is the same as for a freely jointed chain. That is, we define a length l_k by the relation

$$\overline{R^2} = N_k l_k^2 = N \frac{l}{l_k} l_k^2 = (Nl) l_k \qquad (14.2.10)$$

where the last equality arises from the fact that the number of multisegments N_k is less than the number of unit segments N by the factor l/l_k. Nl is the contour length of the chain, and l_k is called the *Kuhn length*.

Note that there is a bit of physical ambiguity in the definition of the Kuhn length because, for a chain of correlated segments, it is never precisely true that the correlation factor can be defined to unity by using larger segments. However, as pointed out above, only a rather short string of persistence lengths suffices for the correlations to become very small. Mathematically, there is no ambiguity since the Kuhn length is defined as the mean square length divided by the contour length. Use of Kuhn segments permits simple models to be used, such as the freely jointed chain or the beads-on-a-string model, that are appropriate for the large-scale properties of long polymer chains.

The angular relations among the segments within a distance of the order of a persistence length are ignored in this approach, but it must be remembered that the multisegment defining the persistence length is not a rigid, linear entity. Rather than regarding the chain as a sequence of larger rigid rods when viewed on a scale larger than that of unit segments, it is physically more appropriate to regard it as a random coil of limp spaghetti. That is, on the scale of persistence lengths, the chain is very flexible.

14.3 Chain length fluctuations

The root mean square end-to-end distance is a good measure of the size of a chain molecule. However, the number of possible chain configurations is very large, many of them having an end-to-end distance far from the root mean square. It is therefore of interest to determine the degree of fluctuation of size from the root mean square length. This is measured by the average difference between the square of the end-to-end distance and the mean square distance. Thus, the relative fluctuation Δ_R is defined as

$$\Delta_R^4 = \frac{\overline{\left(R^2 - \overline{R^2}\right)^2}}{\overline{R^2}^2} = \frac{\overline{R^4} - \overline{R^2}^2}{\overline{R^2}^2} \qquad (14.3.1)$$

so all that is necessary is to compute the fourth moment of the distribution of end-to-end distances. To do this, let us use equation (13.5.19) for the freely jointed chain in the form

$$P(R) = \frac{1}{\sigma^3 \sqrt{8\pi^3}} e^{-R^2/2\sigma^2} \qquad (14.3.2)$$

The fourth moment is given by

$$\overline{R^4} = \frac{4\pi}{\sigma^3 \sqrt{8\pi^3}} \int_0^\infty e^{-R^2/2\sigma^2} R^6 dR = 15\sigma^4 \tag{14.3.3}$$

the standard deviation being related to the mean square distance by

$$\overline{R^2} = 3\sigma^2 \tag{14.3.4}$$

Using (14.3.3) and (14.3.4), equation (14.3.1) gives $\Delta^4 = 2/3$, $\Delta (2/3)^{1/4} = 0.904$. This shows that the fluctuations in end-to-end distance are large.

Let us get the ratio of the probability of having a specified end-to-end distance to that of having the root mean square distance. If we let $R = R_{RMS} + \delta$, then

$$\frac{P(R)}{P(R_{RMS})} = e^{-3/2(2\delta/R_S + \delta^2/R_S^2)} \tag{14.3.5}$$

This gives the relative probability of a fluctuation. A plot of (14.3.5), as a function of $2\delta/R_S$, is shown in figure 14.3. From the curve the probability that a fluctuation in length will occur that deviates from the root mean square value

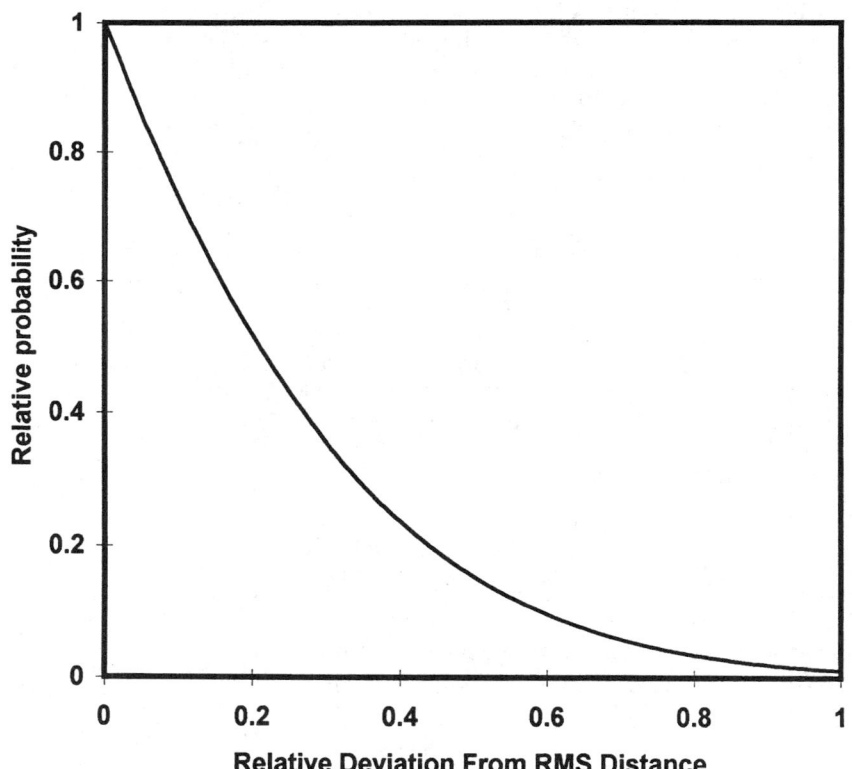

Figure 14.3. Fluctuations for a Gaussian chain. RMS: root mean square.

by 50% is about 15%, again showing that fluctuations from the root mean square distance are substantial.

14.4 Density in a polymer chain

The interactions among segments will be of importance for the partition function. For segments that are adjacent, these interactions can be quite strong. But even segments that are far apart in the chain sequence can approach each other because of coiling, so some measure of the average distance between segments is of interest. This can be obtained as an average density of segments in the volume taken up by the chain. Since a linear measure of the size of the molecule is the root mean square end-to-end distance, an average chain volume is approximately the cube of this. Ignoring unimportant geometrical numerical factors, the volume over which the chain spreads is approximately

$$V_{chain} \approx N^{3/2} l^3 \tag{14.4.1}$$

so the number of segments per unit volume is

$$n_{chain} \approx \frac{1}{N^{1/2} l^3} \tag{14.4.2}$$

Compare this to the hypothetical case in which the chain is tightly coiled, with each segment taking up a volume of l^3 such that the density of segments $\approx l^{-3}$. Then the ratio of the segment density in a random flight chain to that in a tightly coiled chain is $N^{-1/2}$. That is, the chain becomes looser and more spread out with increasing chain size. For a chain consisting of 10,000 segments, for example, the segment density over the volume of the chain is only 0.01 of that in a tightly coiled chain. Most segments, therefore, will not be close to each other and will interact to a negligible degree. Of, course, this is true only for free, isolated chains. In a condensed phase, there will be considerable interaction among segments from different chains. The relatively open structure of chains implies that, in a condensed phase, there is plenty of room for different chains to intertwine and that the random walk results can still be a decent approximate method of describing chain length even though the overall segment density will be much higher than for an isolated chain.

14.5 Partition function of a polymer chain

Consider a polymer chain made up of N monomer units. Each monomer unit consists of a carbon skeleton to which are attached hydrogen atoms and one or more chemical groups, as shown in figure 14.4.

The chemical groups can rotate about the bond angle θ. The rotation is described by the azimuthal angle ϕ. In general, the rotation is not completely free, but is restricted by steric hindrance such that a potential barrier exists at certain values of the relative angles between groups on different monomer units. The potential energy of the chain can be classified as follows:

1. The sum of chemical bond interactions of each atom with atoms adjacent to it: this depends only on the distance between adjacent atoms.

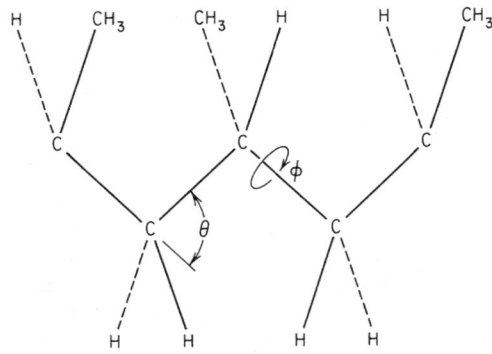

Figure 14.4. Monomer units in a polymer chain.

It includes the minimum of the bond interaction energies and the atomic vibrational energies. To a good approximation, this energy is independent of the chain conformation.

2. The rotational energy arising from the interaction between adjacent groups on the chain: if the rotational barrier is high, this reduces to librational energies. If the barrier is low relative to thermal energy, this becomes the energy of free rotation. The rotational energy arises from physical interactions and is a function of the relative angles between the groups. This is also nearly independent of chain conformation, although the relative rotation between adjacent groups is influenced by the orientation of groups farther away.

3. The physical, van der Waals type of interactions among atoms that are not adjacent to each other: it is convenient to separate this into two parts, interactions among nonadjacent atoms on the same monomer unit and interactions among nonadjacent atoms on different monomer units. The latter interaction is a strong function of chain conformation.

To a first approximation, the energy of a polymer chain can be divided into three parts: the energy of the unit segments, the energy of interaction among adjacent segments, and the energy of interaction among segments that are not adjacent. Let us call these E_1, E_2, E_3, respectively. The first of these can be taken as a constant characteristic of the units making up the chain. The bond angle θ between two adjacent segments can vary (because of thermal fluctuations) over a small range, thereby giving a variation in the distances between the parts of the two segments. This can be approximated by a quadratic function of the distance between the two segments and is a vibrational energy contribution to E_2. There is also a rotational energy contribution arising from the fact that, even though the bond angle can vary only between narrow limits, a segment can rotate through an angle φ about an axis defined by the extension of the preceding segment in the chain. But this rotation is not completely free because, as a segment rotates, its atoms can come close to atoms in the preceding segment. The result is a potential energy of rotation that is a function of the azimuthal rotation angle φ. Depending on the number and nature of the chemical groups attached to the skeletal atoms, the rotational potential can have multiple minima.

The third energy, E_3, is the van der Waals repulsion among atoms on different monomer units and gives rise to the excluded volume. It is often approximated by assigning a single energy of interaction between two segments, thereby replacing interactions among atoms by a kind of average over the atoms in a segment. The partition function for a single chain is then

$$Z_c = \sum_{E_j} \sum_{R,\theta,\phi} W(R)W(\theta,\phi)e^{-E_1/kT}e^{-E_2/kT}e^{-E_3/kT} \tag{14.5.1}$$

The parameters R, θ, ϕ define the configuration of the polymer chain, and E_j denotes the jth quantum state for a given configuration. The number of configurations for a given energy has been factored into two parts that are assumed to be independent. The first, $W(R)$, is the number of ways that the chain can achieve an end-to-end distance R, as given by random walk models. The second is the number of ways that adjacent segments can be oriented relative to each other for a given energy. It will be assumed that the chain has a single bond angle such that $W(\theta,\phi) = W(\phi)$. If the segments can rotate freely about this bond angle, then $W(\theta,\phi) = W(\theta) = 1$. The inner sum is over all configurations for a given energy while the outer sum is over all energy levels.

This is a rather formidable form for the partition function, but it can be readily transformed to the point where simplified models can be applied. The energies in (14.5.1) are taken to be independent such that the partition function can be factored into

$$Z = Z_1 Z_2 Z_3 \tag{14.5.2}$$

in an obvious notation. Our interest is in properties that strongly depend on chain length, so we will focus on Z_3 and write it explicitly as

$$Z_3 = \sum_R W(R)e^{-E_3/kT} \tag{14.5.3}$$

A basic task of polymer statistics is to evaluate this partition function.

14.6 Excluded volume

The models of chain configuration discussed so far are based on random flights for which the segments can interpenetrate. But a polymer chain is more accurately modeled by a self-avoiding flight in which no jump that either brings the particle to a point that was previously visited, or that crosses any previous jump vector, is allowed. This ensures that each segment occupies a volume that is unavailable to other segments. The possible configurations of a chain are therefore more limited than those for a fully random walk. The consequences of this are collectively called the *excluded volume effect*.

It is obvious that there must be a relation between the excluded volume and the size as measured by the root mean square end-to-end distance. An estimate of the root mean square distance, relative to that for a fully random walk, can be obtained from a simple mean field approximation to the partition function by first replacing the sum over states by the most probable state. That is, the partition function (14.5.3) function is approximated by

$$Z = W(\tilde{R})e^{-\tilde{E}/kT} \tag{14.6.1}$$

where $W(\tilde{R})$ is the number of configurations with the most probable end-to-end distance $\tilde{R}$ and $\tilde{E}$ is the total energy of the most probable chain. This most probable distance is not the same as the most probable root mean square distance of a random walk because it includes the effect of the Boltzmann factor of the energy. To get the most probable state, the partition function (14.6.1), will be written for an arbitrary R, and the Helmholtz free energy obtained from this partition function will be minimized with respect to R.

It will be assumed that the number of conformations for an arbitrary R is proportional to the Gaussian probability distribution. That is,

$$W(R) = \frac{W_{tot}}{\sigma^3 \sqrt{8\pi^3}} e^{-R^2/2\sigma^2} \tag{14.6.2}$$

W_{tot} being the total number of possible configurations of the chain. Recall that the general derivation of the Gaussian distribution requires that the jump probability be the same for each jump. This is clearly not true for the first members of the sequence of jumps for a self-avoiding flight. Also, for any number of jumps, the probability that a self-avoiding flight results in a very small value of R is very low because this would require the superposition of many segments. However, for jumps later in the sequence, each jump sees, on the average, a similar environment, so the Gaussian becomes more accurate with increasing N. Since it is only valid for large N in any case, and since the most probable root mean square distance results in a rather open structure (as shown in section 14.4), we can adopt it as an approximate representation even for a self-avoiding walk.

In accord with the discussion of section 14.5, the energy can be separated into two parts: that which is the result of interactions among the segments, and all other energy contributions. Thus, the energy can be written as

$$E = E_o + \Delta E \tag{14.6.3}$$

E_o being the energy in the absence of the excluded volume effect and ΔE being the interaction energy among all segments in the chain. It is at this point that the mean field approximation is made by assuming that the total segment interaction energy can be replaced by the average interaction of one segment with all the others. This average interaction energy is proportional to the average density, $\rho \propto N/R^3$ of segments, so if ε is the average interaction energy per segment, then, again ignoring unimportant geometric factors,

$$\Delta E = K\varepsilon\rho = KN\varepsilon \frac{N}{R^3} \tag{14.6.4}$$

the effective volume of the chain being represented by R^3.

K is a proportionality constant with the dimensions of volume whose value is determined by the fact that the smallest possible increase in density is the reciprocal of the volume of one segment, and for such an increase the energy of interaction of one segment with the rest of the system is ε. That is, $(1/N)(d\Delta E/d\rho) = \varepsilon/(1/l^3)$. Comparing this with the derivative of (14.6.4) shows that $K = l^3$. Then (14.6.4) is

$$\Delta E = l^3 N\varepsilon \left(\frac{N}{R^3}\right) \tag{14.6.5}$$

The partition function (14.6.1) for arbitrary R is then

$$Z = \frac{W_{tot}}{\sigma^3 \sqrt{8\pi^3}} e^{-R^2/2\sigma^2} e^{-E_0/kT} e^{-N^2 l^3 \varepsilon/R^3 kT} \tag{14.6.6}$$

and the Helmoltz free energy is

$$A = -kT \ln Z = \frac{kTR^2}{2\sigma^2} + \frac{\varepsilon N^2 l^3}{R^3} + \text{constant} \tag{14.6.7}$$

where the last term does not depend on R. Differentiating (14.6.7) with respect to R, setting the derivative equal to zero, and solving gives the value of $\tilde{R}$ that minimizes the free energy. That is,

$$\tilde{R} = \left(\frac{\varepsilon}{kT} l^5 N^3\right)^{1/5} \tag{14.6.8}$$

Note that in getting (14.6.8), the standard deviation has been replaced by its random walk value according to $\sigma = (Nl^2/3)^{1/2}$. The excluded volume effect therefore changes the dependence of the end-to-end distance on N from $N^{1/2}$ to the faster $N^{3/5}$. That is, because of the excluded volume, the chain becomes larger as the number of segments increases, relative to the unmodified random flight value. The ratio of the size taking into account excluded volume relative to that for a fully free random walk chain is

$$\frac{\tilde{R}}{\sqrt{\overline{R^2}}} = \frac{(\varepsilon l^5 N^3/kT)^{1/5}}{l\sqrt{N}} = \left(\frac{\varepsilon}{kT}\right)^{1/5} N^{1/10} \tag{14.6.9}$$

Let us revisit the estimate of the segment density of section 14.4, taking into account the excluded volume. Since, from (14.6.7), the volume is proportional to $N^{9/5}$, relative to a tightly coiled chain, the segment density is proportional to $N^{-4/5}$. That is, the chain becomes looser with increasing N faster than for the case of a fully free random walk. This is expected since the mean field parameter is repulsive because it describes the resistance to overlap of segments.

14.7 The force ensemble and chain elasticity

The linear dimensions of free polymer chains are much smaller than the contour length because they are extensively coiled. This means that, if a force is applied to a polymer chain, it can stretch by a large amount. The extreme elasticity of rubbers and gels arises from this fact. An understanding of this phenomenon starts with the study of a single chain subject to a tensile force. The resistance to an external force in a solid polymer has two sources. For a system in which there is no cross-linking, the chains interact through van der Waals forces that provide the reaction to an applied force. This can be quite substantial because the chains are long and tangled up with each other. In chains that are cross-linked in the condensed phase, the specimen is essentially one giant molecule and chains are prevented from slipping past each other by the cross-links. The elastic properties depend on the density of cross-links since the possible extension of the specimen is determined by the average distance between links. If this distance is short, then the chains cannot stretch very much and the system is relatively stiff. If the distance between links is long, then the chains can stretch quite a lot. This is the case for rubber.

Assume that a tensile force F_x is applied to the ends of a polymer chain in the direction of the end-to-end vector **R**. In complete analogy with the pressure partition function (see section 2.16), a force ensemble can be defined that has a force partition function given by

$$Z_F = \sum_R \sum_E W(R_x) e^{-E(R_x)/kT} e^{F_x R_x/kT}$$
$$= \sum_R Z(R_x) e^{F_x R_x/kT} \qquad (14.7.1)$$

F_x is the force exerted *on* the chain, so the sign in the exponential is positive, just as in the magnetic case, where the outer sum is over all possible values of the end-to-end distance and $Z(R_x)$ is the canonical partition function for a chain of length R_x given by

$$Z(R_x) = \sum_E W(R_x) e^{-E(R_x)/kT} \qquad (14.7.2)$$

which is just the canonical partition function (14.5.1) in a slightly different notation. The subscript x indicates that the x-axis of the coordinate system has been chosen to lie along the line of the applied force, and the sum is over all quantum states for a given value of R_x.

Note that the probability that the chain has energy $E(R_x)$ and length R_x is

$$f(E, R_x) = \frac{e^{-E(R_x)/kT} e^{F_x R_x/kT}}{Z_F} \qquad (14.7.3)$$

and the probability that the chain has a length R_x regardless of the energy is

$$f(R_x) = \sum_{E(R_x)} f(E, R_x) \frac{Z(R_x) e^{F_x R_x/kT}}{Z_F} \qquad (14.7.4)$$

The Gibbs free energy for the force ensemble is given by

$$G = -kT \ln Z_F \qquad (14.7.5)$$

and from thermodynamics,

$$G = U - TS - F_x \overline{R_x} \qquad (14.7.6)$$

$$dG = -SdT - \overline{R_x} dF_x \qquad (14.7.7)$$

$$dA = -SdT - F_x d\overline{R_x} \qquad (14.7.8)$$

From this,

$$F_x = \left(\frac{\partial A}{\partial \overline{R_x}}\right)_T = \left(\frac{\partial U}{\partial \overline{R_x}}\right)_T - T\left(\frac{\partial S}{\partial \overline{R_x}}\right)_T \qquad (14.7.9)$$

This separates the force into two parts: the variation of the energy with elongation and the variation of entropy with elongation. From (14.7.7),

$$\overline{R_x} = -\left(\frac{\partial G}{\partial F_x}\right)_T = kT\left(\frac{\partial \ln Z_F}{\partial F_x}\right)_T \qquad (14.7.10)$$

which gives the force–elongation relation if the partition function can be found as a function of the applied force.

Note that for the pressure ensemble, the PV work term had a negative sign, whereas for the force ensemble the work term has a positive sign. The reason for this is that a positive pressure is defined as compressing the system but the tension force is defined as positive if it expands the system.

For a random flight model of the chain, the mean end-to-end distance is zero, but this is obviously not necessarily the case when a force is applied to the chain. In a freely jointed chain there is no excluded volume effect, so $E_3 = 0$. Also, $W(\theta, \phi) = 1$ and we take all the other energy terms to be independent of chain configuration. In addition, these energy terms are assumed to be the energy of one segment times the number of segments. That is, all energy terms except E_1 of section 14.5 are neglected. Then (14.7.1) simplifies to

$$Z_F = z_s^N \sum_{\mathbf{R}} W(R_x) e^{-F_x R_x / kT} \tag{14.7.11}$$

where $z_s = \Sigma_s e^{-es/kT}$ is the partition function for a single segment. Our task is now reduced to finding $W(R_x)$, which is the number of conformations of the chain that give an end-to-end distance of R_x. This can be done as follows.

Each segment of the chain is represented by a vector $\mathbf{r}_i$ of length l. If x_i is the x-component of the ith segment, then

$$R_x = \sum_{i=1}^{N} x_i \tag{14.7.12}$$

Now let us collect all x_i that have the same values into groups and rewrite (14.7.12) as

$$R_x = \sum_{k=1}^{n} N_k x_k \tag{14.7.13}$$

where N_k is the number of segments whose x-component is x_k and n is the number of possible values of x_k. Then (14.7.11) is

$$Z_F = z_s^N \sum_{\{x_k\}} W(x_k) e^{F_x \sum_k N_k x_k / kT} \tag{14.7.14}$$

and $W(x_k)$ is the number of ways that the segments can be distributed to give a set of x_k that add up to R. If N_k is the number of segments with a projection x_k on $\mathbf{R}$, then $W(x_k)$ is just the number of ways of putting N objects in boxes such that N_k are in the kth box. Then

$$Z_F = z_s^N \sum_{\{k\}} \frac{N!}{\prod_k N_k!} e^{F_x \sum_k N_k x_k / k}$$

$$= z_s^N N! \sum_{\{k\}} \prod_{k=1}^{n} \frac{(e^{F_x x_k / kT})^{N_k}}{N_k!} \tag{14.7.15}$$

But the sum over the products is just the expansion of a multinomial according to the multinomial theorem, so

$$Z_F = z_s^N z_\tau^N \tag{14.7.16}$$

where z_τ is defined by

$$z_\tau = \sum_{i=1}^{n} e^{F_x x_i/kT} \qquad (14.7.17)$$

which is the contribution of the force to the partition function.

For the freely jointed chain, the components x_i can take on all values between $-l \le x_i \le +l$. The sum in (14.7.17) can therefore be replaced by an integral and evaluated as follows:

$$\sum_{i=1}^{n} e^{F_x x_i/kT} = \int_{-l}^{+l} e^{F_x x/kT} dx$$

$$= \frac{kT}{F_x}(e^{F_x l/kT} - e^{-F_x l/kT}) \qquad (14.7.18)$$

so (14.7.16) becomes

$$Z_F = \left(z_s \frac{2l}{\tilde{F}_x} \sinh \tilde{F}_x\right)^N \quad \text{(freely jointed chain)} \qquad (14.7.19)$$

where $\tilde{F}_x \equiv F_x l/kT$.

This procedure can also be applied to the case of the restricted jointed chain. If each segment can take on any azimuthal angle relative to the preceding segment, but can only take on longitudinal angles between $-\theta_s$ and $+\theta_s$, then the x_i can take on a continuum of values between $-l\cos\theta_s$ and $+l\cos\theta_s$, so instead of (14.7.18) we get

$$\sum_{i=1}^{n} e^{-F_x x_i/kT} = \frac{kT}{F_x}(e^{F_x l \cos\theta_s/kT} - e^{-F_x l \cos\theta_s/kT}) \qquad (14.7.20)$$

which gives the same form as (14.7.19) except that now

$$\tilde{F}_x \equiv F_x l \cos\theta_s/kT \qquad (14.7.21)$$

The mean length of the chain can now be obtained from (14.7.10) by performing the indicated differentiation of the partition function. This gives

$$\overline{R_x} = Nl\left(\frac{\cosh \tilde{F}_x}{\sinh \tilde{F}_x} - \frac{1}{\tilde{F}_x}\right) \qquad (14.7.22)$$

The factor in the brackets is the Langevin function, which is defined by $L(y) = \coth y - 1/y$, so

$$\overline{R_x} = NlL(\tilde{F}_x) \qquad (14.7.23)$$

As the argument approaches infinity, the Langevin function approaches unity. Therefore, as would be expected, the maximum extension of the chain is equal to the contour length Nl, which is reached only for an infinite force acting on the chain.

For forces that are small relative to l/kT, the force–extension relation is linear, as can be seen from the expansion of the Langevin function as follows:

$$L(\tilde{F}_x) = \frac{1}{\tilde{F}_x} + \frac{\tilde{F}_x}{3} - \frac{\tilde{F}_x^3}{45} + \cdots - \frac{1}{\tilde{F}_x} \quad \text{(for small } \tilde{F}_x\text{)}$$

$$= \frac{\tilde{F}_x}{3} \tag{14.7.24}$$

so

$$\overline{R_x} = \frac{Nl}{3} \tilde{F}_x \tag{14.7.25}$$

but $\tilde{F}_x \equiv F_x l \cos \theta_s / kT$, so

$$\overline{R_x} = \frac{Nl}{3} F_x l \cos \theta_s / kT \tag{14.7.26}$$

For a freely jointed chain with $\theta_s = 0$, the x-component of the mean square end-to-end distance, in the absence of any force, is

$$\overline{R_o^2} = \frac{Nl^2}{3} \tag{14.7.27}$$

so (14.7.25) can be written as

$$\tilde{F}_x = \frac{1}{\overline{R_o^2}} \overline{R_x} \tag{14.7.28}$$

Using (14.7.21) this can also be written as

$$F_x = \frac{kT}{\overline{R_o^2} \cos \theta_s} \overline{R_x} \tag{14.7.29}$$

which shows that for small strains Young's modulus is a linear function of temperature and depends only on the bond angle and root mean square distance of the undeformed chain.

The plot of $\overline{R_x}/NL$ versus $\tilde{F}_x$ in figure 14.5 displays the initial linear region for small forces, with a slope that decreases with increasing force until the maximum extension is reached.

14.8 Elastomers

Elastomers are substances that can be reversibly deformed to a very large extent. Rubber, the prototypical elastomer, can be stretched to several times its original length, the strain being fully reversible, so the material recovers its original dimensions on removal of the force. The molecular structure that is responsible for this consists of long, flexible chains cross-linked into a network. The chains can stretch by uncoiling, and the network structure prevents chain slipping that would result in permanent deformation. A schematic representation of such a structure is shown in figure 14.6.

Consider a system consisting of a macroscopic specimen of rubber in the form of a rectangular parallelepiped of cross-sectional area oriented on a

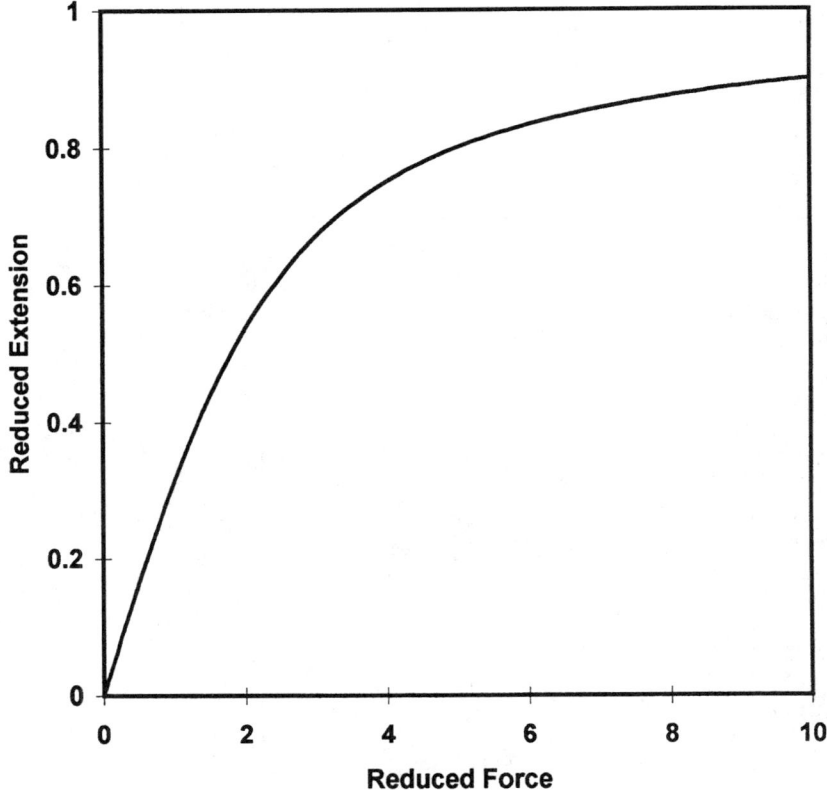

Figure 14.5. Theoretical extension–force curve for a polymer chain.

Cartesian coordinate system such that, in the absence of any force, its length lies along the x-axis with a value L_o. Now apply a tensile force F_x along the x-axis so that the specimen stretches in the x-direction to a new value L. The thermodynamic equations (14.7.6)–(14.7.9) are valid for this system, except that we now replace the mean distance $\overline{R}_x$ by L and therefore

$$G = U - TS - F_x L \tag{14.8.1}$$

$$dG = -SdT - LdF_x \tag{14.8.2}$$

$$dA = -SdT + F_x dL \tag{14.8.3}$$

$$F_x = \left(\frac{\partial A}{\partial L}\right)_T = \left(\frac{\partial U}{\partial L}\right)_T - \left(\frac{\partial S}{\partial L}\right)_T \tag{14.8.4}$$

Another useful thermodynamic relation is the Maxwell reciprocity relation,

$$\left(\frac{\partial S}{\partial L}\right)_T = -\left(\frac{\partial F_x}{\partial T}\right)_L \tag{14.8.5}$$

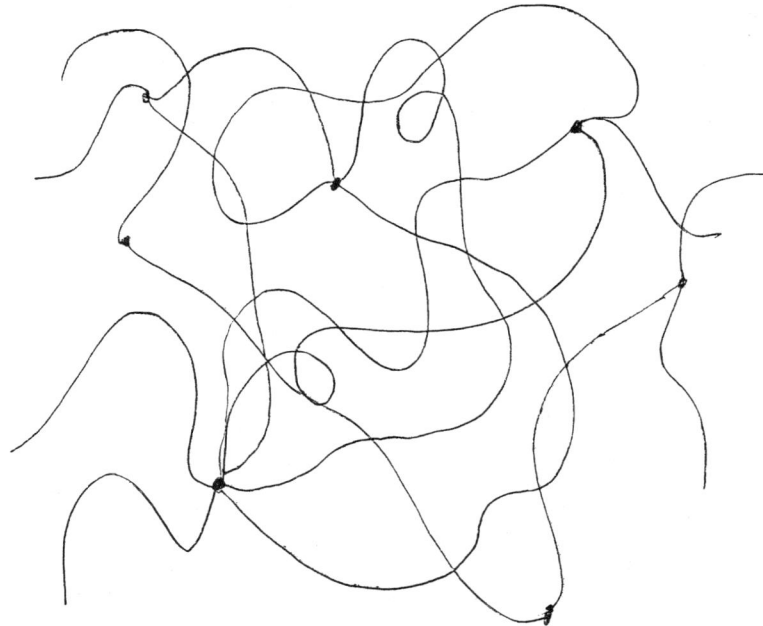

Figure 14.6. Chain structures for elastomers.

which is the analog of equation (1.16.8). (Again, the difference in sign between the pressure and force equations arises from the definition of pressure acting on the system as compression and force acting as tension.)

Using (14.8.5), equation (14.8.4) gives

$$F_x = \left(\frac{\partial U}{\partial T}\right)_T + I\left(\frac{\partial F_x}{\partial T}\right)_L \qquad (14.8.6)$$

By measuring the force as a function of both extension and temperature, the last term in (14.8.6) can be obtained experimentally, and therefore the energy contribution can be derived from data. It is found that the energy contribution is very nearly zero. This is reasonable because the segment density in rubber is comparable to the molecular or atomic density of other condensed systems and the deformation arising from changes in distances among segments is small relative to dimensional changes arising from uncoiling. Therefore, to a good approximation, the force–extension relation for rubber is an entropy effect and is given by

$$F_x = \left(\frac{\partial F_x}{\partial T}\right)_L = -T\left(\frac{\partial S}{\partial L}\right)_T \qquad (14.8.9)$$

the major contribution to this being the conformational entropy of the chains.

The conformational entropy of a single chain is determined by the number of different conformations that have an end-to-end distance R and is proportional to the probability density $P(R)$. The entropy of an ensemble of chains is proportional to the product of the probabilities for all possible values of R, the proportionality constant being the total number of conformations for all possible values of R multiplied by Boltzmann's constant. Upon deforming the system, the probability that a chain has a given value of R changes although the proportionality constant does not. The entropy change is therefore the sum of the entropy changes for each chain:

$$S - S_o = \sum_i N_i \Delta S_i \tag{14.8.10}$$

where N_i is the number of chains with initial length R_i and ΔS_i is the change in conformational entropy of a single chain upon deformation, which is given by

$$\Delta S_i = k \ln \frac{W(R_i^*)}{W(R_i)} \tag{14.8.11}$$

$W(R_i)$ is the number of ways to form a chain with end-to-end distance R_i for the undeformed state, and $W(R^*_i)$ is the corresponding probability in the deformed state. Note that for rubber a chain is defined as the number of segments between two cross-links. The cross-links determine both the number and contour length of chains. Since we are assuming that the chains are Gaussian, both probabilities in (14.8.11) are proportional to Gaussian probabilities according to

$$W(R) \propto P(R) dX dY dZ = \frac{1}{\sigma^3 \sqrt{8\pi^3}} e^{-R^2/2\sigma^2} dX dY dZ \tag{14.8.12}$$

before deformation, and

$$W(R^*) \propto P(R^*) dX^* dY^* dZ^* = \frac{1}{\sigma^3 \sqrt{8\pi^3}} e^{-R^{*2}/2\sigma^2} dX^* dY^* dZ^* \tag{14.8.13}$$

after deformation.

After deformation each point R moves to R^* and $R^2 = X^2 + Y^2 + Z^2$ goes to

$$R^{*2} = X^{*2} + Y^{*2} + Z^{*2} = \lambda_x^2 X^2 + \lambda_y^2 Y^2 + \lambda_z^2 Z^2 \tag{14.8.14}$$

where the λs are defined by

$$\lambda_x = \frac{X^*}{X}, \quad \lambda_y = \frac{Y^*}{Y}, \quad \lambda_z = \frac{Z^*}{Z} \tag{14.8.15}$$

and

$$dX^* = \lambda_x dX, \quad dY^* = \lambda_y dY, \quad dZ^* = \lambda_x dZ \tag{14.8.16}$$

Putting these into (14.8.13) gives the probability in terms of the relative elongations and the original distances as

$$P(R^*) dX^* dY^* dZ^* = \frac{\lambda_x \lambda_y \lambda_z}{\sigma^3 \sqrt{8\pi^3}} e^{-(\lambda_x^2 X^2 + \lambda_y^2 Y^2 + \lambda_z^2 Z^2)/2\sigma^2} dX dY dZ \tag{14.8.17}$$

Therefore, adopting the continuum notation, (14.8.11) can be written as

$$\Delta S_i(R) = k\ln(\lambda_x\lambda_y\lambda_z) + \frac{k}{2\sigma^2}[X^2(1-\lambda_x^2) + Y^2(1-\lambda_Y^2) + Z^2(1-\lambda_z^2)] \quad (14.8.18)$$

Summing these for all M chains according to (14.8.10) gives, in the continuum notation,

$$S - S_o = Mk\ln(\lambda_x\lambda_y\lambda_z) + \frac{k}{2\sigma^2}\Big[(1-\lambda_x^2)\int_X N(X)X^2 dX$$
$$+ (1-\lambda_y^2)\int_Y N(Y)Y^2 dY + (1-\lambda_z^2)\int_Z N(Z)Z^2 dZ\Big] \quad (14.8.19)$$

$N(X)$, $N(Y)$, $N(Z)$ being the number of chains with components of the end-to-end distance given by X, Y, Z, and we have assumed that all chains deform in the same way. The integrals are proportional to mean square lengths. That is, $\int_X N(X)X^2 dX = M\overline{X^2}$, and so on, so (13.8.19) becomes

$$S - S_o = Mk\ln(\lambda_x\lambda_y\lambda_z) + \frac{Mk}{2\sigma^2}\Big[\overline{X^2}(1-\lambda_x^2) + \overline{Y^2}(1-\lambda_y^2) + \overline{Z^2}(1-\lambda_z^2)\Big] \quad (14.8.20)$$

but $\overline{X^2} = \overline{Y^2} = \overline{Z^2} = \overline{R^2}/3$ and $\sigma^2 = R^2/3$. Therefore, (13.8.20) is

$$S - S_o = Mk\ln(\lambda_x\lambda_y\lambda_z) + \frac{Mk}{2}(3 - \lambda_x^2 - \lambda_y^2 - \lambda_z^2) \quad (14.8.21)$$

The entropy change depends only on the deformation and the number of chains.

For rubber, the volume changes are negligible and will be taken to be constant. This means that

$$\lambda_x\lambda_y\lambda_z = 1 \quad (14.8.22)$$

Since both the force and the elongation are taken to be along the x-axis, from (14.8.22) the y- and z-dimensions must shrink according to $\lambda_y = \lambda_z = 1/\sqrt{\lambda_x}$ and (14.8.21) reduces to

$$S - S_o = \frac{Mk}{2}\left(3 - \lambda_x^2 - \frac{2}{\lambda_x}\right) \quad (14.8.23)$$

Now the force–elongation relation can be obtained from (14.8.9) as

$$F_x = -T\left(\frac{\partial S}{\partial L}\right)_T = \frac{MkT}{L_o}\left(\lambda_x - \frac{1}{\lambda_x^2}\right) \quad (14.8.24)$$

It is convenient to divide through by the prefactor on the right-hand side of (14.8.24) to give a reduced force $F_x^{(R)}$ defined by

$$F_x^{(R)} = \left(\lambda_x - \frac{1}{\lambda_x^2}\right) \quad (14.8.25)$$

A plot of this relation is shown in figure 14.7 along with experimental points given by L.R.C. Treloar.[1] The experimental force data were given per initial unit area in kg/cm.[2] These were divided by 3.32 to give a dimensionless reduced force that matches (14.8.25) at intermediate elongation values.

The theory does fairly well up to elongations of five and higher. This is good

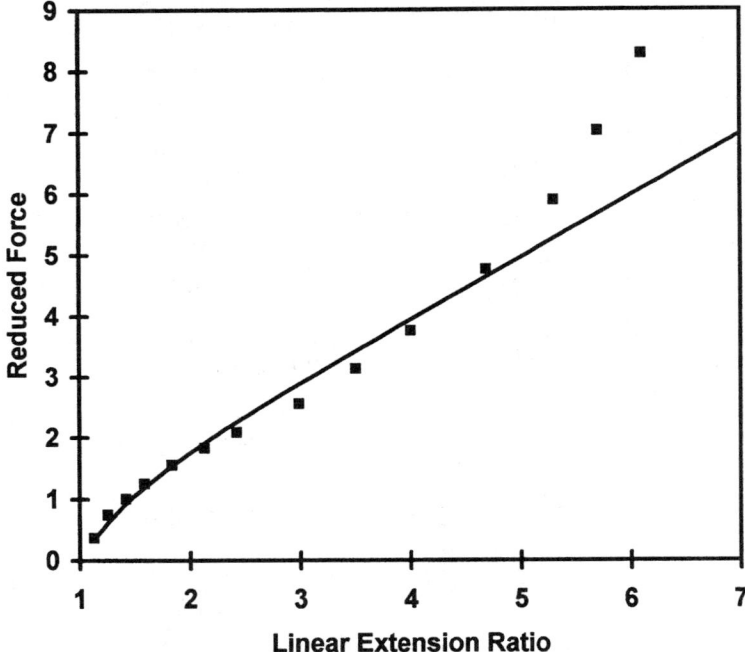

Figure 14.7. Comparison of theory and experiment for the elongation of rubber.

evidence that the stretching of rubber is indeed determined by the entropy of coils, at least at elongations that are not too high. The deviations between theory and experiment arise form several sources, the main ones being the failure of the Gaussian probability for the end-to-end distances of the chains, and possible entanglements of chains that hinder complete uncoiling. Note that rubberlike elastic properties are often found in long chain systems without cross-linking because of entanglements that can act as cross-links if the force is not too high.

For a specimen of initial length L and cross-sectional area A Young's modulus, which is the stress per unit strain, is given by

$$F_Y = \frac{F_x L_o}{A \Delta L} = \frac{F_x}{A(\lambda_x - 1)} \tag{14.8.26}$$

Using (14.8.24) and the definition of the relative elongations, this becomes

$$F_Y = \frac{MkT}{A_o L_o} = \frac{(\lambda_x^2 + \lambda_x + 1)}{\lambda_x} \tag{14.8.27}$$

In the limit of small forces, the relative elongations go to unity and we have

$$F_Y = (F_x \to 0) \frac{3MkT}{V} \tag{14.8.28}$$

Young's modulus depends only on the number of chains per unit volume, and these can be measured from a simple elongation experiment. This is equivalent to determining the average molecular weight of a chain because the density ρ is related to the molecular weight M_W through Avagadro's number N_{av} by

$$\rho = \frac{mass}{V} = \frac{MM_W}{N_{av}V}$$

so

$$F_Y(F_x \to 0) = \frac{3\rho N_{av} kT}{M_W} \quad (14.8.29)$$

This remarkable result states that measurements of moduli can give data on the number of chains per unit volume and their molecular weight.

14.9 The Flory correction

In a network, there is a contribution to the entropy that is proportional to the number of ways that cross-links can be formed. The length of a chain is defined as the distance between cross-links such that, for M chains, the number of cross-links is $M/2$. A cross-link can occur only if two chain segments are close enough together. Following Flory, we assume that, in a cross-link, the segments must be within a volume δV. The number of ways of forming a cross-link is proportional to the probability that two segments are within this volume. But, assuming equal accessibility to all space in the volume V, this probability is proportional to $\delta V/V$. Upon deformation, the volume changes to V^* but δV does not. So for the deformed system, the entropy for a cross-link is proportional to $\delta V/V^*$. Thus, the ratio of the total number of ways of forming $M/2$ cross-links in the deformed system to that in the undeformed system is $(V/V^*)^{M/2} = (\lambda_x\lambda_y\lambda_z)^{-M/2}$ and the corresponding entropy change is $-(M/2)kT\ln(\lambda_x\lambda_y\lambda_z)$
Adding this to (14.8.21) gives the corrected entropy as

$$S - S_o = \frac{Mk}{2}[\ln(\lambda_x\lambda_y\lambda_z) + (3 - \lambda_x^2 - \lambda_y^2 - \lambda_z^2)] \quad (14.9.1)$$

Note that this correction has no effect on the force strain relation for rubber given by (14.8.24) because volume is conserved and the first term is a constant.

14.10 Solutions and gels

Elastomers can absorb surprisingly large amounts of certain liquids without dissolving because dissolution is prevented by the existence of cross-links. Rubber exposed to hydrocarbons, for example, swells considerably. A gel is essentially a solution containing a fraction of polymer that cannot go below a value dictated by the existence of the cross-links. A relatively simple lattice theory of solutions containing polymers has been given by Paul Flory and M.L. Huggins[2] that describes such solutions reasonably well and is applicable to gels.

Let the system be a liquid solvent in contact with a polymer that has absorbed some solvent. If the polymer has no cross-links, then the system may have any composition, depending on the chemical potentials of the components. The possible compositions can range from dilute to highly concentrated solutions. With cross-linking, the possible compositions are restricted since very dilute solutions are prevented from being formed.

The lattice model regards space as being divided into cells of equal size, each of which can accommodate either a solvent molecule or a polymer segment. Any differences in size are ignored, so the molecular fractions of segments and of solvent molecules are equal to their respective volume fractions. The polymer chains are treated as connected strings of beads, each bead taking up the same volume as a solvent molecule. The beads on a string are laid out in a connected set of lattice sites, and all lattice sites not occupied by beads are taken up by solvent molecules. It is assumed that the solution is random in the sense that solvent molecules and polymer segments can occupy all sites with equal probability consistent with the requirements that a chain is a connected set of segments. Also, a mean field approximation is used for the interaction energies. The procedure is to compute the entropy of mixing from the statistical count, and the energy from mean field theory, and get chemical potentials from the resulting free energy. This determines the composition of the system.

Let M_2 be the number of polymer molecules, each having N segments, so that the number of sites occupied by polymer segments is $M_2 N$ and the total number of sites is $M_T = M_1 + NM_2$ where M_1 is the number of sites occupied by solvent molecules. Then the site fractions, also equal to the volume fractions, are given by

$$f_1 = \frac{M_1}{M_T} = \frac{M_1}{M_1 + NM_2} \qquad (14.10.1)$$

$$f_2 = \frac{NM_2}{M_T} = \frac{NM_2}{M_1 + NM_2} \qquad (14.10.2)$$

To get the entropy of mixing, count the number of ways of arranging M_1 solvent and M_2 polymer molecules on the M_T lattice sites. Do this by counting the number of ways of putting the first chain on the lattice, then the second chain, and so on. When all chains have been inserted, then put in the solvent molecules. But after all the chains are in, there is only one way of putting in solvent molecules on the lattice, so only a count for the chains is needed.

Let w_1 be the number of ways of putting in the first chain, w_2 the number of ways of putting in the second chain after the first is in place, and so on. Then the statistical count is

$$W(M_1, M_2) = \frac{1}{M_2!} \prod_{j=1}^{M_2} w_j \qquad (14.10.3)$$

w_j being the number of ways of putting the jth chain on the lattice after $(j + 1)$ chains have been inserted. The $M_2!$ corrects for the indistinguishability of identical polymer chains. The calculations proceeds by expressing w_{j+1} in terms of w_j as follows.

After putting in the jth chain, the number of occupied sites is jN and the fraction of occupied sites at this stage is $\alpha_j = jN/M_T$. Then the first segment of the $(j + 1)$st chain can be placed in the lattice in $(M_T - jN)$ ways since this is the number of empty sites. The second segment can only go into one of the nearest neighbors of the first segment, and assuming a random distribution, this can be done in $z(1 - \alpha_j)$ ways, z being the number of nearest neighbors, since $(1 - \alpha_j)$ is the probability that a site is empty. In laying down the second

LINEAR POLYMER CHAINS

segment, only $(z - 1)$ nearest neighbor sites are available, so this can be done in $(z - 1)(1 - \alpha_j)$ ways. The number of ways of putting in the third segment is a bit more involved because the first two segments can leave either $(z - 1)$ or $(z - 2)$ vacant nearest neighbor sites for the third segment. Thus, the average number of sites not available to the third segment is between 1 and 2. Call this η, so the number of ways of laying down the third segment is $(z - \eta)(1 - \alpha_j)$. Now assume that the number of ways of putting down all subsequent segments is also $(z - \eta)(1 - \alpha_j)$. That is, the number of sites excluded is the same for all segments except for the first. (Actually, it is customary to approximate even further by taking $\eta = 1$ for all segments on the grounds that the theory is quite approximate anyway, but let us carry the theory forward a bit before making this approximation.) The statistical count for putting down $(N - 1)$ segments of the chain on the lattice is then

$$w_{j+1} = (M_T - jN)z(1 - \alpha_j)(1 - \alpha_j)(z - 1)(z - \eta)^{N-3}(1 - \alpha_j)^{N-3} \quad (14.10.4)$$

or, since $(1 - \alpha_j) = (M_T - jN)/M_T$,

$$w_{j+1} = (M_T - jN)^N \frac{z(z-1)}{M_T^2}\left(\frac{z-\eta}{M_T}\right)^{N-3} \quad (14.10.5)$$

Taking logarithms gives

$$\ln w_{j+1} = N \ln(M_T - jN) + (N-3)\ln\frac{z-\eta}{M_T} + \ln\frac{z(z-1)}{M_T^2} \quad (14.10.6)$$

Now take the product of (14.10.6) for all chains (i.e., for $j = 0$ to $j = M_2 - 1$) to get

$$\ln \prod_{i=1}^{M_2} w_i = \sum_{j=0}^{M_2-1} \ln w_{j+1}$$

$$= N \sum_{j=0}^{M_2-1} \ln(M_T - jN) + M_2\left[(N-3)\ln\frac{z-\eta}{M_T} + \ln\frac{z(z-1)}{M_T^2}\right] \quad (14.10.7)$$

For a coordination number of $z = 12$, varying η from 1 to 2 or replacing z by $(z - 1)$ changes the log term by about 5%, so it is sufficiently accurate to write (14.10.7) as

$$\ln \prod_{i=1}^{M_2} w_i = N \sum_{j=0}^{M_2-1} \ln(M_T - jN) + M_2(N-1)\ln\frac{z-1}{M_T} \quad (14.10.8)$$

But

$$\sum_{j=0}^{M_2-1} \ln(M_T - jN) = \sum_{j=0}^{M_2-1} \ln N\left(\frac{M_T}{N} - j\right)$$

$$= \sum_{j=0}^{M_2-1} \ln\left(\frac{M_T}{N} - j\right) + M_2 \ln N$$

and

$$\ln \prod_{j=0}^{M_2-1}\left(\frac{M_T}{N} - j\right) = \ln \frac{\left(\frac{M_T}{N}\right)!}{\left(\frac{M_T}{N} - M_2\right)!} = \ln \frac{\left(\frac{M_T}{N}\right)!}{\left(\frac{M_1}{N}\right)!}$$

Using Stirling's approximation and doing a bit of algebra finally gives

$$\ln \prod_{i=1}^{M_2} w_i = M_T \ln M_T - M_T - M_1 \ln M_1 + M_1$$

$$+ M_2(N-1)\ln\frac{z-1}{M_T} \qquad (14.10.9)$$

so the number of complexions (14.10.3) is

$$\ln W(M_1, M_2) = M_T \ln M_T - M_T - M_1 \ln M_1 + M_1$$

$$- M_2 \ln M_2 + M_2 + M_2(N-1)\ln\frac{z-1}{M_T} \qquad (14.10.10)$$

The entropy of mixing is the entropy of the solution minus the entropy of the pure constituents. But the configurational entropy of pure liquid is zero, so only the entropy of pure polymer is needed. The statistical count for the pure polymer is just (14.10.10) for $M_1 = 0$. Remembering that $M_T = NM_2 + M_1$, we then get

$$\ln W(0, M_2) = NM_2 \ln NM_2 - NM_2$$

$$- M_2 \ln M_2 + M_2 + M_2(N-1)\ln\frac{z-2}{M_2 N} \qquad (14.10.11)$$

Taking the difference between (14.10.11) and (14.10.10) and multiplying by Boltzmann's constant gives the entropy of mixing.

$$\Delta S_m = S(\text{sol}) - S(\text{pure}) = -k(N_1 \ln f_1 + M_2 \ln f_2) \qquad (14.10.12)$$

This is just a random mixing entropy, similar to that for any random solution.

Just as in Bragg–Williams theory, assume nearest neighbor interactions such that $-v_{11}, -v_{22}, -v_{12}$ are the energies of interaction for solvent-solvent nearest neighbors, segment-segment nearest neighbors, and solvent-segment interactions, respectively. Counting the number of contacts is somewhat different than in Bragg–Williams theory because segments are attached to each other. Two of the nearest neighbors to a polymer segment are sure to be occupied by polymer segments, so each segment has $(z-2)f_1$ contacts with a solvent molecule. The energy of solvent-polymer interactions is therefore $NM_2(z-2)f_1 v_{12}$, since NM_2 is the number of polymer segments. Similarly, the energy of segment-segment contacts is $(NM_2/2)(z-2)f_2 v_{22}$, and for solvent-solvent contacts the energy is $(M_1/2)zf_1 v_{11}$. The factor of 1/2 corrects for double counting. Adding these together gives the energy of the solution as

$$E(M_1, M_2) = NM_2(z-2)f_1 v_{12} + \frac{M_1}{2}zf_1 v_{11} + \frac{NM_2}{2}(z-2)f_2 v_{22} \qquad (14.10.13)$$

For pure solvent and pure polymer, respectively, the energies are $(M_1/2)zv_{11}$ and $(M_1/2)zv_{11}$. Subtracting these from (14.10.13) gives the energy of solution formation as

$$E(M_1, M_2) = M_T\left[zf_1 f_2\left(v_{12} - \frac{v_{11}}{2} - \frac{v_{22}}{2}\right) + f_1 f_2(v_{22} - 2v_{12})\right] \qquad (14.10.14)$$

Let

$$v = \left(v_{12} - \frac{v_{11}}{2} - \frac{v_{22}}{2}\right) \qquad (14.10.15)$$

Then

$$E(M_1, M_2) = M_T[zf_1f_2v + f_1f_2(v_{22} - 2v_{12})]$$
$$= M_T\{zf_1f_2v - f_1f_2(v_{11} + 2v)\} \qquad (14.10.16)$$

The energy parameters are all about the same magnitude, and it is customary to neglect the last term in (14.10.16). This recognizes the fact that the theory is approximate and that there is a degree of uncertainty in the definition of the number of nearest neighbor sites. Then, the energy of mixing simplifies to

$$E(M_1, M_2) = M_T z f_1 f_2 v \qquad (14.10.17)$$

The free energy of solution, from (14.10.17) and (14.10.12), is

$$\Delta A_m = \Delta E(M_1, M_2) - T\Delta S_m$$
$$= M_T z f_1 f_2 v + kT(M_1 \ln f_1 + NM_2 \ln f_2) \qquad (14.10.18)$$

The theory developed so far is appropriate to a solution of polymer without cross-links in a solvent. Two modifications to (14.10.18) are needed to apply it to a rubber-solvent gel. The first is that the network is one giant molecule, so $M_2 = 1$ and the last term in (14.10.18) is trivially small and can be taken to be zero. The second is that the entropy of stretching must be included. Note that the site fractions f_1 and f_2 are still volume fractions since the site fraction for the polymer refers to the number of polymer segments.

A gel forms when cross-linking restricts the amount of liquid that can go into solution. The solvent molecules increase the volume by swelling the chains, but because of cross-linking the swelling is the result of stretching of chains and not of pushing chains apart indefinitely. In the solution theory leading to (14.10.18), stretching of the chains is not taken into account and the entropy term is just an entropy of mixing. To apply this to gel formation in cross-linked polymers, the entropy of stretching the chains must be added to the lattice configurational energy of the chains. From (14.9.1) (which includes the Flory correction), this is

$$S - S_o = \frac{M_p k}{2} \ln(\lambda_x \lambda_y \lambda_z) + \frac{M_p k}{2}(3 - \lambda_x^2 - \lambda_y^2 - \lambda_z^2) \qquad (14.10.19)$$

M_p is now the number of chains as defined by the average distance between cross-links. In a cross-linked polymer this is not the same as the number of molecules, an obvious but important fact that modifies the configurational entropy of placing the polymer segments on a lattice.

We assume that the system is isotropic and swells equally in all directions. For such a uniform, homogeneous expansion from dimensions X_p, Y_p, Z_p for the pure polymer to X^*, Y^*, Z^* for the solution

$$\lambda_x \lambda_y \lambda_z = \frac{X^* Y^* Z^*}{X_p Y_p Z_p} = \frac{V^*}{V_p} = \frac{V(\text{gel})}{V(\text{polymer})} = \frac{1}{f_2} \qquad (14.10.20)$$

and

$$\lambda_x = \lambda_y = \lambda_z = \left(\frac{V^*}{V_p}\right)^{1/3} = \left(\frac{1}{f_2}\right)^{1/3} \quad (14.10.21)$$

since, in this model, volume fractions are equal to molecular fractions. The stretching entropy (14.10.19) for the gel is then

$$S - S_o = -\frac{M_2 k}{2} \ln f_2 + \frac{3 M_2 k}{2}\left[1 - \left(\frac{1}{f_2}\right)^{2/3}\right] \quad (14.10.21)$$

Multiplying this by T and subtracting it from (14.10.17) (leaving out the last term) gives the Helmholtz free energy of the gel as

$$\Delta A_m = M_T z f_1 f_2 v + k T M_1 \ln f_1 + \frac{M_2 k T}{2}\left[\ln f_2 - 3 + 3\left(\frac{1}{f_2}\right)^{2/3}\right] \quad (14.10.22)$$

The gel is formed by immersing the rubber in solvent so (14.10.22) is the difference in free energy between gel and solvent. At equilibrium, the chemical potential of solvent in gel and chemical potential of pure solvent must be equal. This means that

$$\mu_1(\text{gel}) - \mu_1(\text{sol}) = \left(\frac{\partial \Delta A_m}{\partial N_1}\right)_{T, M_2} = 0 \quad (14.10.23)$$

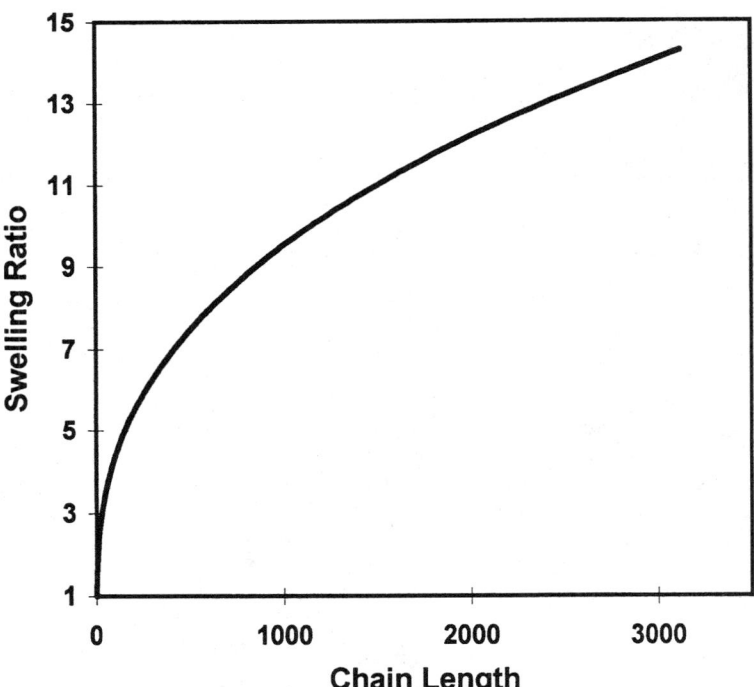

Figure 14.8. Polymer swelling as a function of chain length.

Differentiating (14.10.22) according to (14.10.23) and setting the result equal to zero gives $(zv/kT)f_2^2 + \ln(1-f_2) + f_2 - (f_2/2N) + (1/N)f_2^{1/3} = 0$, or

$$\chi = -\frac{1}{f_2^2}\left[\ln(1-f_2) + f_2 - \frac{1}{N}\left(\frac{f_2}{2} - f_2^{1/3}\right)\right] \qquad (14.10.24)$$

The ratio $\chi \equiv zv/kT$ is called the *interaction parameter*.

For simplicity, no subscript has been added to indicate that f_2 is now the molecular fraction of polymer segments (or volume fraction of polymer) for a gel in equilibrium with solvent. Since f_2 is also the ratio of the volume of dry polymer to the volume of the gel, the swelling ratio is just $1/f_2$. A plot of this swelling ratio as a function of chain length in number of monomer units, assuming a value of 0.5 for the interaction parameter, is shown in figure 14.8. Swelling increases rapidly with increasing chain length at first, but slows down after a chain length of about a thousand segments.

The chain length in (14.10.24) is the distance between cross-links, and its reciprocal is the number of links per chain. The density of links can therefore be obtained from swelling experiments if the interaction parameter is known. Conversely, if the chain length is determined in some other way, such as measuring Young's modulus, then a swelling experiment yields the value of the interaction parameter.

Exercises

14.1 If the correlation angle between adjacent segments of a linear chain molecule is $\pi/4$, get the expansion factor, persistence length, and the Kuhn length (in units of the segment length). Use the long chain limit.

14.2 For a Gaussian chain, what is the ratio of the probability that the end-to-end distance will be twice the root mean square distance; to the probability that it will equal to the root mean square distance?

14.3 For a chain consisting of 20,000 segments, find the ratio of the volume of a Gaussian, taking into account the excluded volume effect, to a tightly coiled chain. Compute this ratio if the mean field interaction energy per segment is $2kT$.

14.4 Given a force F_x acting along the x-direction of a rectangular parallelepiped specimen, show that the mean square deviation in the length along the x-axis is

$$\overline{\Delta R_x^2} \equiv \overline{R_x^2} - \overline{R_x}^2 = -kT\left(\frac{\partial \overline{R_x}}{\partial F_x}\right)_T$$

Find this mean square deviation for an elastomer in which the force–elongation relation is the result of configurational entropy changes of the elastomer coils.

14.5 What is the average molecular weight of the chains in an elastomer whose Young's modulus in the limit of small elongations is 10^6 Pa and whose density is $0.85\,\text{gm/cm}^3$ at $300\,\text{K}$?

14.6 Show that the configurational entropy of a deformed elastomer under linear tensile stress is always less than the entropy of the undeformed system.

Notes

1. The experimental data were read off a curve in Munk (1989, p. 428), which was from Treloar, L.R.G.; 1958; *The Physics of Rubber Elasticity*; Oxford University Press, New York.
2. See Flory (1953), chapters 12 and 13.

15

Vacancies and Interstitials in Monatomic Crystals

15.1 Choice of ensemble

It is generally agreed that the lattice vacancy is the predominant type of point defect in simple metals and rare gas solids. Direct and indirect measurements of the vacancy concentration as a function of temperature exist that support this contention. The concentration of point defects other than vacancies is relatively low in close-packed hexagonal and face-centered cubic structures because their formation requires large lattice distortions. In more open structures, however, such as body-centered cubic and the diamond structure characteristic of semiconductors, interstitials may be important.

Point defects have a number of important effects on crystal properties. They often control the mechanisms of diffusion, they contribute to electrical and thermal resistivity, they play a role in the growth of voids during plastic deformation, and they affect the conductivity of semiconductors. Through their interaction with dislocations and other internal stress sources, they have an effect on the mechanical properties of metals and on microstructure. Also, because of their critical role in diffusion, they are important in the properties of nucleation, growth, and phase transformations. In ionic crystals, they are responsible for the electrical conductivity.

In this chapter, the statistical mechanical theory of a pure monatomic crystal containing vacant lattice sites and atoms in interstitial positions is developed. We first treat crystals in which the monovacancy is the only defect. This allows us to present the theory in its most transparent form. The theory of crystals with more complex point defect systems is a simple extension of this and is treated later.

The theory of point defect concentration is not a classical thermodynamic theory in any sense. As required by the phase rule of Gibbs, the pressure and temperature are sufficient to fix all thermodynamic properties of a one-component system. Point defect theory, therefore, requires statistical mechanics and crystal structure, and the statistical mechanics is needed to account for the very existence of point defects. This is quite a different situation than, for example, the theory of the heat capacity in which statistical mechanics provided a method of computing a true thermodynamic quantity from microscopic considerations. The defect functions are statistical mechanical rather than thermodynamic quantities.

In working out the statistical mechanics of defect crystals, it turns out to be most convenient to use the pressure ensemble. The reason for this is that, in an ensemble in which vacancies or interstitials enter into the definition of state, the volume varies over the members of the ensemble. Also, defect concentration formulas are always expressed in terms of the Gibbs free energy, and the thermodynamic independent variables are temperature and pressure. The equations can be derived in a simple and elegant way from the pressure ensemble, and no unique calculational difficulties arise.

The pressure ensemble represents a system at a temperature T and pressure P with a volume that can vary, although the average volume is fixed. For convenience, the basic pressure ensemble equations are rewritten in the following form:

$$f_j(V_i) = \frac{N_j(V_i)}{X} = \frac{1}{Z_P} e^{-(E_j + PV_i)/kT} \quad (15.1.1)$$

$$\sum_{j,i} f_j(V_i) = 1 \quad (15.1.2)$$

$$Z_P = e^{-G/kT} \quad (15.1.3)$$

In these equations, $N_j(V_i)$ is the number of systems in an ensemble of X members whose volume is V_i and whose energy is E_j. $f_j(V_i)$ is just the probability of occurrence of such a member system in the ensemble. The pressure-ensemble partition function is Z_P, which determines the Gibbs free energy G of the system by equation (15.1.3).

15.2 The vacancy concentration

The problem before us is to compute the concentration of vacancies in a monatomic crystal in equilibrium. To do this, consider an ensemble of X crystals, all of which have the same number N of identical atoms. Each crystal can have any volume V_i and any number of vacancies N_v and can exist in any of a set of quantum states with energy E_j. The probability that a crystal exists in a state characterized by particular values of V_i, N_v, and E_j is given by the pressure ensemble as

$$f_{j,i,v} = \frac{1}{Z_P} e^{-(E_j + PV_i)/kT} \quad (15.2.1)$$

The probability $f(N_v, E_j, V_i)$ that the crystal has the values (N_v, E_j, V_i) is just (15.2.1) multiplied by the number of distinct wave functions for a crystal with N_v vacancies in a state with volume V_i and energy E_j. This degeneracy factor will be written as $w(N_v)\Omega(N_v, E_j, V_i)$. $w(N_v)$ is the number of configurational complexions (statistical count), which is the number of ways of distributing N atoms and N_v vacancies on the lattice, and $\Omega(N_v, E_j, V_i)$ is the degeneracy of a crystal containing N_v vacancies, except for the configurational contribution. Multiplying equation (15.2.1) by this factor gives

$$f(N_v, E_j, V_i) = \frac{1}{Z_P} w(N_v)\Omega(N_v, E_j, V_i) e^{-(E_j + PV_i)/kT} \quad (15.2.2)$$

Summing (15.2.2) over all values of energy and volume for a given number of vacancies gives the probability that the crystal has N_v vacancies as

$$f(N_v) = \frac{1}{Z_P} w(N_v) \sum_{E_j, V_i} \Omega(N_v, E_j, V_i) e^{-(E_j + PV_i)/kT} \qquad (15.2.3)$$

In performing the sum, Ω depends only on the number of vacancies (for a given energy and volume) and not on the configuration of their distribution over the crystal lattice. We also assume that all lattice sites are equivalent in the sense that the energy and volume do not depend on the distance between vacancies. This ignores the possibility of the existence of divacancies, which will be considered in section 15.6.

It is convenient to define a free energy $G\{N_v\}$ by

$$e^{-G\{N_v\}/kT} = \sum_{E_j, V_i} \Omega(N_v, E_j, V_i) e^{-(E_j + PV_i)/kT} \qquad (15.2.4)$$

and to write (15.2.3) as

$$f(N_v) = \frac{1}{Z_P} w(N_v) e^{-G\{N_v\}/kT} \qquad (15.2.5)$$

The free energy defined by (15.2.4) does not include the configurational entropy term arising from the configurational statistical count $w(N_v)$. The use of such a free energy simplifies the development of the theory and will be identified by braces for its argument.

Because the sum of the probabilities is unity and because the free energy of the crystal is related to Z_P by (15.1.3), summation of (15.2.5) over all values of N_v gives

$$e^{-G/kT} = \sum_{N_v} w(N_v) e^{-G\{N_v\}/kT} \qquad (15.2.6)$$

G is the total Gibbs free energy of a crystal and $G\{N_v\}$ is the Gibbs free energy of a crystal containing N_v vacancies except for the configurational contribution of the vacancies.

Equations (15.2.5) and (15.2.6) are the two basic relations of the statistical mechanics of vacancies and enable us to obtain the thermodynamic functions of a crystal containing vacancies as well as the vacancy concentration formula.

As usual, we identify the most probable value with the equilibrium value and therefore maximize (15.2.5) to obtain the equilibrium number of vacancies. Therefore, if we require that

$$\left[\frac{\partial \ln f(N_v)}{\partial N_v}\right]_{N_v = \bar{N}_v} = 0 \qquad (15.2.7)$$

where $\bar{N}_v$ is the number of vacancies at equilibrium, then applying (15.2.7) to (15.2.5) gives

$$\left[\frac{\partial \ln w(N_v)}{\partial N_v}\right]_{N_v = \bar{N}_v} = \frac{1}{kT} \left[\frac{\partial G\{N_v\}}{\partial N_v}\right]_{N_v = \bar{N}_v} \qquad (15.2.8)$$

In performing these differentiations, the number of atoms is held constant so that vacancies are formed by transferring atoms from an internal lattice site to the surface.

The derivative on the right-hand side of equation (15.2.8) is similar to a partial atomic free energy. It is the increase in Gibbs free energy upon adding one vacancy to the crystal, excluding the configurational contribution, and is therefore called the *free energy of vacancy* formation, which we designate by G_v^f:

$$G_v^f = \left(\frac{\partial G\{N_v\}}{\partial N_v}\right)_{N_v=\overline{N}_v} \quad (15.2.9)$$

To get the left-hand side of (15.2.8), we note that the number of ways of putting N atoms and N_v vacancies on $(N + N_v)$ lattice sites is

$$w(N_v) = \frac{(N+N_v)!}{N!N_v!} \quad (15.2.10)$$

Using Stirling's approximation, (15.2.10) gives

$$\ln w(N_v) = (N+N_v)\ln(N+N_v) - N\ln N - N_v \ln N_v \quad (15.2.11)$$

from which

$$\frac{\partial \ln w(N_v)}{\partial N_v} = \ln \frac{N+N_v}{N} \quad (15.2.12)$$

Putting (15.2.9) and (15.2.12) into (15.2.8) gives the vacancy concentration formula:

$$\frac{\overline{N}_v}{N+\overline{N}_v} = e^{-G_v^f/kT} \quad (15.2.13)$$

which shows that the vacancy concentration increases with temperature according to a Boltzmann factor of the free energy of vacancy formation, as expected. The number of vacancies in a crystal is usually much smaller than the number of atoms even near the melting point, so N_v is sometimes neglected in the denominator when writing (15.2.13).

Equation (15.2.13) is often derived more simply by more elementary methods. However, the method used here clearly shows the nature of the quantities involved, provides a sound basis for investigating the thermodynamic functions of a crystal containing vacancies, and is readily generalized to systems containing a more complex array of point defects.

It is clear that the method of finding the equilibrium number of vacancies by maximizing $f(N_v)$ of equation (15.2.5) is completely equivalent to finding the maximum term in the sum (15.2.6) that gives the crystal free energy. In fact, adding the configurational entropy to the configurationless free energy gives the free energy for a crystal containing an arbitrary number of vacancies N_v as

$$G(N_v) = G\{N_v\} - kT \ln w(N_v) \quad (15.2.14)$$

The above method is equivalent to minimizing this expression. That is, the equilibrium number of vacancies is determined by

$$\frac{\partial G(N_v)}{\partial N_v} = 0 \quad (15.2.15)$$

where the derivative refers to a change in the number of vacancies in the crystal interior by transferring atoms to or from the surface. Using (15.2.14) and

(15.2.15) directly saves us from the cumbersome notation that results when the pressure ensemble probability functions are written for complex systems. Therefore, these equations will be used to determine equilibrium defect distribution functions from now on.

15.3 The crystal free energy

The analysis of the thermodynamics of crystals containing vacancies requires an expression for the free energy that contains the vacancy concentration. Such an expression could be derived from (15.2.6) if we knew how to perform the sum over all possible configurations. But, in general, this cannot be done exactly, and it is usual to assume that the sum can be replaced by its largest term and that this term is that for the equilibrium concentration of vacancies. It is, of course, a general rule in statistical mechanics that, for systems containing a large number of atoms, sums over all possible configurations can be replaced by the term with the most probable distribution because the fluctuations from the most probable value are very small (see section 2.17). The present case provides an instructive example because it is particularly easy to perform the sum within the approximation of a Gaussian distribution of fluctuations. The procedure consists of showing that the sum in (15.2.6) can be written as a Gaussian distribution about the equilibrium vacancy concentration, and that the spread of the distribution contributes a negligible amount to the free energy. It is then only necessary to take the nonconfigurational part of the free energy as linear in the vacancy concentration to get the desired result.

We write (15.2.6) in the form

$$e^{-G/kT} = \sum_{N_v} h(N_v) \tag{15.3.1}$$

where $h(N_v)$ is defined by

$$h(N_v) = w(N_v)e^{-G\{N_v\}/kT} \tag{15.3.2}$$

It will now be shown that $h(N_v)$ forms a very narrow Gaussian distribution about the equilibrium value $\overline{N}_v$. Using this distribution, (15.2.6) is then summed to find the crystal free energy.

To get the Gaussian form of (15.3.2), start with the ratio of $h(N_v)$ to its most probable value:

$$\frac{h(N_v)}{h(\overline{N}_v)} = \frac{w(N_v)}{w(\overline{N}_v)} e^{(G\{\overline{N}_v\} - G\{N_v\})/kT} \tag{15.3.3}$$

The deviation from the mean number of vacancies is

$$\Delta N_v = N_v - \overline{N}_v \tag{15.3.4}$$

Take $G\{N_v\}$ to be linear in N_v so that

$$G\{N_v\} = G\{\overline{N}_v\} + \Delta N_v G_v^f \tag{15.3.5}$$

From (15.2.10), the ratio of the statistical counts is

$$\frac{w(N_v)}{w(\overline{N}_v)} = \frac{(N+N_v)!\overline{N}_v!}{(N+\overline{N}_v)!N_v!} \tag{15.3.6}$$

so using (15.3.5) and (15.3.6), equation (15.3.3) becomes

$$\frac{h(N_v)}{h(\overline{N}_v)} = \frac{(N+N_v)!\overline{N}_v!}{(N+\overline{N}_v)!N_v!}\left(e^{-G_v^f/kT}\right)^{\Delta N_v} \qquad (15.3.7)$$

or, since the exponential is given by (15.2.13),

$$\frac{h(N_v)}{h(\overline{N}_v)} = \frac{(N+N_v)!\overline{N}_v!}{(N+\overline{N}_v)!N_v!}\left(\frac{\overline{N}_v}{N+\overline{N}_v}\right)^{\Delta N_v} \qquad (15.3.8)$$

First consider the values of $h(N_v)$ for which ΔN_v is small relative to N_v. With this restriction, it is easy to show that $h(N_v)$ has a Gaussian form by first taking logarithms of (15.3.8), doing a little algebra, and using Stirling's approximation to get

$$\ln\frac{h(N_v)}{h(\overline{N}_v)} = (N+\overline{N}_v)\ln\left(1+\frac{\Delta N_v}{N+\overline{N}_v}\right) - \overline{N}_v\ln\left(1+\frac{\Delta N_v}{\overline{N}_v}\right)$$

$$+\Delta N_v\left[\ln\left(1+\frac{\Delta N_v}{N+\overline{N}_v}\right) - \ln\left(1+\frac{\Delta N_v}{\overline{N}_v}\right)\right] \qquad (15.3.9)$$

Now expand the logarithms with retention of only the first term $[\ln(1 + x) \approx x]$. Equation (15.3.9) then becomes

$$\ln\frac{h(N_v)}{h(\overline{N}_v)} = -\frac{(\Delta N_v)^2}{\overline{N}_v}\frac{N}{N+\overline{N}_v}$$

or, since $\overline{N}_v \ll N$, to a sufficient approximation,

$$\ln\frac{h(N_v)}{h(\overline{N}_v)} = -\frac{(\Delta N_v)^2}{\overline{N}_v}$$

so that

$$h(N_v) = h(\overline{N}_v)e^{-(\Delta N_v)^2/\overline{N}_v} \qquad (15.3.10)$$

which is the Gaussian distribution we were after.

Let us compute the exponential in (15.3.10) for $\Delta N_v/N_v = 10^{-4}$, a value for which the above derivation is certainly valid. Then, since N_v in crystals at temperatures for which the vacancy concentration is detectable is at least of the order of 10^{16}, ΔN_v is about 10^{12}. These values give e^{-10^8} for the exponential in (15.3.10), which is certainly negligible. This shows that (15.3.10) is indeed a very sharply peaked distribution.

All this was assuming ΔN_v is small. What happens if it is not small? We can in fact show that $h(N_v)$ decreases with increasing deviation from the equilibrium N_v whether the deviation is negative or positive, so $h(N_v)$ is even smaller than the above computation for large deviations from equilibrium. To do this, first consider positive values of ΔN_v and rewrite (15.3.8) as

$$\frac{h(N_v)}{h(\overline{N}_v)} = \frac{(N+\overline{N}_v+\Delta N_v)!\overline{N}_v!}{(N+\overline{N}_v)!(\overline{N}_v+\Delta N_v)!}\left(\frac{\overline{N}_v}{N+\overline{N}_v}\right)^{\Delta N_v} \qquad (15.3.11)$$

or

$$\frac{h(N_v)}{h(\overline{N}_v)} = \prod_{j=1}^{\Delta N_v} \frac{(N + \overline{N}_v + j)}{(\overline{N}_v + j)} \left(\frac{\overline{N}_v}{N + \overline{N}_v}\right)^{\Delta N_v} \tag{15.3.12}$$

But every factor in (15.3.12) is less than unity, and the more of them there are, the smaller the value of (15.3.12). This shows that $h(N_v)$ decreases with increasing ΔN_v for all positive values of ΔN_v.

If ΔN_v is negative, we write (15.3.11) in the form

$$\frac{h(N_v)}{h(\overline{N}_v)} = \frac{(N + \overline{N}_v - |\Delta N_v|)! \, \overline{N}_v!}{(N + \overline{N}_v)! (\overline{N}_v - |\Delta N_v|)!} \left(\frac{N + \overline{N}_v}{\overline{N}_v}\right)^{|\Delta N_v|} \tag{15.3.13}$$

or

$$\frac{h(N_v)}{h(\overline{N}_v)} = \prod_{j=1}^{|\Delta N_v|} \frac{(\overline{N}_v - |\Delta N_v| + j)}{(N + \overline{N}_v - |\Delta N_v| + j)} \left(\frac{N + \overline{N}_v}{\overline{N}_v}\right)^{|\Delta N_v|} \tag{15.3.14}$$

Again, each factor in this product is less than unity, and the more factors in the product, the smaller $h(N_v)$. We have therefore shown that $h(N_v)$ decreases with increasing magnitude of ΔN_v, whether the deviation is positive or negative. Since $h(N_v)$ is very small even for small deviations, a negligible error is made if the Gaussian distribution (15.3.10) is adopted for all values of ΔN_v.

Now put (15.3.10) into (15.3.1) to get

$$e^{-G/kT} = h(\overline{N}_v) \sum_{\Delta N_v} e^{-(\Delta N_v)^2/\overline{N}_v} \tag{15.3.15}$$

To evaluate the sum, replace it with an integral as follows:

$$\sum_{\Delta N_v} e^{-(\Delta N_v)^2/\overline{N}_v} = \int_{-\infty}^{\infty} e^{-x^2/\overline{N}_v} dx = \sqrt{\pi \overline{N}_v}$$

and therefore (15.3.15) becomes

$$e^{-G/kT} = h(\overline{N}_v)\sqrt{\pi \overline{N}_v} \tag{15.3.16}$$

From the definition of $h(N_v)$ given by (15.3.2), taking logarithms of (15.3.16) results in

$$\frac{G}{kT} = -\ln w(\overline{N}_v) + \frac{G\{\overline{N}_v\}}{kT} + \frac{1}{2}\ln(\pi \overline{N}_v) \tag{15.3.17}$$

All that remains is to evaluate the first log term. This is easily done from (15.2.11), which gives

$$\ln w(\overline{N}_v) = N \ln\left(1 + \frac{\overline{N}_v}{N}\right) + \overline{N}_v \ln\left(\frac{N + \overline{N}_v}{\overline{N}_v}\right) \tag{15.3.18}$$

The log in the first term on the right can be approximated by the first term in its series expansion and the second term can be expressed in terms of the free energy of vacancy formation by using (15.2.13). Doing this, (15.3.17) and (15.3.18) gives

$$\ln w = \overline{N}_v + \overline{N}_v \frac{G_v^f}{kT} \qquad (15.3.19)$$

and putting this into (15.3.17) gives

$$G = G\{\overline{N}_v\} - \overline{N}_v G_v^f - \overline{N}_v kT \qquad (15.3.20)$$

where the last term in (15.3.17) has been neglected relative to $\overline{N}_v$. This was the term that arose from the fluctuations from the most probable state and is indeed small since it goes as the ratio of the log of the number of vacancies to their number. Taking $G\{N_v\}$ to be linear in the number of vacancies, that is,

$$G\{\overline{N}_v\} = G^o + \overline{N}_v G_v^f \qquad (15.3.21)$$

G^o being the free energy of a vacancy free crystal, we finally get[1]

$$G = G^o - \overline{N}_v kT \qquad (15.3.22)$$

Equation (15.3.21) states that each vacancy contributes an amount to the configurationless free energy $G\{N_v\}$ that is equal to the free energy of vacancy formation. This means that we assume the vacancy concentration to be low enough that the vacancies do not influence each other. Vacancy concentrations are of the order of 10^{-4} atomic percent in simple crystals; this justifies the form of (15.3.21) and the approximations based on assuming $N_v/N \ll 1$ and $\ln N_v \ll N_v$. However, N_v/N was *not* taken to be zero relative to unity in using equation (15.3.18). Doing so would have yielded a zero contribution of vacancies to the crystal free energy rather than $-kT$ per vacancy as displayed in equation (15.3.22).

This remarkably simple result enables us to obtain all the thermodynamic functions of a crystal as a function of vacancy concentration and to derive the differential relations among the vacancy formation quantities. In this connection, it is interesting to note from (15.3.20) that the last two terms represent the configurational contribution to the crystal free energy. Thus, the configurational entropy due to the vacancies is

$$S_v^c = k\overline{N}_v \left(1 + \frac{G_v^f}{kT}\right) \qquad (15.3.23)$$

As will be shown below, the existence of this term means that the crystal entropy is not just the sum of the entropy of a vacancy free crystal and the entropy of vacancy formation.

15.4 Vacancies and thermodynamic functions

The vacancy contributions to all the other thermodynamic functions can be found from equation (15.3.22) through the usual thermodynamic formulas. These formulas define the entropy, volume, energy, enthalpy specific heat at constant pressure, specific heat at constant volume, thermal expansion, and compressibility of defect formation as follows.

The entropy of vacancy formation:

$$S_v^f = -\left(\frac{\partial G_v^f}{\partial T}\right)_P \qquad (15.4.1)$$

The volume of vacancy formation:

$$V_v^f = -\left(\frac{\partial G_v^f}{\partial P}\right)_T \qquad (15.4.2)$$

The energy of vacancy formation:

$$U_v^f = G_v^f + TS_v^f - PV_v^f \qquad (15.4.3)$$

The enthalpy of vacancy formation:

$$H_v^f = U_v^f + PV_v^f = G_v^f + TS_v^f \qquad (15.4.4)$$

The heat capacity at constant pressure of vacancy formation:

$$(C_v^f)_P = \left(\frac{\partial H_v^f}{\partial T}\right)_P \qquad (15.4.5)$$

The specific heat at constant volume of vacancy formation:

$$(C_v^f)_V = \left(\frac{\partial U_v^f}{\partial T}\right)_V \qquad (15.4.6)$$

The thermal expansion of vacancy formation:

$$\alpha_v^f = \frac{1}{V_v^f}\left(\frac{\partial V_v^f}{\partial T}\right)_P \qquad (15.4.7)$$

The compressibility of vacancy formation:

$$\kappa_v^f = \frac{1}{V_v^f}\left(\frac{\partial V_v^f}{\partial P}\right)_T \qquad (15.4.8)$$

These definitions ensure that the defect formation quantities will follow the usual rules of thermodynamics. However, it is not generally true that the defect quantities defined above give the contribution per vacancy to the corresponding crystal quantities. Although this is true for the energy and enthalpy, for example, it is not true for the entropy and the free energy.

To use (15.4.1)–(15.4.8) in conjunction with (15.3.22), we need the derivatives of the vacancy concentration with respect to pressure and temperature. From (15.2.13), and using the definitions of the defect formation quantities, we get

$$\left(\frac{\partial \bar{N}_v}{\partial T}\right)_P = \frac{\bar{N}_v H_v^f}{kT^2} \qquad (15.4.9)$$

$$\left(\frac{\partial \bar{N}_v}{\partial P}\right)_T = \frac{\bar{N}_v V_v^f}{kT} \qquad (15.4.10)$$

From this point on, the bar over the N_v will be dropped for the sake of convenience, with the understanding that all vacancy concentrations in the thermodynamic formulas refer to statistical equilibrium.

Using (15.4.1)–(15.4.10), it is only a matter of a little algebra to derive the crystal thermodynamic functions in the following forms:

The entropy:

$$S = S^o + \frac{N_v}{T}(H_v^f + kT) \qquad (15.4.11)$$

or

$$S = S^o + S_v^c + N_v\, S_v^f \qquad (15.4.12)$$

where S_v^c is the vacancy configurational entropy given by (15.3.23) and the superscript o refers, as usual, to a hypothetical defect-free crystal.

The volume:

$$V = V^o + N_v V_v^f \qquad (15.4.13)$$

The energy:

$$U = U^o + N_v U_v^f \qquad (15.4.14)$$

The enthalpy:

$$H = H^o + N_v H_v^f \qquad (15.4.15)$$

The heat capacity at constant pressure:

$$C_P = C_P^o + N_v\left[(C_v^f)_P + \frac{(H_v^f)}{kT^2}\right] \qquad (15.4.16)$$

The thermal expansion:

$$\alpha = \alpha^o + \frac{N_v V_v^f}{V_o}\left(\alpha_v^f - \alpha^o + \frac{H_v^f}{kT^2}\right) \qquad (15.4.17)$$

The compressibility:

$$\kappa = \kappa^o + \frac{N_v V_v^f}{V_o}\left(\kappa_v^f - \kappa^o + \frac{V_v^f}{kT}\right) \qquad (15.4.18)$$

In deriving (15.4.17) and (15.4.18), terms quadratic in N_v were neglected and the following approximation was used:

$$\left(1 + \frac{N_v V_v^f}{V_o}\right)^{-1} = 1 - \frac{N_v V_v^f}{V_o} \qquad (15.4.19)$$

The heat capacity at constant volume can now be obtained from equation (1.18.18) using (15.4.16)–(15.4.18) by treating all vacancy contributions as small compared to the corresponding crystal quantities, and retaining terms only to the first order in the vacancy concentration. The result is easy to get but rather cumbersome, and it is not reproduced here.

With the above formulas, the effect of vacancies on the thermodynamic functions can be investigated. Note that, only for the volume, the energy and the enthalpy can the vacancy effect be written as an incremental addition of the vacancy formation quantities. For the other thermodynamic functions, the formulas are more complex.

15.5 The vacancy formation functions

The vacancy concentration formula (15.2.13) is often written, with the aid of (15.4.3), as

$$\frac{N_v}{N} = e^{S_v^f/kT} e^{-U_v^f/kT} e^{-PV_v^f/kT} \tag{15.5.1}$$

Where N_v is neglected relative to N in the denominator on the left-hand side. It is this form that is used in the experimental determination of vacancy formation energies and volumes. The formation energy is obtained by measuring a quantity that is proportional to the vacancy concentration as a function of temperature at constant (usually atmospheric) pressure. The formation energy is then obtained from an Arrhenius plot of $\ln N_v$ versus $1/T$. The formation volume is obtained from a plot of $\ln N_v$ versus P at constant temperature.

The applicability of these methods is simple when the formation energy, entropy, and volume are independent of the temperature and pressure, at least within the accuracy of the experiments. The question of the variation of the Gibbs free energy of vacancy formation with temperature or pressure is therefore of considerable importance. If an Arrhenius plot, or a plot of $\ln N_v$ versus P, exhibits a curvature, two causes may be operative. The first is that the temperature or pressure dependencies of the vacancy formation quantities are being reflected in the curvature. The second is that a process other than vacancy formation may be affecting the measurement. Such processes might include, for example, impurity-vacancy binding or divacancy formation. Without some theoretical guide, it is difficult to distinguish between these two possibilities.

It is important to note that a linear Arrhenius plot does not in itself guarantee that U_v^f is independent of temperature. If the vacancy formation energy is linear in the temperature, the Arrhenius plot will still be linear. Likewise, if the volume of vacancy formation is linear in the reciprocal of the pressure, the $\ln N_v$ versus P plot will be linear.

This and the following section develop a theory of the vacancy formation quantities in such a way as to give some physical insight into the factors that determine them. It will also be shown that, for most monatomic crystals, the entropy, energy, and volume of vacancy formation are constants to a good degree of approximation.

To get the Gibbs free energy of formation, and therefore all the other formation quantities, we start with (15.2.9). Since we are dealing with crystals in which the vacancy concentration is small, we will take the derivative in (15.2.9) to be independent of the vacancy concentration. It then becomes equal to the difference in free energy between a crystal having one vacancy and a crystal having no vacancies. Thus,

$$G_v^f = G^1 - G^o \tag{15.5.2}$$

where G^1 is the Gibbs free energy of a crystal containing a single vacancy and G^o is the Gibbs free energy of a perfect crystal.

The reason that it is legitimate to define G^1 and G^o in this way is that $G\{N_v\}$ does not include the configurational contribution to the crystal free energy.

Thus, we can use canonical ensemble theory to calculate the Helmholtz free energy and add to it PV to get

$$G^1 = A^1 + PV \tag{15.5.3}$$

$$G^o = A^o + PV \tag{15.5.4}$$

where the superscript 1 again refers to a crystal containing one vacancy and the superscript o refers to a perfect crystal.

The Helmholtz free energies are given by equation (4.3.13) as

$$A^1 = E_0^1 + kT \sum_{j=1}^{3N} \ln\left(1 - e^{-h\nu_j^1/kT}\right) \tag{15.5.5}$$

$$A^o = E_o^o + kT \sum_{j=1}^{3N} \ln\left(1 - e^{-h\nu_j^o/kT}\right) \tag{15.5.6}$$

in obvious notation.

It is clear from the above that all of the vacancy formation quantities defined in (15.4.1)–(15.4.6) refer to differences between a perfect crystal and a crystal containing one vacancy. That is,

$$S_v^f = S^1 - S^o \tag{15.5.7}$$

$$V_v^f = V^1 - V^o \tag{15.5.8}$$

$$U_v^f = U^1 - U^o \tag{15.5.9}$$

$$H_v^f = H^1 - H^o \tag{15.5.10}$$

$$(C_v^f)_P = C_P^1 - C_P^o \tag{15.5.11}$$

$$(C_v^f)_V = C_V^1 - C_V^o \tag{15.5.12}$$

where, again, the 1 refers to a crystal containing one vacancy and the o to a perfect crystal.

Because of the presence of V_v^f in front of the derivatives in (15.4.7) and (15.4.8), the thermal expansion and compressibility are not given by simple difference formulas, but rather by

$$V_v^f \alpha_v^f = \alpha^1 V^1 - \alpha^o V^o \tag{15.5.13}$$

$$V_v^f \kappa_v^f = \kappa^1 V^1 - \kappa^o V^o \tag{15.5.14}$$

The entire apparatus of canonical ensemble theory and the quasi-harmonic theory of crystals is now available for the computation of the vacancy formation quantities and, therefore, for the vacancy concentration as a function of temperature and pressure.

The Debye temperature represents the maximum normal mode vibration frequency, so for temperatures above Θ_D the $h\nu/kT$ are all small. In the high-temperature limit the formulas (15.5.5) and (15.5.6) can be considerably simplified. Since vacancies exist in appreciable concentrations only at high temperatures, it is sufficiently accurate to adopt the approximation that

$$e^{-h\nu_j^1/kT} = 1 - \frac{h\nu_j^1}{kT} \tag{15.5.15}$$

$$e^{-h\nu_j^o/kT} = 1 - \frac{h\nu_j^o}{kT} \tag{15.5.16}$$

With this approximation, (15.5.5) and (15.5.6) become

$$A^1 = E_o^1 + kT \sum_{j=1}^{3N} \ln\left(\frac{h\nu_j^1}{kT}\right) \tag{15.5.17}$$

$$A^o = E_o^o + kT \sum_{j=1}^{3N} \ln\left(\frac{h\nu_j^o}{kT}\right) \tag{15.5.18}$$

From the above and equations (15.5.2)–(15.5.4), the Gibbs free energy can now be written as

$$G_v^f = (E_o^1 - E_o^o) + kT \sum_{j=1}^{3N} \ln\left(\frac{\nu_j^1}{\nu_j^o}\right) + P(V^1 - V^o) \tag{15.5.19}$$

From the derivation of (15.5.19), the first two terms constitute the Helmholtz free energy of vacancy formation:

$$A_v^f = (E_o^1 - E_o^o) + kT \sum_{j=1}^{3N} \ln\left(\frac{\nu_j^1}{\nu_j^o}\right) \tag{15.5.20}$$

so, using the thermodynamic relations

$$S_v^f = -\left(\frac{\partial A_v^f}{\partial T}\right)_V \tag{15.5.21}$$

$$U_v^f = A_v^f - TS_v^f \tag{15.5.22}$$

gives

$$S_v^f = -k \sum_{j=1}^{3N} \ln\left(\frac{\nu_j^1}{\nu_j^o}\right) \tag{15.5.23}$$

$$U_v^f = E^1 - E^o \tag{15.5.24}$$

and therefore comparing (15.4.3) to (15.5.19) requires that

$$V_v^f = V^1 - V^o \tag{15.5.25}$$

In deriving (15.5.23)–(15.5.25), the physically reasonable assumption is made that the vibration frequencies and zero point energies are explicit functions of the volume only and depend on the temperature and pressure only implicitly through the volume.

To investigate the temperature dependence of the free energy of vacancy formation, we will expand it in a Taylor series, retaining terms to the second order in the temperature and pressure. The origin of the expansion will be taken at zero pressure and the melting point temperature to give

$$G_v^f(P,T) = G_v^f(0,T_m) + P\left(\frac{\partial G_v^f}{\partial P}\right)_{0,T_m} + (T-T_m)\left(\frac{\partial G_v^f}{\partial T}\right)_{0,T_m}$$

$$+ \frac{P^2}{2}\left(\frac{\partial^2 G_v^f}{\partial P^2}\right)_{0,T_m} + \frac{(T-T_m)^2}{2}\left(\frac{\partial^2 G_v^f}{\partial T^2}\right)_{0,T_m}$$

$$+ P(T-T_m)\left(\frac{\partial^2 G_v^f}{\partial T \partial P}\right)_{0,T_m} \tag{15.5.26}$$

The appearance of this equation can be simplified by using the definitions of the defect formation quantities in the preceding section to give

$$G_v^f(P,T) = \tilde{G}_v^f - (T-T_m)\tilde{S}_v^f + P\tilde{V}_v^f - \frac{(T-T_m)^2}{2T_m}(\tilde{C}_v^f)_P$$

$$- \frac{P^2}{2}\tilde{\beta}_v^f\tilde{V}_v^f + P(T-T_m)\tilde{\alpha}_v^f\tilde{V}_v^f \tag{15.5.27}$$

where the tilde signifies values at zero pressure and the melting temperature. The heat capacity at constant pressure in the fourth term was introduced by means of the thermodynamic relation

$$\left(\frac{\partial S_v^f}{\partial T}\right)_P = \frac{(C_v^f)_P}{T} \tag{15.5.28}$$

Experimental measurements of the vacancy concentration as functions of temperature and pressure in face-centered cubic metals show that $U_v^f \cong 1\ eV$, $S_v^f \cong k$, and $V_v^f = V_a/2$, where V_a is the atomic volume. The accuracy of the measurements is not sufficient to detect any temperature or pressure variation from the quadratic terms in (15.5.27). Indeed, if it is assumed that the thermal expansion and compressibility of a crystal containing a vacancy are of the same order as that measured for a real crystal, then a simple calculation shows that the last two terms are negligible relative to PV_v^f up to pressures of about 50,000 atmospheres. Welch[2] has shown that for copper, α_v^f and κ_v^f are actually quite close in value to α_v^o and κ_v^o. His calculations were based on atomistic considerations in which ion-core interaction energies, electron distributions, and lattice vibrations were analyzed for perfect and defect crystals.

An estimate of the fourth term in (15.5.27) can be made from an analysis of the heat capacity of defect formation. The heat capacities at constant pressure and constant volume are related to each other by

$$(C_v^f)_P = (C_v^f)_V + T\left[\frac{V'(\alpha')^2}{\kappa'} - \frac{V^o(\alpha^o)^2}{\kappa^o}\right] \tag{15.5.29}$$

Using (15.5.13) and (15.5.14) and taking the thermal expansion and the compressibility to have about the same values in the perfect and defect crystal, it is easy to show that, to an excellent approximation,

$$(C_v^f)_P = (C_v^f)_V \tag{15.5.30}$$

Applying the high-temperature Debye approximation for the heat capacity in section 4.8 to both the defect and perfect crystal gives

$$C_V^1 = 3Nk\left[1 - \frac{1}{20}\left(\frac{\Theta_D^1}{T}\right)^2\right] \tag{15.5.31}$$

$$C_V^o = 3Nk\left[1 - \frac{1}{20}\left(\frac{\Theta_D^o}{T}\right)^2\right] \tag{15.5.32}$$

Θ_D^1 differs from Θ_D^o because the Debye temperature varies with dilatation. Placing an interior atom on the surface increases the volume of a crystal by one atomic volume. The vacancy formation volume is less than this, however, because the atoms around the empty site relax inward toward the vacancy. This gives rise to a relaxation volume V_R, which is negative; thus, if V_a is the atomic volume,

$$V_v^f = V_a + V_R \tag{15.5.33}$$

The relaxation volume consists of two parts. As the crystal relaxes, any spherical surface surrounding the vacancy and anchored in the atoms sweeps out a volume ΔV^∞. This is equal to V_R only if the crystal is infinite. The distortion around the vacancy induces an elastic stress field throughout the crystal. The crystal surface, however, must be stress free, so image forces give rise to an additional contribution, ΔV^I, to the relaxation volume, which is called the *image volume*. Thus,

$$V_v^f = V_R + \Delta V^\infty + \Delta V^I \tag{15.5.34}$$

ΔV^∞ and ΔV^I can be computed from elasticity theory by methods developed by J.D. Eshelby.[3] Both ΔV^∞ and ΔV^I are negative for vacancies in noble metals. For the copper vacancy $\Delta V^\infty = -0.326 V_a$ and $\Delta V^I = -0.16 V_a$. Of the three terms in (15.5.34), only ΔV^I produces a dilatation, and therefore it is the only volume change that affects the Debye temperature. Using the Gruneisen assumption (see chapter 5), we therefore have

$$\frac{\Theta_D^1 - \Theta_D^o}{\Theta_D^o} = -\gamma \frac{\Delta V^I}{V^o} \tag{15.5.35}$$

Combining (15.5.31), (15.5.32), and (15.5.35), we get

$$C_V^f = C_V^1 - C_V^o = \frac{2Nk}{20}\left(\frac{\Theta_D^1}{T}\right)^2\left(1 - \frac{\gamma \Delta V^I}{V^o}\right) \tag{15.5.36}$$

Since $\Delta V^I \ll V^o$, this can be written as

$$C_V^f = \frac{3k}{10}\left(\frac{\Theta_D^1}{T}\right)^2 \frac{\Delta V^I}{V_a} \gamma \tag{15.5.37}$$

For copper, $\Theta_D/T_m = 0.23$ and $\gamma = 2$, so according to (15.5.37) the vacancy formation heat capacity is about $-0.05k$ at the melting point.

This calculation is very rough, since it ascribes the entire heat capacity of vacancy formation to the elastic image field volume, which is based on linear elasticity theory. The relaxation of atoms right around the vacancy, however, is too large to be described by linear elasticity theory. A more rigorous calculation would start with the atomic interaction force constants for the atoms around the vacancy and proceed to a calculation of the altered vibration frequencies from these force constants. Such calculations have been done in attempts to compute the entropy of vacancy formation. They are fraught with problems, however, and the precise results are sensitive to the details of the models used.

For our purposes, we take $(C_v^f)_V$ to be negligible, since no improvement in our model will increase the value of $-0.05k$ to the point where the fourth term in (15.5.27) needs to be retained, at least for close-packed structures. We will therefore write, for close packed crystals,

$$C_v^f = \tilde{G}_v^f - (T - T_m)\tilde{S}_v^f + P\tilde{V}_v^f \qquad (15.5.38)$$

or

$$G_v^f = \tilde{U}_v^f - T\tilde{S}_v^f + P\tilde{V}_v^f \qquad (15.5.39)$$

so Arrhenius plots can indeed be expected to be linear and yield the energy of vacancy formation, and $\ln N_v$ versus P plots will give the vacancy formation volume.

The validity of (15.5.38) depends on the image volume being small and on the thermal expansion and compressibility of the region around the vacancy not being too greatly different from that of the perfect crystal. Although this seems to be the case in close-packed crystals, it is not true for the more open body–centered cubic structure. In sodium, which has a very small ion core, the relaxation around a vacancy is so large that a vacancy is similar to a small puddle of liquid.

15.6 Vacancies, divacancies, and interstitials

Up to this point, it has been assumed that the vacancy is the only defect in our monatomic crystal. For the noble metals, both experimental and theoretical considerations give the result that the free energy of vacancy formation is considerably less than that of other defects. However, other point defects certainly must exist, and their importance relative to monovacancies can be expected to vary from one material to another. In aluminum, for example, it appears that nearly 40% of the vacant lattice sites are tied up in divacancies at the melting point, whereas in copper at the melting point only about 0.2% of the vacancies are in divacancies.

Most quantitative information about point defects in metals is based on work in face-centered cubic systems. In these systems, only vacancies, and sometimes divacancies, need to be taken into account at equilibrium because the free energy for interstitial formation is so high. But there is no reason to expect these results to carry over to more open structures. In sodium, for example, the ion-core radius is small relative to the nearest neighbor distance, so the free energy of interstitial formation can be expected to be much smaller than in close-packed metals. Unfortunately, there has not been enough research done to identify the nature and number of all the point defects present in such systems, but there are strong indications that the alkali metals have a variety of point defects that do not exist in appreciable numbers in close-packed crystals. In general, the smaller the ratio of the ion radius to the nearest neighbor distance, the greater the variety and complexity of point defects that can be expected.

This section considers the coexistence of vacancies, divacancies, and interstitials in a pure monatomic crystal. The defect concentrations will be determined by a straightforward generalization of (15.2.15):

$$\frac{\partial G(V, 2V, I)}{\partial N_v} = 0$$

$$\frac{\partial G(V, 2V, I)}{\partial N_{2v}} = 0 \qquad (15.6.1)$$

$$\frac{\partial G(V, 2V, I)}{\partial N_I} = 0$$

where N_v, N_{2v} and N_I are the number of vacancies, divacancies, and interstitial atoms, respectively, and $G(V, 2V, I)$ is the Gibbs free energy of a crystal containing N atoms, N_v vacancies, N_{2v} divacancies, and N_I interstitials. This free energy is given by

$$G(V, 2V, I) = G\{V, 2V, I\} - kT \ln w(V, 2V, I) \qquad (15.6.2)$$

where $G\{V, 2V, I\}$ is the Gibbs free energy except for the configurational contribution and $w(V, 2V, I)$ is the number of ways of distributing N atoms, N_v vacancies, N_{2v} divacancies, and N_I interstitials on a lattice of L normal sites and L_I interstitial sites where

$$L = N + N_v + N_{2v} \qquad (15.6.3)$$

and L_I is a small multiple of L, depending on the crystal structure. Note that we have assumed that there is only one kind of site that can be occupied by an interstitial atom. The configurationless free energy is given in terms of a partition function which is completely analogous to (15.2.4).

In (15.6.1), the differentiations are performed while holding the number of atoms constant so that vacancy and divacancy formation processes consist of removing atoms from the interior of the crystal and placing them on the surface, while the formation of an interstitial consists of taking an atom from the surface and placing it in an interior interstitial position.

Equations (15.6.1) and (15.6.2) give the following conditions of defect equilibrium:

$$\frac{\partial \ln w(V, 2V, I)}{\partial N_v} = \frac{G_v^f}{kT} \qquad (15.6.4)$$

$$\frac{\partial \ln w(V, 2V, I)}{\partial N_{2v}} = \frac{G_{2v}^f}{kT} \qquad (15.6.5)$$

$$\frac{\partial \ln w(V, 2V, I)}{\partial N_I} = \frac{G_I^f}{kT} \qquad (15.6.6)$$

where G_v^f, G_{2v}^f, and G_I^f are the free energies of formation of a vacancy, a divacancy, and an interstitial, respectively, defined by

$$G_v^f = \frac{\partial G\{V, 2V, I\}}{\partial N_v} \qquad (15.6.7)$$

$$G_{2v}^f = \frac{\partial G\{V, 2V, I\}}{\partial N_{2v}} \qquad (15.6.8)$$

$$G_I^f = \frac{\partial G\{V, 2V, I\}}{\partial N_I} \tag{15.6.9}$$

G_v^f is the free energy change on bringing an atom from an interior lattice site to the crystal surface, G_{2v}^f is the free energy change on bringing two adjacent atoms from the interior to the surface, and G_I^f is the free energy change on taking an atom from the surface and placing it in an interstitial position.

As usual, we will assume that the concentration of defects is low enough that interactions among them can be ignored and the formation free energies can be taken to be independent of concentration.

The statistical count $w(V, 2V, I)$ is obtained by first counting the number of ways N_{2v} divacancies can be placed on L lattice sites, then counting the number of ways N_v vacancies can be placed on the remaining available lattice sites, and finally counting the number of ways N_I interstitials can be placed on L_I interstitial sites. In performing this count, the restriction is imposed that the different types of defects cannot be nearest neighbors, since this would produce a new kind of defect. Getting the statistical count is straightforward, although it requires some care. A detailed analysis is given in chapter 16, which generalizes our approach to include dilute alloys. The result is that $w(V, 2V, I)$ is the product of three terms:

$$w(V, 2V, I) = w_{2v} w_v w_I \tag{15.6.10}$$

where

$$w_{2v} = \frac{(z/2)^{N_{2v}}}{N_{2v}!} \prod_{m=0}^{N_{2v}-1} (L - 2m) \tag{15.6.11}$$

$$w_v = \frac{[L - (z' + 2)N_{2v}]!}{[L - (z' + 2)N_{2v} - N_v]! N_v!} \tag{15.6.12}$$

$$w_I = \frac{[L_I - z_I N_v - (z_I' + 2)N_{2v}]!}{[L_I - z_I N_v - (z_I' + 2)N_{2v}]! N_I!} \tag{15.6.13}$$

In these equations, z is the number of nearest neighbors to a lattice site, z' is the number of nearest-neighbor lattice sites to a vacancy pair, z_I is the number of nearest-neighbor interstitial sites surrounding a vacancy, and z_I' is the number of interstitial sites that are nearest neighbors to a vacancy pair, defined such that $(z_I' + 2)$ is the number of interstitial sites around a divacancy that are not available to interstitials.

The process of differentiating the statistical count is a bit tedious but straightforward. It is only necessary to use Stirling's approximation and approximations of the form $\ln(1 + x) = x$ for small x, and to take the number of defects as small relative to the number of sites. The product in (15.6.11) is treated in the following way:

$$\sum_{m=0}^{N_{2v}-1} \ln(L - 2m) = \sum_{m=0}^{N_{2v}-1} \ln L(1 - 2m/L) \cong \sum_{m=0}^{N_{2v}-1} \ln L = N_{2v} \ln L$$

Using these approximations gives

$$\ln w(V, 2V, I) = N_v + N_v \ln \frac{L}{N_v} + N_{2v} + N_{2v} \ln \frac{L}{N_{2v}} + N_I + N_I \ln \frac{L}{N_I} \tag{15.6.14}$$

$$\frac{\partial \ln w(V, 2V, I)}{\partial N_v} = \ln \frac{L}{N_v} \tag{15.6.15}$$

$$\frac{\partial \ln w(V, 2V, I)}{\partial N_{2v}} = \ln \frac{zL}{2N_{2v}} \tag{15.6.16}$$

$$\frac{\partial \ln w(V, 2V, I)}{\partial N_I} = \ln \frac{L}{N_I} \tag{15.6.17}$$

Combining these with (15.6.4)–(15.6.6), we get the defect concentrations as

$$\frac{N_v}{L} = e^{-G_v^f/kT} \tag{15.6.18}$$

$$\frac{N_{2v}}{L} = \frac{z}{2} e^{-G_{2v}^f/kT} \tag{15.6.19}$$

$$\frac{N_I}{L} = e^{-G_I^f/kT} \tag{15.6.20}$$

There is a relation between the vacancy and the divacancy concentrations because two vacancies can combine to form a divacancy, and conversely, a divacancy can dissociate into two monovacancies. This process can be represented in the notation of chemical reactions as $V + V \leftrightarrow V_2^1$. The equilibrium constant governing the equilibrium concentrations for this reaction is defined by

$$K(V + V \leftrightarrow V_2) = \frac{N_{2v}/L}{(N_v/L)^2} \tag{15.6.21}$$

From (15.6.18) and (15.6.19), the right-hand side of this is given by

$$\frac{N_{2v}/L}{(N_v/L)^2} = \frac{z}{2} e^{-(G_{2v}^f - 2G_v^f)/kT} \tag{15.6.22}$$

The quantity G_{vv}, defined by

$$G_{vv} = 2G_v^f - G_{2v}^f \tag{15.6.23}$$

is the binding free energy of the divacancy since it is the free energy change upon separating a divacancy into two monovacancies. Equation (15.6.22) is usually written as

$$\frac{N_{2v}}{L} = \left(\frac{N_v}{L}\right)^2 \frac{z}{2} e^{G_{vv}/kT} \tag{15.6.24}$$

The total concentration of vacant sites, N_v^T, is given by

$$N_v^T = N_v + 2N_{2v} \tag{15.6.25}$$

or, using (15.6.18) and (15.6.19),

$$\frac{N_v^T}{L} = e^{-G_v^f/kT}\left[1 + ze^{-(G_v^f + G_{vv})/kT}\right] \tag{15.6.26}$$

422 STATISTICAL MECHANICS OF SOLIDS

Having obtained the equilibrium concentration of defects, we can now derive the crystal free energy in a manner completely analogous to that used in arriving at (15.3.22). If we write the configurationless free energy $G\{V, 2V, I\}$ as linear in the defect concentrations,

$$G\{V, 2V, I\} = G^o + N_v G_v^f + N_{2v} G_{2v}^f + N_I G_I^f \qquad (15.6.27)$$

and put this, along with (15.6.2), into (15.6.14), and then use (15.6.18)–(15.6.20), the result is

$$G\{V, 2V, I\} = G^o - kT\{N_v + N_{2v} + N_I\} \qquad (15.6.28)$$

This is a straightforward generalization of (15.3.22). Just as in section 15.4, we can now get all the thermodynamic crystal functions in a form that displays the defect concentrations. These are listed below.

$$S = S^o + \frac{N_v}{T}(H_v^f + kT) + \frac{N_{2v}}{T}(H_{2v}^f + kT) + \frac{N_I}{T}(H_I^f + kT) \qquad (15.6.29)$$

$$V = V^o + N_v V_v^f + N_{2v} V_{2v}^f + N_I V_I^f \qquad (15.6.30)$$

$$U = U^o + N_v U_v^f + N_{2v} U_{2v}^f + N_I U_I^f \qquad (15.6.31)$$

$$H = H^o + N_v H_v^f + N_{2v} H_{2v}^f + N_I H_I^f \qquad (15.6.32)$$

The heat capacity at constant at constant pressure, the thermal expansion, and the compressibility are just like equations (15.4.16), (15.4.17), and (15.4.18), except that there are three bracketed terms on the right, one for each defect.

15.7 Some numerical results

A variety of experimental methods have been used to determine the values of energies and entropies of defect formation. These include measurement of the electricity resistivity of quenched and irradiated metals, comparison of dilatometric and X-ray measurements of thermal expansion, determination of stored energy release of irradiated and cold-worked specimens, diffusion measurements, and internal friction studies.

The various methods agree in that they give roughly comparable results. The accuracy of any of the results, however, is open to question, despite the fact that they represent a considerable amount of careful and highly competent research. There are two reasons for this. First, defects are present in low concentrations, and very sensitive experimental methods must be used to see their effects. Second, in most materials a rich variety of point defects, as well as more extended defects such as dislocations and grain boundaries, exists. The question of sorting out all the defects and their interactions on the measurements is a very difficult one, so often uncertainty is inherent in the interpretation of the experiments.

Despite the difficulties, valuable information on the defect parameters has been obtained. Table 15.1 shows these parameters for some metals. Most of the values have been reduced to one significant figure. For those values that seem better established, two significant figures are retained.

Figure 15.1 shows the vacancy concentration in copper as a function of temperature according to equation (15.6.18) using the parameters in table 15.1 for

Table 15.1: Point Defect Parameters in Metals

Metal	U_V^f (eV)	S_V^f (eV)	U^W (eV)	S_{2V}^f
Ag	1	—	—	—
Cu	1.05	0.4 k	0.1	—
Au	0.87	0.5 k	—	—
Ni	1.4	1.5 k	0.3	2 k
Al	0.65	0.8 k	0.3	1 k

From an analysis by Seeger A., and H. Mehrer; 1970; "Vacancies and Interstitials in Metals"; *Proceedings of the Julich Conference* (September, 1968); A. Seeger, D. Schumacher, W. Schilling, and J. Diehl, Eds.; North-Holland, Amsterdam.

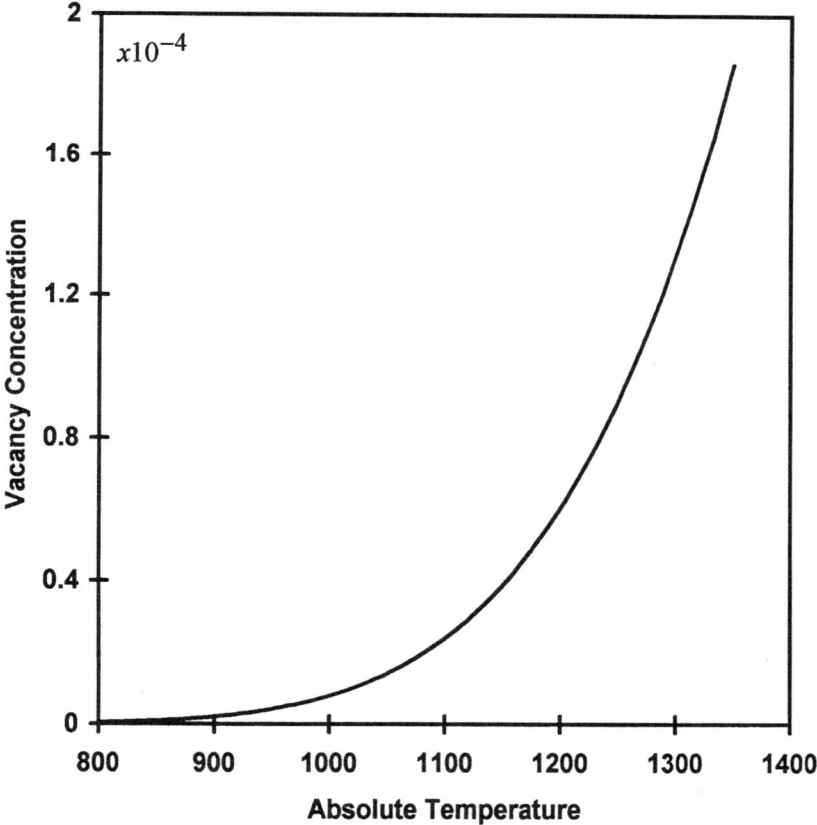

Figure 15.1. Variation of vacancy concentration with temperature in copper.

$G_v^f = U_v^f - TS_v^f$. The atomic fraction of vacancies near the melting point is of the order of 10^{-4}, and it decreases rapidly with temperature. This curve shows why the effects of an equilibrium concentration of vacancies are observable only at high temperatures. Because of the low concentration of monovacancies and the low divacancy binding energy, the divacancy concentration is quite small in copper. Also, because of the high formation energy for interstitials, their concentration is negligible. Thus, in copper at equilibrium, the monovacancy is the predominant point defect.

Since the volume of vacancy formation is positive, the application of pressure decreases the vacancy concentration. This effect is shown in figure 15.2, in which the ratio of the vacancy concentration (at the melting point) at pressure P to that at zero pressure is plotted against pressure according to

$$\frac{N_v(P,T_m)}{N_v(0,T_m)} = e^{-PV_v^f/kT_m} \tag{15.7.1}$$

which is readily obtained from (15.5.1). V_v^f was taken to be one half the atomic volume of copper, and T_m is its melting point.

The percentage vacancy contribution to the heat capacity of copper is displayed in figure 15.3, which was computed from

$$\frac{C_P - C_P^o}{N} = \frac{N_v}{N}\left(\frac{U_v^f}{kT}\right)^2 \tag{15.7.2}$$

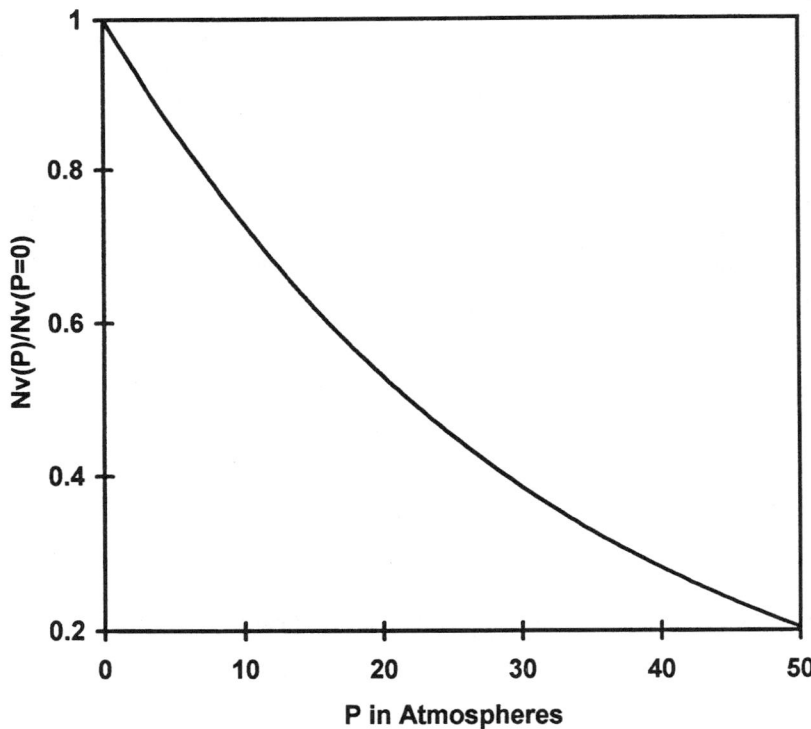

Figure 15.2. Variation of vacancy concentration with pressure in copper.

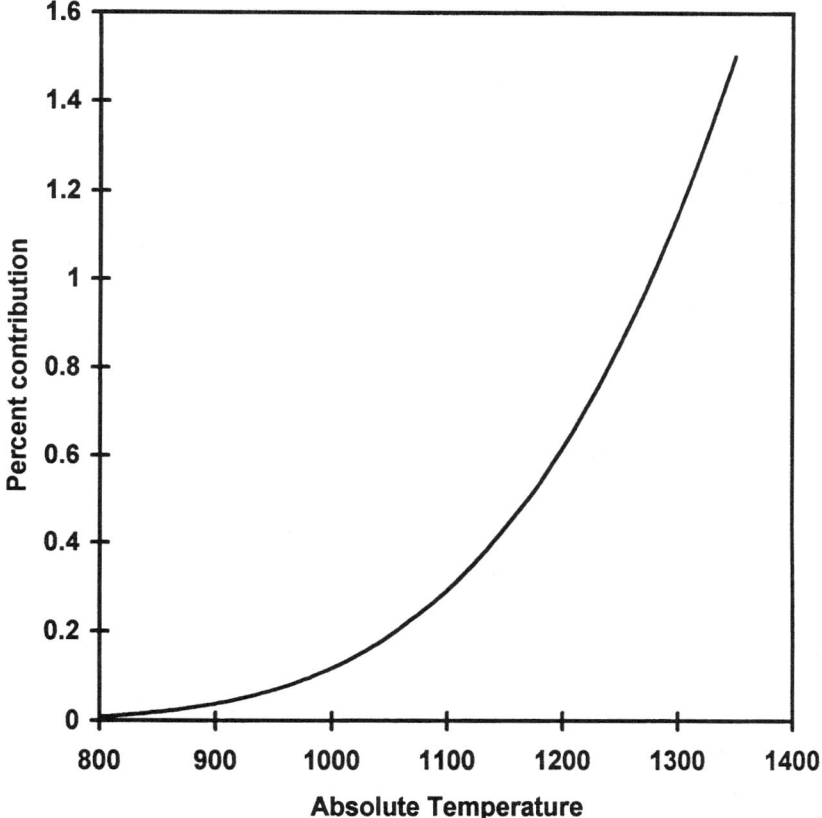

Figure 15.3. Vacancy contribution to the heat capacity of copper.

This equation follows from (15.4.16) since $H_v^f = U_v^f$ at zero pressure and $(C_v^f)_P$ can be shown to be negligible from the calculations in section 15.5. Also, the high-temperature value of $3Nk$ was used for the heat capacity of the perfect crystal, which is sufficiently accurate at temperatures where the vacancy concentrations become important. Note that vacancy formation increases the heat capacity since vacancy creation requires energy.

The contribution of vacancies to the thermal expansion of copper at zero pressure is shown in figure 15.4, which was obtained from equation (15.4.17), in which $(\alpha_v^f - \alpha^o)$ in the right-hand bracket was neglected. Since the thermal expansion of copper is about 70×10^{-6}, we see that the vacancies contribute about 0.07% to the thermal expansion of copper. The contribution to the compressibility is even smaller.

Although these figures show that the contributions of point defects to the thermodynamic properties of pure copper are small, and generally not detectable with any accuracy using available experimental methods, this must not be taken as being true for all metals. The effects are proportional to the defect concentrations, and in many metals they are considerably higher than in copper. As an example, the concentrations of vacancies and divacancies in aluminum are shown as functions of temperature in figure 15.5, and these are

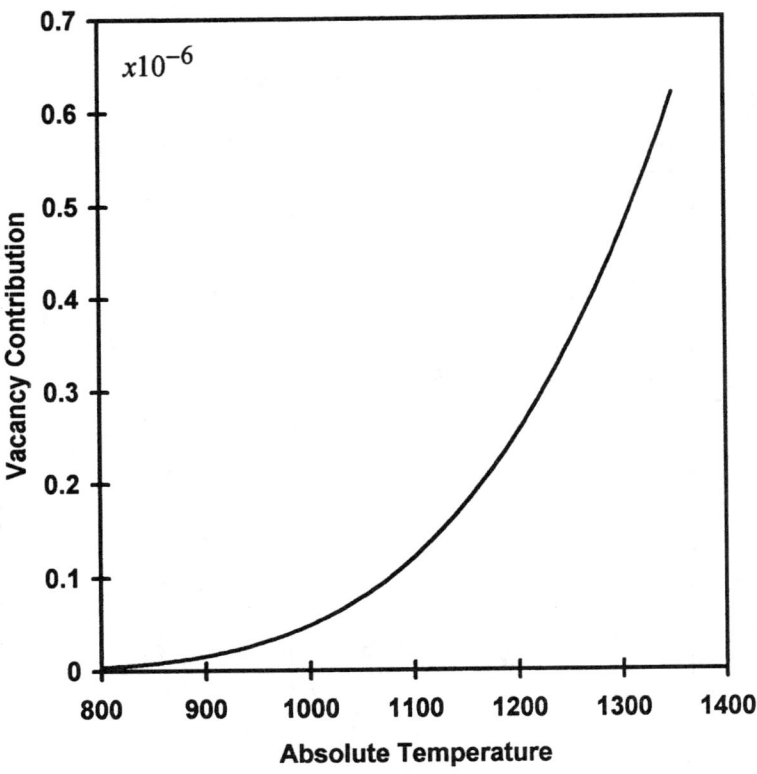

Figure 15.4. Vacancy contribution to thermal expansion in copper.

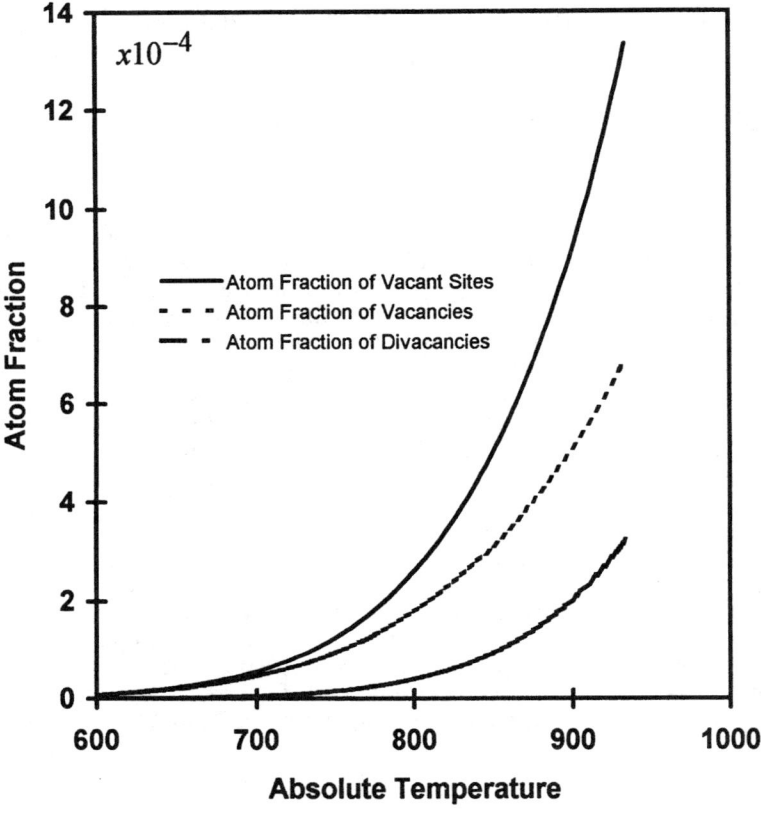

Figure 15.5. Vacancy and divacancy concentrations in aluminum.

much higher than in copper. The total concentration of vacant sites at the melting point is about 10^{-3}, which is a factor of 10 greater than for copper. It is interesting to note that nearly 40% of the vacant sites are tied up in divacancies, so divacancies can by no means be neglected in aluminum. Thus, point defects make significant and measurable contributions to the thermodynamic properties of at least some metals.

Figures 15.6–15.8 display the vacant site contributions to the heat capacity, thermal expansion, and compressibility of aluminum. The percentage increase in heat capacity was computed from (15.7.2). The contributions to the thermal expansion and compressibility were computed from equations just like (15.4.17) and (15.4.18), but containing terms for vacancies and divacancies that are identical in form. The energies of formation for vacancies and divacancies were taken from table 15.1. The volume of formation of a vacancy in aluminum was taken to be half the atomic volume, and the volume of formation of the divacancy was taken to be equal to the atomic volume. This assumes that the relaxation volume around a divacancy is twice that for a vacancy. The vacant site contributions to the heat capacity are much higher than for copper. In fact, the defect contributions are larger than the error in a decent experiment and therefore contribute measurable amounts.

The thermal expansion coefficient of aluminum at 800K is about 10^{-4}, which means that the vacant sites contribute nearly 10%. This is a large effect. The

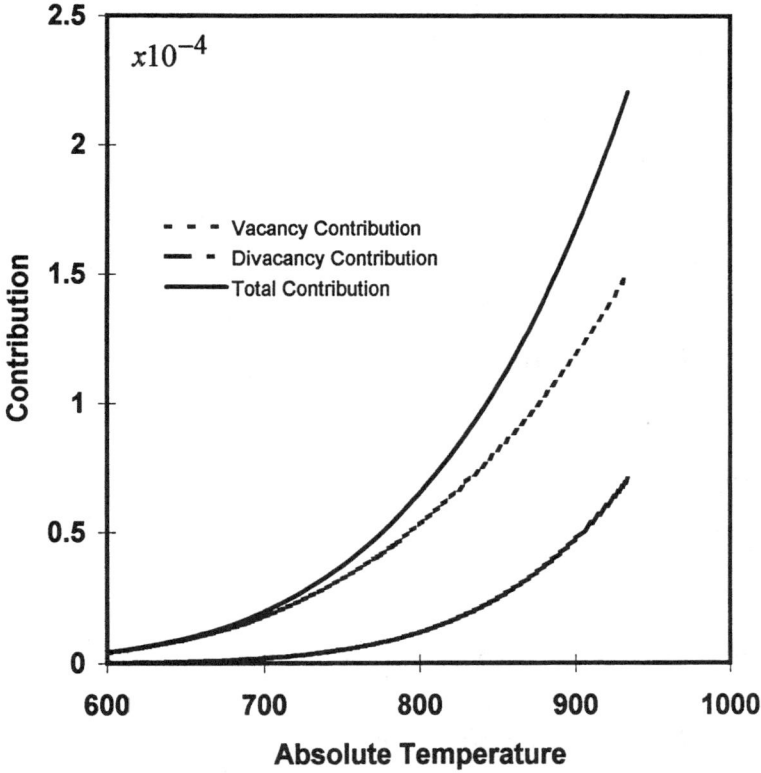

Figure 15.6. Vacant site contribution to heat capacity of aluminum.

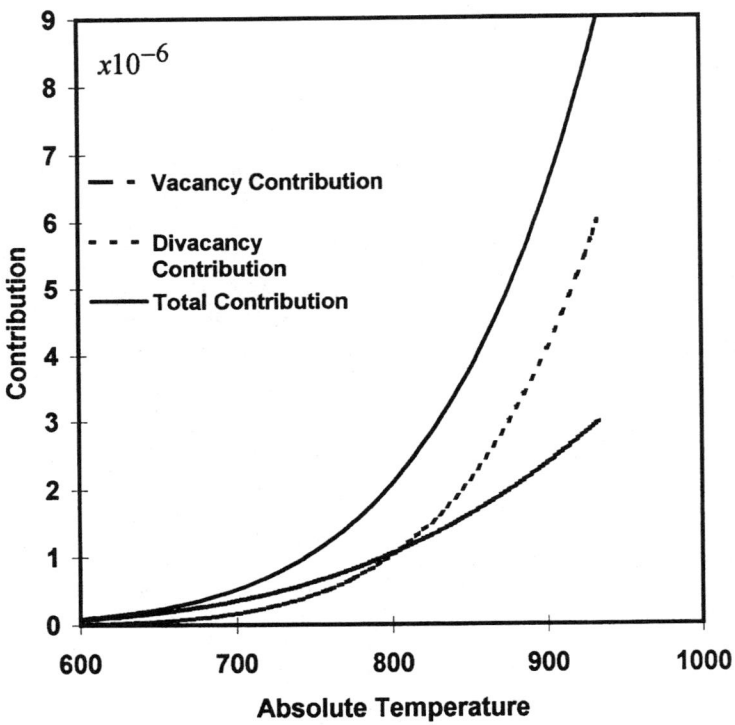

Figure 15.7. Vacant site contribution to thermal expansion of aluminum.

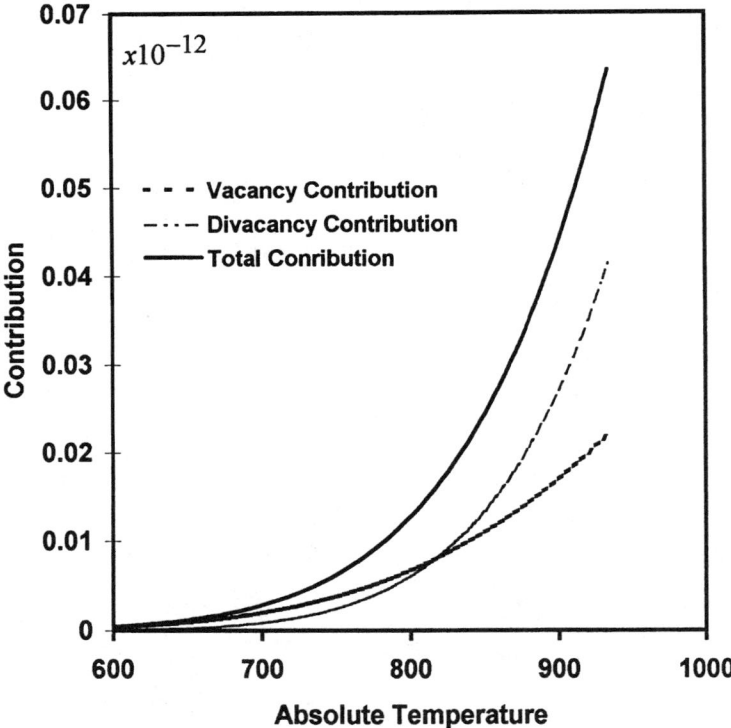

Figure 15.8. Vacant site contribution to compressibility of aluminum.

effect is positive because each vacant site increases the crystal volume by about half an atomic volume.

Vacant sites make the crystal softer so they increase the compressibility. The compressibility of aluminum is about 0.7×10^{-12} cm^2/dyne, so near the melting point the vacant sites add about 10% to the compressibility of aluminum.

Exercises

15.1 What is the fractional concentration of vacancies in aluminum at the melting point? Compare this to the fraction of atoms having "gaslike" properties in the theory of melting given in chapter 10 (table 10.2).

15.2 Assume that the change in vibrational frequencies upon creating a vacancy in a metal can be described by an Einstein model and that one degree of freedom per nearest neighbor atom is altered by the vacancy. For aluminum, what is the fractional change in the vibration frequency of each of the altered degrees of freedom? How does this compare with the estimate of the frequency change for atoms at a surface given in chapter 10?

15.3 What is the average distance between vacancies, in units of half the lattice parameter, at the melting point, in aluminum and in copper?

15.4 Because an atom at a surface has fewer nearest neighbors, the fraction of vacant surface sites is expected to be larger than in the bulk. For copper, compute the atomic fraction of vacancies at a surface, at the melting point, if the energy of vacancy formation at the surface is one half that in the bulk. Ignore entropy terms.

15.5 In the regular solution theory of chapter 10, it was assumed that the crystal energy was a sum of nearest neighbor interactions. Assuming that these interactions are the same in the presence and the absence of vacancies, derive a formula for the formation energy of a vacancy as a function of composition. Ignore the entropy of vacancy formation and all relaxation effects. Assume that every site is surrounded by the average atomic composition.

15.6 For a 50–50 order-disorder alloy, derive the formation energy of a vacancy as a function of the long-range order parameter, assuming each site has an average environment and that the interaction energies are the same in the absence or presence of a vacancy. Ignore entropy terms and all relaxation effects. (Refer to chapter 8.)

Notes

1. There has been considerable interest in this result recently. See Johnson, R.A.; 1994; *Physics Review B; vol. 50*, p. 799; Landsberg, P.T. and S.G. Canagaratna; 1997; *Physics Review B; vol. 55*, p. 5531; and Fahnle, M.; 1998; *Physical Status Solidü; vol 45*, p. R1. These authors were apparently unaware that this result, along with its generalization to dilute alloys containing a variety of point defects (see chapter 16), was published in 1973 by Girifalco, L.A.; (1973); *Statistical Physics of Materials*; John Wiley, New York.

2. Girifalco, L.A., and D.O. Welch; 1967; *Point Defects and Diffusion in Strained Metals*; Gordon and Breach, New York. From Welch's equation 3.22, $\beta_v^f \cong 1.34\beta^0$ and $\alpha_v^f = 0.95\alpha^0$.
3. Eshelby, J.D.; 1956; "The continuum theory of lattice defects", in *Solid State Physics*; F. Seitz and D. Turnbull, Eds.; Academic Press, New York; pp. 79–144.

16

Point Defects in Dilute Alloys

16.1 General comments

This chapter is devoted to an analysis of point defect equilibria in binary alloys for which the concentration of one of the constituents is much smaller than of the other. In such alloys, the minority constituent can itself be regarded as a defect or an impurity.

Our attention will be restricted to one-center and two-center defects, that is, to alterations in the local composition from that of the pure perfect crystal that involve either one atom or an adjacent pair of atoms. The perfect crystal is taken to be one in which all lattice sites are occupied by atoms of the type of the major constituent and all interstitial sites are empty. It will be assumed that all lattice atomic sites are equivalent and that all interstitial sites are equivalent. The results can be easily generalized to crystals in which atoms or interstitial sites are arranged on two or more nonequivalent sublattices, and to crystals containing more than one type of impurity.

The normal lattice on which the atoms in the perfect crystal are distributed will be called the L lattice and contains L sites, while the interstitial lattice is called the L' lattice and contains L' sites. The possible one- and two-center defects will be denoted as follows:

V a vacant lattice site (the monovacancy)
V_2 two adjacent vacant L sites (the divacancy)
B an impurity atom on an L site (the substitutional impurity)
B_2 two impurity atoms on adjacent L sites (the substitutional di-impurity)
BV an impurity atom on an L site adjacent to a vacant L site (the vacancy-lattice impurity complex)
A' a major constituent atom at an L' site (the interstitial)
A'_2 two major constituent atoms on adjacent L' sites (the di-interstitial)
B' an impurity on an L' site (the interstitial impurity)
B'_2 two impurity atoms on adjacent L' sites (the impurity di-interstitial)
$A'B'$ a major constituent atom and an impurity atom on adjacent L' sites (the host-impurity interstitial complex)
$A'V$ a major constituent atom on an L' site adjacent to a vacant site (the interstitial-vacancy complex)
$A'B$ a major constituent atom on an L' site adjacent to an impurity on an L site (the impurity interstitial complex)
BB' an impurity atom on an L site adjacent to an impurity on an L' site (the lattice-interstitial di-impurity)

The first five of these defects occur only on the atomic lattice and will be called *substitutional defects*; the second five occur only on the interstitial lattice and will be called *interstitial defects*; the last three have one center on the atomic lattice and the other on the interstitial lattice and will be called *mixed-lattice defects*.

It is implicit in the definition of these defects that they are surrounded by atoms of the major constituent (A atoms) on the L lattice, and in fact, the immediate environment of these defects should enter into their definitions. This can be seen by the fact that if two vacancies are brought close enough together, they form a divacancy. There is a limit to how close defects can approach each other without losing their identity and being transformed into something else, so there is a minimum number of A atoms surrounding a defect that must be included in its definition. In general, the stronger the lattice distortion around a defect, the larger this minimum number. Thus, in a close-packed structure, the region defining an interstitial will be larger than defining a vacancy because the interstitial distorts the lattice considerably while the atomic displacements around a vacancy are smaller.

Just as in chapter 15, the analysis here is based on the pressure ensemble. However, we now consider a system consisting of the impurity alloy in equilibrium with a gas phase containing A and B atoms. This will permit us to investigate not only the internal defect equilibria, but also the solid-vapor equilibrium and its coupling to the defect equilibria. It turns out that the results can be separated into those that depend on the existence of the vapor and those that do not, and the internal defect equilibria are the same whether or not the vapor is present. This means that the resulting formulas for the internal equilibria are valid even if the impurity has a concentration that is not in equilibrium with its vapor. Thus, much is gained and nothing is lost by using a pressure ensemble that includes the vapor phase. We therefore write the Gibbs free energy G of a crystal containing N_A atoms of type A (host atoms), N_B atoms of type B (impurity atoms) and $N_{(i)}$ defects of type i (as listed above), in equilibrium with a gas containing N_A^o atoms of type A and N_B^o atoms of type B, as

$$e^{-G/kT} = \sum_{N_A, N_B, N_{(i)}} w(N_A, N_B, N_{(i)}) e^{-G\{N_A, N_B, N_{(i)}\}/kT} e^{-G^g(N_A^g, N_B^g)/kT} \quad (16.1.1)$$

In this equation, the total numbers of A atoms and of B atoms are constant. $G\{N_A, N_B, N_{(i)}\}$ is the Gibbs free energy of a crystal containing N_A atoms of type A, N_B atoms of type B, and $N_{(i)}$ defects of type (i), except for the configurational contribution arising from the distribution of atoms and defects on the L and L' lattices. It is defined by a form completely analogous to (15.2.4). $G^g(N_A^o, N_B^o)$ is the free energy of the gas phase.

The equilibrium defect concentrations are again obtained by finding the most probable distribution, which is equivalent to determining the maximum term in the sum of (16.1.1). The Gibbs free energy of the system is

$$G(N_A, N_B, N_{(i)}) = G\{N_A, N_B, N_{(i)}\} + G^g(N_A^g, N_B^g) - kT \ln w[N_A, N_B, N_{(i)}] \quad (16.1.2)$$

in complete analogy with (15.2.14), and the thermodynamic Gibbs free energy of the system is given by (16.1.2) when the $[N_A, N_B, N_{(i)}, N_A^g, N_B^g]$ have their most probable values, which are determined by

$$\delta G[N_A, N_B, N_{(i)}] = 0 \quad (16.1.3)$$

where the variation is taken with respect to changes in the gas and crystal composition. Since the system (crystal + gas) is closed, a change in crystal composition can only take place with a corresponding change in the gas phase. For example, decreasing the number of vacancies in the crystal by unity corresponds to a transfer of an A atom in the gas to a vacant site in the crystal. In general, changing the number of i type defects involves a transfer of A and B atoms to or from the gas phase.

Equation (16.1.3) is valid for any arbitrary variation of crystal composition and can therefore be applied to any specific process of defect formation. For our purposes, it is convenient and instructive to use (16.1.3) in conjunction with specifically defined variations that describe defect formation. Equation (16.1.3) will therefore be applied to (16.1.2) in the usual way to give

$$kT \left.\frac{\partial \ln w}{\partial N_{(i)}}\right|_P = \left.\frac{\partial G\{N_A, N_B, N_{(i)}\}}{\partial N_{(i)}}\right|_P + \left.\frac{\partial G^g(N_A^g, N_B^g)}{\partial N_{(i)}}\right|_P \quad (16.1.4)$$

with the understanding that the derivatives refer to changes in the statistical count and the free energies accompanying a particular process of defect formation. The subscript P specifies the process to which the variation is applied.

16.2 The statistical count for substitutional defects

In this section the statistical count is derived for a dilute alloy containing substitutional defects only. This will make the notation a little easier and make the equations look less complex. The results are then easily generalized to the case containing interstitial and mixed types of defects in an obvious way. Also, in many metals, the formation energy of interstitials is enough higher than that of vacancies that interstitial type defects are not present in large concentrations.

We are given a lattice of L equivalent sites containing N_A A atoms, N_B B atoms, N_V monovacancies, N_{2V} divacancies, N_{2B} impurity pairs, and N_{BV} impurity-vacancy complexes. The alloy is dilute and therefore

$$N_B \ll N_A$$

$$N_V \ll N_A$$

$$N_{2V} \ll N_A \quad (16.2.1)$$

$$N_{2B} \ll N_A$$

$$N_{BV} \ll N_A$$

The total number of lattice sites is

$$L = N_A + N_B + N_v + 2N_{2v} + 2N_{2B} + 2N_{Bv} \quad (16.2.2)$$

The total number of B atoms and of vacant sites is

$$N_B^T = N_B + 2N_{2B} + N_{Bv} \quad (16.2.3)$$

$$N_v^T = N_v + 2N_{2v} + N_{Bv} \quad (16.2.4)$$

The way to get the statistical count is to remember that it is the number of states for which the crystal has a given energy spectrum. Thus, it can be computed by starting with a crystal having an allowed distribution of defects and counting the number of permutations among them that do not change the energy. A method for doing this was proposed by A. B. Lidiard and R. E. Howard,[1] but their result depends on the order in which the calculation for different defects is made, and the statistical count contains some configurations in which the number of di-defects are not properly computed. In the limit of small concentrations, the errors introduced by these factors vanish, so correct formulas are obtained for dilute alloys with a small number of defects. Nevertheless, it is desirable to have a method of counting the number of complexions that removes these errors.

Such a method is readily developed by first defining "core fields" for the defects as the number of sites a defect must occupy to retain its identity, as discussed in section 16.1. Let Z_1 be the number of such sites for one-center defects (vacancy, impurity) and Z_2 be the number of such sites for two-center defects (di-vacancy, di-impurity, vacancy-impurity complex). In the simplest case, where the defect is defined by requiring that only nearest neighbors to the defect must be occupied by A atoms, $Z_1 = z + 1$ and $Z_2 = z' + 2$, where z is the number of nearest neighbors to a lattice site and z' is the number of nearest neighbors to a pair of adjacent sites. The total number of sites pre-empted by the defects is

$$\Delta \equiv Z_1 N_B + Z_1 N_v + Z_2 N_{2B} + Z_2 N_{2v} + Z_2 N_{Bv} \qquad (16.2.5)$$

The number of complexions is just the number of ways the defects can be moved about in the crystal without having any overlap of their core fields.

Now consider a single impurity pair. If all other defects are held in fixed positions, the number of configurations that can be generated by moving the pair, without changing the crystal energy, is just the number of ways of distributing the pair on the available sites, which is $(z/2)(L - \Delta)$. This is obtained from the fact that the first atom of the pair has $(L - \Delta)$ sites available to it and that atom is always attached to another that can go on any of the z nearest neighbors to the first atom. This gives $z(L - \Delta)$ possible configurations. But the two atoms are identical, and exchanging them does not give a new configuration, so the result must be divided by 2. Since the number of di-impurities is N_{2B}, the total number of configurations generated by the permutation of impurity pairs is $[(z/2)(L - \Delta)]^{N_{2B}}$. This, however, clearly includes configurations that differ only by an interchange of impurity pairs and therefore must be divided by $N_{2B}!$ to give the number of configurations of indistinguishable impurity pairs (for each configuration of all other kinds of defects) as $(1/N_{2B}!)[(z/2)(L - \Delta)]^{N_{2B}}$.

There is obviously a factor of this kind for every type of defect, so the total statistical count is

$$W = \frac{[z(L-\Delta)/2]^{N_{2B}} [z(L-\Delta)/2]^{N_{2v}} (L-\Delta)^{N_B} (L-\Delta)^{N_v} [z(L-\Delta)]^{N_{BV}}}{N_{2B}! N_{2v}! N_{Bv}! N_v! N_B!} \qquad (16.2.6)$$

This can be reduced to a more compact form by labeling the defects with the running index (i) so that $N_{(i)}$ is the number of defects of type i ($i = 2B, 2_v, B_v, B, v$) and defining a "rotational factor" ω_i, which is unity for one-center defects, z for two-center defects consisting of unlike species, and $z/2$ for pairs consisting of identical species. ω_i is called the *defect rotational factor* because it describes the degree of indistinguishability produced by rotating the defect.

If Stirling's approximation for the factorials is used in the form

$$N! = N^N e^{-N} \tag{16.2.7}$$

equation (16.2.6) can now be written as

$$w = \prod_{(i)} e^{N_i} \left[\frac{\omega_i (L - \Delta)}{N_{(i)}} \right]^{N_{(i)}} \tag{16.2.8}$$

where the product is over all defects of type i. Equation (16.2.8) is obviously a general result in that any defect, whether of the substitutional, interstitial, or mixed type, contributes a factor to the statistical count of the form

$$w_{(i)} = e^{N_i} \left[\frac{\omega_i (L - \Delta)}{N_{(i)}} \right]^{N_{(i)}} \tag{16.2.9}$$

In applying this formula, it is only necessary to determine the number of lattice sites available to the defect and the rotational factor ω_i.

The derivatives of the statistical count will be needed in the following section. They are given by

$$\frac{\partial \ln w_{(i)}}{\partial N_{(i)}} = \ln \frac{\omega_i (L - \Delta)}{N_{(i)}} \tag{16.2.10}$$

16.3 Defect concentration formulas for substitutional defects

Now that the statistical count has been determined, the defect concentrations can be obtained from equation (16.1.4). In applying (16.1.4), it is only necessary to specify the process for making the defect by transfers of atoms to or from the gas phase.

To create a vacancy, either an A atom or a B atom can be removed from its lattice position in the crystal and transferred to the vapor. If an A atom is transferred, then (16.1.4) is written as

$$kT \left. \frac{\partial \ln w}{\partial N_v} \right|_{A \to V} = \left. \frac{\partial G\{\ \}}{\partial N_v} \right|_{A \to V} + \left. \frac{\partial G^g}{\partial N_{(i)}} \right|_{A \to V} \tag{16.3.1}$$

where the arguments in the free energies have been omitted for convenience.

The derivative of the statistical count for this process only involves the vacancy contribution, so from (16.2.10),

$$\left. \frac{\partial \ln w}{\partial N_v} \right|_{A \to V} = \frac{\partial \ln w_v}{\partial N_v} = \ln \frac{(L - \Delta)}{N_v} \tag{16.3.2}$$

The first term on the right of (16.3.1) is the free energy change on forming a vacancy by removing an A atom from the crystal and transferring it to the vapor phase. This free energy will be denoted by $G^f_{v(A)}$. The right-hand side of (16.3.1) will therefore be written as

$$\left. \frac{\partial G\{\ \}}{\partial N_v} \right|_{A \to V} + \left. \frac{\partial G^g}{\partial N_v} \right|_{A \to V} = G^f_{v(A)} + \mu_A \tag{16.3.3}$$

where μ_A is the chemical potential of component A in the gas phase. That is,

$$\left.\frac{\partial G^g}{\partial N_v}\right|_{A \to V} = \frac{\partial G^g}{\partial N_A} \equiv \mu_A$$

because the vacancy formation process adds an A atom to the vapor at constant temperature and pressure. It is evident that $G^f_{v(A)} + \mu_A \equiv G^f_{vA}$ is the free energy of vacancy formation as it is usually defined. This can be seen by writing the derivative in (16.3.3) as

$$\left.\frac{\partial G\{\ \}}{\partial N_v}\right|_{A \to V} = G\{N_A - 1, N_v + 1\} - G\{N_A, N_v\}$$

where $G\{N_A, N_v\}$ is the free energy of a crystal containing N_A A atoms and N_v vacancies before the transfer and $G\{N_A - 1, N_v + 1\}$ is the free energy of the crystal after the transfer that then contains one less A atom and one more vacancy. However, since at equilibrium μ_A is also the chemical potential of the crystal, $G\{N_A - 1, N_v + 1\} + \mu_A = G\{N_A, N_v + 1\}$, which is the free energy of a crystal containing N_A A atoms and $N_v + 1$ vacancies. Therefore,

$$\left.\frac{\partial G\{\ \}}{\partial N_v}\right|_{A \to V} + \mu_A = G\{N_A, N_v + 1\} - G\{N_A, N_v\}$$

This is just the configurationless free energy of formation at constant number of atoms as it is usually defined. Equations (16.3.2) and (16.3.3) can now be combined according to (16.3.1) to get

$$\frac{N_v}{L - \Delta} = \lambda_A^{-1} e^{-G^f_{v(A)}/kT} \tag{16.3.4}$$

where λ_A is the absolute activity defined by $\lambda_A = e^{\mu_A/kT}$.

We adopt the notational convention that parentheses in the subscript of the defect formation quantities indicate that the defect is formed by transfers to the vapor phase. The quantities without parentheses in the subscript indicate defect formation by transfers to the surface.

Now consider the process of forming a vacancy by removing a B atom from the crystal and transferring it to the vapor. This increases the number of vacancies and decreases the number of B atoms, so changes in the statistical counts of both the vacancies and the B atoms must be accounted for. Therefore, for this process, (16.2.8) gives

$$\left.\frac{\partial \ln w}{\partial N_v}\right|_{B \to V} = \frac{\partial \ln w_v}{\partial N_v} - \frac{\partial \ln w_B}{\partial N_B} = \ln \frac{N_B}{N_v} \tag{16.3.5}$$

Since

$$\left.\frac{\partial G^g}{\partial N_v}\right|_{B \to V} = \frac{\partial G^g}{\partial N_B} = \mu_B$$

the change in the Gibbs free energy for this process is

$$\left.\frac{\partial G\{\ \}}{\partial N_v}\right|_{B \to V} + \mu_B = G^f_{v(B)} + \mu_B \tag{16.3.6}$$

where μ_B is the chemical potential of B.

POINT DEFECTS IN DILUTE ALLOYS 437

For the process under consideration, (16.1.4) is

$$kT\frac{\partial \ln w}{\partial N_v}\bigg|_{B\to V} = \frac{\partial G\{\ \}}{\partial N_v}\bigg|_{B\to V} + \frac{\partial G^g}{\partial N_v}\bigg|_{B\to V} \quad (16.3.7)$$

so (16.3.5)–(16.3.7) give

$$\frac{N_v}{N_B} = \lambda_B^{-1} e^{-G_{v(B)}^f/kT} \quad (16.3.8)$$

where $\lambda_B = \exp(\mu_B/kT)$ is the absolute activity of the impurity.

The definition of the vacancy formation free energy in this case is completely analogous to that in (16.3.3), so $[G_{v(B)}^f + \mu_B]$ is the free energy change on removing a B atom from the interior of the crystal and placing it on the surface, leaving behind a vacant site.

The equilibrium concentration of B atoms can be obtained by combining (16.3.4) and (16.3.8) to give

$$\frac{N_B}{L-\Delta} = \frac{\lambda_B}{\lambda_A} e^{-[G_{v(A)}^f - G_{v(B)}^f]/kT} \quad (16.3.9)$$

This equation determines the equilibrium solubility of B in A for a crystal in equilibrium with its vapor and shows that there is a close connection between solubility and defect formation. Since we are restricted to dilute solutions, $G_{v(A)}^f$ must always be sufficiently greater than $G_{v(B)}^f$ to satisfy the condition that $N_B \ll L$.

To get the equilibrium concentration of divacancies, consider the process in which two vacancies come together to form a divacancy, leaving the number of all other defects and the composition of the gas unchanged. In this case,

$$\frac{\partial \ln w}{\partial N_v}\bigg|_{2V\to V_2} = -2\frac{\partial \ln w_v}{\partial N_v} + \frac{\partial \ln w_{2v}}{\partial N_{2v}} \quad (16.3.10)$$

because the process destroys two vacancies and creates a divacancy. Only the free energy of the crystal changes in this process since the gas is unaffected, so the application of (16.1.4) gives

$$-2\frac{\partial \ln w_v}{\partial N_v} + \frac{\partial \ln w_{2v}}{\partial N_{2v}} = \frac{1}{kT}\frac{\partial G\{\ \}}{\partial N_{2v}}\bigg|_{2V\to V_2} \quad (16.3.11)$$

The derivative on the right-hand side of (16.3.11) is the free energy of binding for a divacancy because it represents the free energy difference of a bound pair and a separated pair of vacancies. It will be written as

$$G^w \equiv \frac{\partial G\{\ \}}{\partial N_{2v}}\bigg|_{2V\to V_2} \quad (16.3.12)$$

Placing this in (16.3.11) and evaluating the derivatives of the statistical count gives

$$\frac{N_{2v}}{L-\Delta} = \frac{z}{2}\left(\frac{N_v}{L-\Delta}\right)^2 e^{G^w/kT} \quad (16.3.13)$$

438 STATISTICAL MECHANICS OF SOLIDS

In a similar way, if we go through the process of forming a di-impurity from two isolated impurities, we get

$$\frac{N_{2B}}{L-\Delta} = \frac{z}{2}\left(\frac{N_B}{L-\Delta}\right)^2 e^{G^{BB}/kT} \qquad (16.3.14)$$

where G^{BB} is the binding free energy for an impurity pair.

Finally, if the process of forming an impurity-vacancy complex by bringing together an isolated vacancy and an isolated B atom in the crystal is applied to (16.1.4), the result is

$$\frac{N_{Bv}}{L-\Delta} = \frac{z}{2}\left(\frac{N_B}{L-\Delta}\right)\left(\frac{N_v}{L-\Delta}\right) e^{G^{Bv}/kT} \qquad (16.3.15)$$

This is an appropriate place to point out that the lack of precision in the definition of the number of sites excluded from the statistical count has very little consequence. As long as all concentrations are small, Δ is small relative to the total number of lattice sites and can usually be neglected.

All the equilibrium concentration formulas for substitutional defects have now been obtained. The numbers of one-center defects can be eliminated from the expressions for the two-center defects by using (16.3.4) and (16.3.9). A summary of the defect concentration formulas is given below, where the atomic concentrations $N_{(i)}/(L-\Delta)$ are written as $C_{(i)}$ to abbreviate the notation:

$$C_v = \lambda_A^{-1} e^{-G^f_{v(A)}/kT} \qquad (16.3.16)$$

$$C_{2v} = \frac{z}{2} C_v^2 e^{-G^w/kT} \qquad (16.3.17)$$

$$C_{2v} = \lambda_A^{-2} e^{-G^f_{2v}/kT} \qquad (16.3.18)$$

$$C_{Bv} = z C_B C_v e^{-G^{Bv}/kT} \qquad (16.3.19)$$

$$C_{2B} = \frac{z}{2} C_B^2 e^{-G^{BB}/kT} \qquad (16.3.20)$$

$$C_B = \frac{\lambda_B}{\lambda_A} e^{-G^s_B/kT} \qquad (16.3.21)$$

$$C_{Bv} = z \frac{\lambda_B}{\lambda_A^2} e^{-G^f_{Bv}/kT} \qquad (16.3.22)$$

$$C_{2v} = \frac{z}{2}\left(\frac{\lambda_B}{\lambda_A}\right) e^{-G^f_{2B}/kT} \qquad (16.3.23)$$

In these equations, we have defined the following:

The free energy of divacancy formation:

$$G^f_{2v} = 2 G^f_{v(A)} - G^w \qquad (16.3.24)$$

The free energy of impurity atom dissolution:

$$G^s_B = G^f_{v(A)} - G^f_{v(B)} \qquad (16.3.25)$$

The free energy of impurity-vacancy complex formation:

$$G^f_{Bv} = G^s_B + G^f_{v(A)} - G^{Bv} \qquad (16.3.26)$$

The free energy of di-impurity formation:

$$G^f_{2B} = 2G^s_B - G^{BB} \qquad (16.3.27)$$

Equations (16.3.16)–(16.3.23) were obtained for a dilute binary alloy in equilibrium with its vapor. However, only the last three of these equations contain parameters that depend on the vapor-solid equilibrium. The first five equations describe the internal defect equilibria and are valid even if the crystal is not in equilibrium with its vapor, provided only that the crystal composition is a constant. In this case, of course, C_B is not given by (16.3.21) but is given in terms of the total impurity concentration, the concentration of di-impurities, and vacancy-impurity complexes as $C_B = C_B^T - C_{Bv} - 2C_{2B}$. This partial equilibrium is quite common and represents a metal containing a small amount of a second, nonvolatile component whose vapor pressure is never high enough to alter the crystal composition. Equations (16.3.16)–(16.3.20) can therefore be taken to describe a system that is only in internal equilibrium, or also in equilibrium with its vapor. If external equilibrium is in effect, the concentrations involving B atoms are given by (16.3.21)–(16.3.23).

Note that the formation free energies were defined relative to the gas phase, since we were considering a crystal in equilibrium with its vapor. The vapor-crystal equilibrium is often neglected in defect studies, and the formation free energies are then defined by forming the defect by transferring atoms to or from the surface. The two definitions lead to the same defect concentration formulas when only internal equilibria are considered. This is ensured by the presence of the absolute activities in (16.3.16)–(16.3.20).

The total concentration of vacant sites and of B atoms is given by

$$C_v^T = C_v + 2C_{2v} + C_{Bv} \qquad (16.3.28)$$

$$C_B^T = C_B + 2C_{2B} + C_{Bv} \qquad (16.3.29)$$

or,

$$C_v^T = C_v\left(1 + zC_v e^{G^w/kT} + zC_B e^{G^{Bv}/kT}\right) \qquad (16.3.30)$$

$$C_B^T = C_B\left(1 + zC_B e^{G^{BB}/kT} + zC_v e^{G^{Bv}/kT}\right) \qquad (16.3.31)$$

It is sometimes necessary to compute C_B and C_{BB} from the total amount of B in the alloy. This is readily done by solving (16.3.31) to obtain

$$C_B = A\left[(1 + BC_B^T)^{1/2} - 1\right] \qquad (16.3.32)$$

where

$$A \equiv \frac{1 + zC_v e^{G^{Bv}/kT}}{2ze^{G^{BB}/kT}} \qquad (16.3.33)$$

and

$$B \equiv \frac{2}{A\left(1 + ze^{G^{Bv}/kT}\right)} \qquad (16.3.34)$$

This gives C_B, and then C_{2B} can be computed from (16.3.20).

When the impurity concentration is low enough, the computation can be simplified since C_B can be replaced by C_B^T in the bracketed term of (16.3.31), provided the di-impurity binding free energy is not too large. Then, solving for C_B gives

$$C_B \cong C_B^T \left(1 + zC_B^T e^{G^{BB}/kT} + zC_v e^{G^{Bv}/kT}\right)^{-1} \qquad (16.3.35)$$

Once C_B is computed from (16.3.32) or (16.3.35), the other defect concentrations can be obtained from (16.3.16)–(16.3.20).

16.4 Internal equilibria for substitutional defects

For a monatomic metal with a nonvolatile impurity, the point defect concentrations are governed by (16.3.16)–(16.3.20), along with (16.3.32), which relates the internal defect equilibria to the total amount of impurity. From (16.3.17), (16.3.19), and (16.3.20), we have

$$\frac{C_{2v}}{C_v^2} = \frac{z}{2} e^{G^{vv}/kT} \qquad (16.4.1)$$

$$\frac{C_{Bv}}{C_B C_v} = z e^{G^{Bv}/kT} \qquad (16.4.2)$$

$$\frac{C_{2B}}{C_B^2} = \frac{z}{2} e^{G^{BB}/kT} \qquad (16.4.3)$$

These equations have the form of equilibrium constants for chemical reactions. In fact, a reaction can be defined in which two vacancies come together to form a divacancy, along with its converse in which a divacancy dissociates into two vacancies. In chemical reaction notation this is represented by

$$V + V \leftrightarrow V_2 \qquad (16.4.4)$$

and the right-hand side of (16.4.1) is seen to be just the equilibrium constant for this reaction. Similarly, two impurity atoms can form a di-impurity, or a di-impurity can dissociate. The corresponding reaction is

$$B + B \leftrightarrow B_2 \qquad (16.4.5)$$

with (16.4.3) as its equilibrium constant. The vacancy-impurity reaction is

$$B + V \leftrightarrow BV \qquad (16.4.6)$$

and its equilibrium constant is given by (16.4.2).

This scheme clearly shows the interdependence of the defect concentrations. For example, adding B atoms to the crystal shifts the reaction (16.4.6) to the right according to (16.4.2), thereby increasing the number of impurity-vacancy complexes and the total number of vacant sites. Also, adding B atoms shifts the reaction (16.4.5) to the right, creating more impurity pairs. A rise in temperature, however, decreases the ratio of impurity pairs to single impurity atoms since the equilibrium constant defined by (16.4.3) decreases with increasing temperature.

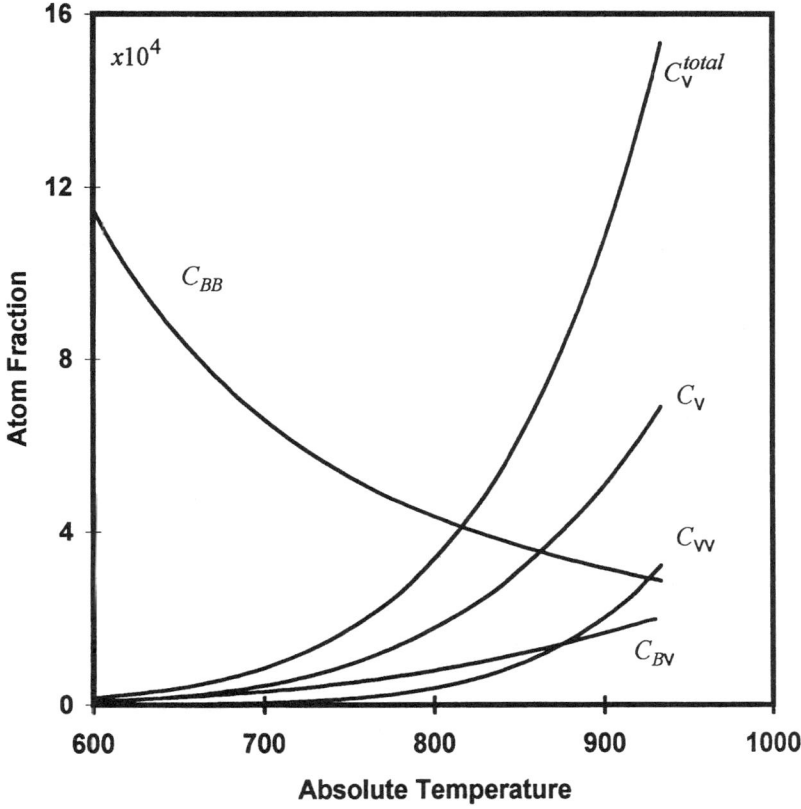

Figure 16.1. Defect concentrations in impure aluminum.

The addition of impurities has a significant effect on the defect equilibria. In figure 16.1 the defect concentrations in aluminum are plotted for a total impurity concentration of 10^{-3} atomic fraction. For these calculations, the data in table 15.1 were used, and it was assumed both the divacancy and di-impurity binding energies were 0.2 eV and that the binding entropies were zero. The figure shows that between 10% and 20% of the impurity is tied up in vacancy pairs and that between 3% and 12% are in vacancy-impurity complexes, depending on the temperature. The impurity has also increased the total number of vacant sites. In fact, at low temperatures, the majority of vacant sites are in impurity-vacancy complexes. Even at high temperatures, the impurity has increased the number of vacant sites by over 10%.

The magnitude of the effect of impurities on the defect equilibria depends, of course, on the impurity concentration, but also on the values of the binding energies. These are not known with any accuracy, but they range from 0 to about 0.5 eV.

16.5 Quenched-in resistivity of dilute binary alloys

The electrical resistivity of a metal has its origins in deviations from the perfect crystal structure, which results in the scattering of electrons. Since point defects in metals scatter electrons, the electrical resistivity can be used to study defect properties.

The electrical resistivity is a function of temperature, impurity content, and defect concentration. At normal temperatures, the effect of temperature is large since many lattice vibrations of all wavelengths are excited. At low temperatures, however, only a few phonons of long wavelength are active. Thus, the resistivity decreases with decreasing temperature. Because of the existence of impurities and defects, the resistivity retains a measurable value even at very low temperatures. It is this temperature-independent intrinsic resistivity that is of interest for defect studies.

For purposes of illustration, consider a pure metal containing only vacancies and dislocations. We assume the dislocation density to be independent of temperature. Now consider the following experiment. The metal is maintained at some high temperature T_q long enough to ensure that defect equilibrium is established. The temperature of the metal is then suddenly changed to a lower temperature T_f. The resistivity is then measured at liquid helium temperatures. It will be assumed that T_f is low enough to immobilize the defects and that the rate of quenching from T_q to T_f is so rapid that all defects present at T_q are present at T_f. The electrical resistivity of the quenched specimen can then be written as

$$\rho(T_q \to T_f) = \rho(D) + \rho_v(T_q) \tag{16.5.1}$$

where $\rho(D)$ is the resistivity arising from the fixed defects such as grain boundaries and $\rho_v(T_q)$ is the specific resistivity per atomic fraction of vacancies at temperature T_q. Phonon contributions do not appear since we are making measurements at liquid helium temperatures. [Whatever residual phonon contribution exists can be lumped into $\rho(D)$ if all measurements are made at the same temperature.]

The vacancy concentration is characteristic of that at the initial temperature T_q. Now if another identical metal sample is held at T_f long enough to ensure defect equilibrium, its vacancy concentration will be $C_v(T_f)$ and the resistivity, measured at liquid helium temperature, will be

$$\rho(T_f) = \rho(D) + \rho_v(T_f) \tag{16.5.2}$$

We assume that T_f is low enough that vacancies have a low mobility and do not anneal out while bringing the specimen to liquid helium temperature. Taking the difference between (16.5.2) and (16.5.1) gives

$$(\Delta\rho)_{T_q} = \rho(T_q \to T_f) - \rho(T_f)$$
$$= \rho_v[C_v(T_q) - C_v(T_f)] \tag{16.5.3}$$

$(\Delta\rho)_{T_q}$ is called the *quenched-in resistivity*. Note that T_f can be chosen to have a value for which $C_v(T_f)$ is negligible. The quenched-in resistivity is then proportional to the vacancy concentration at the quench temperature T_q, and substituting the vacancy concentration formula into (16.5.3) gives

$$(\Delta\rho)_{T_q} = \rho_v e^{-G_v^f/kT} \tag{16.5.4}$$

Now replace G_v^f by $U_v^f - TS_v^f$ and take logarithms. The result is

$$\ln(\Delta\rho)_{T_q} = \ln\rho_v + \frac{S_v^f}{k} - \frac{U_v^f}{kT} \tag{16.5.5}$$

Therefore, the vacancy formation energy can be obtained from experimental data by plotting $\ln(\Delta\rho)_{T_q}$ versus $1/T$ and computing the slope. In general, ρ_v is not known with any accuracy, so S_v^f cannot be obtained from the intercept of such a plot. Thus, while the quenched-in resistivity can give defect formation energies, it leads to unreliable results for total defect concentrations unless it is combined with some other kind of data.

The idealized experiment described above is difficult in practice, the major problem being the inability to achieve an infinitely fast quench. Thus, in a real experiment, the defect concentration is not characteristic of that at temperature T_q since some of the defects are lost as the specimen cools. Nevertheless, by performing the experiment at several quenching rates and extrapolating the results to infinite quenching rates, significant data can be obtained. In fact, careful experiments have been done on pure gold and on gold containing known amount of impurity. It is therefore worthwhile to consider the case of the dilute binary alloy, using the results of section 16.3.

For a dilute substitutional alloy, the point defect contribution to the resistivity is the sum of contributions from vacancies, divacancies, solute atoms, vacancy-solute complexes, and solute-solute pairs. The obvious generalization of (16.5.3) for this alloy is that the defect and impurity contributions to the resistivity at the temperatures T_q and T_f are given by

$$(\Delta\rho)_{T_q} = \rho(T_q \to T_f) - \rho(T_f) \tag{16.5.6}$$

with

$$\rho(T_q \to T_f) = \rho_v C_v(T_q) + \rho_{2v} C_{2v}(T_q) + \rho_{Bv} C_{Bv}(T_q) + \rho_B C_B(T_q) + \rho_{2B} C_{2B}(T_q) \tag{16.5.7}$$

$$\rho(T_f) = \rho_v C_v(T_f) + \rho_{2v} C_{2v}(T_f) + \rho_{Bv} C_{Bv}(T_f) + \rho_B C_B(T_f) + \rho_{2B} C_{2B}(T_f) \tag{16.5.8}$$

In this equation, the defect concentrations are given by (16.3.16)–(16.3.20) and ρ_v, ρ_{2v}, ρ_{Bv}, ρ_B, ρ_{2B} are the specific resistivities (per atom fraction) of vacancies, di-vacancies, vacancy-impurity complexes, and impurity pairs, respectively. It is obvious that for an impure metal, a plot of $\ln(\Delta\rho)_{T_q}$ versus $1/T$ is not linear.

Careful measurement of the quenched-in resistivity have been made by J. Bass on pure gold and on gold containing tin or silver as impurities.[2] These experiments were analyzed according to (16.5.6)–(16.5.8) in conjunction with (16.3.16)–(16.3.20).[3] Since a number of the parameters in the above equations were unknown, the best fit of the equations to the data had to be obtained by a computer-programmed parametric curve-fitting process. The deviation from a linear Arrhenius plot produced by a small amount of tin was remarkable. In fact, detailed analysis showed that the large deviation from linearity at low temperature is the result of the existence of tin-tin pairs. The values for the binding energies that were consistent with the data are $U^{vv} = 0.2$ eV, $U^{BB} = 0.2$, and $U^{Bv} = 0.5$ eV. Clearly, careful and detailed quenching experiments can give valuable results.

16.6 Some general theory

In this section, the basic theory of defect concentrations in a dilute alloy is summarized. The results obtained above for substitutional defects are generalized to include all the substitutional, interstitial, and mixed-lattice defects described in section 16.1.

In section 16.2, it was pointed out that any defect, whether of the substitutional, interstitial, or mixed-lattice type, contributes a factor to the statistical count given by equation (16.2.9). Accordingly, for the general case,

$$w = \prod_i w_{(i)} = \prod_i e^{N_{(i)}} \left[\frac{\omega_{(i)}(L-\Delta)}{N_{(i)}} \right]^{N_{(i)}} \tag{16.6.1}$$

where the product is now taken over all types of mono- and di-defects. This equation looks as if it applies only to crystals for which the number of interstitial sites equals the number of atomic lattice sites because the same L appears in every factor. However, it is easily applicable to the more general case in which L and L' are not equal by letting L always be the number of substitutional sites and appropriately redefining the rotational factors $\omega_{(i)}$.

Examination of equations (16.3.16)–(16.3.23) shows that all defect concentrations can be written in the following two equivalent forms:

$$C_{(i)} = \omega_{(i)} \lambda_{(i)} e^{-G_{(i)}^f/kT} \tag{16.6.2}$$

$$C_{(i)} = \omega_{(i)} e^{-G^{(i)}/kT} \tag{16.6.3}$$

These formulas are also correct for interstitial and mixed-lattice defects. The $G_{(i)}^f$ are the defect-free energies of formation referring to the vapor phase while $G^{(i)}$ are defect-free energies of formation referring to the crystal surface. The $\lambda_{(i)}$ are products of the absolute activities, and the $\omega_{(i)}$ are the rotational factors, generalized when necessary to account for the inequality of L and L'. The interpretations of the $\omega_{(i)}$, $\lambda_{(i)}$, $G_{(i)}^f$, and $G^{(i)}$ are obtained in precisely the same way for interstitial and mixed-lattice defects as for substitutional defects. Binding free energies also exist in the general case that are analogous to those already defined for substitutional one- and two-center defects.

Not all of the possible one-center and two-center defects will be important in all alloys. The concentration of each type of defect depends on its free energy of formation, and it is necessary to know the free energies in order to know which defects predominate in a particular system. For close-packed metals, the vacancy is the major intrinsic defect, and many impurities enter the lattice substitutionally. However, there is strong evidence that cadmium dissolves in lead by the so-called dissociative mechanism, in which the cadmium exists in both substitutional and interstitial sites, despite the fact that lead has the face-centered cubic structure. Furthermore, it appears that there is a strong attraction between an interstitial cadmium impurity and a vacancy. In lead that contains small amounts of cadmium impurity, therefore, the defects that must be considered are V, V_2, B, B_2, B', B'_2, BV, $B'V$, and BB'. Of these, it is probably safe to assume that B_2, B'_2, and BB' are in much smaller concentrations than are any of the others. It is of interest to write out the concentration formulas for the dissociative mechanism explicitly since diffusion experiments exist that have been interpreted by this mechanism. Applying equation (16.6.2) gives

$$C_V = \lambda_A^{-1} e^{-G_{V(A)}^f/kT} \tag{16.6.4}$$

$$C_{2V} = \frac{Z}{2} \lambda_A^{-2} e^{-G_{2V}^f/kT} \tag{16.6.5}$$

$$C_B = \frac{\lambda_B}{\lambda_A} e^{-G_B^S/kT} \tag{16.6.6}$$

$$C_{B'} = \lambda_B e^{-G_{B'}^S/kT} \tag{16.6.7}$$

$$C_{Bv} = z\frac{\lambda_B}{\lambda_A^2} e^{-G_{Bv}^f/kT} \tag{16.6.8}$$

$$C_{B'v} = z\frac{\lambda_B}{\lambda_A} e^{-G_{B'v}^f/kT} \tag{16.6.9}$$

These equations represent full equilibrium in which the impurity concentration is determined by the metal vapor in contact with the crystal. To obtain the internal equilibrium equations, consider the following reactions:

$$V + V \leftrightarrow V_2 \tag{16.6.10}$$

$$B + V \leftrightarrow BV \tag{16.6.11}$$

$$B' + V \leftrightarrow B'V \tag{16.6.12}$$

$$B \leftrightarrow B' + V \tag{16.6.13}$$

The first three of these reactions represent the usual formation of di-defects from monodefects. The last reaction represents the transfer of the impurity among the substitutional and interstitial sites. Applying the law of mass action to the above reactions gives

$$\frac{C_{2v}}{C_v^2} = K_1 \tag{16.6.14}$$

$$\frac{C_{Bv}}{C_B C_v} = K_2 \tag{16.6.15}$$

$$\frac{C_{B'v}}{C_{B'} C_v} = K_3 \tag{16.6.16}$$

$$\frac{C_{B'} C_v}{C_B} = K_4 \tag{16.6.17}$$

K_1, K_2, K_3, and K_4 are equilibrium constants that are readily obtained by substituting (16.6.4)–(16.6.9) into the left-hand sides of (16.6.14)–(16.6.17). These equations, along with the vacancy concentration formula, completely determine the conditions of internal equilibrium, and it is easy to show that

$$C_v = e^{-G_{vA}^f/kT} \tag{16.6.18}$$

$$C_{2v} = \frac{z}{2} C_v^2 e^{G^W/kT} \tag{16.6.19}$$

$$C_{Bv} = zC_B C_v e^{G^{Bv}/kT} \tag{16.6.20}$$

$$C_{B'v} = zC_{B'} C_v e^{G^{B'v}/kT} \tag{16.6.21}$$

$$\frac{C_{B'} C_v}{C_B} = \lambda_A e^{-[G_{B'}^S + G_{v(B)}^f]/kT} \tag{16.6.22}$$

These equations are valid for both the full-scale equilibrium and the case in which no metal atoms can be transferred to the vapor such that only internal equilibrium exists.

For the case of full-scale equilibrium, it is clear that we have arrived at the statistical thermodynamics of the solubility of slightly soluble additions to a monatomic crystal. In fact, the total concentration of solute G_B^T is given by

$$C_B^T = C_B + C_{B'} + C_{Bv} + C_{B'v} \quad (16.6.23)$$

where the terms on the right-hand side are given by equations (16.6.5)–(16.6.9). These equations contain chemical potentials that are determined by the partial pressures of A and B in the gas phase, so they describe the effect of pressure on solubility. Let us derive this relation explicitly for the case in which the concentrations of pairs can be ignored such that the solute exists only as atoms at regular lattice or interstitial sites. Then, using (16.6.6) and (16.6.7), (16.6.23) becomes

$$C_B^T = \frac{\lambda_B}{\lambda_A} e^{-G_B^s/kT} + \lambda_B e^{-G_{B'}^s/kT} \quad (16.6.24)$$

If the vapor is an ideal gas, then the absolute activities are given in terms of the partial pressures P_A and P_B as [see equation (3.4.25)]

$$\lambda_A = \frac{P_A}{kT} \Lambda_A^3 \quad (16.6.25)$$

$$\lambda_B = \frac{P_B}{kT} \Lambda_B^3 \quad (16.6.26)$$

where Λ_A and Λ_B are the thermal wavelengths for gas A and gas B, respectively.

For interstitial solution, the solubility is proportional to the partial pressure of the solute, but for substitutional solution, the solubility depends on the ratio of the partial pressures of solute and solvent because a vacant A site must be created to accommodate a solute atom.

It must be stressed that metals and alloys are often not in equilibrium with their vapor phases, and in those cases the impurity concentration is a constant independent of pressure. It is then related to the internal equilibria through equations (16.6.20)–(16.6.22). These, along with (16.6.23), give four equations in four unknowns that must be solved to get the individual concentrations C_B, $C_{B'}$, C_{Bv}, and $C_{B'v}$. If we include the effects of impurity pairs, which were neglected in the above discussion, the situation becomes more complex and seven simultaneous equations must be solved to get the individual concentrations of defects involving the impurity.

16.7 Thermodynamics of the dilute alloy

The contribution of all one- and two-center defects to the thermodynamic functions can be obtained by a generalization of the method used for vacancies in chapter 15. This method consists of writing the free energy of the crystal as a sum of a configurationless part and the contribution from the configurational entropy. The defect concentration formulas are then substituted into the configurational contribution, and finally, the configurationless free energy is expressed as a linear function of defect formation free energies.

The crystal free energy is

$$G_c[N_A, N_B, N_{(i)}] = G\{N_A, N_B, N_{(i)}\} - kT \ln w[N_A, N_B, N_{(i)}] \quad (16.7.1)$$

This is a generalization of equation (15.2.14). The configurationless free energy on the right-hand side of (16.7.1) is the same as that in (16.1.2) and is determined from the following partition function of the pressure ensemble:

$$e^{-G\{N_A, N_B, N_{(i)}\}/kT} = \sum_{i,j} e^{-(E_j + PV_i)/kT} \quad (16.7.2)$$

where the sum is over all energy states and over all possible volumes of the crystal.

Now substitute (16.6.1) into (16.7.1) to get

$$G_c[N_A, N_B, N_{(i)}] = G\{N_A, N_B, N_{(i)}\} - kT \sum_i N_{(i)}$$

$$- kT \sum_i N_{(i)} \ln\left[\frac{\omega_{(i)}(L-\Delta)}{N_{(i)}}\right] \quad (16.7.3)$$

But from (16.6.2),

$$\ln\left(\frac{\omega_{(i)}}{C_{(i)}}\right) = \ln\left[\frac{\omega_{(i)}(L-\Delta)}{N_{(i)}}\right] = \frac{G^f_{(i)}}{kT} - \ln \lambda_{(i)} \quad (16.7.4)$$

so (16.7.3) becomes

$$G_c(N_A, N_B, N_{(i)}) = G\{N_A, N_B, N_{(i)}\} - \sum_i N_{(i)} G^f_{(i)}$$

$$+ kT \sum_i N_{(i)} \ln \lambda_{(i)} - kT \sum_i N_{(i)} \quad (16.7.5)$$

At this point, we wish to express the configurationless free energy as a linear function of the number of defects, choosing the perfect crystal containing only A atoms as the reference state. Starting with a perfect crystal, each vacancy that is formed increases the free energy by $(G^f_{v(A)} - \mu_A)$ since the number of A atoms is not changed by vacancy formation, and $G^f_{v(A)}$ represents a free energy for removal of the A atom to the vapor phase. The free energy to add a B atom to the perfect crystal is $(G^S_B - \mu_A)$ since G^S_B is the free energy to exchange a B atom in the gas phase with an A atom in the crystal, and μ_A is the free energy required to maintain the number of A atoms constant. In a similar way, the change in free energy upon adding defects of all the types can be obtained from the definitions of the formation energies $G^f_{(i)}$. When this is done, and $G\{N_A, N_B, N_{(i)}\}$ is written as a linear function of these free energy changes, it is found that the first three terms in (16.7.5) reduce to $G^0\{N_A\}$, the configurationless free energy of a perfect crystal containing N_A atoms of type A. (Note that this is just the total free energy of the perfect crystal since in this case the statistical count is unity.) The defect contributions to $G\{N_A, N_B, N_{(i)}\}$ are canceled by the second and third terms in (16.7.5). Therefore, we can write (16.7.5) as

$$G_c(N_A, N_B, N_{(i)}) = G^0\{N_A\} - kT \sum_i N_{(i)} \quad (16.7.6)$$

This equation is a generalization of equations (15.3.22) and (15.6.28).

The thermodynamic functions can now be obtained from (16.7.6) by application of formulas having the same form as equations (15.4.1)–(15.4.8). The results are

$$S = S^0 + \sum_i \frac{N_{(i)}}{T}[H^{(i)} + kT] \qquad (16.7.7)$$

$$V = V^0 + \sum_i N_{(i)} V^{(i)} \qquad (16.7.8)$$

$$U = U^0 + \sum_i N_{(i)} U^{(i)} \qquad (16.7.9)$$

$$H = H^0 + \sum_i N_{(i)} H^{(i)} \qquad (16.7.10)$$

$$C_P = C_P^0 + \sum_i N_{(i)} \left[C_P^{(i)} + \frac{H^{(i)^2}}{kT^2} \right] \qquad (16.7.11)$$

$$\alpha = \alpha^0 + \sum_i \frac{N_{(i)} V^{(i)}}{V_0} \left[\alpha^{(i)} - \alpha^0 + \frac{H^{(i)}}{kT^2} \right] \qquad (16.7.12)$$

$$\kappa = \kappa^0 + \sum_i \frac{N_{(i)} V^{(i)}}{V_0} \left[\kappa^{(i)} - \kappa^0 + \frac{V^{(i)}}{kT} \right] \qquad (16.7.13)$$

These are generalizations of equations (15.4.11)–(15.4.18). In these equations, the left-hand sides represent crystal properties obtained from $G_c[N_A, N_B, N_{(i)}]$ by the usual thermodynamic formula. The thermodynamic functions labeled (i) are derived from the defect formations free energies $G^{(i)}$ by equations that are totally analogous to the ordinary relations of thermodynamics.

In the equations for the thermal expansion and compressibility, terms to the second order in the defect concentrations were neglected.

Exercises

16.1 A planar trivacancy in a face-centered cubic metal consists of three nearest-neighbor vacant sites on a (111) plane. Write the statistical count for this trivacancy as a function of trivacancy concentration. Also get the statistical count if the tri-defect consists of two vacancies and one impurity atom, all being nearest neighbors.

16.2 For a pure monatomic crystal in equilibrium with its vapor, derive the explicit relation for the difference between the free energy to create a vacancy by transferring an atom to the gas phase and transferring an atom to the crystal surface, as a function of temperature and vapor pressure. Assume that the vapor is a monatomic ideal gas.

16.3 Henry's law states that, for a dilute solution in equilibrium with its vapor, the partial pressure of the solute component, at constant temperature, is directly proportional to its concentration. Show that this follows from the theory of point defects in dilute alloys and find the proportionality constant. Hint: use the formula for the impurity concentration for an alloy in equilibrium with its vapor and assume the vapor is an ideal gas.

Notes

1. Lidiard, A.B.; 1960; "The Influence of Solutes on Self-Diffusion in Metals"; *Philosophical Magazine*; vol. 5, p. 1171. Howard, R.E., and A.B. Lidiard; 1964; "Matter Transport in Solids"; *Report on Progress in Physics*; vol 27, p. 161.
2. Bass, J.; 1965; "Quenched Resistance in Dilute Gold-Tin, Gold-Silver Alloys"; *Physics Review*; vol. 137, p. A765.
3. Liu, G.C.T., L.A. Girifalco, and R. Maddin; 1969; "Quenched in Electrical Resistivity of Dilute Binary Alloys"; *Physical Status Solidi*; vol. 31, p. 303. See also Girifalco, L.A; 1973; *Statistical Physics of Materials*; John Wiley, New York.

17

Diffusion in Simple Crystals

17.1 The empirical laws of diffusion

The phenomenological description of diffusion is embodied in Fick's two laws, which are empirical statements that relate the diffusive flow of matter to concentration gradients. To illustrate these laws, consider a single-phase dilute binary alloy in which the constituents are inhomogeneously distributed and let C be the concentration of the minor constituent. The concentration is a function of position and time and the concentration gradient induces a flow of matter. Fick's first law states that the flux of the diffusing species in a given direction has a magnitude proportional to the concentration gradient in that direction. That is, if J_1 is the flux along the x_1 axis, then

$$J_1 = -D_1 \frac{\partial C}{\partial x_1} \tag{17.1.1}$$

where D_1 is a proportionality factor called the *diffusion coefficient*. The negative sign expresses the fact that diffusion occurs from regions of high concentration to regions of low concentration.

If (x_1, x_2, x_3) are the points in a Cartesian coordinate system with unit vectors $(\mathbf{i}_1, \mathbf{i}_2, \mathbf{i}_3)$, then we would expect that an equation similar to (17.1.1) to hold for each of the three directions along the coordinate axes. For isotropic systems such as gases and liquids, this turns out to be the case. Furthermore, for such systems the diffusion coefficient is experimentally found to be the same in all directions. In nonhomogeneous systems, however, such as crystals with a low degree of symmetry, the diffusion coefficient is not the same in all directions. Also, the flux in one direction can depend on the flux in other directions. These results can be described by a generalized Fick's first law as

$$J_1 = -D_{11} \frac{\partial C}{\partial x_1} - D_{12} \frac{\partial C}{\partial x_2} - D_{13} \frac{\partial C}{\partial x_3} \tag{17.1.2}$$

Similarly, the fluxes in the other two directions of the Cartesian coordinate system are given by

$$J_2 = -D_{21} \frac{\partial C}{\partial x_1} - D_{22} \frac{\partial C}{\partial x_2} - D_{23} \frac{\partial C}{\partial x_3} \tag{17.1.3}$$

$$J_3 = -D_{31} \frac{\partial C}{\partial x_1} - D_{32} \frac{\partial C}{\partial x_2} - D_{33} \frac{\partial C}{\partial x_3} \tag{17.1.4}$$

These equations can be written in a compact form by defining the flux vector **J** as

$$\mathbf{J} = J_1 \mathbf{i}_1 + J_2 \mathbf{i}_2 + J_3 \mathbf{i}_3 \tag{17.1.5}$$

and the diffusion dyadic **D** as

$$\begin{aligned}\mathbf{D} = &\ D_{11}\mathbf{i}_1\mathbf{i}_1 + D_{12}\mathbf{i}_1\mathbf{i}_2 + D_{13}\mathbf{i}_1\mathbf{i}_3 \\ &+ D_{21}\mathbf{i}_2\mathbf{i}_1 + D_{22}\mathbf{i}_2\mathbf{i}_2 + D_{23}\mathbf{i}_2\mathbf{i}_3 \\ &+ D_{31}\mathbf{i}_3\mathbf{i}_1 + D_{32}\mathbf{i}_3\mathbf{i}_2 + D_{23}\mathbf{i}_2\mathbf{i}_3\end{aligned} \tag{17.1.6}$$

(see appendix 8). Then the equations (17.1.2)–(17.1.4) give the generalized Fick's first law as

$$\mathbf{J} = -\mathbf{D} \cdot \nabla C \tag{17.1.7}$$

Fick's second law is obtained from (17.1.7) by application of the equation of continuity, which states that

$$\frac{\partial C}{\partial t} + \nabla \cdot \mathbf{J} = 0 \tag{17.1.8}$$

This is just an expression of the law of the conservation of matter and states that any change in concentration in a volume element is the result of the difference in matter flow in and out of the volume element.

Combining (17.1.7) and (17.1.8) gives a generalized form of Fick's second law:

$$\frac{\partial C}{\partial t} = \nabla \cdot \mathbf{D} \cdot \nabla C \tag{17.1.9}$$

Note that if the diffusion dyadic is independent of position, then (171.9) reduces to

$$\frac{\partial C}{\partial t} = \sum_{r,s=1}^{3} D_{rs} \frac{\partial^2 C}{\partial x_r \partial x_s} \tag{17.1.10}$$

The set of quantities D_{rs} is the diffusion tensor and is equivalent to the diffusion dyadic.

As shown in appendix 8, it is always possible to refer a crystal dyadic to principle axes so that the dyadic has only diagonal components D_1, D_2, D_3. When this is done, the diffusion dyadic has the form

$$\mathbf{D} = D_1 \mathbf{i}_1 \mathbf{i}_1 + D_2 \mathbf{i}_2 \mathbf{i}_2 + D_3 \mathbf{i}_3 \mathbf{i}_3 \tag{17.1.11}$$

In isotropic and cubic systems this takes a particularly simple form since $D_1 = D_2 = D_3 = D$. If D is a constant independent of position and time, the simplification is even greater and Fick's laws become

$$\mathbf{J} = -D\nabla C \tag{17.1.12}$$

$$\frac{\partial C}{\partial t} = \nabla^2 C \tag{17.1.13}$$

Recall that the theory of random flights yielded equation (13.8.6) for the change in time of the probability distribution function for the atomic displacement.

This is fully equivalent to equation (17.1.13). An alternate, and somewhat more general, connection between diffusion and random flight is given in sections 17.2 and 17.3.

The value of these results lies primarily in the fact that they can be used to compute the distribution of the diffusing species in space and time. Thus, if the initial boundary conditions are known, (17.1.13) can be solved for $C(\mathbf{r}, t)$. Also, it is possible to set up experimental conditions corresponding to particularly simple solutions of (17.1.13), thereby making it possible to determine the diffusion coefficient from experiment.

The purpose of theory at this point is to provide a derivation of Fick's laws from statistical-kinetic theory and to relate the diffusion coefficient to the atomistic mechanisms responsible for diffusion. First, the method of transition probabilities will be used to show that the diffusion coefficient is proportional to the mean square of atomic displacements per unit time. Then the mean square displacements will be related to the defect concentration and to the atomic jump frequencies. The detailed treatment is restricted to simple systems but is sufficient to illustrate the general principles of the statistical mechanical theory of diffusion.

17.2 Transition probabilities and Fick's laws

Consider a volume element $d\mathbf{r}$ centered about a point defined by the position vector $\mathbf{r}$ in a material medium. The diffusion through this volume element can be described by counting the number of particles of the diffusion species entering and leaving the volume element per unit time. To do this, define a conditional transition probability density per unit time $\Lambda(\mathbf{r}|\mathbf{R})$ such that $\Lambda(\mathbf{r}|\mathbf{R})d\mathbf{R}dt$ is the probability that, if a particle is at $\mathbf{r}$, it will move a distance $\mathbf{R}$ into the volume element $d\mathbf{R}$ during the time dt. Since the number of particles in $d\mathbf{r}$ is $C(\mathbf{r}, t)d\mathbf{r}$, then the number of particles leaving $d\mathbf{r}$ in time dt is

$$\int_{\mathbf{R}} C(\mathbf{r}, t)d\mathbf{r}\Lambda(\mathbf{r}|\mathbf{R})d\mathbf{R}dt \qquad (17.2.1)$$

To get the number of particles entering the volume element $d\mathbf{r}$, consider a volume element $d\mathbf{R}$ about a point $(\mathbf{r} - \mathbf{R})$. The number of particles moving from this volume into $d\mathbf{r}$ during time dt is the number of particles in $d\mathbf{R}$ times the probability that a particle will move from $(\mathbf{r} - \mathbf{R})$ to $\mathbf{r}$, which is

$$C(\mathbf{r} - \mathbf{R}, t)\Lambda(\mathbf{r} - \mathbf{R}|\mathbf{R})d\mathbf{R}d\mathbf{r}dt \qquad (17.2.2)$$

Integrating this over all $\mathbf{R}$ gives the number of particles entering the volume element $d\mathbf{r}$ in time dt as

$$\int_{\mathbf{R}} C(\mathbf{r} - \mathbf{R}, t)\Lambda(\mathbf{r} - \mathbf{R}|\mathbf{R})d\mathbf{R}d\mathbf{r}dt \qquad (17.2.3)$$

The difference between (17.2.3) and (17.2.1) gives the increase in the number of particles in $d\mathbf{r}$, during time dt, as a result of particles moving out of $d\mathbf{r}$ and particles moving into $d\mathbf{r}$. If this difference is divided by $d\mathbf{r}$ and dt, the result is the rate of increase of the concentration at $\mathbf{r}$ at time t. Therefore,

$$\frac{\partial C}{\partial t} = \int_{\mathbf{R}} [C(\mathbf{r} - \mathbf{R}, t)\Lambda(\mathbf{r} - \mathbf{R}|\mathbf{R}) - C(\mathbf{r}, t)\Lambda(\mathbf{r}|\mathbf{R})]d\mathbf{R} \qquad (17.2.4)$$

Now let us assume that the transition probability is very small for large $\mathbf{R}$ compared to the distances over which diffusion is measured. This is certainly true for crystals in which diffusion takes place via jumps of atomic dimensions. It is even true for gases and liquids in which the magnitude of $\mathbf{R}$ is of the order of the mean free path. Therefore, Fick's laws can be obtained from (17.2.4) by expanding the first term in a Taylor series and retaining only the first two terms as follows:

$$C(\mathbf{r}-\mathbf{R},t)\Lambda(\mathbf{r}-|\mathbf{R}) = C(\mathbf{r},t)\Lambda(\mathbf{r}|\mathbf{R}) - \Lambda(\mathbf{r}|\mathbf{R})\mathbf{R}\cdot\nabla C + \frac{1}{2}\Lambda(\mathbf{r}|\mathbf{R})(\mathbf{R}\cdot\nabla)^2 C \quad (17.2.5)$$

Now assume that the transition probability is independent of position $\mathbf{r}$. This will be true for the cases we are now considering. Then, substituting (17.2.5) into (17.2.4) gives

$$\frac{\partial C}{\partial t} = -\int_\mathbf{R} \Lambda(\mathbf{r}|\mathbf{R})\mathbf{R}\cdot\nabla C d\mathbf{R} + \frac{1}{2}\int_\mathbf{R} \Lambda(\mathbf{r}|\mathbf{R})(\mathbf{R}\cdot\nabla)^2 C d\mathbf{R} \quad (17.2.6)$$

Remember that the concentration is evaluated at $\mathbf{r}$ and is independent of $\mathbf{R}$. Also, since the transitional probabilities are assumed to be independent of position, it is natural to take them to be the same for migrations $\mathbf{R}$ and $-\mathbf{R}$ in opposite directions. Then the first integral in (17.2.6) vanishes and we get

$$\frac{\partial C}{\partial t} = \frac{1}{2}\int_\mathbf{R} \Lambda(\mathbf{r}|\mathbf{R})(\mathbf{R}\cdot\nabla)^2 C d\mathbf{R} \quad (17.2.7)$$

Let the components of $\mathbf{R}$ and $\mathbf{r}$ in a Cartesian coordinate system be (X_1, X_2, X_3) and (x_1, x_2, x_3), respectively. Then

$$(\mathbf{R}\cdot\nabla)^2 C = \frac{1}{2}\sum_{i,j=1}^{3} X_i X_j \frac{\partial^2 C}{\partial x_i \partial x_j} \quad (17.2.8)$$

and (17.2.7) can then be written as

$$\frac{\partial C}{\partial t} = \frac{1}{2}\sum_{i,j=1}^{3} \overline{X_i X_j} \frac{\partial^2 C}{\partial x_i \partial x_j} \quad (17.2.9)$$

where

$$\overline{X_i X_j} = \int_\mathbf{R} X_i X_j \Lambda(\mathbf{r}|\mathbf{R}) d\mathbf{R} \quad (17.2.10)$$

These are the mean values of the quadratic migration distance products per unit time, the averages being taken over all possible migration distances.

Equation (17.2.9) has precisely the form of Fick's second law for the case of constant diffusion coefficients, and comparison with (17.1.10) shows that the components of the diffusion dyadic are given in terms of the migration distance averages as

$$D_{rs} = \frac{1}{2}\overline{X_r X_s} \quad (17.2.11)$$

If the coordinate axes are chosen to be the principle axes of diffusion, then only the diagonal terms in (17.2.11) are nonzero and equation (17.2.9) becomes

$$\frac{\partial C}{\partial t} = \frac{1}{2}\sum_{i=1}^{3} \overline{X_i^2}\frac{\partial^2 C}{\partial x_i^2} \qquad (17.2.12)$$

and the principle diffusion coefficients are given by

$$D_i = \frac{1}{2}\overline{X_i^2} \qquad (17.2.13)$$

Equation (17.2.12) is a fundamental relationship that connects the diffusion coefficient to the mean square particle displacement per unit time. It is this result that enables the macroscopic diffusion coefficient to be understood in terms of atomic mechanisms.

When all the D_i are equal (as in isotropic systems or cubic crystals) (17.2.12) becomes

$$\frac{\partial C}{\partial t} = \frac{\overline{X^2}}{2}\nabla^2 C \qquad (17.2.14)$$

where $\overline{X^2} = \overline{X_1^2} = \overline{X_2^2} = \overline{X_3^2}$. The diffusion coefficient is often expressed in terms of $\overline{R^2}$ instead of $\overline{X^2}$. Since

$$\overline{R^2} = \overline{X_1^2} + \overline{X_2^2} + \overline{X_3^2} \qquad (17.2.15)$$

we have, for an isotropic or cubic medium,

$$\overline{R^2} = 3\overline{X^2} \qquad (17.2.16)$$

and we write (17.2.14) as

$$\frac{\partial C}{\partial t} = D\nabla^2 C \qquad (17.2.17)$$

where

$$D = \frac{1}{6}\overline{R^2} \qquad (17.2.18)$$

Note that the statistical interpretation of Fick's first law follows directly from a comparison of (17.2.17) to the equation of continuity. For the cubic or isotropic case, we immediately get

$$\mathbf{J} = -D\nabla C \qquad (17.2.19)$$

with D given by (17.2.18).

Let us remember that, in deriving these results, the conditional transition probability was assumed to be isotropic and independent of position. This is a strong assumption, and since it is necessary if the theory is to yield Fick's laws, it is worthwhile to pause and consider its physical implications. Clearly, a constant transition probability implies that the material is homogeneous over the distances at which diffusion is measured. This means that there must be no external fields and no temperature gradients and that variations in concentrations are small enough so as to have a negligible effect. These are the kinds of systems for which Fick's laws hold. If fields or temperature gradients exist, or if the variations in concentrations are sufficiently large, then the transition probability depends on position. The diffusion coefficients are then also functions of position. Furthermore, the gradient terms in the Taylor expansion of the transition probability no longer integrate to zero, and the diffusion equations then have linear terms that are proportional to fields and to gradients of the concentration and temperature. The theory then becomes somewhat more complex.

17.3 Atomic jumps and the diffusion coefficient

Diffusion in crystals takes place by discrete atomic jumps in which an atom moves from one lattice site to another. The simplest diffusion system consists of a dilute interstitial impurity in a cubic metal, as exemplified by carbon in iron. In such a system, an impurity atom at an interstitial site spends most of its time executing small vibratory motions about its mean position. Occasionally, however, the impurity atom acquires a large amount of energy as the result of local thermal fluctuations and jumps to an adjacent interstitial site. Consequently, the impurity atom wanders through the crystal in a tortuous path that is the sum of a large number of random jumps.

The motion of a vacancy takes place in a similar fashion, the elementary vacancy jump consisting of the movement of an atom adjacent to the vacant site. In tracer movement by a vacancy mechanism, the tracer atom moves by jumping into an adjacent vacancy. In all these cases, the length of the displacement vector over a period of time is the sum of elementary vectors of the same length. Thus, if **R** is the overall displacement of a diffusing entity per unit time, then

$$\mathbf{R} = \sum_{i=1}^{\Gamma} \mathbf{r}_i \quad (17.3.1)$$

where $\mathbf{r}_i$ is the elementary jump vector for the ith jump and Γ is the number of jumps per unit time.

The diffusion coefficient is proportional to the square of the displacement, so we need to square (17.3.1) to get

$$R^2 = \sum_{i,j=1}^{\Gamma} \mathbf{r}_i \cdot \mathbf{r}_j \quad (17.3.2)$$

Now rewrite this by separating out the terms for which $i = j$:

$$R^2 = \sum_{i=1}^{\Gamma} \mathbf{r}_i^2 + \sum_{i,j \neq i}^{\Gamma} \mathbf{r}_i \cdot \mathbf{r}_j \quad (17.3.3)$$

The second term on the right can be written in terms of partial sums as

$$\sum_{i \neq j}^{\Gamma} \mathbf{r}_i \cdot \mathbf{r}_j = 2 \sum_{i=1}^{\Gamma-1} \mathbf{r}_i \cdot \mathbf{r}_{i=1} + 2 \sum_{i=1}^{\Gamma-2} \mathbf{r}_i \cdot \mathbf{r}_{i=2} + \ldots \quad (17.3.4)$$

The validity of (17.3.4) is readily seen by writing out the terms on the right-hand side as a square array and regrouping terms. The factor of 2 arises because the sum is over both i and j. Using (17.3.4), equation (17.3.3) becomes

$$R^2 = \sum_{i=1}^{\Gamma} r_i^2 + 2\sum_{j=1}^{\Gamma-1}\sum_{i=1}^{\Gamma-j} \mathbf{r}_i \cdot \mathbf{r}_{i+j} \tag{17.3.5}$$

Now let us restrict ourselves to cubic monatomic crystals for which all the jump vectors have the same magnitude r. Then the first sum is just r times the number of jumps per unit time Γ, and the second sum consists of terms given by

$$\mathbf{r}_i \cdot \mathbf{r}_{i+j} = r^2 \cos\theta_{i,i+j} \tag{17.3.6}$$

where $\theta_{i,i+j}$ is the angle between the ith and the $(i=j)$th jump. Equation (17.3.5) can therefore be written as

$$R^2 = \Gamma r^2 + 2r^2 \sum_{j=1}^{\Gamma-1}\sum_{i=1}^{\Gamma-j} \cos\theta_{i,i+j} \tag{17.3.7}$$

In this equation R is the magnitude of the displacement of a single atom per unit time. To obtain the mean square displacement that appears in the transition probability theory, it is only necessary to average (17.3.7) over a large number of atoms, each atom making Γ jumps per unit time. That is,

$$\overline{R^2} = \Gamma r^2 f \tag{17.3.8}$$

where

$$f = 1 + \frac{2}{\Gamma} \sum_{j=1}^{\Gamma-1}\sum_{i=1}^{\Gamma-j} \overline{\cos\theta_{i,i+j}} \tag{17.3.9}$$

It is obvious that we have been treating diffusion as a random flight just as in chapter 13, for the special case of solid state diffusion, by a somewhat different route. The method of getting from (13.2.11) to (13.2.19) is clearly completely valid for the present case, and f is the correlation factor given by equation

$$f = \frac{1+q}{1-q} = \frac{1+\overline{\cos\theta}}{1-\overline{\cos\theta}} \tag{17.3.10}$$

For solid state diffusion, the correlation factor is determined by the crystal structure and the diffusion mechanism. In general, computation of f is difficult, but for simple diffusion mechanisms it is a constant near unity.

For diffusion of an interstitial impurity in a cubic crystal, it is clear that $\overline{\cos\theta} = 0$, since for every possible atomic jump in a given direction, there is also a possible jump in the opposite direction. In this case, therefore, $f = 1$, and diffusion is said to occur by an uncorrelated random flight of the impurity atoms. Physically, this means that successive jumps of an interstitial impurity are completely independent. This is a consequence of the fact that the jump frequency Γ is much smaller than the vibration frequencies of the crystal normal modes. An impurity therefore stays at an interstitial site long enough to lose all memory of its preceding jump. Clearly, if we regard a vacancy as a diffusion entity, its migration is also uncorrelated and $f = 1$ for vacancy diffusion.

Diffusion of a substitutional tracer atom by a vacancy mechanism is another matter, however. After a tracer atom jumps into a vacant site, the vacancy is still next to the tracer atom. Furthermore, the time it takes the vacancy to move away from the tracer is comparable to the time it takes the tracer to move back into the vacancy. Therefore, the probability that a tracer will move back to the site it has left is greater than the probability that it will move to some other site. This means that the jump probabilities are not equal in all directions; the jumps are correlated because the direction of a jump depends on the direction of the preceding jump and $\cos\theta$ is not zero. Detailed calculations show that $f = 0.72$ for the body-centered cubic structure and $f = 0.78$ for the face-centered cubic structure.[1]

Combining (17.3.8) with (17.2.18) shows that the diffusion coefficient is given by

$$D = \frac{f}{6} r^2 \Gamma \qquad (17.3.11)$$

For the case of interstitial diffusion, $r = r_I$, the distance between interstitial sites; $f = 1$, and $\Gamma = \Gamma_I$ is the jump frequency for interstitial atoms. Therefore, we write the diffusion coefficient as

$$D_I = \frac{1}{6} r_I^2 \Gamma_I \qquad \text{(interstitial diffusion)} \qquad (17.3.12)$$

A similar formula holds for the diffusion of vacancies:

$$D_v = \frac{1}{6} r_L^2 \Gamma_v \qquad \text{(vacancy diffusion)} \qquad (17.3.13)$$

In this equation, r_L is the distance between nearest neighbor sites and Γ_v is the jump frequency for a vacancy.

For self-diffusion by a vacancy mechanism, $\Gamma = \Gamma_v C_v$, where C_v is the atom fraction of vacant sites. This is so because Γ is the actual number of jumps an atom makes per second. Of course, an atom cannot move unless a vacancy is next to it, so the jump frequency for an atom into a vacancy, Γ_v, must be multiplied by the probability C_v that a site is vacant. Thus, for self-diffusion

$$D_s = \frac{f}{6} r_L^2 \Gamma_v C_v \qquad \text{(self-diffusion)} \qquad (17.3.14)$$

It is found experimentally that diffusion coefficients in crystals have an Arrhenius-type dependence on temperature and pressure of the form

$$D = D_0 e^{-(Q^* + PV^*)/kT} \qquad (17.3.15)$$

where D_o, Q^*, and V^* are constants. A major purpose of the application of statistical mechanics to diffusion is the derivation of this experimental result, and the interpretation of the pre-exponential factor D_o and of the heat and volume of activation Q^* and V^*. All that remains to achieve this goal is to develop the theory of the jump frequency.

17.4 The jump frequency in one dimension

During an atomic jump, an atom passes from one equilibrium position to another. It is clear that, in the process of doing this, the atom meets strong repulsive forces from its neighbors, so it must surmount an energy barrier. The migrating atom only occasionally acquires enough energy to climb this barrier.

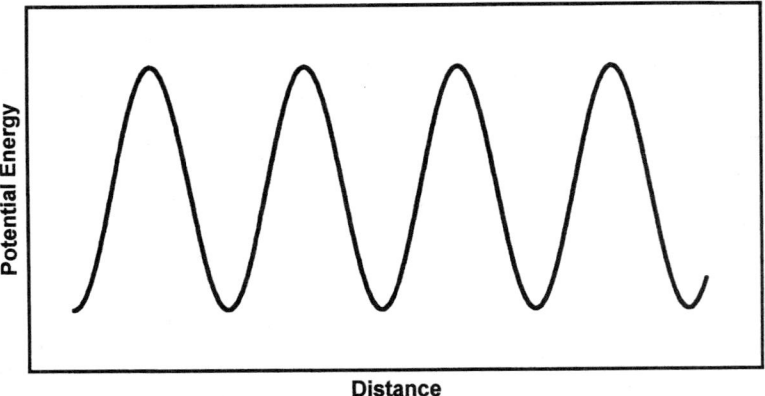

Figure 17.1. Diffusion of a particle in a one-dimensional periodic potential.

This energy is a local thermal fluctuation consisting of the coming together of phonons of sufficient energy and the correct directionality to move the atom over the barrier. To introduce the method of dealing with this situation, we consider a one-dimensional analog of the jump process as shown in figure 17.1.

This represents a single particle of mass m moving in one dimension in a one-dimensional periodic potential. The single particle represents the diffusing atom, and the periodic potential is the analog of the interaction energy of this atom with the rest of the crystal. The equilibrium positions of the particle are at the minima of the periodic potential, and the difference between the maximum energy E_M and the minimum energy E_o is the activation barrier. The difference $(E_M - E_o)$ is taken to be large enough that most of the time the particle vibrates about the bottom of the well in accord with Hooke's law. When the particle acquires an energy equal to or greater than $(E_M - E_o)$ by thermal interaction with its surroundings, it climbs out of the well and moves over the barrier to enter a new position adjacent to its old one. We want to know how often the particle will do this, and the answer will be sought by using classical statistical mechanics.

Assume that the top of each barrier is flat enough that a very small, but nonzero, distance l can be defined such that the potential energy is very nearly constant over this distance. When crossing through this distance l the particle will have a mean velocity $\bar{v}$ given by classical statistical mechanics as

$$\bar{v} = \frac{\int_0^\infty v e^{-mv^2/kT} dv}{\int_0^\infty e^{-mv^2/kT} dv} \tag{17.4.1}$$

Performing the integrations gives

$$\bar{v} = \sqrt{\frac{kT}{2\pi m}} \tag{17.4.2}$$

Out of some long time interval τ, the particle spends most of its time near the bottom of the well and a small amount of time in one of the regions l at the top of a barrier. Let τ_B be the time it spends near an energy minimum and

DIFFUSION IN SIMPLE CRYSTALS

$\tau(l)$ be the time it spends in one of the regions at the top of a barrier. Also, let $\bar{\tau}$ be the average time it takes the particle to pass through the distance l. Then $[\tau(l)]/\bar{\tau}$ is just the total number of crossings the particle makes from one well to another in the total time τ. Dividing this number by τ then gives the number of jumps per unit time, which is just the definition of the jump frequency. Therefore,

$$\Gamma = \frac{\tau(l)}{\bar{\tau}\tau} \tag{17.4.3}$$

But the mean velocity is

$$\bar{v} = \frac{l}{\bar{\tau}} \tag{17.4.4}$$

so (17.4.3) becomes

$$\Gamma = \frac{\bar{v}\tau(l)}{l\tau} \tag{17.4.5}$$

Furthermore, τ is very nearly equal to τ_B since a jump is a relatively rare occurrence. Using this fact and (17.4.2), equation (17.4.5) becomes

$$\Gamma = \sqrt{\frac{kT}{2\pi m}} \frac{\tau(l)}{l\tau_B} \tag{17.4.6}$$

The ratio $\tau(l)/\tau_B$ can be obtained from the basic axiom of statistical mechanics that states that time averages are equal to ensemble averages, so the time a system spends in any group of states is proportional to the partition function for those states. Therefore,

$$\frac{\tau(l)}{\tau_B} = \frac{\sum_{(l)} e^{-E(l)/kT}}{\sum_{(B)} e^{-E(B)/kT}} \tag{17.4.7}$$

where the numerator is the partition function when the particle is in the region l at the top of the barrier and the denominator is the partition function when the particle is in the region around the bottom of the energy well. In the classical limit, the sums become integrals over coordinates and momenta. The integrals over momenta are the same for both partition functions, so (17.4.7) becomes, in the classical limit,

$$\frac{\tau(l)}{\tau_B} = \frac{\int_{x_M - 1/2}^{x_M + 1/2} e^{-\phi(x)/kT} dx}{\int_{-\infty}^{\infty} e^{-\phi(x)/kT} dx} \tag{17.4.8}$$

where x_M is the position of the maximum. $\phi(x)$ is the potential energy of the particle as a function of the distance measured from the position of the energy minimum. In the region l, this energy is very nearly constant. That is,

$$\phi(x) = E_M \quad \text{(in region } l\text{)} \tag{17.4.9}$$

so the integral of the numerator of (17.4.8) becomes

$$l e^{-E_M/kT} \tag{17.4.10}$$

In the region near the bottom of the well, the particle is harmonically bound to the energy minimum such that

$$\phi(x) = E_0 + \frac{B}{2}x^2 \quad \text{(near the minimum)} \tag{17.4.11}$$

where B is the Hooke's law force constant.

The integral in the denominator of (17.4.8) should be taken over the bottom of the well and part of the way up its walls. However, the exponential factor rapidly approaches zero as $\phi(x)$ becomes large, so a negligible error is introduced if the integration is taken from $-\infty$ to ∞. Using (17.4.11), the denominator in (17.4.8) then becomes

$$e^{-E_0/kT} \int_{-\infty}^{\infty} e^{-Bx^2/2kT} dx = e^{-E_0/kT} \sqrt{\frac{2\pi kT}{B}} \tag{17.4.12}$$

Substituting (17.4.10) and (17.4.12) into (17.4.8) gives

$$\frac{\tau(l)}{\tau_B} = e^{-(E_M - E_0)/kT} \sqrt{\frac{2\pi kT}{B}} \tag{17.4.13}$$

Combining this with (17.4.6) gives the jump frequency formula as

$$\Gamma = \frac{1}{2\pi} \sqrt{\frac{B}{m}} e^{-(E_M - E_0)/kT} \tag{17.4.14}$$

But the pre-exponential factor is just the vibration frequency of the particle at the bottom of the well, so (17.4.14) can be put in the more usual form

$$\Gamma = \nu e^{-U_m/kT} \tag{17.4.15}$$

where

$$U_m \equiv E_M - E_0 \tag{17.4.16}$$

is called the *activation energy for migration*.

The physical interpretation of (17.4.15) is straightforward. The particle vibrates near the bottom of the well ν times per second. To climb out of the well it must acquire an energy equal to or greater than the well height U_m. The probability that the particle can do this is given by the Boltzmann factor of the activation energy. The product of this probability with the vibration frequency gives the jump frequency.

This one-dimensional example illustrates the application of statistical mechanics to the theory of the jump frequency. In real crystals, of course, atomic migration is a many-body process involving the motion of many atoms. The migrating atom acquires its energy from the motion of other atoms. Also, as the atom approaches the energy barrier, it interacts with other atoms that move as a result of the interaction. The following section takes up the jump frequency in a many-body system.

17.5 Many-body theory of the jump frequency

The many-body theory of the jump frequency can be developed as a generalization of the method of the preceding section. The system under

DIFFUSION IN SIMPLE CRYSTALS 461

consideration is a single vacancy in an otherwise perfect monatomic crystal, and the temperature is taken to be high enough that the semiclassical version of statistical mechanics can be used when needed.[2]

In accord with the semiclassical description, the system is represented by a point in configuration phase space. This space is defined by the coordinates of the system so that at any particular time the configuration phase point has the coordinates $(q_1, q_2, q_3, \ldots, q_{3N})$. Most of the time the atoms vibrate about their mean positions, and the harmonic approximation can be used to express the potential energy of the system as

$$\phi(q) = \phi_0 + \frac{1}{2} \sum_{i=1}^{3N} m\omega_i^2 q_i^2 \qquad (17.5.1)$$

where ω_i is the angular frequency of the ith normal mode whose coordinate is q_i, m is the mass of an atom, and ϕ_o is the potential energy of the system when all atoms are at their equilibrium positions.

Occasionally, however, an atom next to the vacancy will acquire enough energy to jump into it. In the process of jumping, the atom will pass through an energy barrier separating the initial and final equilibrium positions. The midpoint is a critical position; if the atom reaches this midpoint with a nonzero velocity, it will move into the vacancy, leaving a vacancy behind, and an atomic jump will have occurred. The coordinates of all the atoms when the migrating atom is at the critical position will be called the *activated state for atomic migration*. Actually, there is not just one activated state; there is an entire ensemble of them since many "midpoint configurations" are consistent with the requirement that the migrating atom be in a proper position with sufficient velocity to move into the vacancy. The ensemble of activated states is just a subensemble of the complete ensemble of the system.

It can be expected that the activated states are all close together in the same sense that the normal states are all close together. The activated state of lowest energy will be that in which the migrating atom is midway between its initial and final position, and all other atoms are at rest at equilibrium positions. Of course, these equilibrium positions will be different than those for the normal states because the migrating atom at the top of the energy barrier interacts strongly with the surrounding atoms and pushes them to new positions.

In terms of the configuration phase space, the migration process can be described as follows: the phase point representing the system spends most of its time executing small motions around a region whose potential energy is given by (17.5.1). That is, it moves through the ensemble of normal states. Occasionally, however, the phase point leaves that $3N$-dimensional well to jump into a similar adjacent well. In doing this, it passes through a region of configuration phase space representing the subensemble of activated states. If we let (q_1, q_2, q_3) be the coordinates of the migrating atom and choose the coordinate system such that q_1 is along a line joining the initial and final positions, it is clear that the configuration region corresponding to the activated state will have a very small extension along the q_1-axis. Call this extension δ and let $\bar{v}_1$ be the average velocity of the migrating atom along the q_1 when it is in the activated region. Then the mean time the phase point spends in the activated region during a migration is given by

$$\bar{\tau} = \frac{\delta}{\bar{v}_1} \qquad (17.5.2)$$

Now let $\tau(\delta)$ be the total time the phase point spends in the activated region out of a long time τ. Then $\tau(\delta)/\bar{\tau}$ is the number of times an atom crosses the activated region during the time τ, and dividing this by the total time τ gives the jump frequency for a vacancy as

$$\Gamma_v = \frac{\tau(\delta)}{\tau\tau} \tag{17.5.3}$$

or, using (17.5.2),

$$\Gamma_v = \frac{\bar{v}_1}{\delta}\frac{\tau(\delta)}{\tau} \tag{17.5.4}$$

Replacing $\bar{v}_1$ by $\overline{p_1}/m$, where $\overline{p_1}$ is the average momentum of the moving atom along the migration direction in the activated region, gives

$$\Gamma_v = \frac{\overline{p_1}}{m\delta}\frac{\tau(\delta)}{\tau} \tag{17.5.5}$$

The average momentum is readily obtained from equation (2.14.26) as

$$\overline{p_1} = \frac{1}{(2\pi mkT)^{1/2}}\int_0^\infty p_1 e^{-p_1^2/2mkT}dp_1 \tag{17.5.6}$$

The integration is taken from zero to infinity because we are averaging only over the positive momenta that take the moving atom across the barrier. Performing the integration gives

$$\overline{p_1} = m\left(\frac{kT}{2\pi m}\right)^{1/2} \tag{17.5.7}$$

Thus, equation (17.5.5) now reads

$$\Gamma_v = \left(\frac{kT}{2\pi m}\right)^{1/2}\frac{1}{\delta}\frac{\tau(\delta)}{\tau} \tag{17.5.8}$$

Just as in the one-dimensional case, we now invoke the equivalence of time and ensemble averages and replace the ratio of times in (17.5.8) by a ratio of partition functions. Now, however, the pressure canonical ensemble will be introduced for this purpose because the volumes of the member systems of the ensemble are not the same in the normal and activated states. Also, the final results can then be easily expressed in terms of Gibbs free energies, thereby facilitating the treatment of the effect of pressure on diffusion. Therefore,

$$\frac{\tau(\delta)}{\tau} = \frac{Z_P^*}{Z_P} \tag{17.5.9}$$

where Z_P^* is the pressure canonical partition function for the activated region and Z_P is the pressure canonical partition function for the normal crystal. These are given by

$$Z_P = e^{-G/kT} = \sum_{E,V} e^{-(E+PV)/kT} \tag{17.5.10}$$

$$Z_P^* = e^{-G^*/kT} = \sum_{E,V}{}^* e^{-(E+PV)/kT} \tag{17.5.11}$$

The sum in (17.5.10) is over all volumes and energy states of the normal crystal, but the sum in (17.5.11) is taken only over the activated states. G is the Gibbs free energy of the normal crystal and G^* is the Gibbs free energy of the activated state as defined by the sum in (17.5.11). It is convenient to separate the pressure term from these free energies and to write

$$G = A + PV \tag{17.5.12}$$

$$G^* = A^* + PV^* \tag{17.5.13}$$

where A, V and A^*, V^* are the Helmholtz free energy and volume of the normal state and the activated state, respectively.

Using (17.5.8)–(17.5.13), the vacancy jump frequency is now

$$\Gamma_v = \left(\frac{kT}{2\pi m}\right)^{1/2} \frac{1}{\delta} e^{-PV_v^m} \frac{e^{-A^*/kT}}{e^{-A/kT}} \tag{17.5.14}$$

where

$$V_v^m = V^* - V \tag{17.5.15}$$

is called the *volume of vacancy migration*.

Equation (17.5.14) can be put into a simpler and more useful form by evaluating the exponentials of the Helmholtz free energies using canonical ensemble theory. This states that

$$e^{-A/kT} = \sum_i e^{-E_i/kT} \tag{17.5.16}$$

$$e^{-A^*/kT} = \sum_i{}^* e^{-E_i/kT} \tag{17.5.17}$$

The sum in (17.5.16) is taken over all states of the normal crystal, while the sum in (17.5.17) is taken only over all activated states. In the semiclassical approximation, (17.5.16) becomes (see section 2.14)

$$e^{-A/kT} = \frac{1}{N!\Lambda^{3N}} \int e^{-\phi(q)/kT} dq \tag{17.5.18}$$

[Remember that Λ is the thermal wavelength defined in equation (2.14.21).] Assuming that the crystal is harmonic, the quadratic potential of (17.5.1) can be substituted into the integral of (17.5.18) to give $\int e^{-\phi(q)/kT} dq = e^{-\phi_0/kT} \int e^{-m\omega_1^2/2kT} e^{-m\omega_2^2/2kT} \ldots dq_1 dq_2 \ldots$ This multidimensional integral separates into a product of $3N$ one-dimensional integrals all of the form

$$\int_{-\infty}^{\infty} e^{-m\omega_i^2/2kT} = \left(\frac{2\pi kT}{m\omega_i^2}\right)^{1/2} \tag{17.5.19}$$

$$\int e^{-\phi(q)/kT} dq = e^{-\phi_0/kT} \prod_{i=1}^{3N} \left(\frac{2\pi kT}{m\omega_i^2}\right)^{1/2} \tag{17.5.20}$$

and substituting this into (17.5.18) gives

$$e^{-A/kT} = \frac{1}{\Lambda^{3N}} e^{-\phi_0/kT} \prod_{i=1}^{3N} \left(\frac{2\pi kT}{m\omega_i^2}\right)^{1/2} \tag{17.5.21}$$

Equation (17.5.17) can be treated in a similar way by writing it in the semi-classical approximation as

$$e^{-A^*/kT} = \frac{1}{\Lambda^{3N}} \int {}^* e^{-\phi^*(q)/kT} dq \qquad (17.5.22)$$

where the integration is now carried out over all coordinates in the activated region of phase space and $\phi^*(q)$ is the potential energy in that region. Written out explicitly, the integral in equation (17.5.22) is

$$\int {}^* e^{-\phi^*(q)/kT} dq = \int_{x_1-\delta/2}^{x_1-\delta/2} (3N-1) \int_{-\infty}^{\infty} e^{-\phi^*(q_1, q_2 \ldots)/kT} dq_1 dq_2 dq_3 \ldots \qquad (17.5.23)$$

The integral over dq_1 is taken only over a length δ since this was defined as the thickness of the activated region along the direction of migration; x_1 is the value of q_1 at the minimum activated configuration. Since δ is small and we will later take the limit as $\delta \to 0$, the potential $\phi^*(q)$ will be taken to be independent of q_1 over the length δ. The integral over q_1 is therefore readily performed and (17.5.23) becomes

$$\int {}^* e^{-\phi^*(q)/kT} dq = \delta \int_{-\infty}^{\infty} (3N-1) \int_{-\infty}^{\infty} e^{-\phi^*(q_1, q_2 \ldots)/kT} dq_2 dq_3 \ldots \qquad (17.5.24)$$

The harmonic approximation will again be used to express the potential energy as a function of coordinates in the activated region as

$$\phi^*(q) = \phi_o^* + \frac{1}{2} \sum_{i=1}^{3N-1} m\omega_i^{*} q_i^2 \qquad (17.5.25)$$

where ϕ_o^* is the potential energy when the migrating atom is midway between its initial and final positions and all other atoms are at the mean positions in the activated region. The ω_i^* are normal mode frequencies of the activated state. Using (17.5.25) in (17.5.24) and evaluating the integrals gives

$$\int {}^* e^{-\phi^*(q)/kT} dq = \delta e^{-\phi_o^*/kT} \prod_{i=1}^{3N-1} \left(\frac{2\pi kT}{m\omega_i^{*2}} \right)^{1/2} \qquad (17.5.26)$$

so (17.5.22) becomes

$$e^{-A^*/kT} = \frac{\delta}{N! \Lambda^{3N}} e^{-\phi_o^*/kT} \prod_{i=1}^{3N-1} \left(\frac{2\pi kT}{m\omega_i^{*2}} \right)^{1/2} \qquad (17.5.27)$$

Now combine (17.5.27), (17.5.21), and (17.5.14). The result is

$$\Gamma_v = e^{-(U_v^m + PV_v^m)/kT} \frac{\prod_{i=1}^{3N} v_i}{\prod_{i=1}^{3N-1} v_i^*} \qquad (17.5.28)$$

where

$$U_v^m \equiv \phi_o^* - \phi_0 \qquad (17.5.29)$$

is the vacancy migration energy and the angular frequencies have been replaced by the ordinary frequencies $\nu = \omega/2\pi$.

Another form of (17.5.28) can be written by making use of the identities

$$\prod_{i=1}^{3N} \nu_i = \left(\frac{kT}{h}\right)^{3N} \exp\left(\frac{1}{k}\sum_{i=1}^{3N} k \ln \frac{h\nu_i}{kT}\right) \tag{17.5.30}$$

$$\prod_{i=1}^{3N} \nu_i^* = \left(\frac{kT}{h}\right)^{3N-1} \exp\left(\frac{1}{k}\sum_{i=1}^{3N-1} k \ln \frac{h\nu_i}{kT}\right) \tag{17.5.31}$$

The reason for doing this is that it can easily be shown (see problem 4.5 of chapter 4) that the high-temperature approximation for the entropy of a harmonic system is

$$S = 3Nk - k\sum_{j=1}^{3N} \ln\left(\frac{h\nu_j}{kT}\right) \tag{17.5.32}$$

Thus, (17.5.30) and (17.5.31) give

$$\prod_{i=1}^{3N} \nu_i = \left(\frac{kT}{h}\right)^{3N} e^{-S/k} \tag{17.5.33}$$

$$\prod_{i=1}^{3N-1} \nu_i^* = \left(\frac{kT}{h}\right)^{3N-1} e^{-S^*/k} \tag{17.5.34}$$

where S and S^* are the entropies of the normal and activated states, respectively. Equation (17.5.28) therefore becomes

$$\Gamma_v = \frac{kT}{h} e^{-(U_v^m + PV_v^m - TS_v^m)/kT} \tag{17.5.35}$$

where

$$S_v^m = S^* - S \tag{17.5.36}$$

is the entropy of activation. The Gibbs free energy of vacancy migration is defined by

$$G_v^m = U_v^m + PV_v^m - TS_v^m \tag{17.5.37}$$

so an alternative form for the jump frequency is

$$\Gamma_v = \frac{kT}{h} e^{-G_v^m/kT} \tag{17.5.38}$$

This derivation for the jump frequency was carried out for vacancy motion for the sake of definiteness. But obviously, similar results can be obtained in a similar manner for other diffusion mechanisms. In particular, the jump frequency for an interstitial is

$$\Gamma_I = \frac{kT}{h} e^{-G_I^m/kT} \tag{17.5.39}$$

G_I^m being the Gibbs free energy of motion for an interstitial, defined in complete analogy with G_v^m.

It is important to note that the free energies of motion are not free energies in the usual sense. G is the free energy of the crystal and has its usual meaning. However, G^* is defined in terms of restricted partition sums. Furthermore, it refers to a system having one less degree of freedom than the crystal in its normal state. This missing degree of freedom is responsible for the factor kT/h in the jump frequency formulas.

Often, it is the temperature dependence of the jump frequency that is of major interest. It is then more convenient to retain the jump frequency in the form (17.5.28) rather than (17.5.38) since the free energy of motion is linear in the temperature. The appearance of the formula is simplified by defining an effective frequency by

$$\tilde{v}_v = \frac{\prod_{i=1}^{3N} v_i}{\prod_{i=1}^{3N-1} v_i^*} \tag{17.5.40}$$

so that (17.5.28) becomes

$$\Gamma_v = \tilde{v}_v e^{-(U_v^m - PV_v^m)/kT} \tag{17.5.41}$$

Similarly, for interstitial diffusion, we can write

$$\Gamma_I = \tilde{v}_I e^{-(U_I^m - PV_I^m)/kT} \tag{17.5.42}$$

where U_I^m and V_I^m are the energy and volume of interstitial migration.

It must be stressed that this treatment of the jump frequency is based on equilibrium statistical mechanics. The activated state is one of the states in the ensemble representing the complete equilibrium, and in computing the jump frequency, we have just counted the frequency with which the system moves from one set of states to another in the ensemble. In applying this theory to diffusion in a concentration gradient, it is assumed that the deviation from equilibrium caused by the nonzero gradients is not sufficient to seriously affect the equations derived on an equilibrium basis. The theory developed here, therefore, does not touch the fundamental question of irreversible processes, and in those cases in which irreversibility is of prime importance, such as diffusion in a temperature gradient, the present theory may need modification. In many cases, however, particularly when the diffusing material is present in small quantities or gradients are small, the theory is quite satisfactory.

Migration energies are often of the order of an electron volt, and effective frequencies are of the order of 10^{13} cycles per second. This means that an interstitial or a vacancy makes about 10^8 jumps per second at 1000K. This number decreases rapidly with decreasing temperature.

17.6 The diffusion coefficient

The results obtained so far are sufficient to give the diffusion coefficient as an explicit function of temperature and pressure. Substituting (17.5.41) and (17.5.42) into (17.3.12)–(17.3.14) gives

$$D_I = \frac{1}{6} \tilde{v}_I r_I^2 e^{-(U_I^m + PV_I^m)/kT} \tag{17.6.1}$$

$$D_v = \frac{1}{6}\tilde{v}_v r_L^2 e^{-(U_v^m + PV_v^m)/kT} \tag{17.6.2}$$

$$D_s = \frac{f}{6}\tilde{v}_v r_L^2 e^{-(U_v^m + PV_v^m)/kT} C_v \tag{17.6.3}$$

According to equation (15.5.1) the vacancy concentration is a Boltzmann factor containing formation energy, pressure, and entropy of vacancy formation. All three of the diffusion coefficients can then be written in the experimentally observed form given by (17.3.15) where now the pre-exponential factors, heats and volumes of activation are as follows.

For interstitial diffusion:

$$D_0 = \frac{1}{6}\tilde{v}_I r_I^2$$

$$Q^* = U_I^m \tag{17.6.4}$$

$$V^* = V_I^m$$

For vacancy diffusion:

$$D_0 = \frac{1}{6}\tilde{v}_v r_L^2$$

$$Q^* = U_v^m \tag{17.6.5}$$

$$V^* = V_v^m$$

For self diffusion:

$$D_0 = \frac{f}{6}\tilde{v}_v r_L^2 e^{S_v^m}$$

$$Q^* = E_v^m + U_v^m \tag{17.6.6}$$

$$V^* = V_v^m + V_v^f$$

This formulation of the diffusion coefficient has been successful in treating diffusion data in a large number of systems. The energy of activation can be determined from diffusion experiments carried out at a series of temperatures since the slope of $\ln D$ versus $1/T$ gives Q^*. The volume of activation is determined using data on the diffusion coefficient as a function of pressure, but at a constant temperature, from a plot of $\ln D$ versus P. Precise diffusion experiments are difficult to perform, but a number of systems have been carefully studied. For cubic metals, the pre-exponential factor is of the order of unity (ranging from 0.05 to 2), while the energies of activation are usually several electron volts and the volumes of activation are roughly equal to the atomic volume.

Exercises

17.1 For vacancy migration in a simple crystal, take the migration energy to be 1 eV and the migration volume to be 10^{-23} cm^3. Compute the vacancy jump frequency for a vacancy at 500K and at 1000K at zero pressure and at 50,000 atm. Take the effective frequency to be 10^{13}/sec.

17.2 If the effective frequency for vacancy migration is 10^{13}/sec, compute the entropy of vacancy migration at 1000K and 1500K.

17.3 Assume that self-diffusion in a simple metal takes place by both a vacancy and a divacancy mechanism. Show that the effective activation energy, defined as the slope of the Arrhenius plot, is the weighted average of the individual activation energies. Assume the pressure is zero.

17.4 Assume that the migration energy for self-diffusion in aluminum is the same for vacancies and di-vacancies and that the activation energy for diffusion by the vacancy mechanism is 28.8 kcal/mole. What is the activation energy for self-diffusion by the di-vacancy mechanism if the divacancy binding energy is 0.3 eV?

Notes

1. See chapter 3 in Manning (1968) and appendix E in Borg and Dienes (1988).
2. This method was first presented by Vineyard, G.H.; "The Frequency Factors and Isotope Effects in Solid State Rate Prucesses"; 1957; *Journal of Physical Chemical Solutions;* vol. 3, p. 121.

Appendix 1

Combinatorial Problems in Statistical Mechanics

A. Ensemble statistics

In ensemble statistics, it is necessary to compute the number of complexions of the ensemble for a given distribution of the X member systems among the possible states of a system. A distribution is defined by the set of integers $\{N_j\} = (N_1, N_2, \ldots, N_j \ldots)$ such that N_j is the number of member systems in the jth quantum state. The number of complexions is just the number of ways of realizing this distribution.

Since the members of the ensemble are macroscopic systems, they are distinguishable from one another. Therefore, we need to count the number of ways of arranging X distinguishable systems such that N_1 are in state 1, N_2 are in state 2, ..., N_j are in state j, and so on, with the condition that the total number of systems is

$$X = \sum_j N_j \tag{A.1.1}$$

Clearly, this is equivalent to putting X marbles in boxes such that N_1 are in the first box, N_2 are in the second box, and so on, without regard to the order of the arrangement of marbles in a particular box. Call this number W.

First, let us show that the number of ways of putting the marbles in the boxes in such a way that the ordering of the marbles *is* taken into account is $X!$. For this case, each box would have a set of compartments that hold one marble each, and a permutation of the marbles among compartments would count as a new complexion. If the marbles are put in position one at a time, the first marble could go in any compartment and could therefore be placed in X ways. The second marble could be placed in only $(X - 1)$ ways, since one compartment is already occupied, and the total number of ways of placing the two marbles is $X(X - 1)$. The third marble can be placed in $(X - 2)$ ways, and the total number of ways of placing three marbles is $X(X - 1)(X - 2)$. This process is continued until all X marbles are placed, giving $X!$ as the total number of ways of arranging X marbles in boxes in such a way that, within a box, each ordering of the arrangement of marbles is a different complexion. But we are not really interested in this problem; we want W, in which the arrangement within the boxes is neglected. Obviously, W is smaller than $X!$, and if we multiply W by the number of ways of ordering N_1 marbles in the first box (which is $N_1!$), the number of ways of ordering N_2 marbles in the second box (which is $N_2!$), and so on, we will get the total number of possible permutations that include ordering (which is $X!$). That is, $WN_1!N_2!\ldots = X!$, so

$$W = \frac{X!}{\prod_j N_j!} \tag{A.1.2}$$

This is the number of complexions for an ensemble for a given distribution of member systems among states.

B. Maxwell–Boltzmann statistics

In particle statistics, all particle states are divided into groups such that the jth group consists of states with nearly the same energy in a range $\Delta\varepsilon_j$ centered on ε_j where $\Delta\varepsilon_j \ll \varepsilon_j$. We want to calculate the total number of complexions of the system of N particles when N_1 are in the first group, N_2 are in the second group, and so on. In the Maxwell–Boltzmann case, the particles are distinguishable and there is no limit to the number of particles in each group. The desired number of complexions is denoted by W_{MB}.

First, calculate the number of ways of putting N particles in groups such that there are N_j in the jth group. This is obviously just like the ensemble case, and the result is

$$\frac{N!}{\prod_j N_j!} \tag{A.1.3}$$

But there is a difference between the present problem and that for ensembles. In the ensemble problem all states in a j-group were identical, whereas in the present problem all states in a j-group are different, although they all have very nearly the same energy. Also, there is no limit to the number of particles that can go into a particular state in any j-group. To get the right number of complexions, (A.1.3) must therefore be multiplied by the number of ways of arranging N_j particles among the number of states in each j-group. If the jth group has ω_j states, then N_j particles can be put in these states in $\omega_j^{N_j}$ ways. This is so because one particle can be put in any one of ω_j states; a second particle can also be placed in any of the ω_j states since any number of particles can occupy a particular state and the number of ways of placing the two particles is ω_j^2. Continuing this process, N_j particles can be put in the ω_j states in $\omega_j^{N_j}$ ways. Therefore, to get W_{MB}, multiply (A.1.3) by $\omega_1^{N_1}$, $\omega_2^{N_2}$, and so on. That is,

$$W_{MB} = N! \prod_j \frac{\omega_j^{N_j}}{N_j!} \tag{A.1.4}$$

C. Fermi–Dirac statistics

Again, N particles are to be distributed among particle states that are divided into groups such that the jth group contains N_j particles, and we want to compute the number of ways of doing this. And again, the energies of the states in the jth group are in a range $\Delta\varepsilon_j$ centered on the energy ε_j. But now the particles are not distinguishable and there can be at most only one particle in a given state.

First, consider one group, say, the jth, that has ω_j states, and calculate the number of ways of putting N_j particles in this group of states. If the particles were distinguishable, this would be just $\omega_j!/(\omega_j - N_j)!$ because the first particle could be placed in ω_j ways, the second in $(\omega_j - 1)$ ways, and so on, and the last in $(\omega_j - N_j + 1)$ ways. The product of all these gives

$$\omega_j(\omega_j - 1)(\omega_j - 2)\ldots(\omega_j - N_j + 1) = \frac{\omega_j!}{(\omega_j - N_j)!} \quad (A.1.5)$$

But the particles are *not* distinguishable, and the above expression must be corrected to get the right number of complexions. This correction consists of dividing (A.1.5) by $N_j!$ because that many of the permutations of the particles among the states count as the same arrangement when the particles are indistinguishable. The number of ways of putting N_j particles in the j-group for the Fermi–Dirac case is therefore

$$\frac{\omega_j!}{N_j!(\omega_j - N_j)!} \quad (A.1.6)$$

To obtain the total number of complexions for all j-groups, just take the product of (A.1.6) over all j. There is no need to worry about permutations from one j-group to another because the particles are indistinguishable and the N_j are fixed numbers, so any interchange of particles among the groups does not give a new complexion. The desired result, the number of complexions for Fermi–Dirac particles, is therefore

$$W_{\text{FD}} = \prod_j \frac{\omega_j!}{N_j!(\omega_j - N_j)!} \quad (A.1.7)$$

D. Bose–Einstein statistics

Bose–Einstein particles are indistinguishable and there is no limit to the number of particles that can be in any one state. We want the number of ways of distributing N such particles among states so that there are N_j in the group that has ω_j states. Just as in the Fermi–Dirac case, first compute the number of complexions for one group, labeled j, and then take the product over all j. This procedure is valid because the particles are indistinguishable.

Now pick out a group labeled j. Each distribution of the N_j particles among the ω_j states in this group can be represented by the following scheme:

	3, 5	5	·	1, 2, 9	23	...	
$k =$	1	2	3	4	5	6 ...	

The vertical lines separate the particle states, and the numbers between the lines label the particles in each state. The states are numbered by the index k, which runs from unity to ω_j, the total number of states in the group. Thus, the distribution represented above is the one in which particles 3 and 5 are in the first state, particle 7 is in the second state, there are no particles in the third state, particles 1, 2, and 9 are in the fourth state, particle 23 is in state 5, and so on. Clearly, if the particles were distinguishable, the number of complexions would be the number of ways of permuting the integers 1 to N_j and the vertical lines separating the states. This would give all possible distributions of distinguishable particles.

There are a total of ω_j vertical lines, but in our scheme, the first one must be kept in place so that we do not have any particles that are not in some state. Therefore, only $(\omega_j - 1)$ lines can be permuted. Therefore, the total number of objects (lines plus integers) to be permuted is $(N_j + \omega_j - 1)$ and the total number of permutations is $(N_j + \omega_j - 1)!$. This is greater than the number of complexions for the jth group for two reasons. First, the permutation of a set of symbols consisting of a line and the integers following it with another such set does not alter the combined distribution in the jth group. (E.g., in the distribution shown above, permuting the vertical line and the integers 1, 2, 3 following it as a set, with the vertical line and the integer 23 as a set, does not give a new distribution of indistinguishable particles.) There are $(\omega_j - 1)$ such sets of symbols, so our first result must be divided by $(\omega_j - 1)!$. Second, permutations of the N_j particles with each other leaves everything unchanged because the particles are indistinguishable, so we must divide through by $N_j!$. The number of ways of putting the N_j particles in the jth group is therefore $(N_j + \omega_j - 1)!/N_j!(\omega_j - 1)!$ Taking the product over all j gives the number of complexions for the Bose–Einstein case as

$$W_{BE} = \prod_j \frac{(N_j + \omega_j - 1)!}{N_j!(\omega_j - 1)!} \qquad (A.1.8)$$

Appendix 2

The Method of Undetermined Multipliers

Lagrange's method of undetermined multipliers is used frequently in physical problems to determine maxima or minima of certain functions subject to subsidiary conditions. In this appendix, a brief description of the mathematics involved is presented without encumbering the analysis with physical applications.

We are given a function F of a number of variables $y_1, y_2, \ldots$, and so on:

$$F = F(y_1, y_2, \ldots) \tag{A.2.1}$$

The variables are themselves functions of a set of parameters $x_1, x_2, \ldots$, and so on. That is,

$$\begin{aligned} y_1 &= y_1(x_1, x_2, \ldots) \\ y_2 &= y_2(x_1, x_2, \ldots) \\ y_3 &= y_3(x_1, x_2, \ldots) \\ &\ldots \end{aligned} \tag{A.2.2}$$

Now we want to find the functional form of equations (A.2.2) that gives F a stationary value (makes F maximum or minimum). Throughout the search for this functional form, the xs are taken to be given and to remain constant.

If F is to be stationary, then any variation of F, resulting from a variation in the ys, must be zero. That is,

$$\delta F = \frac{\partial F}{\partial y_1} \delta y_1 + \frac{\partial F}{\partial y_2} \delta y_2 + \ldots = \sum_{i=1}^{r} \frac{\partial F}{\partial y_i} \delta y_i = 0 \tag{A.2.3}$$

where r is the total number of ys. If the y_i were all independent, each coefficient of δy_i would have to be zero because the variations δy_i are completely arbitrary. The problem would then be solved. All we need do is set each partial derivative equal to zero. We are much more interested in the case in which the y_i are not completely independent, but in which some functions of the y_i and the x_i exist that must also be stationary. Let us assume that we have two such functions G_1 and G_2 whose form is known and that we write as

$$G_1 = G_1(y_1, y_2, \ldots; x_1, x_2, \ldots) \tag{A.2.4}$$

$$G_2 = G_2(y_1, y_2, \ldots; x_1, x_2, \ldots) \tag{A.2.5}$$

We require G_1 and G_2 to be constants, and they are therefore stationary with respect to a variation in the ys, so

$$\delta G_1 = \sum_{i=1}^{r} \frac{\partial G_1}{\partial y_i} \delta y_i = 0 \tag{A.2.6}$$

$$\delta G_2 = \sum_{i=1}^{r} \frac{\partial G_2}{\partial y_i} \delta y_i = 0 \tag{A.2.7}$$

The y_i are not independent because of the relations (A.2.4) and (A.2.5), which give us two relations among the y_j. But this lack of complete independence can be taken into account in the following manner. Multiply the variations in G_1 and G_2 by the constants a_1 and a_2 to get the obvious relations

$$a_1 \delta G_1 = 0$$

$$a_2 \delta G_2 = 0 \tag{A.2.8}$$

If these are added to δF, we have

$$\delta F + a_1 \delta G_1 + a_2 \delta G_2 = 0 \tag{A.2.9}$$

This is obviously true for *any* constants a_1 and a_2. This means that (A.2.3), (A.2.6), and (A.2.7) can be combined according to (A.2.9) to get

$$\sum_{i=1}^{r} \left(\frac{\partial F}{\partial y_i} + a_1 \frac{\partial G_1}{\partial y_i} + a_2 \frac{\partial G_2}{\partial y_i} \right) \delta y_i = 0 \tag{A.2.10}$$

and this equation will be true regardless of the values of a_1 and a_2, so we can give these constants any value we choose. Now let us take a_1 and a_2 to have values that satisfy the simultaneous equations

$$\frac{\partial F}{\partial y_1} + a_1 \frac{\partial G_1}{\partial y_1} + a_2 \frac{\partial G_2}{\partial y_1} = 0 \tag{A.2.11}$$

$$\frac{\partial F}{\partial y_2} + a_1 \frac{\partial G_1}{\partial y_2} + a_2 \frac{\partial G_2}{\partial y_2} = 0 \tag{A.2.12}$$

the derivatives being evaluated at the values of y_1 and y_2 that make F stationary. Once this is done, the first two terms in (A.2.10) vanish and y_1 and y_2 are fixed. Equation (A.2.10) then becomes

$$\sum_{i=3}^{r} \left(\frac{\partial F}{\partial y_i} + a_1 \frac{\partial G_1}{\partial y_i} + a_2 \frac{\partial G_2}{\partial y_i} \right) \delta y_i = 0 \tag{A.2.13}$$

The derivatives, of course, all being evaluated at values of y_i for which F is stationary. But now all the remaining y_i are independent because there were only two conditions on the y_i and two of the y_i have been fixed. This means that all of the coefficients of δy_i in (A.2.13) are zero and we have for all i

$$\frac{\partial F}{\partial y_i} + a_1 \frac{\partial G_1}{\partial y_i} + a_2 \frac{\partial G_2}{\partial y_i} = 0 \tag{A.2.14}$$

This is the solution to our problem. Since F, G_1, and G_2 are known functions of the y_i, and since G_1 and G_2 contain the parameters x_j, (A.2.14) can be solved for y_i in terms of the x_j. These solutions still contain a_1 and a_2, but these multipliers can be removed by substituting the solutions $y_i = y_i(x_j; a_1, a_2)$ into (A.2.4) and (A.2.5) and solving for a_1 and a_2. In actual applications to physical problems, the undetermined multipliers often have important physical interpretations, so they are usually retained in the functional form of the y_i. The generalization to any number of subsidiary conditions with corresponding multipliers $a_1, a_2, a_3, \ldots$ is obvious.

In the theory of the canonical ensemble, F corresponds to $\ln W\{N_i\}$, the y_i correspond to the N_i, and the x_j correspond to the E_j. The subsidiary functions G_1 and G_2 correspond to the number of systems in the ensemble and to the total energy of the ensemble.

The above procedure tells us the conditions for the function F to be stationary, but it does not tell us whether it is at a maximum or a minimum. But this decision can be made by recognizing that, if a function has a maximum, then in the immediate neighborhood of the maximum it approximates a parabola that is concave downward while in the neighborhood of a minimum it approximates a parabola that is concave upward. Therefore, if the function is expanded to the second order in a Taylor series about its stationary value, then it is at a maximum if the coefficient of the quadratic term is negative and at a minimum if the coefficient of the quadratic term is positive. Applying this to our case, the function we want to investigate is $(F + a_1 G_1 + a_2 G_2)$ since this is the function whose stationary value is given by (A.2.14). Call this function L such that

$$L = L(y_i) = F + a_1 G_1 + a_2 G_2 \tag{A.2.15}$$

Denote the values of y_i for which L is stationary by y_i^s. At these values of the y_i, equation (A.2.14) holds and we will call L^s the stationary value of (A.2.15). Now expand (A.2.15) in a Taylor series about the stationary value up to the second order, remembering that the linear terms vanish because of (A.2.14). The result is

$$L = L^s + \frac{1}{2} \sum_{i,j} \left(\frac{\partial^2 L}{\partial y_i \partial y_j} \right)_s (y_i - y_i^s)(y_j - y_j^s) \tag{A.2.16}$$

We restrict ourselves to functions L that are sums of functions for each y_i. That is, we take L to have the form

$$L = \sum_i f_i(y_i) \tag{A.2.17}$$

where each f_i is a function of only one y_i. This is the type of function met with in the application of Lagrange's undetermined multipliers to the problem of finding the most probable distribution. The mixed derivatives in (A.2.16) then vanish and we have

$$L = L^s + \frac{1}{2} \sum_i \left(\frac{\partial^2 L}{\partial y_i^2} \right)_s (y_i - y_i^s)^2 \tag{A.2.18}$$

So the recipe is simple: find the second derivative of L at the stationary point. If it is negative, the point is a maximum; if it is positive, the point is a minimum.

In statistical mechanics, the functions G_1 and G_2 are usually linear in the y_i (i.e., in ensemble theory both the number of systems in the ensemble and the ensemble energy are linear in N_j, the number of systems in each state j). Equation (A.2.18) then simplifies further because the second derivatives of G_1 and G_2 vanish and (A.2.18) reduces to

$$L = L^s + \frac{1}{2}\sum_i \left(\frac{\partial^2 F}{\partial y_i^2}\right)_s (y_i - y_i^s)^2 \qquad \text{(A.2.19)}$$

In general, different second derivatives for different is can have different signs, so our functions (L or F) can have a maximum with respect to some variables and a minimum with respect to others. This would be the situation at a saddle point. For our applications, the derivatives generally have the same sign for all i.

Appendix 3
Stirling's Approximation

If N is a large, positive integer, Stirling's approximation states that $N!$ is approximately given by

$$\ln N! = N \ln N - N \qquad (A.3.1)$$

We can show that this is approximately correct by using the relation between a sum and an integral. Since $\ln x$ is a monotonic increasing function of x, then

$$\sum_{j=1}^{N} \ln j < \int_0^N \ln x\, dx < \sum_{j=1}^{N} \ln(j+1) \qquad (A.3.2)$$

These inequalities can be made obvious by graphing the function $\ln x$ and comparing it to the summed areas of the unit stepwise divisions representing the sums in (A.3.2). From (A.3.2) it follows that

$$\int_0^{N-1} \ln x\, dx < \sum_{j=1}^{N} \ln j < \int_0^N \ln x\, dx \qquad (A.3.3)$$

Performing the integrals, and recognizing that the sum in the middle is $\ln N!$, we get

$$(N-1)[\ln(N-1) - 1] < \ln N! < N(\ln N - 1) \qquad (A.3.4)$$

If unity is neglected relative to N, equation (A.3.1) follows immediately. It is trivial to show that the outside terms in (A.3.4) differ by a quantity of order $\ln N$. For very large numbers, the logarithm is much smaller than the number, so the greater N, the more accurate is (A.3.1).

The following table shows that Stirling's approximation is a good one for remarkably small values of N:

N	$\ln N!$	$N \ln N$
50	148	146
100	363	360
200	864	860
300	1415	1411
400	2000	1997
500	2611	2607
600	3242	3238

However, there are times when (A.3.1) is not sufficiently accurate even for large values of N because of cancellations occurring in ratios of factorials. It is then necessary to use the following more accurate formula:

$$\ln N! = \left(N + \frac{1}{2}\right)\ln N - N + \frac{1}{2}\ln 2\pi + \frac{1}{12N} \qquad (A.3.4)$$

This approximation neglects terms of order N^{-3} and smaller. Derivations of this formula can be found in *The Mathematics of Physics and Chemistry* by Henry Margenau and George M. Murphy (1956) and *Elements of Pure and Applied Mathematics* by Harry Lass (1957).

Appendix 4

Sums and Integrals

In this appendix, we evaluate certain sums and integrals that are useful in statistical mechanics. The sums are

$$S_1 = \sum_{j=0}^{\infty} x^j = \frac{1}{1-x} \qquad x < 1 \tag{A.4.1}$$

$$S_2 = \sum_{j=0}^{\infty} j x^j = \frac{x}{(1-x)^2} \qquad x < 1 \tag{A.4.2}$$

$$S_3 = \sum_{j=1}^{\infty} \frac{(-1)^{j-1}}{j^2} = \frac{\pi^2}{12} \tag{A.4.3}$$

$$S_4 = \sum_{j=1}^{\infty} \frac{1}{j^2} = \frac{\pi^2}{6} \tag{A.4.4}$$

$$S_5 = \sum_{j=1}^{\infty} \frac{1}{j^4} = \frac{\pi^4}{90} \tag{A.4.5}$$

$$S_6 = \sum_{j=0}^{\infty} \frac{(j+C)!}{j!} x^j = \frac{C!}{(1-x)^{C+1}} \qquad C = \text{integer} \tag{A.4.6}$$

Equation (A.4.1) is just the geometric series and can be proven by starting with the partial sums

$$S_1(n) = \sum_{j=0}^{n} x^j = 1 + x + x^2 + \ldots x^n \tag{A.4.7}$$

Multiply this by x to get

$$x S_1(n) = x \sum_{j=0}^{n} x^j = x + x^2 + \ldots + x^{n+1} \tag{A.4.8}$$

Now subtract (A.4.8) from (A.4.7) and solve for $S_1(n)$. The result is

$$S_1(n) = \sum_{j=0}^{n} x^j = \frac{1-x^{n+1}}{1-x} \tag{A.4.9}$$

Since $x < 1$, taking the limit of (A.4.9) as $n \to \infty$ gives (A.4.1). Also, differentiating (A.4.9) gives

$$x\frac{dS_1(n)}{dx} = \sum_{0}^{n} jx^j = \frac{1-x^{n+1}}{(1-x)^2} - \frac{(n+1)x^n}{(1-x)} \tag{A.4.10}$$

Equation (A.4.2) can be obtained from this by taking the limit for infinite n, but it is easier to differentiate (A.4.1) with respect to x to get

$$\frac{dS_1}{dx} = \sum_{j=0}^{\infty} jx^{j-1} = \frac{1}{(1-x)^2} \tag{A.4.11}$$

from which (A.4.2) follows immediately.

Equations (A.4.3) and (A.4.4) can be obtained by expanding the function $f(x) = x^2$ in a Fourier series in the interval $-\pi$ to $+\pi$. The result is

$$x^2 = \frac{\pi^2}{3} + 4\sum_{j=1}^{\infty} \frac{(-1)^j}{j^2} \cos jx \tag{A.4.12}$$

Letting $x = 0$ in this equation gives (A.4.3), and letting $x = \pi$ gives (A.4.4).

A similar procedure works for (A.4.5). Expand $f(x) = x^4$ in a Fourier series to get

$$x^4 = \frac{\pi^4}{5} + 8\pi^2 \sum_{j=1}^{\infty} \frac{(-1)^j}{j^2} \cos jx - 48 \sum_{j=1}^{\infty} \frac{(-1)^j}{j^4} \cos jx \tag{A.4.13}$$

Letting $x = \pi$, (A.4.13) becomes

$$\pi^4 = \frac{\pi^4}{5} + 8\pi^2 \sum_{j=1}^{\infty} \frac{1}{j^2} - 48 \sum_{j=1}^{\infty} \frac{1}{j^4} \tag{A.4.14}$$

Replacing the first sum on the right by $\pi/6$ according to (A.4.4) and solving for the second sum gives (A.4.5)

To evaluate (A.4.6), note that it has the form of a Taylor expansion, so a function $F(x)$ exists such that

$$F(x) = \sum_{j=0}^{\infty} \frac{1}{j!} \left(\frac{d^j F}{dx^j}\right)_0 x^j \tag{A.4.15}$$

with the derivatives evaluated at $x = 0$ being given by

$$\left(\frac{d^j F}{dx^j}\right)_0 = (j+C)! \tag{A.4.16}$$

It is easily verified that the function whose derivatives at $x = 0$ are given by (A.4.16) is

$$F(x) = \frac{C!}{(1-x)^{C+1}} \tag{A.4.17}$$

because $d^jF/dx^j = (C+j)!/(1-x)^{C+j+1}$, so $(d^jF/dx^j)_0 = (C+j)!$. The sum S_6 of equation (A.4.6) is therefore just the function $F(x)$ of equation (A.4.17).

Now we want to prove the following:

$$A_1 = \int_{-\infty}^{\infty} e^{-ax^2} dx = \sqrt{\frac{\pi}{a}} \qquad (A.4.18)$$

$$A_2 = \int_0^{\infty} x^2 e^{-ax^2} dx = \frac{1}{4}\sqrt{\frac{\pi}{a^3}} \qquad (A.4.19)$$

$$A_3 = \int_0^{\infty} \sqrt{x} e^{-ax} dx = \frac{1}{2}\sqrt{\frac{\pi}{a^3}} \qquad (A.4.20)$$

$$A_4 = \int_0^{\infty} x^{3/2} e^{-ax} dx = \frac{3}{4}\sqrt{\frac{\pi}{a^5}} \qquad (A.4.21)$$

$$A_5 = \int_0^{\infty} x^{2n} e^{-jx} dx = \frac{2n!}{j^{2n+1}} \qquad (A.4.22)$$

In (A.4.22), j and n are positive integers.

Equation (A.4.18) is just the Gauss integral and is obtained by first changing variables to $u = xa^{1/2}$ to get

$$A_1 = \frac{1}{\sqrt{a}} \int_{-\infty}^{\infty} e^{-u^2} du \qquad (A.4.23)$$

which can be squared to give

$$A_1^2 = \frac{1}{a} \int_{-\infty}^{\infty} e^{-u^2} du \int_{-\infty}^{\infty} e^{-v^2} dv \qquad (A.4.24)$$

Now transform to polar coordinates so that $r^2 = u^2 + v^2$, $dudv = 2\pi rdr$, which transforms (A.4.24) to

$$A_1^2 = \frac{2\pi}{a} \int_0^{\infty} e^{-r^2} rdr \qquad (A.4.25)$$

Changing variables to $y = r^2$, this becomes

$$A_1^2 = \frac{\pi}{a} \int_{-\infty}^{\infty} e^{-y} dy = \frac{\pi}{a} \qquad (A.4.26)$$

from which (A.4.18) follows.

Equation (A.4.19) is obtained from (A.4.18) by differentiating A_1 with respect to a:

$$\frac{dA_1}{da} = -\int_{-\infty}^{\infty} x^2 e^{-ax^2} dx = -\frac{1}{2}\sqrt{\frac{\pi}{a^3}} \qquad (A.4.27)$$

which is just (A.4.19) since the integrand is an even function.

To prove (A.4.20), transform the integration variable to $u = x^{1/2}$ so that

$$A_3 = 2\int_0^{\infty} u^2 e^{-u^2} du \qquad (A.4.28)$$

Comparing this with (A.4.19) shows that (A.4.20) is correct.
Differentiation of (A.4.20) gives

$$\frac{dA_3}{da} = \int_0^\infty x^{3/2} e^{-ax} dx = -\frac{3}{4}\sqrt{\frac{\pi}{a^5}} \tag{A.4.29}$$

which proves (A.4.21).

The proof of (A.4.22) starts with the variable transformation $y = jx$ so that

$$A_5 = \frac{1}{j^{2n+1}} \int_0^\infty y^{2n} e^{-y} dy \tag{A.4.30}$$

Equation (A.4.22) then follows from a successive integration by parts. That is,

$$\int_0^\infty y^{2n} e^{-y} dy = -y^{2n} e^{-y} \Big|_0^\infty + 2n \int_0^\infty y^{2n-1} e^{-y} dy$$
$$= 2n \int_0^\infty y^{2n-1} e^{-y} dy$$
$$= 2n(2n-1) \int_0^\infty y^{2n-2} e^{-y} dy$$
$$\cdots$$
$$= 2n! \int_0^\infty e^{-y} dy = 2n!$$

Putting this result in (A.4.30), we get (A.4.22).

Two integrals in the Debye theory of crystals are

$$A_6 = \int_0^\infty \frac{x^3 dx}{e^x - 1} = \frac{\pi^4}{15} \tag{A.4.31}$$

$$A_7 = \int_0^\infty \frac{x^4 dx}{(e^x + e^{-x} - 2)} = \int_0^\infty \frac{x^4 e^{-x} dx}{(1 - e^{-x})^2} = \frac{4\pi^4}{15} \tag{A.4.32}$$

Equation (A.4.31) can be obtained by using the series expansion

$$\frac{1}{e^x - 1} = e^{-x} + e^{-2x} + e^{-3x} \ldots \tag{A.4.33}$$

so that

$$A_6 = \sum_{n=1}^\infty \int_0^\infty x^3 e^{-nx} dx \tag{A.4.34}$$

Now let $y = nx$ so that (A.4.33) can be written as

$$A_6 = \sum_{n=1}^\infty \frac{1}{n^4} \int_0^\infty y^3 e^{-y} dy \tag{A.4.35}$$

The integral can be evaluated by successive integration by parts:

$$\int_0^\infty y^3 e^{-y} dy = -y^3 e^{-y} \Big|_0^\infty + 3\int_0^\infty y^2 e^{-y} dy$$
$$= 3\int_0^\infty y^2 e^{-y} dy = 6\int_0^\infty y e^{-y} dy$$
$$= 6\int_0^\infty y e^{-y} dy = 6$$

Thus, (A.4.35) becomes

$$A_6 = 6\sum_{n=1}^\infty \frac{1}{n^4} \tag{A.4.36}$$

The sum was evaluated above and is given by (A.4.5). Using this result in (A.4.36), we recover (A.4.31).

Equation (A.4.32) can be obtained by integration of (A.4.31) by parts. Thus,

$$\frac{\pi^4}{15} = \int_0^\infty \frac{x^3 dx}{e^x - 1} = \frac{x^4}{4(e^x - 1)}\Big|_0^\infty + \frac{1}{4}\int_0^\infty \frac{x^4 dx}{(e^x - 1)^2}$$
$$= \frac{1}{4}\int_0^\infty \frac{x^4 dx}{(e^x - 1)^2} \tag{A.4.37}$$

which reduces to (A.4.31).

Appendix 5

Fermi Integrals

If $g(\varepsilon)$ is a monotonically increasing function of ε whose value is zero when ε is zero, and $f(\varepsilon)$ is the Fermi function defined by

$$f(\varepsilon) = \frac{1}{e^{(\varepsilon-\mu)/kT}+1} \quad (A.5.1)$$

then the Fermi integral is defined by

$$I = \int_0^\infty g(\varepsilon)f(\varepsilon)d\varepsilon \quad (A.5.2)$$

or

$$I = -\int_0^\infty F(\varepsilon)\frac{\partial f}{\partial \varepsilon}d\varepsilon \quad (A.5.3)$$

Equation (A.5.3) is obtained from (A.5.2) by an integration by parts, and $F(\varepsilon)$ is defined by

$$F(\varepsilon) = \int_0^\varepsilon g(x)dx \quad (A.5.4)$$

Since $g(\varepsilon)$ is given, $F(\varepsilon)$ is a known function. Examples of Fermi integrals are those with $g(\varepsilon)$ given by $\sqrt{\varepsilon}$ and $\varepsilon^{3/2}$, which are used in getting the Fermi energy and energy as a function of temperature for a gas of free electrons.

From the form of the Fermi function, the derivative $\partial f/\partial \varepsilon$ is practically zero for all energies except in the vicinity of the Fermi energy, near $\varepsilon = \mu$. This means that a rapidly converging series can be obtained by expanding $F(\varepsilon)$ in a Taylor series about μ. Start with the Taylor expansion:

$$F(\varepsilon) = \sum_{r=0}^\infty \frac{1}{r!}(\varepsilon-\mu)^r F_r(\mu) \quad (A.5.5)$$

$F_r(\mu)$ being the rth derivative of $F(\varepsilon)$ evaluated at $\varepsilon = \mu$. That is,

$$F_r(\mu) = \left(\frac{\partial^r F}{\partial \varepsilon^r}\right)_{\varepsilon=\mu} \quad (A.5.6)$$

Equation (A.5.3) can now be written as

$$I = -\sum_{r=0}^{\infty} \frac{F_r(\mu)}{r!} I_r \tag{A.5.7}$$

I_r being defined by

$$I_r = \int_0^{\infty} (\varepsilon - \mu)^r \frac{\partial f}{\partial \varepsilon} d\varepsilon \tag{A.5.8}$$

Now let us work on this integral. From the definition of the Fermi function, the derivative is

$$\frac{\partial f}{\partial \varepsilon} = -\frac{e^{(\varepsilon-\mu)/kT}}{kT[1+e^{(\varepsilon-\mu)/kT}]^2} \tag{A.5.9}$$

Define a variable z by

$$z = \frac{\varepsilon - \mu}{kT} \tag{A.5.10}$$

so that (A.5.9) becomes

$$\frac{\partial f}{\partial \varepsilon} = -\frac{e^z}{kT(1+e^z)^2} \tag{A.5.11}$$

Using this in (A.5.8) and putting the result in (A.5.7) gives

$$I = -\sum_{r=0}^{\infty} (kT)^r \frac{F_r(\mu)}{r!} J_r \tag{A.5.12}$$

where the J_r are integrals defined by

$$J_r = \int_{-\mu/kT}^{\infty} \frac{z^r e^z dz}{(1+e^z)^2} \tag{A.5.13}$$

First note that for all reasonable temperatures, μ/kT is large, and since the integrand is very small for large z, the lower limit can be replaced by $-\infty$ with practically no loss of accuracy. Next note that if r is odd, the integrand in (A.5.13) is antisymmetric, while if r is even, it is symmetric. This is so because the derivative of the Fermi function is symmetric about z:

$$\frac{e^z}{(1+e^z)^2} = \frac{1}{(e^z + e^{-z} + 2)} = \frac{e^{-z}}{(1+e^{-z})^2} \tag{A.5.14}$$

as can be verified with a little algebra. This means that the integrals J_r all vanish if r is odd, whereas if r is even, so $r = 2n$ where n is any positive integer, we can write

$$J_{2n} = \int_{-\infty}^{\infty} \frac{z^{2n} e^z dz}{(1+e^z)^2}$$

$$= 2\int_{0}^{\infty} \frac{z^{2n} e^z dz}{(1+e^z)^2}$$

$$= 2\int_{0}^{\infty} \frac{z^{2n} e^{-z} dz}{(1+e^{-z})^2} \tag{A.5.15}$$

The integral in the last of these equations can be related to known functions by expanding the denominator and integrating term by term. According to the binomial theorem,

$$\frac{1}{(1+e^{-z})^2} = \sum_{j=1}^{\infty} (-1)^{j-1} j e^{-(j-1)z} \tag{A.5.16}$$

so (A.5.15) becomes

$$J_{2n} = 2\sum_{j=1}^{\infty} (-1)^{j-1} j \int_0^{\infty} z^{2n} e^{-jz} dz \tag{A.5.17}$$

The integrals are standard forms given by (A.4.21) and therefore (A.5.17) reduces to

$$J_{2n} = 2(2n)! \sum_{j=1}^{\infty} \frac{(-1)^{j-1}}{j^{2n}} \tag{A.5.18}$$

This is what we were after because the sums can be evaluated. The sum for $n = 1$ has already been given in appendix 4 by (A.4.3) and has the value $\pi^2/12$. Putting this in (A.5.18) for $n = 1$, we get

$$J_2 = \frac{\pi^2}{3} \tag{A.5.19}$$

Thus, we can write (A.5.7) to an approximation including the terms for $r = 0$ and $r = 2$ as

$$I = -F_0(\mu) + F_2(\mu) \frac{(\pi k T)^2}{6} \tag{A.5.20}$$

This is ordinarily sufficiently accurate. A more general treatment starts with recognizing that the sums in (A.5.16) are closely related to the Riemann zeta function, which has been extensively studied and whose values are known. Numerical values of all the integrals J_{2n} can then be obtained.

The widespread use of personal computers provides an alternate method of obtaining the expansion coefficients in (A.5.7) since it is a simple matter to evaluate the integral J_{2n} numerically. Doing this for $n = 1$ to 4 gives the following results:

n	$J_{2n}/2n!$
1	1.64494
2	1.89406
3	1.97099
4	1.99137

Note that these coefficients are all of the order of unity, so the rapid convergence of (A.5.11) is not the result of rapidly decreasing values of the expansion coefficients, but is due to the fact that the ratio of kT to the Fermi energy is small. The functions $g(\varepsilon)$, and therefore $F(\varepsilon)$, are usually small powers of ε, so when their derivatives are evaluated at $\varepsilon = \mu$, the ratio kT/μ then becomes the expansion variable. Since this ratio is of the order of 0.05 or less even for temperatures above 1000 K, rapid convergence is assured.

Appendix 6

Kirkwood's Second Moment

The second moment of the energy in Kirkwood's order-disorder theory is defined by

$$M_2 = \frac{1}{g(R)} \sum_k (W_k - \overline{W})^2 \qquad (A.6.1)$$

For a 50-50 AB alloy, the second moment can be expressed in terms of the long-range order parameter R as

$$M_2 = \frac{Nzv^2}{8}(1 - R^2)^2 \qquad (A.6.2)$$

To show that this is the case, we follow the derivation given by Nix and Shockley (*Review of Modern Physics*; vol. 10, pp. 1–71; 1938). Let the index i represent a pair of nearest neighbor sites and define a parameter p_i such that $p_i = 1$ whenever both sites of the pair are occupied by A atoms and $p_i = 0$ otherwise. (Note that one of these sites is always an α site while the other is always a β site.) For any configuration of the crystal, the total number of AA pairs is therefore

$$Q_{AA} = \sum_i p_i \qquad (A.6.3)$$

where the sum is taken over all pairs in the crystal.

From the definitions in chapter 8, the energy of the kth configuration is

$$W_k = w_o - vQ_{AB} \qquad (A.6.4)$$

and putting this in (A.6.1) gives

$$M_2 = \frac{v^2}{g(R)} \sum_k (Q_{AB} - \overline{Q}_{AB})^2 \qquad (A.6.5)$$

Since the sum divided by the number of configurations is just the average over all configurations, (A.6.5) is written more concisely as

$$M_2 = v^2 \overline{(Q_{AB} - \overline{Q}_{AB})^2} \tag{A.6.6}$$

Taking the supra average gives

$$M_2 = v^2 \left(\overline{Q_{AB}^2} - \overline{Q}_{AB}^2 \right) \tag{A.6.7}$$

and remembering that the number of unlike pairs is $Q_{AB} = Q - Q_{AA} - Q_{BB} = Q - 2Q_{AA}$, this becomes

$$M_2 = 4v^2 \left(\overline{Q_{AA}^2} - \overline{Q}_{AA}^2 \right) \tag{A.6.8}$$

Using (A.6.3) in (A.6.8) gives the second moment in terms of the p_i as

$$M_2 = 4v^2 \sum_i \sum_j \left(\overline{p_i p_j} - \overline{p}_i \overline{p}_j \right) \tag{A.6.9}$$

For a particular pair of sites labeled i, the probability that the α site contains an A atom is just

$$r_\alpha = \frac{N_{A\alpha}}{N_A} = \frac{(1+R)}{2} \tag{A.6.10}$$

and the probability that the β site contains an A atom is

$$w_\beta = \frac{N_{A\beta}}{N_B} = \frac{(1-R)}{2} \tag{A.6.11}$$

The probability that p_i is unity is therefore $(r_\alpha w_\beta)$, and the probability that p_i is zero is $(1 - r_\alpha w_\beta)$. The average value of p_i is therefore

$$\overline{p}_i = \overline{p}_j = r_\alpha w_\beta = rw \tag{A.6.12}$$

Note that r_α is the probability that the α site is rightly occupied (A on α) and that this is equal to the probability that the β site is rightly occupied (B on β). Similarly, w_β is the probability that a β site is wrongly occupied (A on β). That is, $r_\alpha = r_\beta = r$ and $w_\alpha = w_\beta = w$ are the probabilities that a site is rightly or wrongly occupied, respectively.

A little more work is needed to get the average of the square of the p_i so that the double sum in (A.6.9) can be computed. To this end, note that there are four different kinds of terms:

1. i and j represent the same pair of sites;
2. i and j represent two pairs, each having the same α site but different β sites;
3. i and j represent two pairs, each having the same β site but different α sites; and
4. i and j represent two pairs with no sites in common.

For each of these types of terms, we need to know their number for which the average of the product $p_i p_j$ and the product of the average p_i are unity. The product of the averages is always given from (A.6.12) as $r^2 w^2$, so we only need to now concentrate on computing the average of the product.

For type 1 sites, the total number of pairs is Q and the probability that p_i (and therefore $p_i p_j$, since the same sites are in both sums) is unity, is just rw

(i.e., the probability that A is on α and A is on β) so for pairs i, j of type 1, the contribution to (A.6.9) is

$$M_2(1) = 4v^2 Q(rw - r^2 w^2) \qquad (A.6.13)$$

where the first term is from the average of the product and the second term is from the product of the averages [see equation (A.6.12)].

Now consider pairs of sites of type 2 in which the pairs have a common α site. For these pairs, there are a total of $N/2$ α sites. For each of these α sites there are $z(z-1)$ ways of choosing an adjacent β site. The total number of terms in the double sum over these types of sites is then $z(z-1)N/2 = (z-1)Q$. Of these, we want the fraction for which $p_i p_j = 1$. But this is just the probability that the α site is rightly occupied, times the probability that two adjacent β sites are wrongly occupied. That is, $p_i p_j = 1$ for a fraction $r_\alpha w_\beta w_\beta = rw^2$ of the sites, so for sites of type 2, the contribution to (A.6.9) is

$$M_2(2) = 4v^2 Q(z-1)(rw^2 - r^2 w^2) \qquad (A.6.14)$$

The same analysis for type 3 sites gives

$$M_2(3) = 4v^2 Q(z-1)(r^2 w - r^2 w^2) \qquad (A.6.15)$$

Again, the first terms in (A.6.14) and (A.6.15) arise from the average of the product, while the second terms are from the product of the averages.

The remaining sites are of type 4, in which none of the sites overlap. Clearly $p_i p_j$ is unity for sites of this type only if all four sites are occupied by A atoms. Consider the α sites. The number of these is $N/2$, and $rN/2$ of them contain A atoms (rightly occupied), so the probability that a site is rightly occupied is just r. But the probability that another α contains an A atom is $[(rN/2) - 1]/[(N/2) - 1]$ because one of the α sites is already occupied by an A atom. The probability that there are A atoms on both α sites is therefore

$$\frac{r\left(\frac{rN}{2} - 1\right)}{\frac{N}{2} - 1} \qquad (A.6.16)$$

Going through the same analysis for the β sites requires that both β sites be wrongly occupied for $p_i p_j$ to be unity and the probability that this is the case is given by $w[(wN/2) - 1]/[(N/2) - 1]$.

Multiplying these last two expressions together gives the fraction of terms for which $p_i p_j = 1$ in the double sum for sites of type 4, and multiplying the result by Q^2, the number of terms in the double sum, gives the contribution of pairs of type 4 to (A.6.9) as

$$M_2(4) = 4v^2 Q^2 \left[\frac{r^2 w^2 (1 - 2/rN)(1 - 2/wN)}{(1 - 2/N)^2} - r^2 w^2 \right] \qquad (A.6.18)$$

This expression can be considerably simplified by doing a little algebra and dropping terms in N^{-2} relative to terms in N^{-1} and terms in N^{-1} relative to unity. The result is

$$M_2(4) = 4zQv^2(2r^2 w^2 - rw) \qquad (A.6.19)$$

Now let us add (A.6.13), (A.6.14), (A.6.15), and (A.6.19) to get the total contribution to (A.6.9) so that we finally get the remarkably simple equation for the second moment:

$$M_2 = 4Qv^2r^2w^2 \qquad (A.6.20)$$

In arriving at this, we used the fact that $r + w = 1$ and $Q = zN/2$. Using the expressions (A.6.10) and (A.6.11) for r and w in terms of the long-range order parameter R, we recover equation (A.6.2), which is the desired result.

Appendix 7

The Generalized Lattice Gas

Just as for the simple lattice gas that is equivalent to the Ising model, divide the system into cells such that, at most, only one molecule can occupy a given cell and a cell can either be occupied or empty. Let the total number of molecules be N and the total number of sites be M. Also, define a parameter that describes the occupancy of a cell as $e_j = 0$ if the jth cell is empty, and $e_j = 1$ if the jth cell contains a molecule. Note that

$$\sum_j e_j = N \qquad (A.7.1)$$

Instead of restricting ourselves to nearest neighbor interactions, let the potential energy of interaction of two atoms in two cells labeled i and j be $-v_{ij}$. That is, it is still assumed that the system can be described by pairwise central interactions, but these include all pairs and not just nearest neighbors. Note that v_{ij} is a constant. We also assume that there is a binding energy of an atom to a cell given by $-v_j^0$ and that this can be different for every cell. If either of the two cells i and j are empty, then the interaction energy and binding energy are both zero, so the Hamiltonian of the system is

$$H\{e_i e_j\} = -\frac{1}{2}\sum_{i \neq j} v_{ij} e_i e_j - \sum_j v_j^0 e_j \qquad (A.7.2)$$

The first term on the right is a double sum over all i and j that are not equal and is multiplied by $1/2$ to avoid double counting. The argument $\{e_i e_j\}$ is attached to the Hamiltonian to remind us that, because the molecules interact, the energy depends on the particular distribution of the molecules in the cells. To simplify the notation we account for the fact that a molecule does not interact with itself by defining $v_{jj} = 0$.

This model is particularly useful for analyzing the distribution of atoms in such systems as impurities in crystals, at grain boundaries or at dislocations, or in stress fields, as well as the relative concentration of atoms in solutions. In such problems, the statistical average of the occupation of cells by atoms is of critical importance. It then turns out that it is convenient to choose a specific site labeled k and to separate the Hamiltonian into two parts: one that contains k and one that does not. Equation (A.7.2) is therefore written as

$$H = H(k) + H(\neq k) \qquad (A.7.3)$$

where

$$H(k) = -\sum_{j} v_{kj} e_k e_j - e_k v_k^0 \quad (A.7.4)$$

and

$$H(\neq k) = -\frac{1}{2} \sum_{i,j(\neq k)} v_{ij} e_i e_j - \sum_{j(\neq k)} v_j^0 e_j \quad (A.7.5)$$

Equation (A.7.4) is the Hamiltonian for the interaction of a molecule with cell k and with all other molecules in the rest of the crystal (which is zero if the site is unoccupied) and (A.7.5) is the potential energy of all other molecules in the rest of the crystal, except for their interaction with the molecule in cell k. Let us shorten the notation to reflect these definitions. Then, the energy for the kth cell is

$$H(k) = -e_k v(k) \quad (A.7.6)$$

where $v(k)$ is defined by comparing (A.7.6) and (A.7.4).

Now construct the grand canonical partition function, using (A.7.6) and (A.7.5) for the Hamiltonian

$$Q = \sum_N \sum_{\{e_i e_j\}}^{\Omega} e^{\beta e_k v(k)} e^{-\beta H(\neq k)} e^{\beta N \mu} \quad (A.7.7)$$

μ being the chemical potential. As usual, sums over possible states are indicated by braces.

The inner sum is over all possible distributions of the molecules in the cells, there being one such distribution for each set $\{e_i e_j\}$. The total number of distributions is just the number of ways of putting N molecules in M cells, so

$$\Omega = \frac{M!}{N!(M-N)!} \quad (A.7.8)$$

The outer sum is over all possible numbers of molecules, up to the number of sites, but since M is very large (infinity in the thermodynamic limit), the sum is over all positive integers.

Our aim is to get the thermodynamic probability that the cell k is occupied. This is just the statistical mechanical average of the occupation index e_k and given by

$$\overline{e_k} = \sum_{\{e_i e_j\}} e_k f\{e_i e_j\} \quad (A.7.9)$$

where $f\{e_i e_j\}$ is the grand canonical distribution function given by

$$f\{e_i e_j\} = \frac{1}{Q} e^{\beta e_k v(k)} e^{-\beta H(\neq k)} e^{\beta N \mu} \quad (A.7.10)$$

so written out explicitly, (A.7.9) is

$$\overline{e_k} = \frac{\sum\limits_{\{e_j e_j\}} e_k e^{\beta e_k v(k)} e^{-\beta H(\neq k)} e^{\beta N \mu}}{\sum\limits_{\{e_j e_j\}} e^{\beta e_k v(k)} e^{-\beta H(\neq k)} e^{\beta N \mu}} \quad (A.7.11)$$

If only nearest neighbor molecules interact then all terms that do not contain k cancel out, the only surviving terms being sums over the possible values of e_k, so (A.7.11) reduces to

$$\overline{e_k} = \frac{\sum\limits_{\{e_k\}} e_k e^{\beta e_k v(k)} e^{\beta e_k \mu}}{\sum\limits_{\{e_k\}} e^{\beta e_k v(k)} e^{\beta e_k \mu}} \quad (A.7.12)$$

Since the possible values of e_k are zero and unity, this reduces to

$$\overline{e_k} = \frac{e^{\beta v(k)} e^{\beta \mu}}{e^{\beta v(k)} e^{-\beta \mu} + 1}$$

$$= \frac{1}{1 + e^{-\beta[\mu + v(k)]}} \quad (A.7.13)$$

This has the same form as the Fermi–Dirac distribution because each "particle" (cell) can exist in only one of two states.

In this form, the lattice gas model neglects any kinetic energy contributions to the partition function, which, because the molecules are bound to cells, will consist of vibrations. This is not easy to do in the general case because a factor must be included in the partition function that contains the normal modes v_n. This factor is

$$\sum\limits_{\{e_j e_j\}} e^{-\beta \Sigma_n (1+1/2) h v_n} \quad (A.7.14)$$

In general, the sum in the exponential is *not* a sum over lattice cells, so the separation of terms between those that include a cell k and those that do not is not possible. But in the special case of the Einstein approximation, such a separation can be made because then each vibrational mode is attached to a molecule. The vibrational energy is then included by adding the term $\Sigma_{\alpha=1}^{3}(1 + 1/2)hv_k^\alpha$ to the binding energy of the molecule to the cell. The molecule may have different frequencies in different directions, as would be the case when it is bound to a surface or a dislocation line; v_k° is then reinterpreted to include the vibrational term. There are cases in which this is an excellent approximation. One of these is when the number of molecules is much less than the number of cells, so there is no coupling among the molecules and the molecules do vibrate independently. Another is when the molecules are tightly bound such that the vibrational spectrum can be divided into two parts: a vibration within cells and vibrational modes among cells. This, for example, would be the case for chemical adsorption on a surface. In both cases, however, it is still assumed that the cells only provide sites for occupancy and have no physical properties other than a binding energy. That is, it is assumed that any vibrational energy associated with the cells is the same whether or not the cells are occupied. This is often a reasonable assumption, especially at high temperatures.

Appendix 8

Dyadics and Crystal Symmetry

A.8.1 Dyadic algebra

A dyadic is an operator whose properties are most easily understood by writing it as the juxtaposition of two vectors. Thus, given two vectors **A** and **B**, the dyadic **D** is an operator defined by

$$\mathbf{D} = \mathbf{AB} \tag{A.8.1.1}$$

A is called the *antecedent* of the dyadic, and **B** is called the *consequent*.

If **A** and **B** are written in terms of the unit vectors $\mathbf{i}_1$, $\mathbf{i}_2$, $\mathbf{i}_3$ in a Cartesian coordinate system so that

$$\mathbf{A} = A_1\mathbf{i}_1 + A_2\mathbf{i}_2 + A_3\mathbf{i}_3 \tag{A.8.1.2}$$

and

$$\mathbf{B} = B_1\mathbf{i}_1 + B_2\mathbf{i}_2 + B_3\mathbf{i}_3 \tag{A.8.1.3}$$

then the dyadic **D** = **AB** is

$$\begin{aligned}\mathbf{D} = \mathbf{AB} = &\, A_1B_1\mathbf{i}_1\mathbf{i}_1 + A_1B_2\mathbf{i}_1\mathbf{i}_2 + A_1B_3\mathbf{i}_1\mathbf{i}_3 \\ &+ A_2B_1\mathbf{i}_2\mathbf{i}_1 + A_2B_2\mathbf{i}_2\mathbf{i}_2 + A_2B_3\mathbf{i}_2\mathbf{i}_3 \\ &+ A_3B_1\mathbf{i}_3\mathbf{i}_1 + A_3B_2\mathbf{i}_3\mathbf{i}_2 + A_3B_3\mathbf{i}_3\mathbf{i}_3\end{aligned} \tag{A.8.1.4}$$

The pairs of unit vectors $\mathbf{i}_r\mathbf{i}_s$ are called *dyads*. The components of the dyadic are defined by

$$D_{rs} = A_rB_s \tag{A.8.1.5}$$

Two dyadics are equal if their corresponding components are equal. Thus, **D** = **D**′ means that $D_{rs} = D'_{rs}$. **AB** is called the *conjugate dyadic* of **BA** and is often designated by a subscript c. Thus, if **D** = **AB**, then its conjugate is $\mathbf{D}_c = \mathbf{BA}$.

The sum of two dyadics is found by adding their components. Thus, **F** = **D** + **E** means $F_{rs} = D_{rs} + E_{rs}$. A multiplier on the left of a dyadic is called a *prefactor*, while a multiplier on the right is called a *postfactor*. Pre- and postfactors can be scalars, vectors, or dyadics, and the multiplication can be scalar multiplication, scalar or vector products, or dyadic multiplication.

A dyadic operating on a scalar simply multiplies the dyadic as follows:

$$\mathbf{AB}a = a\mathbf{AB} \qquad (A.8.1.6)$$

Both the pre- and postdot products of a dyadic with a vector are vectors. That is,

$$\mathbf{C} \cdot \mathbf{AB} = (\mathbf{C} \cdot \mathbf{A})\mathbf{B} = \mathbf{B}(\mathbf{C} \cdot \mathbf{A}) \qquad (A.8.1.8)$$

$$\mathbf{AB} \cdot \mathbf{C} = \mathbf{A}(\mathbf{B} \cdot \mathbf{C}) = (\mathbf{B} \cdot \mathbf{C})\mathbf{A} \qquad (A.8.1.9)$$

where $(\mathbf{C} \cdot \mathbf{A})$ and $(\mathbf{B} \cdot \mathbf{C})$ are scalar products of $\mathbf{C}$ and $\mathbf{A}$ and of $\mathbf{B}$ and $\mathbf{C}$, respectively.

Both the pre- and postcross products of a vector with a dyadic are dyadics. Thus,

$$\mathbf{AB} \times \mathbf{C} = \mathbf{A}(\mathbf{B} \times \mathbf{C}) \qquad (A.8.1.10)$$

$$\mathbf{C} \times \mathbf{AB} = (\mathbf{C} \times \mathbf{A})\mathbf{B} \qquad (A.8.1.11)$$

Dyadic algebra is therefore noncommutative. It is clear from the above that all of dyadic algebra and analysis follows directly from the rules of vector algebra and analysis.

If both the pre- and postdot product of a dyadic $\mathbf{D} = \mathbf{AB}$ is formed with two vectors $\mathbf{f}$ and $\mathbf{g}$, the result is a scalar. That is,

$$\mathbf{f} \cdot \mathbf{D} \cdot \mathbf{g} = \mathbf{f} \cdot \mathbf{AB} \cdot \mathbf{g} = (\mathbf{f} \cdot \mathbf{A})(\mathbf{B} \cdot \mathbf{g})$$
$$= \sum_{i,j} f_i A_i B_j g_j = \sum_{i,j} D_{ij} f_i g_j \qquad (A.8.1.12)$$

This equation illustrates the economy of dyadic notation in that it bypasses the need to identify coordinate axes by subscripts.

The last sum in (A.8.1.12) is called a quadratic form. Such quantities arise frequently in physical applications, particularly in the theory of harmonic vibrations.

A dyadic $\mathbf{D}$ is symmetric if $D_{rs} = D_{sr}$ and antisymmetric if $D_{rs} = -D_{sr}$. In general, a dyadic is neither symmetric nor antisymmetric, as is evident from equation (A.8.1.4). However, the sum

$$D_{sr}^{+} \equiv D_{rs} + D_{sr} \qquad (A.8.1.13)$$

is always symmetric, and the difference

$$D_{sr}^{-} \equiv D_{rs} - D_{sr} \qquad (A.8.1.14)$$

is always antisymmetric.

Note that the D_{sr} are the components of the dyadic that is conjugate to $\mathbf{D}$ so that the (A.8.1.13) and (A.8.1.14) are equivalent to

$$\mathbf{D}^{+} = \mathbf{D} + \mathbf{D}_c \qquad (A.8.1.15)$$

$$\mathbf{D}^{-} = \mathbf{D} - \mathbf{D}_c \qquad (A.8.1.16)$$

Obviously,

$$\mathbf{D} = \frac{1}{2}\{\mathbf{D}+\mathbf{D}_c\} + \frac{1}{2}\{\mathbf{D}-\mathbf{D}_c\} \qquad (A.8.1.17)$$

so a dyadic is always easily decomposed into symmetric and antisymmetric parts.

In component form, (A.8.1.13) and (A.8.1.14) give

$$D_{rs} = \frac{1}{2}(D_{sr}^+ + D_{sr}^-) \qquad (A.8.1.18)$$

which is equivalent to (A.8.1.17). Note that it is only for symmetric dyadics that the order of writing the two vectors that define the dyadic is immaterial, because in general, $D_{rs}(=A_rB_s)$ and $D_{sr}(=A_sB_r)$ are not equal.

The symmetry or antisymmetry of a dyadic is preserved under a rotation of axes. That is, if a dyadic with components D_{rs} is symmetric (or antisymmetric) in a coordinate system defined by the orthogonal unit vectors $\mathbf{i}_r$, then it is also symmetric (or antisymmetric) in a coordinate system with orthogonal vectors $\mathbf{e}_n$, which are the result of the linear coordinate transformation

$$\mathbf{e}_n = \sum_r a_{nr}\mathbf{i}_r \qquad (A.8.1.19)$$

To show this, we need to recall some results from the theory of linear transformations.

The a_{nr} are the direction cosines between the two sets of unit vectors since the dot products of the two sets of vectors are

$$\mathbf{e}_n \cdot \mathbf{i}_s = \sum_r a_{nr}\mathbf{i}_r \cdot \mathbf{i}_s = \sum_r a_{nr}\delta_{rs} = a_{ns} \qquad (A.8.1.20)$$

A vector $\mathbf{F}$ is expressed in the two-coordinate systems in terms of its components and the unit vectors in the two systems as

$$\mathbf{F} = \sum_j F_j \mathbf{i}_j = \sum_i F_i' \mathbf{e}_i \qquad (A.8.1.21)$$

Substituting (A.8.1.19) into (A.8.1.21) gives

$$\mathbf{F} = \sum_j F_j \mathbf{i}_j = \sum_{i,j} F_i' a_{ij} \mathbf{i}_j \qquad (A.8.1.23)$$

so equating coefficients of the unit vectors gives the transformation equation for the vector components as

$$F_j = \sum_i a_{ij} F_i' \qquad (A.8.1.24)$$

It is also easy to show that

$$\mathbf{i}_n = \sum_r a_{rn} \mathbf{e}_r \qquad (A.8.1.25)$$

and

$$F_j' = \sum_i a_{ji} F_i \qquad (A.8.1.26)$$

The dyadic **D** is

$$\mathbf{D} = \sum_{r,s} D_{rs} \mathbf{i}_r \mathbf{i}_s \quad (A.8.1.27)$$

and substituting (A.8.1.25) into (A.8.1.27) gives the components of the transformed dyadic as

$$\mathbf{D}' = \sum_{r,s,n,m} D_{rs} a_{nr} a_{ms} \mathbf{e}_n \mathbf{e}_m \quad (A.8.1.28)$$

or, in component form,

$$D'_{nm} = \sum_{r,s} a_{nr} a_{ms} D_{rs} \quad (A.8.1.29)$$

If $D_{rs} = D_{sr}$, then interchanging the subscripts r and s in the sum in (A.8.1.29) shows that $D'_{nm} = D'_{mn}$. Similarly, if $D_{rs} = -D_{sr}$, then interchanging subscripts in the sum shows that $D'_{nm} = -D'_{mn}$. The symmetry properties are preserved under a rotation of axes.

The close relationship between dyadics and matrices is evident from equation (A.8.1.4). In fact, dyadics are made formally equivalent to 3 × 3 square matrices and to second rank tensors by defining a rule for multiplying dyadics by treating their components as the components of matrices and then applying the rules of matrix multiplication. The entire algebra and calculus of dyadics then follows from their representation in component form.

A.8.2 Principle axes

A symmetric dyadic has the important property that a coordinate system can always be found in which only the diagonal components of the dyadic are nonzero. That is, if a dyadic is

$$\mathbf{D} = \sum_{r,s} D_{rs} \mathbf{i}_r \mathbf{i}_s \quad (A.8.2.1)$$

where the unit vectors define an arbitrary Cartesian coordinate system, and if

$$D_{rs} = D_{sr} \quad (A.8.2.2)$$

then another Cartesian system, defined by the unit vectors $\mathbf{e}_1, \mathbf{e}_2, \mathbf{e}_3$ can always be found such that

$$\mathbf{D}' = \sum_{m,n} D'_{mn} \mathbf{e}_m \mathbf{e}_n \delta_{mn} = \sum_m D'_{mm} \mathbf{e}_m \mathbf{e}_m \quad (A.8.2.3)$$

This can be demonstrated by starting with the fact that the sets of unit vectors $\mathbf{i}_n$ and $\mathbf{e}_n$ are related by (A.8.1.19) and (A.8.1.25). Not all the transformation coefficients a_{rs} are independent. Because the unit vectors are orthogonal in both coordinate systems, (A.8.1.25) gives

$$\mathbf{i}_r \cdot \mathbf{i}_s = \sum_{m,n} a_{rm} a_{sn} \mathbf{e}_m \cdot \mathbf{e}_n = \delta_{rs} = \sum_{m,n} a_{rm} a_{sn} \delta_{mn} \quad (A.8.2.4)$$

from which it follows that

$$\sum_m a_{rm}a_{sm} = \delta_{rs} \qquad (A.8.2.5)$$

Since this equation is invariant with respect to an r, s interchange, only six of these nine equations among the transformation coefficients are independent.

To show that a transformation of the form (A.8.1.25) always exists that transforms a symmetric dyadic (A.8.2.1) into the diagonal form (A.8.2.3), carry out the transformation by substituting (A.8.1.25) into (A.8.2.1) to get

$$\mathbf{D}' = \sum_{r,s}\sum_{m,n} a_{rm}a_{sn}D_{rs}\mathbf{e}_m\mathbf{e}_n \qquad (A.8.2.6)$$

Then, if (A.8.2.6) is to have the diagonal form (A.8.2.3), corresponding components must be equal and we should have

$$D'_{mn}\delta_{mn} = \sum_{r,s} a_{rm}a_{sn}D_{rs} \qquad (A.8.2.7)$$

There are 12 unknowns that must be determined if (A.8.2.7) is to be valid: the three components D'_{mn} and the nine transformation coefficients, the original components D_{rs} and the original coordinate system being regarded as given. Equation (A.8.2.7) represents a set of nine equations of which only six are independent because of the symmetry requirement of equation (A.8.2.2). When added to the six independent equations of (A.8.2.5), we have 12 independent equations to find our 12 unknowns. Thus, a diagonal form can always be found for a symmetric dyadic. The axes defined by the $\mathbf{e}_n$ are called the *principal axes*. This result is often stated by saying that a symmetric dyadic can always be transformed to its principle axes.

The isomorphism of dyadic to matrix algebra is clear, and the problem of finding the diagonal form of a dyadic is equivalent to that of diagonalizing a 3×3 matrix to find its eigenvalues. The above development shows that diagonalization is possible for any dyadic (or matrix) provided that it is symmetric. If in addition none of the eigenvalues of the matrix are equal, then diagonalization is possible even for nonsymmetric dyadics.

A.8.3 Dyadics in crystals

In material media, dyadics couple two vectors that denote properties of, or processes in, the material. The diffusion equation, for example, can be written as

$$\mathbf{J} = -\mathbf{D}\cdot\nabla C \qquad (A.8.3.1)$$

where the diffusion dyadic $\mathbf{D}$ connects the flux $\mathbf{J}$ of an atomic or molecular species to the gradient of the concentration (see chapter 17). Such dyadics are properties of the material, and if the material is a crystal, they must reflect the crystal symmetry. The effects of crystal symmetry are readily found by subjecting the dyadic to the crystal symmetry operations.

First, consider a dyadic $\mathbf{D}$ that is expressed in its principle axes as

$$\mathbf{D} = \sum_{r=1}^{3} D_{rr}\mathbf{e}_r\mathbf{e}_r \qquad (A.8.3.2)$$

and define its component in an arbitrary direction by

$$\mathbf{D}_n = \mathbf{n} \cdot \mathbf{D} \tag{A.8.3.3}$$

$\mathbf{n}$ being a unit vector in the given direction, which is related to the unit vectors $\mathbf{e}_r$ of the principle axes by

$$\mathbf{n} = n_1 \mathbf{e}_1 + n_2 \mathbf{e}_2 + n_3 \mathbf{e}_3 \tag{A.8.3.4}$$

n_1, n_2, and n_3 being the direction cosines of the vector $\mathbf{n}$ referred to the principle axes.

Combining (A.8.3.2) and (A.8.3.4) according to (A.8.3.3) gives

$$\mathbf{D}_n = (n_1 \mathbf{e}_1 + n_2 \mathbf{e}_2 + n_3 \mathbf{e}_3) \cdot \sum_{r=1}^{3} D_{rr} \mathbf{e}_r \mathbf{e}_r = \sum_{r=1}^{3} n_r D_{rr} \mathbf{e}_r \tag{A.8.3.5}$$

The utility of this result is readily illustrated applying it to equation (A.8.3.1). That is, the component of flux in the $\mathbf{n}$ direction is

$$\mathbf{n} \cdot \mathbf{J} = -\mathbf{n} \cdot \mathbf{D} \cdot \nabla C = -\sum_{r=1}^{3} n_r D_{rr} \mathbf{e}_r \cdot \nabla C \tag{A.8.3.6}$$

In particular, consider a medium for which all D_{rr} are equal to the same constant D. Then (A.8.3.6) is

$$\mathbf{n} \cdot \mathbf{J} = -D \sum_{r=1}^{3} n_r \mathbf{e}_r \cdot \nabla C \tag{A.8.3.7}$$

But the sum in (A.8.3.7) is just the unit vector $\mathbf{n}$, so we have

$$\mathbf{n} \cdot \mathbf{J} = -D\mathbf{n} \cdot \nabla C \tag{A.8.3.8}$$

so the flux in a particular direction is proportional to the component of the gradient of concentration in that direction.

Equation (A.8.3.8) states that a medium for which the principle components of diffusion are all the same has the same diffusion coefficient in all directions. That is, the medium is isotropic. Note that while the above was applied to diffusion, similar results hold for any property or process in which two vectors are linked by a dyadic. That is, in isotropic systems, dyadics have the same component in all three principle directions and are the same in all Cartesian coordinate systems.

This is not generally true for crystals. To see the effect of crystal symmetry, start with a cubic crystal, take the coordinate axes to be along the three cubic crystallographic directions, and write out the dyadic as

$$\begin{aligned}\mathbf{D} = & D_{11}\mathbf{i}_1\mathbf{i}_1 + D_{12}\mathbf{i}_1\mathbf{i}_2 + D_{13}\mathbf{i}_1\mathbf{i}_3 \\ & + D_{21}\mathbf{i}_2\mathbf{i}_1 + D_{22}\mathbf{i}_2\mathbf{i}_2 + D_{23}\mathbf{i}_2\mathbf{i}_3 \\ & + D_{31}\mathbf{i}_3\mathbf{i}_1 + D_{32}\mathbf{i}_3\mathbf{i}_2 + D_{33}\mathbf{i}_3\mathbf{i}_3 \end{aligned} \tag{A.8.3.9}$$

Cubic symmetry requires that rotation by 90° about any of the three unit vectors must leave the dyadic unchanged. For example, if a rotation about $\mathbf{i}_3$ is performed such that $\mathbf{i}_1$ goes to $\mathbf{i}_2$ and $\mathbf{i}_2$ goes to $-\mathbf{i}_1$, or if a rotation about $\mathbf{i}_2$ brings $\mathbf{i}_1$ to $\mathbf{i}_3$ and $\mathbf{i}_3$ goes to $-\mathbf{i}_1$, the dyadic must have the same components before

APPENDIX 8 501

and after the rotations. That is, **D** is invariant with respect to either of the transformations:

$$\begin{aligned}\mathbf{i}_1 &\to \mathbf{i}_2 \\ \mathbf{i}_2 &\to -\mathbf{i}_1\end{aligned} \qquad \text{(A.8.3.10)}$$

and

$$\begin{aligned}\mathbf{i}_1 &\to \mathbf{i}_3 \\ \mathbf{i}_3 &\to -\mathbf{i}_1\end{aligned} \qquad \text{(A.8.3.11)}$$

Now apply the first of these to (A.8.3.9) by simply replacing $\mathbf{i}_1$ by $\mathbf{i}_2$ and $\mathbf{i}_2$ by $-\mathbf{i}_1$ to get

$$\begin{aligned}\mathbf{D} = &D_{11}\mathbf{i}_2\mathbf{i}_2 - D_{12}\mathbf{i}_2\mathbf{i}_1 + D_{13}\mathbf{i}_1\mathbf{i}_3 \\ &- D_{21}\mathbf{i}_1\mathbf{i}_2 + D_{22}\mathbf{i}_1\mathbf{i}_1 - D_{23}\mathbf{i}_1\mathbf{i}_3 \\ &+ D_{31}\mathbf{i}_3\mathbf{i}_2 1 D_{32}\mathbf{i}_3\mathbf{i}_1 + D_{33}\mathbf{i}_3\mathbf{i}_3\end{aligned} \qquad \text{(A.8.3.12)}$$

But the dyadic is invariant with respect to this rotation, so the coefficients of corresponding dyads in (A.8.9) and (3.12) must be equal. This gives

$$\begin{aligned}D_{11} &= D_{22}; & D_{12} &= -D_{21}; & D_{13} &= D_{23} \\ D_{21} &= -D_{12}; & D_{22} &= D_{11}; & D_{23} &= -D_{13} \\ D_{31} &= -D_{32}; & D_{32} &= D_{31}; & D_{33} &= D_{33}\end{aligned} \qquad \text{(A.8.3.13)}$$

Combining some of these equations simplifies the set (A.8.3.13) to

$$\begin{aligned}D_{11} &= D_{22}; \quad D_{12} = -D_{21} \\ D_{13} &= D_{31} = D_{23} = D_{32} = 0\end{aligned} \qquad \text{(A.8.3.14)}$$

The diffusion dyadic (A.8.3.9) therefore reduces to

$$\mathbf{D} = D_{11}\mathbf{i}_1\mathbf{i}_1 + D_{11}\mathbf{i}_2\mathbf{i}_2 + D_{33}\mathbf{i}_3\mathbf{i}_3 + D_{12}\mathbf{i}_1\mathbf{i}_2 + D_{21}\mathbf{i}_2\mathbf{i}_1 \qquad \text{(A.8.3.15)}$$

Now apply the rotation (A.8.3.11) to (A.8.3.15). The result is

$$\mathbf{D} = D_{11}\mathbf{i}_3\mathbf{i}_3 + D_{11}\mathbf{i}_2\mathbf{i}_2 + D_{33}\mathbf{i}_1\mathbf{i}_1 + D_{12}\mathbf{i}_3\mathbf{i}_2 + D_{21}\mathbf{i}_2\mathbf{i}_3 \qquad \text{(A.8.3.16)}$$

and equating coefficients of the dyads in (A.8.3.15) and (A.8.3.16) shows that $D_{33} = D_{11}$ and $D_{12} = D_{21} = 0$. Therefore, in cubic crystals, the dyadic becomes

$$\mathbf{D} = D(\mathbf{i}_1\mathbf{i}_1 + \mathbf{i}_2\mathbf{i}_2 + \mathbf{i}_3\mathbf{i}_3) \qquad \text{(A.8.3.17)}$$

where $D_{11} = D_{22} = D_{33} \equiv D$ is the same for all directions. The dyadic in a cubic crystal is isotropic just as in a noncrystalline medium, and any set of Cartesian coordinates defines a set of principle axes.

Now take a hexagonal crystal and choose a set of axes such that $\mathbf{i}_1$ and $\mathbf{i}_2$ are in the basal plane and $\mathbf{i}_3$ is perpendicular to the basal plane. The crystal is symmetric with respect to a 60° degree rotation about $\mathbf{i}_3$ in the basal plane, so the dyadic is invariant with respect to a rotation defined by

$$\begin{aligned}\mathbf{i}_1 &\to \frac{1}{2}\mathbf{i}_1 + \frac{\sqrt{3}}{2}\mathbf{i}_2 \\ \mathbf{i}_2 &\to -\frac{\sqrt{3}}{2}\mathbf{i}_1 + \frac{1}{2}\mathbf{i}_2\end{aligned} \qquad \text{(A.8.3.18)}$$

On making the interchange in (A.8.3.9) according to (A.8.3.18) and equating coefficients as before, we get

$$\mathbf{D} = D_{11}\mathbf{i}_1\mathbf{i}_1 + D_{11}\mathbf{i}_2\mathbf{i}_2 + D_{33}\mathbf{i}_3\mathbf{i}_3 \quad (A.8.3.19)$$

provided we take $D_{12} = D_{21}$.

In a hexagonal crystal the dyadic has two independent components, one associated with the basal plane and the other with the direction perpendicular to the basal plane. The principle axes consist of any two perpendicular axes in the basal plane and a third axis perpendicular to the basal plane.

Note that in arriving at (A.8.3.19) we assumed that the dyadic was symmetric while no such assumption was needed for the cubic case because it was isotropic. Equation (A.8.3.19) is therefore correct only for symmetric dyadics. For dyadics that are not symmetric, results such as (A.8.3.19) are valid only for the symmetric part of the dyadic. But crystal dyadics are usually symmetric. Consider a crystal dyadic that is antisymmetric with components D_{rs}^-. Since $D_{rs}^- = -D_{sr}^-$, there are no diagonal components and the antisymmetric dyadic has the form

$$\mathbf{D}^- = D_{12}^-\mathbf{i}_1\mathbf{i}_2 + D_{13}^-\mathbf{i}_1\mathbf{i}_3 + D_{21}^-\mathbf{i}_2\mathbf{i}_1 + D_{23}^-\mathbf{i}_2\mathbf{i}_3 + D_{31}^-\mathbf{i}_3\mathbf{i}_1 + D_{32}^-\mathbf{i}_3\mathbf{i}_2 \quad (A.8.3.20)$$

and assume that the unit vectors can be chosen such that the crystal has a reflection plane defined by the coordinate axes $\mathbf{i}_2$, $\mathbf{i}_3$ that leaves the crystal invariant. That is, changing $\mathbf{i}_1$ into $-\mathbf{i}_1$ leaves the crystal dyadic unchanged. Then (A.8.3.20) becomes

$$\mathbf{D} = -D_{12}^-\mathbf{i}_1\mathbf{i}_2 - D_{13}^-\mathbf{i}_1\mathbf{i}_3 - D_{21}^-\mathbf{i}_2\mathbf{i}_1 + D_{23}^-\mathbf{i}_2\mathbf{i}_3 - D_{31}^-\mathbf{i}_3\mathbf{i}_1 + D_{32}^-\mathbf{i}_3\mathbf{i}_2 \quad (A.8.3.21)$$

and equating coefficients in equations (A.8.3.19) and (A.8.3.21) gives

$$\begin{aligned} D_{12}^- = -D_{12}^- = 0 \\ D_{13}^- = -D_{13}^- = 0 \end{aligned} \quad (A.8.3.22)$$

Similarly, if there is a reflection across the $\mathbf{i}_1$, $\mathbf{i}_2$ plane that leaves the crystal invariant, we find that

$$D_{23}^- = -D_{32}^- = 0 \quad (A.8.3.23)$$

and the antisymmetric crystal dyadic vanishes.

Application of similar procedures to other crystal systems shows that the results for tetragonal and trigonal systems are just like those for the hexagonal. Dyadics in systems of lower symmetry have three independent principle components.

A.8.4. Symmetry of the diffusion dyadic

There is no general proof that all crystal dyadics must be symmetric. It can be shown that thermodynamics requires that dyadics that represent equilibrium thermodynamic properties must be symmetric to satisfy the law of conservation of energy, but diffusion is not an equilibrium process. However, the symmetry of the diffusion dyadic can be demonstrated by remembering that in a diffusion experiment, it is always the divergence of the flux that is measured rather than the flux itself. All that we can measure is the difference in the amount of material entering and leaving a volume element. This is illustrated by the fact that it is always Fick's *second* law that is used to analyze diffusion experiments.

With this in mind, let us separate the diffusion dyadic into its symmetric and antisymmetric parts and write

$$\mathbf{D} = \mathbf{D}^+ + \mathbf{D}^- \tag{A.8.4.1}$$

and put this into Fick's second law to get

$$\frac{\partial C}{\partial t} = \nabla \cdot \mathbf{D}^+ \cdot \nabla C + \nabla \cdot \mathbf{D}^- \cdot \nabla C \tag{A.8.4.2}$$

The components of $\mathbf{D}^-$ are D^-_{mn}, and since $D^-_{mn} = -D^-_{nm}$, all diagonal components vanish. Then if we expand the second term on the right of (A.8.4.2), we find that it vanishes identically because the order of taking a second-order derivative is immaterial and the off-diagonal elements are antisymmetric. Fick's second law then reduces to

$$\frac{\partial C}{\partial t} = \nabla \cdot \mathbf{D}^+ \cdot \nabla C \tag{A.8.4.3}$$

This equation tells us that only the symmetric part of the diffusion dyadic contributes to the change in concentration with time and the antisymmetric part is unobservable, so we may as well take the diffusion dyadic to be symmetric. Then the symmetry consideration given above for a symmetric dyadic holds for the diffusion dyadic.

Additional Readings

The literature on the topics treated in this book is enormous. The number of books on thermodynamics or statistical mechanics alone is staggering. Below, I offer some selections that readers would find useful for further studies.

Chapter 1

Callen, Herbert B.; (1961); *Thermodynamics*; John Wiley, New York.
Fermi, Enrico; (1936); *Thermodynamics*; Prentice-Hall, New York; reprinted by Dover Publications, 1956.
Gaskell, David R.; (1995); *Introduction to the Thermodynamics of Materials*; Taylor and Francis, Washington, D.C.
Glasstone, Samuel; (1947); *Thermodynamics for Chemists*; van Nostrand, New York.
Guggenheim, E.A.; (1949); *Thermodynamics*; North Holland, Amsterdam; Interscience, New York.
Landau, L.D., and E.M. Lifshitz; (1958); *Statistical Physics*; Pergamon Press, London.
Margenau, Henry, and George M. Murphy; (1956); *The Mathematics of Physics and Chemistry*; chapter 1; van Nostrand; New York.
Slater, J.C.; (1939); *Introduction to Chemical Physics*; chapters 1 and 2; McGraw-Hill, New York.
van Ness, H.C.; (1969); *Understanding Thermodynamics*; McGraw-Hill, New York; reprinted by Dover Publications, 1983.
Zemanski, Mark W.; (1957); *Heat and Thermodynamics*; McGraw-Hill, New York.

Chapters 2 and 3

Bowley, Roger, and Mariana Sanchez; (1996); *Introductory Statistical Mechanics*; Clarendon Press, Oxford.
Chandler, David; (1987); *Introduction to Modern Statistical Mechanics*; Oxford University Press, Oxford.
Eyring, H., D. Henderson, B.J. Stover, and E.M. Eyring; (1964); *Statistical Mechanics and Dynamics*; Wiley, New York.
Fowler, R.H.; (1936); *Statistical Mechanics*; Cambridge University Press, Cambridge.

Fowler, R.H., and E.A. Guggenheim; (1930); *Statistical Thermodynamics*; 2nd printing with corrections, 1949; Cambridge University Press, Cambridge.
Girifalco, L.A.; (1973); *Statistical Physics of Materials*; John Wiley, New York.
Hill, T.L.; (1960); *Statistical Mechanics*; Addison-Wesley, Reading, Mass.
Kittel, Charles, and Herbert Kroemer; (1980); *Thermal Physics*; Freeman, San Francisco.
Landau, L.D., and E.M. Lifshitz; (1958); *Statistical Physics*; Pergamon Press, London.
Mayer, J.E., and M.G. Mayer; (1952); *Statistical Mechanics*; John Wiley, New York.
McDonald, D.K.C.; (1963); *Introductory Statistical Mechanics for Physicists*; John Wiley, New York.
McQuarrie, Donald A.; (1976); *Statistical Mechanics*; Harper & Row, New York.
Reif, F.; (1965); *Fundamentals of Statistical and Thermal Physics*; McGraw-Hill, New York.
Schrodinger, E.; (1952); *Statistical Thermodynamics*; Cambridge University Press, Cambridge; Reprinted by Dover Publications, 1989.
Tolman, R.C.; (1938); *Statistical Mechanics*; Oxford University Press, Cambridge University Press, Cambridge.
Wannier, G.H.; (1966); *Statistical Physics*; John Wiley, New York.

Chapters 4 and 5

The texts cited above contain much material on the harmonic theory and equation of state of crystals. Additional readings are:.
Born, M., and K. Huang; (1954); *Dynamical Theory of Crystal Lattices*; Clarendon Press, Oxford.
Callaway, Joseph; (1974); *Quantum Theory of the Solid State*; chapter 1; Academic Press, New York.
de Launay, Jules; (1956); "The Theory of Specific Heats and Lattice Vibrations"; in *Solid State Physics, vol. 2*; F. Seitz, and D. Turnbull, Eds.; Academic Press, New York, pp. 219–303.
Gschneider, Karl A.; (1964); "Physical Properties and Interrelationships of Metallic and Semimetallic Elements"; in *Solid State Physics, vol. 16*; F. Seitz, and D. Turnbull, Eds.; Academic Press, New York, pp. 276–426.
Weinreich, Gabriel; (1965); *Solids: Elementary Theory for Advanced Students*; John Wiley, New York.

Chapters 6 and 7

Blatt, Frank J.; (1968); *Physics of Electronic Conduction in Solids*; McGraw-Hill, New York.
Raimes, S.; (1963); *The Wave Mechanics of Electrons in Solids*; North-Holland, Amsterdam.
Wilson, A.H.; (1965); *The Theory of Metals*, 2nd ed., Cambridge University Press, Cambridge.

Chapter 8

Fowler, R.H., and E.A. Guggenheim; (1956); *Statistical Thermodynamics*; chapter 13; Cambridge University Press, Cambridge.

Guttman, L.; (1956); "Order-Disorder Phenomenon in Metals"; in *Solid State Physics, vol. 3*; F. Seitz, and D. Turnbull, Eds.; Academic Press, New York, pp. 146–243.

Muto, T., and Y. Takagi; (1955); "The Theory of Order-Disorder in Alloys"; in *Solid State Physics, vol. 1*; F. Seitz, and D. Turnbull, Eds.; Academic Press, New York, pp. 193–282.

Chapter 9

Cusack, N.; (1958); *The Electrical and Magnetic Properties of Solids*; Longmans, Green, London.

Fowler, R.H.; (1936); *Statistical Mechanics*, 2nd ed.; chapter 12; Cambridge University Press, Cambridge.

Chapters 10 and 11

Binney, J.J., N.J. Dowrick, A.J. Fisher, and M.E.J. Newman; (1993); *The Theory of Critical Phenomena: An Introduction to the Renormalization Group*; Clarendon Press, Oxford.

Chaiken, P.M., and T.C. Lubensky; (1995); *Principles of Condensed Matter Physics*; Cambridge University Press, Cambridge.

Goldenfeld, Nigel; (1992); *Lectures on Phase Transitions and the Renormalization Group*; Addison-Wesley, New York.

Goodstein, David L.; (1975); *States of Matter*; Prentice-Hall, Englewood Cliffs, N.J.; reprinted by Dover Publications, 1985.

Ma, Shang-Keng; (1976); *Modern Theory of Critical Phenomena*; Addison-Wesley, New York.

Plischke, Michael, and Birger Bergersen; (1994); *Equilibrium Statistical Physics*, 2nd ed.; chapters 3–6; World Scientific, Singapore.

Stanley, H. Eugene; (1971); *Introduction to Phase Transitions and Critical Phenomena*; Oxford University Press, Oxford.

Yeomans, J.M.; (1992); *Statistical Mechanics of Phase Transitions*; Clarendon Press, Oxford.

Chapter 12

Adamson, Arthur W.; (1967); *Physical Chemistry of Surfaces*, 2nd ed.; Interscience, New York.

Davis, W. Ted; (1996); *Statistical Mechanics of Phases, Interfaces and Thin Films*; VCH Publishers, New York.

Sutton, A.P., and R.W. Balluffi; (1995); *Interfaces in Crystalline Materials*; Clarendon Press, Oxford.

Zangwill, Andrew; (1988); *Physics at Surfaces*; Cambridge University Press, Cambridge.

Chapter 13

Chandrasekhar, S.; (1943); "Stochastic Problems in Physics and Astronomy"; *Reviews of Modern Physics; vol. 15*, No. 1, pp. 1–89; Reprinted in *Selected Papers on Noise and Stochastic Processes*," Dover Publications, 1954. pp. 3–92.

Weiss, G.H.; (1994); *Aspects and Applications of the Random Walk*; North-Holland, Amsterdam.

Manning, J.R.; (1968); *Diffusion Kinetics for Atoms in Crystals*; chapter 2; van Nostrand, New York.

Reif, F.; (1965); *Fundamentals of Statistical and Thermal Physics*; chapter 1; McGraw-Hill, New York.

Chapter 14

Flory, Paul J.; (1953); *Principles of Polymer Chemistry*; Cornell University Press, Ithaca, N.Y.

Flory, Paul J.; (1969); *Statistical Mechanics of Chain Molecules*; Interscience, New York.

Hill, T.L.; (1960); *Statistical Thermodynamics*; chapters 13 and 21; Addison-Wesley, Reading, Mass.

Mark, James E.; (1993); "The Rubber Elastic State"; in *Physical Properties of Polymers*; American Chemical Society, Washington, D.C.; pp. 3–59.

Munk, Petr; (1989); *Introduction to Macromolecular Science*; John Wiley, New York.

Sun, S.F.; (1994); *Physical Chemistry of Macromolecules*; John Wiley, New York.

Chapters 15–17

Boltaks, B.I.; (1963); *Diffusion in Semiconductors*; Academic Press, New York; first published in Moscow, 1961; translated from the Russian in 1963 by J.I. Carasso.

Borg, Richard J., and G.J. Dienes; (1988); *An Introduction to Solid State Diffusion*; Academic Press, New York.

Damask, A.C., and G.J. Dienes; (1963); *Point Defects in Metals*; Gordon and Breach, New York.

Flynn, C.P.; (1972); *Point Defects and Diffusion*; Clarendon Press, Oxford.

Girifalco, L.A.; (1964); *Atomic Migration in Crystals*; Blaisdell Publishing, New York.

Girifalco, L.A., and David Welch; (1967); *Point Defects and Diffusion in Strained Metals*; Gordon and Breach; New York.

Manning, J.R.; (1968); *Diffusion Kinetics for Atoms in Crystals*; van Nostrand Co., New York.

Nowick, A.S., and J.J. Burton; (1975); *Diffusion in Solids*; Academic Press, New York.

Shewmon, P.G.; (1963); *Diffusion in Solids*; McGraw-Hill, New York.

Tuck, Brian; (1974); *Introduction to Diffusion in Semiconductors*; IEE Monograph Series 16; Stevenage Peter Peregrinus Ltd.

van Beuren, H.G.; (1960); *Imperfections in Crystals*; North-Holland, Amsterdam.

Index

absorbing barrier, 361–363
absorption, 288
acceptors, 163, 167, 168, 171
activated state for atomic migration, 461, 466
   normal modes for, 464
activity, 342
   and impurities, 437
   and point defects, 436, 437, 439
   and solid solubility, 446
adhesion and cohesion, 328–333
   energy of adhesion, 328, 330
   energy of surface cohesion, 329, 331, 333
adiabatic walls, 3, 4
adiabatic process, 6, 7
adsorption, 288, 321
   activation energy for, 326
   differential heat of, 326
   of dissociated molecules, 331
   energy of, 335, 336, 339
   free energy of, 325
   integral heat of, 328
   isosteric heat of, 326, 327
   and mobile layer, 339–340
   multilayer, 340–345
   thermodynamics of, 325–328
adsorption isotherm, 325, 326
   for mobile monolayer, 340
   for multilayer adsorption, 340–345
aluminum, crystal properties and vacancies, 428
aluminum, defect concentrations in, 427
aluminum, total vacancy concentration in, 427
alloying energy, 284
alloys, dilute binary, 431, 432, 439
   diffusion in, 450
   thermodynamics of, 446–448
anharmonicity, 280
   and Gruneisen assumption, 129–134
anharmonic potential, 136
antiferromagnet, 235, 245
   critical temperature for, 250
   internal field in, 249
   mean field theory for, 248–251
   magnetization in, 250
   Neel temperature, 251
Arrhenius law for diffusion, 457
atomic jumps, 455–457
   and diffusion, 453, 455–457
   energy barrier for, 457, 458, 461
   and thermal fluctuations, 458
atomic jump vector, 455
atomic levels and band formation, 159–160

band theory, 159–162
   and electrical conduction, 160
band width, 159
β-brass, 203, 204
BET isotherm, 340–345
binding energy of electrons to impurities, 162, 163
binding free energy of two center defects, 438–439, 440, 444
binodal curve, 269, 270
block spins, 306, 308, 309
Bohr magneton, 236, 239, 242, 246, 247, 294
Boltzmann transport equation, 181, 183
   and Lorentz force, 183
Bose-Einstein statistics, 72, 75, 83, 255

Bose-Einstein statistics (*cont.*)
  and grand canonical ensemble, 81–83
  and multilayer adsorption, 341
  and phonons, 119–121
Bosons, 70
Bragg-Williams theory, 209, 213, 222, 223, 225, 227, 262, 291, 375, 398
  and ferromagnets, 246, 247
  and regular solutions, 282
  and thermodynamic quantities, 217–219
Brillouin function, 237, 238, 239, 240, 242, 244, 247, 250
Brownian motion, 369
bulk modulus, 29, 127, 131, 133, 134, 139
  of crystals, 131–132
  of fluid near critical point, 291
  of free electrons, 154

Carnot cycle, 11–12
canonical ensemble, 41, 42–44
  and Helmholtz free energy, 52
  for order-disorder alloys, 209
  partition function, 49
  and thermodynamics, 48–53
canonical pressure ensemble, 41, 63–64, 403, 432
  and atomic migration, 462
  and Gibbs free energy, 64
  partition function, 64
  and point defects, 403, 404, 447
characteristic function for random flight, 363, 367–368
chemical potential, 20–21
  in Debye model, 271
  free electrons, 144, 146
  in gels, 400
  and Gibbs adsorption isotherm, 324
  grand canonical ensemble, 63
  of ideal gas, 76, 79, 272, 336
  and impurity concentration in semiconductors, 174–175
  and interfaces, 322, 339
  of liquids, 275
  and Maxwell construction, 267–268
  and particle statistics, 75, 144
  and solubility, 446
  and Thomas-Fermi theory, 156
chemisorption, 289, 326, 335, 337
Clapeyron equation, 260, 271, 272
  for surface-gas equilibrium, 327
Clausius-Clapeyron equation, 261, 272
  for surface-gas equilibrium, 327
collision derivative, 181, 182, 184
communal entropy, 276–277
complexions, 42, 43, 276
  in Bose-Einstein statistics, 72
  for canonical ensemble, 42
  for defect crystals, 433–437, 444
  and entropy, 54, 69–71
  for extrinsic semiconductors, 167–169
  for Fermi-Dirac statistics, 71
  for general point defects, 444
  for grand canonical ensemble, 61
  for Maxwell-Boltzmann statistics, 72
  and particle statistics, 71–72
  substitutional defects, 433–435, 436, 437
compressibility, 29, 127, 131, 134, 135, 141
  analogy to susceptibility, 291
  of crystals, 131, 132
  of fluid near critical point, 291
  of free electrons, 154
  point defect contribution to, 412, 422
condensation and physical adsorption, 343
conduction band, 160, 163, 165, 166, 167
conformational entropy of elastomers, 391
contact angle, 322
continuity equation, 451, 454
contour length in polymer chains, 379, 385, 388
  in rubber, 392
continuous string, 99–100
copper-gold, 202–203
cooperative phenomena, 204, 235
copper, crystal properties and vacancies, 424–425
copper, vacancy concentration in, 424
correlation angle, 352, 353, 377
correlation energy, 125
correlation factor, 352, 353, 354,
  for diffusion, 456, 457
correlation function, 297, 298, 300
  and expansion factor, 378
  for Ising model, 301, 302
  and surfaces, 330

correlation length, 298, 299,
  for Ising model, 301, 302
  and renormalization, 308
critical exponents, 289, 290, 306,
    311, 316
  and correlation length, 298
  for ferromagnet or order-disorder
    alloy, 293, 311
  relations among, 311, 316
  for surface tension, 333–335
  for van der Waals model, 292
critical opalescence, 292
critical point, 273, 288, 289–293,
    302, 303, 305, 310, 316
  for ferromagnets and order-
    disorder alloys, 294
  for surface tension, 333–335
  in van der Waals model, 265, 266,
    269
critical temperature
  in antiferromagnets, 250, 251
  in β-brass, 232
  in Bragg-Williams theory, 219, 220,
    223, 224
  and correlation function, 297
  comparison by various methods,
    229
  ferromagnetic, 246
  in Landau theory, 293
  second moment approximation,
    223, 224, 225
  for surface tension, 333–335
  in van der Waals model, 266
cross links, 385, 389, 395, 399
  and contour length, 392
  and gels, 395, 396
crystal energy, 126, 128, 136, 137,
    139
  for nearly free electron metals,
    125
  universal energy curve for, 126–
    128
Curie temperature, 243, 244, 247
Curie-Weiss law, 243, 244, 251

de Broglie relation, 84
Debye energy and heat capacity
  functions, 113, 132, 271
Debye model, 108, 270, 278, 416
  chemical potential in, 271, 281
  experimental tests of, 116–118
  frequency distribution, 111, 133,
    274
  Gruneisen assumption and, 134–
    136

high temperature approximation,
    113–114
low temperature limit, 114–115
Debye temperature, 111, 133, 136,
    274, 278, 279, 280, 281, 414,
    417
  relation to Einstein temperature,
    116
  in order-disorder alloys, 210
Debye T^3 law, 115, 152
decimation, 304
  and critical point, 305
  and renormalization, 304, 307
  and partition function, 305
defect concentration formulae, 406,
    421, 438, 440
defect equilibria, 440
  effect of impurities on, 421, 441
defect rotational factor, 434, 444
density of states, 77–78
  in conduction bands, 164
  for electrons and holes, 166
  of free electrons, 145, 146
  of ideal gas, 77, 78
  in k-space, 85
  in a magnetic filed, 240
  in momentum space, 85
  in semiconductors, 166
  for standing elastic waves, 109–
    111
  two dimensional, 87
  representations of, 84–86
  for valence bands, 164
  in velocity space, 85
detailed balanced, 55
dielectric constant in silicon and
    germanium, 163
diamagnetism, 234
diathermic walls, 4
di-defects, 434, 438
diffusion, 403
  Fick's first law, 450, 451
  Fick's second law, 451
  heat of activation, 457
  of interstitial impurity, 455, 457
  in order-disorder alloys, 204
  and point defects, 403
  point source, 371
  principle axes of, 454
  and random flight, 352, 349, 451,
    452
  tracer atoms, 456
  and transition probability, 452–455
  of vacancies, 455, 456
  volume of activation, 457

diffusion coefficient, 273, 450, 451, 466–467
  and activation energy, 457
  and atomic jumps, 455–457
  interstitial diffusion, 457
  and mean square displacement, 454
  self diffusion, 457
  temperature and pressure dependence, 457
  vacancy diffusion, 457
diffusion dyadic, 351
  and migration distances, 453
diffusion equation and random flight, 371–372
diffusion in a liquid, Gaussian model, 367
diffusion tensor, 451
dislocations, 143, 288, 321, 361, 403, 422, 442
distribution function. *See also* probability distribution
  Bose-Einstein, 72, 75
  for electrons and holes, 163, 164, 196, 197
  for electrons in external fields, 180
  Fermi-Dirac, 71, 73, 75, 155
  for free electrons, 146, 147, 148
  Maxwell-Boltzmann, 69, 72, 73, 74, 75
  rate of change of, 181, 183
  semi-classical, 73, 75, 197
divacancies, 405, 418, 420, 432, 433, 440
  in aluminum, 426
  concentration of, 437
  free energy of binding, 421, 437
donors, 163, 167, 168, 169, 171
drift derivative, 182
drift velocity, 182, 184
Dulong-Petit law, 152

effective mass, 144, 153, 161, 164, 196
Einstein function, 105, 108, 109
  superposition of, 107–108
Einstein heat capacity model, 106–107
Einstein oscillators, 107–108
Einstein temperature, 107
  relation to Debye temperature, 116
elastic continuum model, 108
  frequency distribution, 109–111
elasticity of polymer chain, 385–389
elasticity of rubber, 389–395

  and entropy, 391–393, 399
  Flory correction for, 395
elastic string, 99–100
elastomers, 389–395
electrical conductivity, 152, 160, 177, 179, 186, 187, 188, 190, 195
  in intrinsic semiconductors, 196–201
  in extrinsic semiconductors, 200
  in metals, 188
  and point defects, 403, 441
  temperature dependence in metals, 442
  temperature dependence in semiconductors, 198, 199, 200
electric field induced by temperature gradient, 189
electronic heat capacity, 152–153
electron-ion interaction, 162–163
electron scattering, 179, 180, 441
electron spin in extrinsic semiconductors, 168, 169
electron transport equations, 184
energy gap, 160, 163, 165
energy levels in solids, 159–161
  impurities in semiconductors and, 162
  crystal vibrations, 101
  particle in a box, 77, 145
ensemble, 41–42
  of activated states, 466
ensemble parameters and thermodynamics, 52, 53
enthalpy, 10
entropy, 13
  and equilibrium conditions, 14
  of free electrons, 151, 152
  high temperature approximation for harmonic crystal, 465
  and number of complexions, 69–71
  in order-disorder alloys, 203
  and randomness, 54
  statistical mechanical, 50, 53, 54, 62, 70
  statistical mechanical and the second law, 54–56
  time dependence of, 55
entropy of mixing, 284, 396, 398, 399
equal a priori probabilities, 43
equation of state, 28–31
  of copper, 129
  free electron gas, 153–155

and heat capacity relations, 28, 29
high temperature approximation for solids, 132
ideal gas, 80
low temperature approximation for solids, 132
of Mie and Gruneisen, 131
two dimensional ideal gas, 88
and universal energy curve for solids, 128, 129, 132
equilibrium, 1–2
  absolute, 1, 24
  and ensemble parameters, 46
  conditions of, 22
  and entropy, 14
  and free energy, 14–16
  internal, 1, 432, 439, 445, 446
  metastable, 24
  partial, 2
equilibrium constants
  for extrinsic semiconductors, 171–172
  and order-disorder alloys, 227
  for point defects, 440–441
equipartition of energy, 106
ergodic hypothesis, 41, 273, 350, 459, 462
Euler's theorem, 25, 323
exchange energy, 125
excluded volume in polymers, 383–385
exclusion principle, 160
expansion factor, 378
Eyring model of liquids, 273, 275

Fermi-Dirac statistics, 71, 73, 75, 83, 145, 152, 168
  for extrinsic semiconductors, 169–170
  and grand canonical ensemble, 81–83
  and phonons, 119–121
  and surface adsorption, 341
Fermi distribution, 145–148
  in an electric or magnetic field, 178, 179
  in a temperature gradient, 179
Fermi energy, 146, 150, 152, 156, 165, 195
  impurity concentration and, in semiconductors, 173–175
  in intrinsic semiconductors, 166, 171
  in a magnetic field, 241
  and screening distance, 159
  temperature dependence of, 151, 154, 155, 190
Fermi function, 145, 149, 187, 241
  for impurity levels, 170
  for holes, 164
Fermi integral, 149, 150, 166, 189, 190, 195
Fermi level (see Fermi energy)
Fermi momentum, 155, 156
fermions, 70
ferrimagnetism, 236
ferromagnetism, 234, 235, 245, 249, 251, 288
  critical point, 293
  domains, 235
  ground state, 252
  heat capacity data, 247
  internal field, 242
  mean field theory, 242–244, 247
  partition function for, 296
  spontaneous magnetization, 244, 246
Fick's laws of diffusion, 450, 451, 454, 455
  and transition probability, 452–455
film pressure, 325
fixed points and renormalization, 304, 313, 314, 315
fluctuations, 64–67
  deviations from most probable distribution, 66
  and correlation length, 295–298
  and diffusion, 455
  and electrical conduction in semiconductors, 160
  and energy, 65
  and magnetic susceptibility, 296
  and polymer chain length, 379–380
  and vacancy concentration, 407–408
flux of electrons, 180, 186
flux equations for electrons, 185–186
  and Hall effect, 193
  for holes, 196
force ensemble, 386
Fourier transform and random flight, 363, 364
Fourier's law, 38
free electrons, 125, 143, 148, 155, 157, 177
  average energy of, 148, 151
  chemical potential for, 147, 151
  compressibility of, 154
  density of states for, 145

514 INDEX

free electrons (cont.)
  energy distribution of, 145
  entropy of, 152
  equation of state for, 154
  in external fields, 177–179
  in extrinsic semiconductors, 167–172
  heat capacity of, 151, 152
  in intrinsic semiconductors, 163–167
  temperature dependence of energy, 151
  in a temperature gradient, 179
  velocity of, in external fields, 178
free energy and equilibrium, 14–16
free volume, 262, 265, 274
frequency variation with volume, 130, 137
frequency distribution in the harmonic model, 103–104
fusion, 260

gels, 385, 395–400
  chemical potential in, 400
  interaction parameter, 400
Gibbs adsorption isotherm, 323–324, 325
Gibbs-Duhem equation, 26, 258, 267
Gibbs free energy, 15
  and equilibrium, 16
Gibbs-Helmholtz equations, 17
Gibbs reference surface, 319, 320
Girifalco-Good equation, 331, 347
grain boundary, 345, 361, 422, 442
grand canonical ensemble, 41, 60–64
  and multilayer adsorption, 341
  and particle statistics, 81–84
  partition function, 63, 82, 342
grand canonical pressure ensemble, 41
Gruneisen assumption, 139–134
  and Debye model, 134–136
Gruneisen constant, 130, 135, 136, 137, 139, 417
Gruneisen equation, 131
Gruneisen parameter, 131, 137–139, 141
  theory of, 137–139
  values of, 140
gyromagnetic ratio, 236

Hall angle, 193
Hall coefficient, 192, 196, 200
  in semiconductors, 200
Hall effect, 191–196

harmonic model, 92, 101, 137, 270
  Debye model, 108–112
  and diffusion, 461, 464
  Einstein model, 26
harmonic oscillator, 94, 97, 101
heat capacity, 9, 17, 31, 18, 93, 131
  of antiferromagnets, 251
  at constant pressure, 134
  and equation of state, 28–31
  Einstein model, 106–107
  of ferromagnets, 247
  of free electrons, 151, 152
  and harmonic model, 104–105
  high temperature limit for harmonic crystal, 105–106
  low temperature limit for harmonic crystal, 106, 151
  in metals, 152
  vacancy contribution to, 424, 427
heat, 4, 8
  mechanical equivalent of, 3–4
heat flow equation, 191–192
Helmholtz free energy, 15
  and equilibrium, 15
  and partition function, 53
  relation to ensemble parameter, 53
holes, 160, 162, 163, 165, 196
  and electrical conductivity, 198
  energy distribution for, 164, 165, 166

ideal gas, 2, 76, 261, 262, 271
  density of states for, 77, 78
  energy of, 79, 154
  energy levels for, 77
  entropy, 78, 80
  equation of state, 2, 80
  Gibbs free energy of, 76, 78
  partition function for, 76, 78, 274, 276
  temperature, 2–3
  two dimensional, 87–88, 328, 339
impurity, 431, 432
  diffusion of, 361
  dissociative mechanism for cadmium in lead, 444
  energy levels in semiconductors, 162
  pairs, 433, 440
  segregation of, 345–346
interfaces, 318, 319, 320
  and adhesion, 328
  and concentration, 320
  energy of, 329, 333
  free energy of, 319, 323, 331

thermodynamics of, 322–325
interfacial phase, 320
interfacial tension, 321, 322, 332, 333
interstitials, 403, 418
　diffusion of, 455
　migration of, 466
intrinsic resistivity, 442
ionization energy, 163, 200
ionization reactions in semiconductors, 171–173
Ising chain, 298–302, 299, 305, 312, 316
　renormalization of, 302–305
Ising model, 235, 245, 247, 252, 288, 289, 290, 292, 303, 312
　and fluctuations, 295
　Landau expansion for, 293
　partition function for, 298–301
isothermal process, 6
irreversible process, 5, 7

jellium model, 157
jump frequency, 371, 455, 456, 465
　effective frequency in, 466
　for an interstitial, 465
　many-body theory of, 460–466
　one-dimensional model, 457–460
jump probability, 350, 351, 359, 351, 352, 355, 359
　and drift displacement, 369
　Fourier transform for, 363, 364
　for freely jointed flight, 364
　and Gaussian distribution, 369, 370, 384
　Gaussian model, 367
　vacancy mechanism, 457

Kadanoff construction, 305–311, 312, 313
Kirkwood expansion, 316
　and Landau theory, 292–295
Kirkwood method for order-disorder alloys, 213, 225, 298
　second moment approximation, 222, 227, 293
Kuhn length, 379

Lagrangian multipliers, method of, 44–45, 61
　for extrinsic semiconductors, 169
　for grand canonical ensemble, 61
Lame constants, 135
Landau theory of ordering, 292–295
Langevin function, 239, 388

Langmuir adsorption isotherm, 335–339, 344, 346
Laplace's equation, 157
lattice gas, 288, 289, 337, 345, 346
lattice renormalization, 306, 308, 309
　and correlation length, 308
　and partition function, 309
Legendre transformation, 19
Lennard-Jones potential, 125, 263, 332
librations, 261, 261
　in polymer chains, 382
Lindemann equation, 280
Linear chain, 93–99
　equations of motion, 93, 94
　frequencies in, 97–99
liquid state, 272–276
Lorenz number, 192

magnetic domains, 235
magnetic effects and electron transport, 185
magnetic moment, 236
magnetic susceptibility, 234, 291
　analogy to compressibility, 32
　critical point for, 289
　divergence of, 296
　and fluctuations, 295–296
magnetic systems
　ordering in, 235
　thermodynamics of, 31, 31
magnetization, 234, 235, 236, 245, 246, 247, 291, 292
　classical limit for paramagnet, 238
　critical point for, 289
　experimental data for iron, nickel, cobalt, 247
　and internal field, 242
　for free electrons, 242
　for paramagnetism, 237, 238, 240
　saturation for ferromagnet, 243
　saturation for paramagnet, 238
　and spin waves, 252, 255, 256
magnons, 255
Markoff method for random flight, 363–367
mass action laws in extrinsic semiconductors, 171–173
Maxwell distribution of velocities, 86–87
Maxwell equal area construction, 267, 268
Maxwell reciprocity relations, 27, 29, 390
McLean isotherm, 346

McLeod, equation, 334
mean field theory, 242–244, 247, 248, 252, 290, 291, 292, 295, 375
  in polymers, 383
  in gels, 396
mean free path, 188
  and diffusion, 453
  for electrons and holes in semiconductors, 199
mean velocity of migrating atom, 458, 461
mechanical equivalent of heat, 3–4, 8
melting, 273, 281–282
  and crystal vibrations, 277–281
  heat of, 273, 282
  and vacancies, 273
melting point, 261, 278, 280, 281
metastable states, 24, 268, 269
microcanonical ensemble, 41
microscopic reversibility, 55
Mie-Gruneisen equation of state, 131
mobile monolayers, 339–340
monolayer adsorption, 335–339
Monte Carlo, 316, 317

Neel temperature, 251
normal modes, 101, 130, 131, 136, 277, 414, 456
  activated state, 464
  in linear chain, 93–100
  in order-disorder alloys, 210
  at a surface, 280–281
n-type conductor, 163

order-disorder alloys
  analogy with ferromagnet, 235, 245, 246, 291
  configurational energy, 211, 213, 216, 217, 226
  configurational entropy, 216
  configurational states, 213
  critical point for, 289
  domains in, 235
  heat capacity of, 216, 223, 296
  moment expansion of free energy, 215
  moment expansion of partition function, 214
  partition function, 209–210, 213, 214, 226
  thermodynamic quantities and degree of order, 217, 220

order-disorder transition, 203, 204, 217, 288
  critical temperature for, 204, 293
  and heat capacity, 204, 210, 222
  heat capacity data, 231
  and physical properties, 204
order-disorder structures, 202–203
ordering energy, 203, 211, 226
order parameter, 205, 206, 207, 208, 209, 212
  and average energy, 215
  experimental data, 230
  and free energy, 216
  temperature dependence of, 216, 219, 220, 222, 223

parachor, 335
paramagnetism, 235, 236–240, 251
  of free electrons, 240–242
  partition function for, 236, 237
paramagnetic susceptibility, 240–242
  of free electrons, 242
  in antiferromagnet, 251
particle statistics and thermodynamics, 74–76
partition function, 46
  for adsorbed molecule, 336, 339, 340
  configurational, 59, 263, 264
  and force ensemble, 386
  ideal gas, 59, 274, 276
  Ising model, 299–300, 303, 304
  for a liquid, 273
  for a ferromagnet, 296
  for order-disorder alloys, 209–210
  for paramagnetism, 236, 237
  for polymer chains, 381–383
  and renormalized lattice, 309, 314
  semiclassical, 57, 58, 263
  two dimensional ideal gas, 88
  van der Waals model, 265
periodic boundary conditions, 95, 96
persistence length, 376, 377, 379
  and polymer chain size, 378
phase diagram, 259,
  for regular solution, 286
  for van der Waals fluid, 270
phase equilibria
  gas-liquid, 262, 268, 270, 271, 272, 291, 294, 295, 344
  gas-solid, 266, 270
  liquid-solid, 272
  second order, 244, 251, 292
  stability conditions for, 23–24

thermodynamics for one
  component system, 258–262
triple point, 258
two dimensional, 325
phase rule, 22, 23, 258, 403
phase space, 56, 461, 464
phonon gas, 119–121
photons, 119, 121, 255
physical adsorption, 326
point defects, 157, 159, 275, 403,
    406, 418
  in aluminum, 441
  binding free energy, 438–439, 440,
    444
  and chemical potential, 436
  core fields of, 434
  and electrical resistivity, 422
  and electron scattering, 179
  equilibrium, 419, 421
  experimental results, 422–429
  formation, 433, 436
  formation quantities, 416, 444
  general theory of, 443–446
  interstitial, 432, 433
  internal equilibria, 432, 439, 446
  mixed lattice type, 432
  and quenched-in resistivity, 443
  statistical count for, 420, 433
  substitutional, 432, 433
  substitutional concentration
    formulas, 435–440
  and thermodynamic functions, 422
  two-center defects, 434, 438
  types of, 431–432
Poisson's equation, 155, 157
Poisson's ratio, 135, 156
polymer chains, 349, 354, 357, 361
  density in, 381
  elasticity of, 385–389
  entropy of, 376
  excluded volume in, 383–385
  freely jointed, 367, 377, 379, 388,
    389
  partition function of, 381–383, 385
  and random flight, 350, 375, 383
  and self similarity, 370
pressure ensemble. *See* canonical
  pressure ensemble
probability distribution. *See also*
  distribution function
  canonical, 43–45
  in coordinates or momenta, 60,
    155
  definitions, 47–48
  in extrinsic semiconductors, 170
  free electrons, 145
  grand canonical, 61, 81
  semiclassical, 57
p-type conductor, 163

quantum statistics, conditions for,
    75, 144
quasi-chemical method, 225–230,
    299
  and critical temperature, 229
  equilibrium constant, 227
  free energy in, 228
  free energy of reaction, 227
  order parameter temperature
    dependence, 228, 229
quenched-in resistivity, 441–443
  and vacancy formation energy,
    443

radial distribution function, 263, 264
random flight, 351, 352
  and Bernoulli (binomial)
    distribution, 356, 359
  and central limit theorem, 367–
    369
  and characteristic function, 363
  and diffusion, 456
  and diffusion equation, 371–372
  and Gaussian distribution, 359,
    360, 367, 369
  general solution, 367–369
  on a lattice, 354–361
  and polymer chain, 375, 383
  and probability distribution
    function, 358, 359
  self-avoiding, 375, 384
random solution, 283
  entropy of, 284
reciprocal lattice, 100–101
reciprocity relations of Maxwell, 27,
    29, 390
reflecting barrier, 361–363
regular solution theory of binary
    alloys, 282–286
relaxation, 417
  around point defects, 275
relaxation time for electrons , 183,
    184, 187, 188
  and crystal vibrations, 188
  in semiconductors, 199, 200
renormalization, 292, 316
  and block spins, 306, 308, 309
  and one-dimensional Ising model,
    302–305
  and recursion relations, 305, 316

renormalization group, 311–313, 315
    and scaling, 313–316
renormalization operator, 312, 313
renormalization theory, 303
    and fixed points, 304
    transformation matrix, 315
renormalized lattice, 306, 308, 309, 310, 312, 313, 314
resistivity. *See also* electrical conductivity
    temperature dependence in metals, 188
response functions, 296
reversible process, 6
rotational barrier in polymers, 382
rotations, 361–362
rubber, 385, 389, 391, 395
    entropy effect in stretching of, 391, 392
    Flory correction, 395
    force-elongation relation, 393–394

Sacker-Tetrode equation, 80
scaling, 292, 298, 308, 309, 310, 311, 313, 315, 316
    and renormalization group, 313–316
    and self similarity, 370
screening, 143, 159
segregation at interfaces, 345–346
segregation energy, 346
self-similarity, 370–371, 376
semiclassical approximation, 56–60, 197
    conditions for, 75, 144
    partition function for, 58–59
semiconductors, 145, 159, 160
    extrinsic, 167–172, 200
    impurity levels in, 162–163
    intrinsic, 163–167, 166
    and mass action law, 171–173
    two-band model, 163
significant structures, method of, 272
solid solution, 282, 284, 285, 286, 446
solid surface, 280
solubility, 437
    free energy of, 438
    statistical thermodynamics of, 446
solutions of polymers, 395–400
spinodal curve, 269
spin operator, 252
spin waves, 251–256
    frequency dispersion of, 254–255
spontaneous process, 5

spreading coefficient, 322
state, 1, 34
    microscopic and macroscopic, 36–37
    thermodynamic, 16, 35–36
state distribution, 42
state function, 2, 6, 26
statistical count (*see* complexions)
    entropy and, 54, 69–71
    for point defects, 444
    substitutional defects, 433–435, 436, 437
steric hindrance, 351, 381
sublattice, 202, 204, 206, 207, 212, 213, 235, 249
sublimation, 260, 261, 270–272
    heat of, 272
substitutional defects, 435–440
    internal equilibria for, 440–441
surface area from BET isotherm, 344
surface concentration, 324
surface energy, heterogeneity of, 345
surface free energy, 323, 325
surface melting, 280–281
surface phase, 327
    Clapeyron equation for, 327
surface pressure, 88, 325, 328
surface tension, 321, 324, 325
    critical point and critical exponent for, 333–335
    and the parachor, 335

temperature, 2–3
    absolute, 3, 12
    empirical, 2
    ideal gas, 2
    relation to ensemble parameter, 52
    statistical mechanical, 52, 62
thermal conductivity in metals, 152, 179, 188–192
thermal energy, 131, 134
thermal expansion, 29, 30, 93, 131, 134, 136, 139, 271
    and vacancies, 425, 427
thermal wavelength, 59, 78, 263, 463
thermodynamic efficiency, 11, 12
thermodynamic potentials, 16, 19, 20
thermodynamic stability, 23, 24
Thomas-Fermi equation, 156
    linearized, 157
Thomas Fermi theory, 155–159
    and point defects, 157
time averages, 40
transition probability, 55, 180, 181
    and diffusion, 452, 455, 456

and Fick's laws, 452–455
triple point, 258

uncertainty principle, 57
Universal energy curve, 126, 138, 139, 278
  scaling factor for, 127, 128, 139
  and vibrational amplitudes, 279
universality class, 316

vacancies, 403, 418, 436
  concentration formula, 406
  contribution to crystal free energy, 410
  formation volume, 275
  Gibbs free energy of formation, 275, 276
  and liquid theory, 273, 274, 275, 276, 277
  and melting, 277
  relaxation around, 418
  and thermodynamic functions, 410–413
vacancy concentration, 403, 404–407, 413
  in aluminum, 426
  in copper, 423,
  and diffusion coefficient, 457, 467
  and Gaussian distribution, 407–409
  total, 421, 427, 433
  in liquids, 282
vacancy diffusion, 456
vacancy formation, 402–407
  configurational entropy of, 410
  energy from quenched-in resistivity, 443
  entropy of, 410, 418
  free energy of, 406, 410, 418, 436, 437
  thermodynamic functions of, 410–411
vacancy formation functions, 413–418
  and canonical ensemble, 414
  and quasi-harmonic crystal theory, 414

temperature and pressure dependence, 413–416
vacancy mechanism for diffusion, 456–457
vacancy migration, 461
  entropy of, 465
  free energy of, 465
  volume of, 463
valence band, 160, 163, 164–166, 167, 198
van der Waals equation, 262, 265
  reduced form of, 266
van der Waals isotherm, 267
  and Maxwell construction, 268
van der Waals model, 262–270, 272
  and critical point, 265, 266, 269
  Helmholtz free energy for, 265
  partition function for, 265
vaporization, 260, 261, 273
  heat of, 273
vibrations
  amplitudes of, 278, 280
  in crystals, 92, 93, 129, 134, 136, 152
  elastic string, 99–100
  and electrical conductivity, 167, 188
  linear chain, 99
  in liquids, 281
  and melting, 277–281
  in order-disorder alloys, 210

Weiss effective mean field, 242
wetting, 322
Wiedemann-Franz law, 188, 192
Wigner-Seitz radius, 125, 127, 139
work, 6–8

Young's equation, 322
Young's modulus, 389, 394
  and polymer molecular weight, 395, 400

zero point energy, 131, 132, 133, 141, 271, 272, 281, 415